WHALES AND DOLPHINS OF THE SOUTHERN AFRICAN SUBREGION

(with apologies to the porpoise)

Shepherd's beaked whale – adult

Shepherd's beaked whale – calf

Pygmy sperm whale

Atlantic spotted dolphin

Spectacled porpoise – male

Southern right whale

Whales and Dolphins of the Southern African Subregion

Peter B. Best
Mammal Research Institute, University of Pretoria, South Africa

Illustrations by Pieter A. Folkens

CAMBRIDGE
UNIVERSITY PRESS

We would like to express our heartfelt thanks to the following organisations whose financial contributions have assisted in making this publication possible:

De Beers
A DIAMOND IS FOREVER

DE BEERS MARINE

mazda
Wildlife Fund
The Driving Force
Behind Conservation

CAMBRIDGE UNIVERSITY PRESS
Cambridge, New York, Melbourne, Madrid, Cape Town, Singapore, São Paulo, Delhi

Cambridge University Press
The Water Club, Beach Road, Granger Bay, Cape Town 8005, South Africa

www.cambridge.org
Information on this title: www.cambridge.org/9780521897105

First published 2007

Reproduction by Resolution Colour (Pty) Ltd, Cape Town
Printed in Singapore by Tien Wah Press (Pte) Ltd
Sponsors' and Collectors' editions bound by Graphicraft, Cape Town

Project manager: Alison Paulin
Editor: Douglas van der Horst
Proofreader: Joy Clack
Indexer: Ethne Clarke
Monochrome illustrations: Val Myburgh
Maps: John Hall

ISBN-13 978-0-521-89710-5 (Standard Edition)

ISBN-13 978-0-521-89712-9 (Collectors' Edition)

ISBN-13 978-0-521-89711-2 (Sponsors' Edition)

Cover image: Heaviside's dolphin off Namibian coast, Dionys Moser
Frontispiece: (not to scale) Shepherd's beaked whale – adult and calf, *Tasmacetus shepherdi*; Pygmy sperm whale, *Kogia breviceps*; Atlantic spotted dolphin, *Stenella frontalis*; Spectacled porpoise – male, *Phocoena dioptrica*; Southern right whale, *Eubalaena australis*

If you want to know more about this book or any other Cambridge University Press publication, phone us at +27 (0)21 412-7800, fax us at +27 (0)21 419-8418 or send an e-mail to capetown@cambridge.org

CONTENTS

CONTRIBUTORS

ISABELLE J. ANSORGE is a lecturer in the Oceanography Department at the University of Cape Town. Her main interests are to understand and quantify the critical influence the circulation and processes in the Southern Ocean have on the ocean environment and productivity of subantarctic islands such as the Prince Edward Islands. Dr Ansorge is responsible for all sea-going training and it is intended that through this training, all students completing a degree in Oceanography, whether undergraduate or postgraduate, receive hands-on experience in operating oceanographic instrumentation and in the analysis of all hydrographic data collected.

JOHANN R. E. LUTJEHARMS is a physical oceanographer who holds the chair of Ocean Climatology at the University of Cape Town. Educated at this university and at the University of Washington, Seattle, he has a wide-ranging interest in many aspects of the oceans adjacent to southern Africa. To this end he has collaborated with marine biologists, meteorologists, chemical oceanographers, climatologists, maritime industrialists and geological oceanographers. For this work he has received the Alexander von Humboldt Preis, the John D. Gilchrist medal, the Havenga medal, the Fridjof Nansen medal, as well as the John F. W. Hershel medal. He is the author of an acclaimed book, *The Agulhas Current* (Springer, 2006).

WILLIAM F. PERRIN is the Senior Scientist for Marine Mammals at the Southwest Fisheries Science Center of the US NOAA National Marine Fisheries Service in La Jolla, California. He obtained his PhD from UCLA in 1972 and specializes in the systematics, ecology and conservation of whales, dolphins and porpoises. He works with the International Whaling Commission, the World Conservation Union, and the Convention on Migratory Species. With two colleagues he edited the *Encyclopedia of Marine Mammals* (Academic Press, 2002).

FOREWORD

IT WAS WITH considerable surprise that I received the request to write the Foreword to this impressive volume, but one that I am very honoured to accept. Peter Best was one of the first cetacean biologists I met almost 30 years ago, and one of those who has most influenced my thinking, not only on cetacean conservation and biology, but also on the privilege and burden faced by scientists when dealing with a group of mammals that represent perhaps the world's most charismatic megafauna.

Consciously or subconsciously, by his behaviour in international fora such as the Scientific Committee of the International Whaling Commission, he has shown me that the most productive and honourable role a scientist can play is to be thorough, questioning and objective, rather than taking the more political role of an unquestioning lobbyist. The implications of a conclusion reached using one's 'best' scientific judgement should never influence one's presentation of the results – I sometimes fear that the honesty and integrity of such a view is being lost given the many environmental problems facing the world and the need to 'act'. Peter's behaviour throughout his long and productive career shows that the best, most lasting and most effective conservation action comes from strong, objective science.

From his UK beginnings, Peter has evolved into the leading cetologist of southern Africa, and it is more than fitting that he has written the definitive volume of southern African cetaceans. In a publishing world where knowledge and experience of cetaceans sometimes seem to be the least important criteria for writing a popular book on the subject, it is indeed a delight to see a true expert in the field put so much effort into what, despite the ever-expanding research into all aspects of cetacean biology, is destined to be a classic reference for many years to come. In what clearly is a labour of love, Peter has consulted in his own inimitable, sometimes grumpy but always fair way, all of the available literature and distilled this into a concise, accurate and occasionally provocative account of the wide range of cetacean species in this fascinating area of the world's oceans.

The number of references cited is indeed a testament both to his thoroughness and to his belief that the reader should be able to trace the source of any statements made, and if necessary question the conclusions reached. The species accounts are simply excellent, from their considerations of taxonomy and general biology through to their discussion of conservation status. The volume also contains many gems of Peter's own original work and careful thinking. As just one of many new ideas, rare in such encyclopaedic tomes, I would cite his possible link of strandings of beaked whales to whalers' use of ASDIC (a form of sonar) – a good example of taking modern findings and extending them to past situations.

In a diplomatic written style not always present in his blunt verbal (sometimes audible only to blue whales) interventions, I was impressed by his rather gentle 'questioning of the wisdom of attempting a global assessment' of any species – when he and many others have spent their careers emphasising the importance of population level rather than species level assessments. Despite the authoratitive, scholarly style of the volume, there are intriguing and occasional glimpses of Peter's dry humour; however, even as a not-so-good Irish Catholic, I hesitate to draw attention to Peter's questioning of the local Cape Town community with respect to the restorative powers of ambergris for problems of erectile dysfunction!

It would be quite remiss of me to complete this Foreword without reference to the outstanding illustrations and photographs to be found in this volume. The illustrations of the species can only be described as excellent – in choosing Pieter Folkens, Peter Best clearly chose the best. The photographs too, whether of skeletal material, archive material or of animals in the field, have been carefully and expertly chosen.

This volume is without doubt a major contribution to cetacean literature – we should all be grateful.

GREG DONOVAN
Head of Science, *International Whaling Commission*
Editor, *Journal of Cetacean Research and Management*

THE MAMMAL RESEARCH INSTITUTE

University of Pretoria

THE MAMMAL RESEARCH INSTITUTE (MRI) was founded at the University of Pretoria in 1966, and is the only research institute dedicated specifically to the uniquely diverse mammalian fauna of Africa. Based on Main Campus in the eastern suburbs of Pretoria, the MRI is presently involved in research on mammals ranging in size from bats to whales, and in habitats from the Antarctic pack ice to the Serengeti Plains. The MRI's mission is to maintain its long-standing international reputation for undertaking and communicating high-quality research on the biology and ecology of African mammals. This mission is pursued with a focus on issues of particular relevance to conserving Africa's indigenous mammals and their habitats while promoting sustainable human development.

The MRI is embedded within the Department of Zoology and Entomology and provides an institutional identity to the mammal-oriented researchers in the department. Coordinating existing research, and identifying important topics for future research, are the responsibilities of the Director of the MRI, who is normally also the Austin Roberts Professor of Mammalogy. To date there have been four directors: Professor 'Waldo' Meester (1966–1971), Professor John Skinner (1972–1998), Professor Johan du Toit (1999–2005) and Professor Elissa Cameron (2006–). MRI projects are undertaken and/or overseen by MRI Associates, of which there are normally about seven in the Department of Zoology and Entomology, with an additional four Extraordinary Professors and two Research Fellows from other institutions in South Africa and abroad. Young researchers represent the life-blood of the MRI, and at any one time there are usually about five Postdoctoral Fellows, 20 PhD students and 20 MSc students busy on their individual research projects. The cumulative output of research publications is impressive for an institute of this relatively small size: over the past five years an average of 50 peer-reviewed (ISI-listed) scientific papers were published per year with the MRI cited as the institutional affiliation of at least one of the authors of each paper.

Apart from a small annual allocation from the University of Pretoria to cover office expenses, funds for the activities of the MRI are raised on a competitive basis by MRI Associates from a wide range of granting agencies in South Africa and abroad. In addition, the MRI manages several collaborative projects based in South Africa, the two largest such collaborations being with the University of California at Berkeley and the University of Cambridge. International academic links extend beyond research, however, and student exchange agreements are in place with a number of universities in Europe and North America. Through an outreach project the MRI also places two graduates per year in Tanzania, where they serve as university lecturers in wildlife ecology.

Students at the University of Pretoria have the opportunity of taking modules in mammalogy from the third year of their BSc degree, but they only occupy workspace in the MRI when they register for MSc or PhD degrees. Even then, because most MRI projects are field-based, our students are scattered far and wide, with some only able to communicate with their supervisors by e-mail or telephone for much of the time that they are occupied with data collection. There is nevertheless a lively and cosmopolitan assemblage of postgraduate students to be found in the MRI at most times, with a core group derived from the one-year intensive MSc degree in African mammalogy. For this programme a limited number of places (10 or fewer) are offered each year to graduates with a good honours degree (or four-year bachelor degree) in biology with a major in zoology. The appeal of spending a year immersed in learning about the biology, ecology, and conservation of Africa's mammals draws applicants from all over the world, and the most successful often stay on to continue with PhD studies. Information on the MSc programme in African Mammalogy, and all other activities of the MRI, can be found on the MRI website:
http://www.up.ac.za/academic/zoology/MRI.html

The publication of the authoritative *Mammals of the Southern African Subregion* has been one of the major achievements of the MRI. The first edition, under the authorship of the late Dr Reay Smithers, was published in 1983, and subsequent editions have appeared in 1990 (co-authored by Prof. John Skinner and Reay Smithers) and 2005 (co-authored by Profs John Skinner and Christian Chimimba). The latter edition was published by Cambridge University Press, and the present volume continues this relationship between Cambridge University Press and the MRI. Although *Whales and Dolphins of the Southern African Subregion* is a stand-alone publication, it has been produced very much in the style of the last edition of *Mammals of the Southern African Subregion*.

ACKNOWLEDGEMENTS

IN A COMPILATION such as this, which represents the distillation of a very large number of influences, experiences and published and unpublished observations by a substantial number of people over a protracted period, it is difficult to do justice to all the contributors. If I have omitted someone from the following list it is surely an oversight on my part and not a reflection on their contribution, and I ask them to accept my apologies.

My grateful thanks are due to the following individuals and institutions (in no particular order):

Caroline Weir of Ketos Ecology and BP for allowing me access to unpublished observations off Angola (including incidental sightings made by Alex Vogel and Ian Nicolson); Tara Pelembe and Emma Bennett for information on what occurs off Ascension and St Helena islands respectively; Violaine Drouot for similar information from Réunion; Mike Griffin for assistance with Namibian cetacean records; Bill Perrin for answering numerous queries about obscure dots on maps; Gabbie Luyt for providing personal details of the dolphins at Sea World; Shelton Dudley for correcting statistics of dolphin catches in shark nets off KwaZulu-Natal; James Mead and Charley Potter for providing access to specimens at the Smithsonian Institution for photography; Sylvia van Zyl and Gill Watson for similar privileges at Bayworld, Port Elizabeth; Jimmy Glass for acts beyond the call of duty in securing invaluable specimens from Tristan da Cunha; Ann Bowles and Walter Munk for arranging and allowing me access to original sightings from the 1991 ATOC expedition to Heard Island; Cherry Allison and Marion Hughes for extracting records from the IWC's sighting data base; Michael Meÿer and a veritable army of colleagues – Keiko Sekiguchi, Meredith Thornton, Simon Elwen, Colin Potts, Derek Ohland, Desray Reeb, Sidney Kannemeyer, Blake Abernethy, Peter Heather-Noon, Mike Combrink, Deon Kotze and Ken Findlay – who assisted in the grim and greasy business of retrieving stranded material and preparing skeletons.

The writing of this book would not have been possible without funding provided by the Island Foundation, USA, and the Mammal Research Institute Book Fund, South Africa (at the instigation of Johan du Toit). Rina Krynauw and Alieya Haider, chief librarians at the South African Museum and the John D. Gilchrist libraries respectively, were extremely generous and patient in handling my many literature requests.

The team at Cambridge University Press – designer Sandra Hall, editor Douglas van der Horst, cartographer John Hall, proofreader Joy Clack and especially commissioning editor Alison Paulin – did an outstanding job in ensuring quality while maintaining impossible deadlines: my sincerest thanks to you all.

I would like to express my indebtedness to individuals who have had a profound influence on my career, some of them unfortunately no longer living: my parents Kenneth and Marjorie, for allowing me to pursue the career I wanted; Tim Potts, my biology teacher at school, for encouraging me to set my sights higher; Chris Ash, who introduced me to the world of whales and whaling and was a most understanding first employer; Bob Rand, for tolerating many a churlish insult in the process of breaking me in as a scientist; John Bannister, a friend indeed and an invaluable personal adviser over many decades; Graham Ross, a cherished and highly respected colleague, whose emigration in 1989 was a severe loss to southern African cetology; John Skinner, for offering me a lifeline when the South African government dispensed with my services; and Johan du Toit, for bringing me back to earth with a bump.

Finally, none of this would have been possible without the life-support system provided by my wife and soul-mate Maggi. She has suffered my long periods away in the field or at meetings, and endured my frequent returns home 'in bad odour', not without complaint but always with good humour. To her I owe more than it is possible to put into words, and I dedicate this volume to her.

INTRODUCTION

WHALES, DOLPHINS AND porpoises all belong to the mammalian order Cetacea. They are the only mammals to spend their entire lives in water and have exchanged a layer of hair for a thick coating of insulating blubber. They have streamlined their body shape by losing all external trace of ears or hind limbs, and are able to propel their body using horizontally flattened tail flukes (their forelimbs having been relegated to paddle-shaped flippers for steering and stability). They are divided into mysticetes, or those with whalebone (horny baleen plates in the upper jaw), and odontocetes, or toothed whales (although dentition may be reduced to a single pair of teeth in the lower jaw).

Worldwide there are just over 80 species of cetaceans. The uncertainty about the exact number is caused by several factors. For even the 'better known' large whale species, of which many tens of thousands of individuals were killed by the whaling industry, only a tiny fraction are represented in museum collections, presumably because of the onerous task of preparing a skull or skeleton of such a large animal and then finding the space to store it. Consequently, no large series of skulls or skeletons exists for any single population of the great whales.

Other cetaceans are oceanic by nature, rarely venturing into inshore waters, and because they are relatively small in size they did not attract the attention of whalers. Such animals may only make themselves known to science when they strand on a coastline, and this has to happen near somebody who is keen enough to record the event and collect some material (and not eat the specimen). As a result, the science of cetacean taxonomy was handicapped for many years by the relative lack of comparative material and its slow rate of accumulation. It has taken the advent of modern genetic techniques to revolutionise the situation: DNA can be extracted from old museum specimens and other existing collections of biological material, and it can also be obtained from living cetaceans at sea through biopsies. Data from a relatively large series of individuals can now be collected in a reasonably short period and placed in an overall phylogenetic relationship.

However, as with so many other advances in science, all this new knowledge has created almost as many questions as it has answered. Consequently, while we might have been satisfied in the past to talk about one species of Bryde's whale worldwide, we now think there might be two or three. Genetics now tells us that there are three species of right whale (when some of us thought there might only be one), yet we battle to tell them apart morphologically. As more and more information accumulates about individual variation within and between populations, it seems inevitable that much of this will lead to the formulation of new cetacean subspecies or even species.

One of the regions of the world where our knowledge about cetacean fauna is weakest is southern Africa, despite the fact that this is a region of high oceanographic productivity and biological diversity that was once the site of the largest whaling ground outside the Southern Ocean. It is also home to about 63% of the world's whale and dolphin species (not to forget the porpoise), and contains one of the key areas for beaked whale diversity in the world. Attributing reasons for this ignorance is clearly beyond the scope of this book, but they can roughly be classified as political and socio-economic.

By displaying the region's rich cetacean fauna and summarising what we know and do not know about it, this book hopes to stimulate interest in this truly fascinating group of animals and encourage the accumulation of more knowledge about them. Only in this way can their conservation be truly assured.

About the book

As with many regional faunal guides, the geographical limits are largely arbitrary. In this case, the subregion is defined as extending from the equator to the edge of Antarctica, and between longitude 20° W and 80° E, or roughly from the middle of the Atlantic Ocean to the middle of the Indian Ocean. While these boundaries may not correspond to key oceanographic features, they have been chosen to be as broad as possible to make the addition of new species to the fauna list for the region less likely, and to cover as many destinations as possible that the sea-going traveller from southern Africa might visit.

For each species, illustrations are provided of the whole animal, including a side view (all species) and a belly or back view where appropriate. If colour varies with age, there is also an illustration of the juvenile. As photographs of whole cetaceans, particularly the larger ones, are difficult to obtain, we have been particularly fortunate to acquire the services of Pieter Folkens, the renowned marine mammal illustrator. Photographs of live animals are included where possible in order to give an impression of the 'jizz' of the species at sea, while the standardised views of the skull are intended to assist in identifying specimens for which no information on external appearance is available.

A word about the skull photographs. Those for which I was responsible were taken with a Nikon D70 camera and an 18–70 mm 1:3.5-4.5G ED lens. Readers should be aware that because of the wide-angle nature of this lens, I was able to capture even the largest blue whale skull from a reasonable distance, but not without incurring some degree of distortion. Many of the large whale skulls were also photographed through a barricade of metal and other supports, and although the repro house has done an admirable job of removing them from the images, those skulls that have been extensively worked in this way have been annotated as 'retouched'. Mainly as the result of the mechanics of handling the mandibles of large whales, they have been photographed from the lingual side in mysticetes, but the labial side in odontocetes. On occasions where the inappropriate mandible has been photographed for the corresponding lateral view of the skull, its image has been reversed (and noted as such).

The distribution maps provided for most species include actual sighting positions as well as a 'best impression' swath of colour to suggest what the range might be. In this way it is hoped that the reader will get a better feel for the reliability of the distribution shown. Strandings have been considered where appropriate but their distribution treated with some circumspection given that it may include animals outside their normal range. The sources of all location data are held by the MRI Whale Unit.

Each species account contains the following components:

- Appearance
- Length:weight relationship
- Skeletal data
- Distribution and migrations
- External commensals and parasites
- Food and feeding behaviour
- Reproduction and growth
- Behaviour and vocalisations
- Conservation status
- References

In compiling these accounts, priority was given to the results of data collected within the subregion, and only if information was deemed inconclusive or incomplete were data from other populations provided.

Nothing is more annoying than trying to figure out the origin of statements made in general texts, so an attempt has been made to cite a source for each statement of significance. To make these citations as unobtrusive as possible they have been indicated by means of superscript numerals. The names of the authors and the dates of publication appear at the end of each species account, and the full details of each source are given in the Bibliography at the end of the book.

Most species accounts contain a citation to 'MRI Whale Unit data'. This refers principally to unpublished data compiled by the Unit, but also includes compilations of published and unpublished data, some (such as the IWC's DESS data set) not originally collected by the Unit.

A word of explanation is needed about the units of measurement that appear in the book. Although metric units have been preferred throughout, occasionally non-metric measures such as nautical miles, imperial tons or even feet have crept in. These are there because the original (usually historical) source used them, and conversion might have been misleading. For instance, large whales taken by the industry were traditionally measured to the nearest foot, and this has been reflected in much of the scientific literature arising therefrom. Converting the results to an accuracy of more than one decimal point of a metre could give a spurious impression of accuracy.

To somewhat leaven the dough of scientific material, boxes containing brief discussions of topics of particular interest that could not be covered in any detail in the species accounts are scattered throughout the text. The choice and content of these may tend to reflect the author's particular areas of interest or bias, and for this he makes no apology.

PETER B. BEST
Cape Town
September 2007

Key to museum acronyms

MNM	Maldives National Museum, Maldives
NHMZD	Natural History Museum, London, England
NMNZ/TMP	Museum of New Zealand, Te Papa Tongarewa, Wellington, New Zealand
PEM	Bayworld, Port Elizabeth, South Africa
QMJM	Queensland Museum, Brisbane, Queensland, Australia
USNM	Smithsonian Institution, Washington DC, United States of America
ZM	Iziko South African Museum, Cape Town, South Africa

MAP OF THE SUBREGION

THE SOUTHERN AFRICAN subregion has been defined for the purposes of this book as the area extending from the equator to the edge of the Antarctic continent, and between the longitudes of 20° W and 80° E.

Shown below is the map of this subregion that is used within the species accounts to indicate distribution, but providing relevant nations, islands, towns, cities and other geographical features. For oceanographic features such as currents and fronts, the reader should consult Figs 2 to 4 on pages 8, 9 and 10. Anglicised versions of names have been used throughout, as these are how they generally appear in the original publications. An attempt has been made to include all localities mentioned in the text, but for reasons of space it may not have been possible to accommodate all of them on the map.

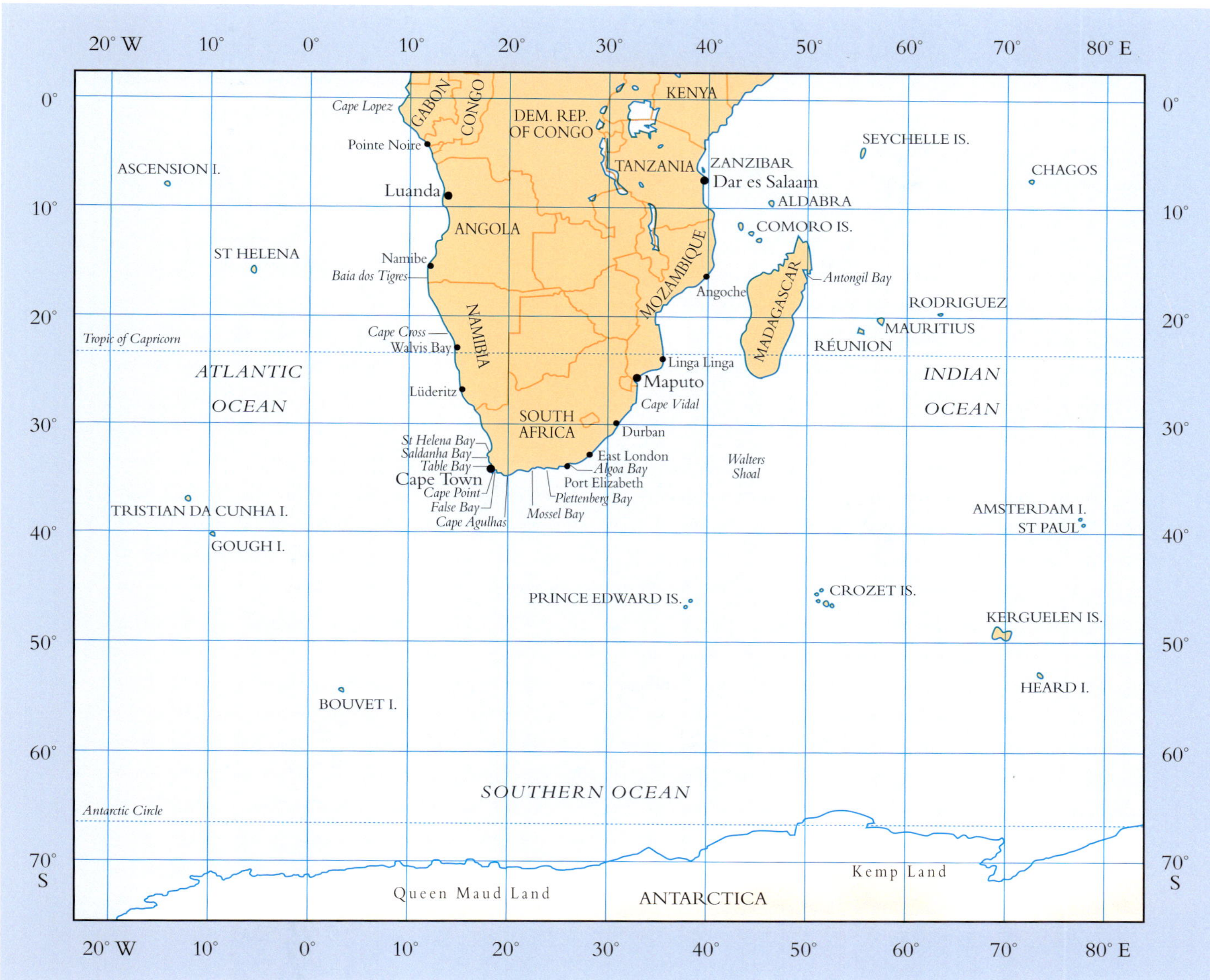

FIG. 1 Map of the southern African subregion as used to depict distribution, showing place names and geographical features.

THE CETACEAN ENVIRONMENT OFF SOUTHERN AFRICA

Isabelle J. Ansorge and Johann R. E. Lutjeharms
Department of Oceanography, University of Cape Town, South Africa

THE OCEAN REGION extending from the tropics to Antarctica and including the south-west Indian and the south-east Atlantic oceans encompasses the widest possible variety of circulatory regimes. Many of these may affect cetacean distribution, including currents,[1] fronts,[2] vertical stratification, biological productivity and even faint traces of chemical and physical constituents of sea water that might aid in navigation.

Along the eastern seaboard of southern Africa the circulation is dominated by the greater Agulhas Current system,[3, 4] the largest western boundary current in the Southern Hemisphere (Fig. 3) (western boundary here meaning the western boundary of the anti-clockwise gyre in the South Indian Ocean). It is a narrow, intense current that extends to the ocean floor and is fed by a complex of sources, perturbations of which dramatically influence its behaviour. Deriving its waters from the tropics and subtropics it is low in nutrients and therefore in biological productivity. The circulation on the western side of the subcontinent is very different. Here the Benguela Current is a wide, shallow movement of water towards the equator, an eastern boundary current[5] (Fig. 2). On its landward side there is an extensive coastal upwelling system, driven by the local winds.[6] This water is cold, rich in nutrients and supports a very productive ecosystem.

Moving south from these subtropical regimes, one enters the Southern Ocean. This ocean region is characterised by zonal fronts, extending roughly along lines of latitude[7] (Fig. 4). The northernmost of these, and the intrinsic border to the Southern Ocean, is the Subtropical Convergence, a region where there is a very large meridional gradient in both temperature and salinity. This front is more intense south of Africa than anywhere else in the world's oceans. Further south other ocean fronts are found, such as the Antarctic Polar Front, but these do not have the same intense gradients. Along and between these fronts the largest current in the world's oceans – the Antarctic Circumpolar Current – carries its waters from west to east. The southern part of the Southern Ocean undergoes marked seasonal changes. A large part is covered by ice in winter, which causes a freshening of the surface layer in summer when this pack ice melts.

The Benguela Current system

The Benguela region encompasses the western coastline, roughly from Cape Town to southern Angola, and is largely characterised by wind-driven coastal upwelling of cool, sub-thermocline water[5] (Fig. 2). The Benguela Current system consists of a coastal branch, modulated by local weather processes, and an offshore oceanic flow, which forms the eastern limb of the South Atlantic subtropical gyre.[5, 8] The atmospheric high pressure cell situated over the south-east Atlantic provides the driving force for the Benguela Current regime.[8] In the south Benguela region, the south-east trade winds are responsible for the offshore transport of near-surface layers, resulting in a belt of coastal upwelling (see below) along the South African and Namibian coasts. The strongest winds are found where the coastline changes orientation from broadly north-westerly to more northerly, at Lüderitz and Cape Frio, and the weakest between Möwe Bay and Walvis Bay, where the coastline is concave.

The coastal upwelling area of the Benguela ecosystem extends between 14° S and 37° S, where its northern boundary has been shown to oscillate in a band between 14° S and 17° S[9, 10] (Fig. 2). There is a well-defined mean flow in the Benguela Current that is mostly confined near the African continent, with typical speeds of 10–30 cm/sec.[5] On its western side, however, flow is more transient and is characterised by large eddies shed from the retroflection of the Agulhas Current (Fig. 3), which are probably responsible for the spatial variations in transport.[11] In the south the current has a width of 200 km, but as it flows northward it widens rapidly to 750 km in the north.[12]

Upwelling regime

Upwelling, the process whereby the surface water is removed offshore by wind action and cold, nutrient-rich water is drawn to the surface, plays an important role in the oceanography and fisheries of the Benguela region. The phenomenon occurs all along the coast of south-western Africa, but is most localised in distinct cells characterised by recurring patches of cold water at the surface (Fig. 2). The area at Lüderitz (27° S to 28° S) presents a permanent and active upwelling cell[5, 13] that divides the Benguela system physically[5] and biologically[14] into two distinct areas, the northern and southern Benguela. Wind stress is greatest here and sea temperatures are lowest.[5, 10] Strong upwelling occurs here for 95% of the year,[10, 15] with a slight maximum in spring and minimum in autumn. It has been suggested that the surface features associated with the upwelling system are largely confined to within 300 km of the coast.[10]

There are high levels of inter-annual variability in the upwelling regions. From sea surface temperature measurements the existence of a weak 8±10-year cycle between alternating cool and warm periods has been found,[16] the cool periods being associated with enhanced cold-water upwelling and vice versa for warm periods. Sea surface temperature may vary 5 °C or more from the long-term mean. Inter-annual variability in sea temperature and salinity,[17] sea level,[18] winds[19] and atmospheric pressure,[20] all occur in the south-east Atlantic, although relatively few long-term studies have been undertaken.

Longshore fronts and cold filaments

Well-developed longshore thermal fronts exist between 18° S and 34° S. These fronts demarcate the seaward boundary of upwelled water. South of Lüderitz, a single front is usually well defined, and although spatially and temporally variable, coincides approximately with the edge of the continental shelf.[5] Further north, the surface definition of the front is more diffuse and multiple fronts are evident on occasion. Upwelling filaments are characteristic of these offshore fronts and exist as westward surface streamers of cold water typically 50 km wide and extending beyond the normal offshore extent of the upwelling cell.[21] These fronts have a typical lifespan from a few days to weeks, and are generally perpendicular to the coast. Although the influence of the upwelling regime may be between 150 and 200 km wide on average, the filamentous mixing area may extend up to 625 km offshore.[10, 21] Upwelling filaments tend to be highly-convoluted, eddy-shedding regions, and show extensive seaward penetrations at major upwelling cells such as Cape Frio and Lüderitz/Walvis Bay.[5, 10]

Poleward undercurrent

The most significant discovery during the past two and a half decades has been of a poleward undercurrent to the Benguela system[5] lying 100–150 km offshore at a depth of 200–400 m. On the shelf the undercurrent can be found at 50–100 m, and between Cape Point and Cape Cross (Namibia) it exhibits a seasonal trend in strength and occurrence. South of 33° S, where the shelf is narrow, the flow is not seasonally dependent, and average net poleward flow of only 5 km/day has been observed.[6] In contrast, north of 33° S, where the shelf broadens, the flow is variable over periods of weeks, alternating between the equator and the poles. This poleward undercurrent may provide clues to the occurrence of oxygen-depleted shelf water, which may be advected southwards from the tropical south-east Atlantic along the edge of the shelf.[22]

While the coastal upwelling area of the Benguela ecosystem shares many of the generic characteristics of other eastern boundary currents, it is unique in that it is bordered at both northern and southern ends by warm water systems, the Angola Current and the Agulhas Current respectively.[5] These borders are known to be highly dynamic with prominent 'pulsing' implications for the local ecosystem.

Angola Current system

The westward flow in the 15–20° S latitudinal band forms the confluence of the Benguela Current and the southerly flank of a large clockwise gyre west of Angola. The poleward Angola Current is a fast (40–50 cm/sec.), narrow and stable flow of warm (more than 24 °C), saline water that reaches a depth of 250–300 m and meanders laterally and vertically over both the continental shelf and the slope of the Angolan coast.[23–26] This water mass gradually becomes colder and less saline as it travels further south.[26] In addition, the Angola Current displays a degree of seasonality, retreating north-westwards in winter and spring to be replaced by slightly cooler waters (20–26 °C). This periodic south-east advance and north-west retreat of the Angola Current seems to be linked to the intensity of upwelling off the Namibian coast.[27, 28]

Another prominent feature associated with this Current is a regionally fixed, cyclonic gyre known as the Angola Dome lying at approximately 12° S, 4–5° E.[26] The Angola Dome is characterised by a decrease in salinity, compared to the surrounding water, that may result from the entrainment and vertical mixing of low-salinity Congo River water from the surface layer. The Angola Dome is less prominent during winter months,[29] and its width and extension depend on the intensity of the south-easterly trade winds.[30]

Angola-Benguela front

The Angola-Benguela frontal zone is the convergence between the south-flowing, warm Angola Current and the north-flowing, cold Benguela Current.[9] This frontal region also demarcates the water of the warm, nutrient-poor Angola Current and the cold, nutrient-rich Benguela upwelling regime, creating a transition zone between the tropical ecosystem in the north and the upwelling-driven ecosystem in the south.[26] This front is a permanent feature at the sea surface, maintained throughout the year in a narrow latitudinal band between 14° S and 6° S, with a general west to east orientation at the coast.[27] It is characterised by an average sea surface temperature of 23.6 °C in the northern region and 20.6 °C in the southern boundary at 16° S. Salinity data suggest the front may extend to a depth of 200 m, but it is particularly marked at the surface by a temperature gradient reaching 4 °C per 1° of latitude.[9] It is termed a frontal zone because often two or three parallel fronts can be observed.[6] The front fluctuates seasonally in latitude, seaward extent, temperature and strength, being furthest south in summer and early autumn and furthest north in winter and early spring. This migration has been associated with the seasonal movement of the Intertropical Convergence Zone in the equatorial Atlantic (a region where the north-east and south-east trade winds converge, forming an often continuous band of clouds).[31]

Temperatures associated with the front exhibit a similar pattern of seasonal change. At the coast, the front is least defined in winter, a period coinciding with the coolest water off Namibia and the stronger northward flow of Benguela water. It is most clearly developed during summer and autumn, when the warm-flowing southward Angola Current is at its maximum and coastal upwelling is at its weakest. On average, the front reaches its southernmost position in near-coastal zones and penetrates seawards up to distances of about 250 km, although traces of it can be found up to 1 000 km offshore. The maximum seaward extent is in spring and summer.[9, 27]

Benguela Niños

At times an extreme southward migration of the Angola-Benguela front results in large intrusions of warm, saline tropical water.[5, 17, 32] These intermittent anomalous warm events obstruct coastal upwelling by depressing the thermocline and stabilising the water column.[33] Extremely warm, nutrient-poor waters replace the cold, nutrient-rich upwelling water, with drastic consequences for productivity in the northern Benguela ecosystem. Thus, well-known warm years (1934, 1963, 1984 and 1995) are characterised by a reduction of upwelling in the northern Benguela and increased rainfall over the Namibian hinterland.[5] From a number of similarities between the 1963 and 1984 events and El Niño-Southern Oscillation events in the Pacific, the term 'Benguela Niño' has been suggested to describe them.[34] Similarities included apparent connections to anomalous atmospheric conditions in the tropical Atlantic in terms of relaxed westerlies along the equator, poleward propagation of tropical water along the Namibian coast, suppression of phytoplankton production, displacement of fish stocks, and a reduction in fish catches.[32] Long-term sea level records have shown that anomalously high sea levels correspond to these years of strong positive anomalies in sea temperature along the entire south-west African coast.[18] Furthermore, it has been suggested that Benguela Niño signals propagate polewards from the equatorial Atlantic similar to El Niño events, although not as intensely.[34]

Seasonality

The large-scale atmospheric weather systems over the south Atlantic display a seasonal pattern of northward and southward migration, and the effect of this seasonal cycle together with changes in insolation is manifested in the upper layer of the ocean.[5] Off the Namibian coastline there is a distinct seasonal cycle in surface and upper layer temperature, with a maximum in March and a minimum in August/September, the overall range in sea surface temperature being about 4–6 °C.[5] In the southern Benguela the upwelling season coincides with the period of maximum insolation, and accordingly the range in seasonal average temperature is only 1–2 °C. At the boundaries of the upwelling system, the water column becomes strongly stratified during the austral summer, particularly so over the eastern Agulhas Bank and off Angola, north of the Angola-Benguela front.

In the south, the boundary at the Agulhas retroflection likewise displays apparent seasonality, being furthest south in autumn. The Agulhas retroflection area is one of the global 'hot spots' in terms of variability of temperature and water movements.

Agulhas Current system

The Agulhas Current system can be thought of as extending from the Comoro Islands in the northern mouth of the Mozambique Channel (Fig. 3) to the Subtropical Convergence in the south.[35] The longitudinal borders are more difficult to define. The sources of the system include the whole South Indian Ocean and the throughflow between the Indonesian archipelago, whereas its outflow goes both into the South Atlantic Ocean and towards Australia in the South Indian Ocean. The behaviour of this system is also greatly affected by the underlying bottom topography.

The most obvious element of this bathymetry is the presence of the large land mass of Madagascar. It has a narrow shelf on its eastern side that stabilises the East Madagascar Current flowing along this coastline. Furthermore, the presence of Madagascar prevents the formation of a strong western boundary current in the Mozambique Channel. South of Madagascar there is a very shallow ridge, the Madagascar Ridge, that may play a decisive role in the movement of water in that region. From the coastline of Africa a similar, but deeper ocean ridge, the Mozambique Ridge, extends southwards, running parallel to the Madagascar Ridge. It may inhibit the flow of bottom water, but at lesser depths it may also affect the movement of deep-sea eddies. The edge of the continental shelf of eastern Africa is mostly narrow, allowing the Agulhas Current to flow quite close to the coast. However, at certain places there are offsets in the coast, such as at the Delagoa Bight, and here lee eddies may be formed (Fig. 3). South of Africa the continental shelf widens to form the Agulhas Bank. The Agulhas Current continues to follow the shelf edge, thus moving further offshore as it progresses southward along this wider shelf.[36] One of the final outflows of the Agulhas Current, the Agulhas Return Current, has to negotiate a number of shallower features on its way eastwards. These include the Agulhas Plateau (Fig. 3), which causes this current to carry out wide meanders in the direction of the equator.

The sources of the Agulhas Current proper come from an extensive recirculation in the south-west Indian Ocean, from water originating east of Madagascar, and from water passing through the Mozambique Channel. The circulation in the Mozambique Channel is dominated by eddies that form at the narrows of the channel and move steadily southward. These eddies may extend to great depths and may be anti-cyclonic (anti-clockwise) or cyclonic. The former seem to be stronger and longer lasting. As these eddies move past the shelf edge they may stimulate the formation of lee eddies. These eddies have upwelling at their centres and may thus enhance biological productivity. One such eddy is found in the Delagoa Bight[37] at the mouth of the Mozambique Channel.

To the east of Madagascar the South Equatorial Current carries water from east to west. As it reaches the eastern coast of Madagascar it splits (Fig. 3), one branch going north and the other south. The northern branch is eventually reunited with the rest of the South Equatorial Current and moves past the northern tip of Madagascar. The southern branch rapidly increases in strength until it reaches the southernmost part of Madagascar. Here it may either retroflect and flow eastward (as the South Indian Ocean Countercurrent) or move over the Madagascar Ridge. In this process it may break up into an assortment of eddies[38, 39] that subsequently drift into the Mozambique Channel or towards the Agulhas Current.[40, 41] At the point where the southern branch of the East Madagascar Current leaves the coast there is a well-developed upwelling cell[42] that has been associated with the presence of blue whales in spring.[43]

The Agulhas Current proper becomes fully constituted only somewhere along the north-east coast of South Africa. It is an intense current, about 100 km wide, and some of its surface water moves at speeds of more than 2 m/sec. Its northern part, namely as far downstream as Port Elizabeth (Fig. 3), is extremely stable, meandering less than 15 km to either side.[1] This stability is interrupted at irregular intervals by a singular meander, the Natal Pulse,[44] the lateral dimensions of which grow as it travels downstream. On its inshore side it causes rapid but short-lived changes in the direction of the coastal currents.

The depth of the Agulhas Current at any one location can vary over a period of a few months from the surface to 2 000 m or all the way to the bottom. Its landward border is sharply defined by a sudden increase in speed and temperature if one travels across it in a seaward direction.[45] At a depth of about 1 000 m the current is underlain by a countercurrent[46] that carries water towards the equator and may extend as one continuum into the Mozambique Channel. Between Durban and St Lucia, to the north, there is a wider and partly sheltered shelf region, the Natal Bight, while at St Lucia itself there is an intense upwelling cell[42] that brings nutrient-rich water onto the shelf and may control the whole ecosystem of the Bight.

By contrast the southern Agulhas Current meanders much more widely. Along the eastern side of the Agulhas Bank it develops shear edge eddies[47] and attendant plumes of warm water. The latter may penetrate far onto the Bank. Similar to the situation in the Natal Bight, there is a distinct upwelling cell at Port Alfred[48] where the Agulhas Current leaves the narrow shelf. Here nutrient-rich water is believed to be drawn up onto the shelf and spreads over the whole Agulhas Bank. The circulation on the western side of the Agulhas Bank may be more complicated.

(continued on page 11)

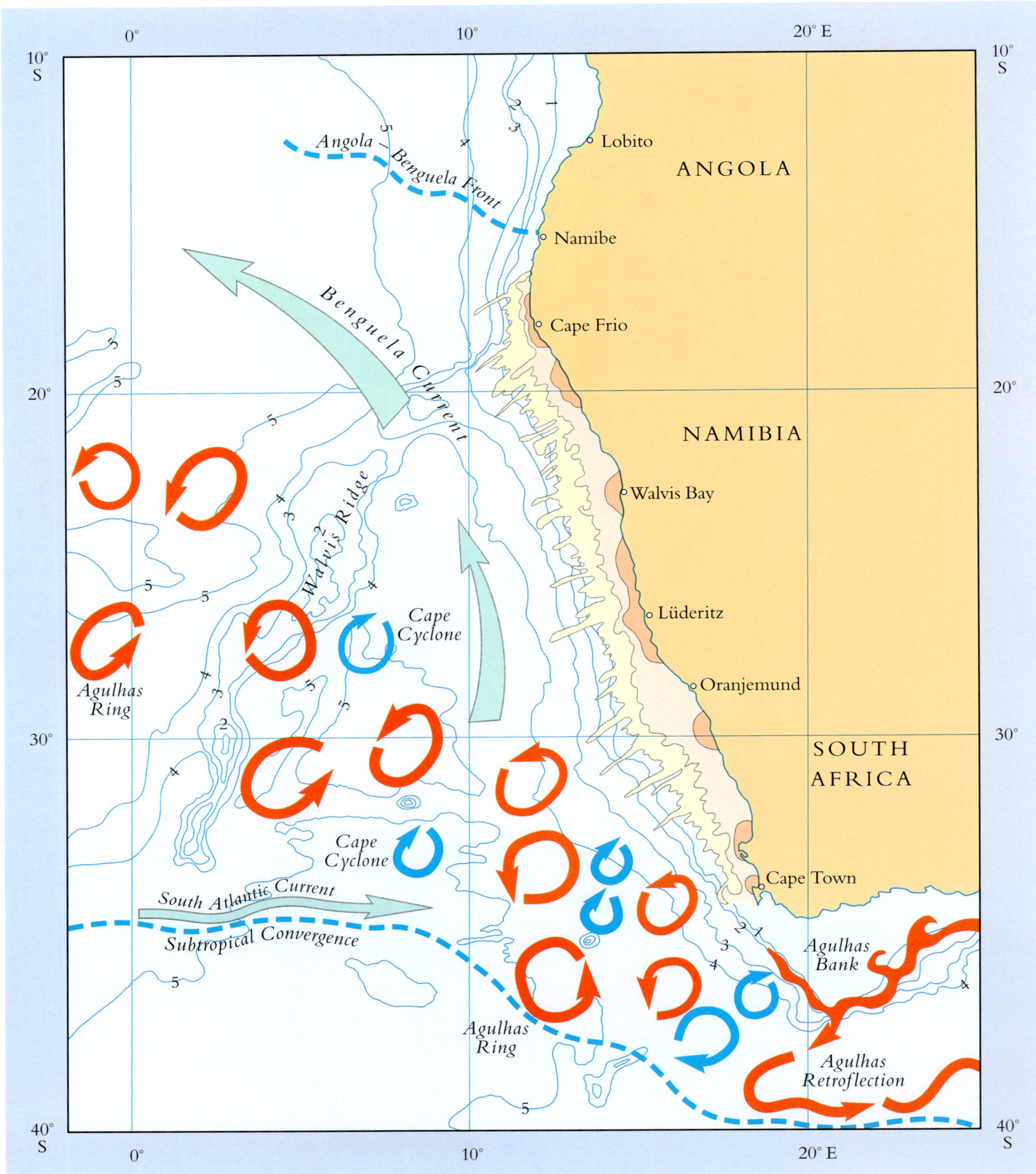

FIG. 2 A pictorial representation of the fronts and currents of the Benguela system in the south-east Atlantic Ocean. Intense currents, such as the Agulhas Current and Agulhas rings, are shown by solid arrows whereas general background drifts, such as that of the Benguela Current and the South Atlantic Current, are indicated by light blue arrows. Fronts are shown by broken lines. All these are superimposed on the bottom topography, which is given in kilometres. The continental shelf shallower than 1 km is the lightest shade of blue. The Agulhas Current with its warm (red) water, sheds Agulhas rings (red) at the Agulhas retroflection, which then move north-westward into the South Atlantic Ocean. They are interspersed by cold cyclones (blue) with a clockwise spinning direction. The wind-driven upwelling along the coast is not directly related to the Benguela Current. It consists of three parts: site-specific upwelling cells next to the coast (dark pink), a contiguous upwelling zone (light pink), and a seaward band of cold filaments, wisps and eddies (yellow).

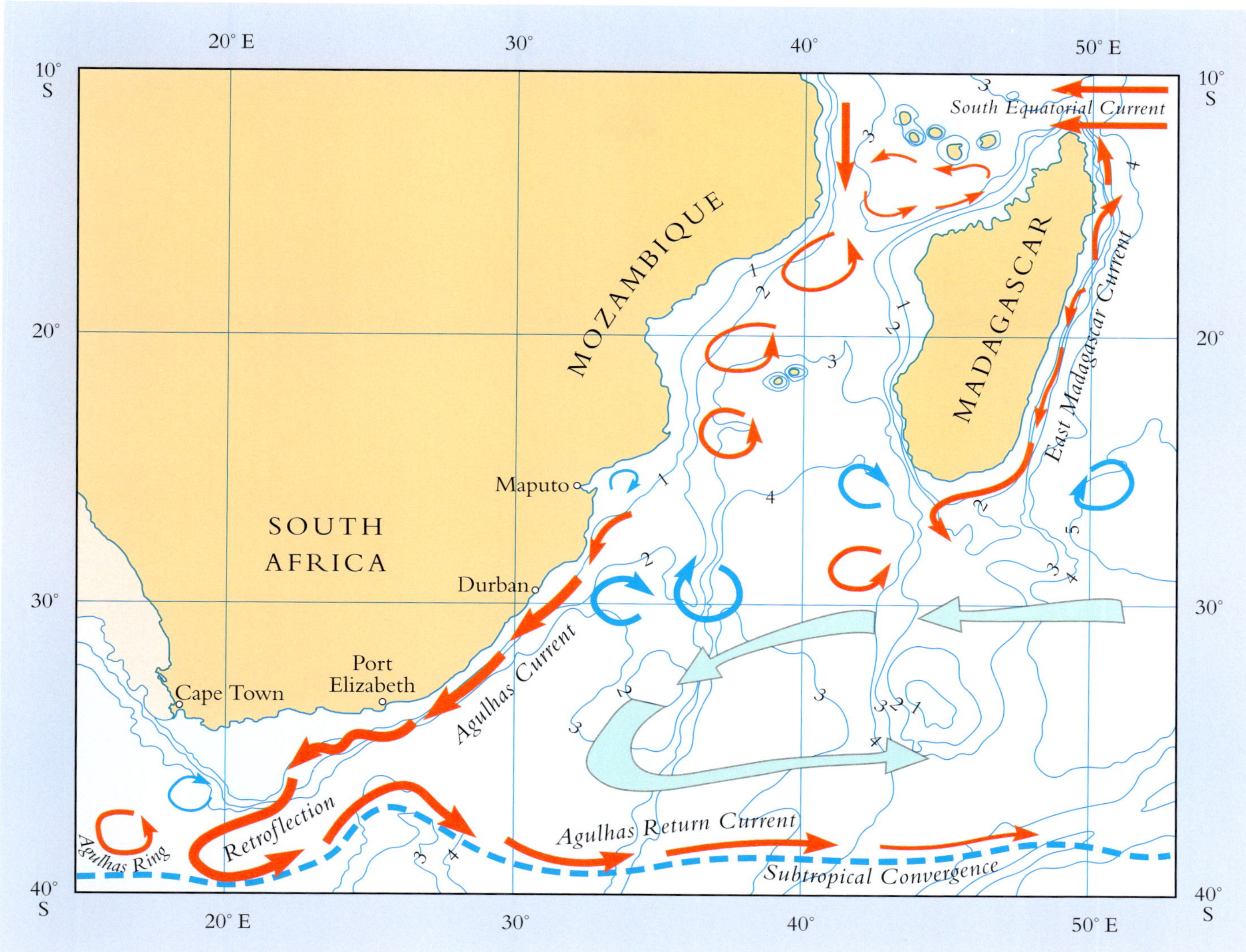

FIG. 3 A pictorial representation of the front and currents of the greater Agulhas Current system in the south-west Indian Ocean. Intense currents, such as the Agulhas and East Madagascar Currents, as well as Agulhas rings, are shown by solid arrows, whereas the general background flow is indicated by light blue arrows. The front is shown by a broken line. All these are superimposed on the bottom topography, which is given in kilometres. The continental shelf shallower than 1 km is the lightest shade of blue. Coastal upwelling along the south-west coast is indicated by a light pink shade. The Agulhas Current with its warm (red) water, sheds Agulhas rings (red) at the Agulhas retroflection. Warm eddies are also formed in the Mozambique Channel and south of Madagascar, while clockwise rotating eddies (blue) arrive from the east and may also be formed south of Madagascar and near Maputo.

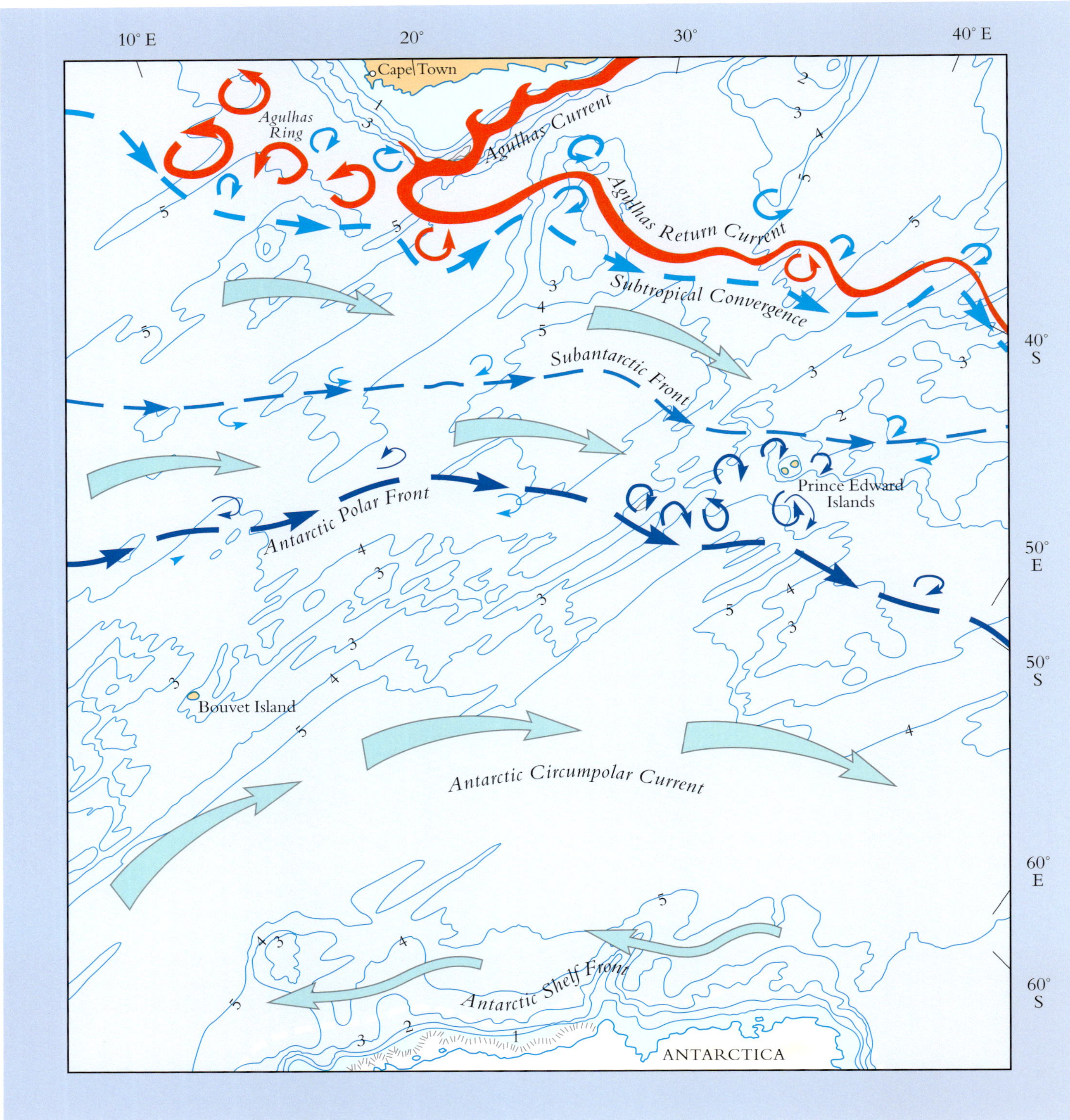

FIG. 4 A pictorial representation of the fronts and currents in the ocean region between Africa and Antarctica. South of the Subtropical Convergence is the Southern Ocean. Intense currents, such as the Agulhas Return Current and Agulhas rings, are shown by solid arrows, whereas general background drift, such as that of the Antarctic Circumpolar Current, is indicated by light blue arrows. Fronts are shown by broken lines with arrows indicating the concentrated movement at these fronts. All these are superimposed on the bottom topography, which is given in kilometres. The continental shelf shallower than 1 km is the lightest shade of blue, whereas the edge of the shelf ice is shown by a picket fence. The Agulhas Current with its warm (red) water, sheds Agulhas rings (red) at the Agulhas retroflection. They are interspersed by cold cyclones (blue) spinning clockwise. Eddies at fronts have different spinning directions. Note in particular the preponderance of eddies in the vicinity of the Prince Edward Islands. Note also that the map projection is not rectangular.

On the whole it is believed to be towards the equator,[49] which can be seen by the occasional northward advection of a filament of warm surface water from the landward edge of the Agulhas Current. However, there is evidence that the intermittent presence of a cyclonic lee eddy off this shelf slope may cause occasional reversals of flow, with water being carried towards the pole instead.

Once the Agulhas Current has passed the southern tip of the African continental shelf, it turns back on itself in a tight loop, called the Agulhas retroflection[50] (Fig. 3). This configuration has been shown to be unstable and at irregular intervals the loop occludes, forming independent Agulhas rings that move off into the South Atlantic. During this process large intrusions of cold subantarctic surface water move into the South Atlantic[51] between the newly formed rings and the newly constituted Agulhas retroflection loop. This contrast in warm and cold water, or nutrient-poor and nutrient-rich water, may be reflected in the presence of cetaceans.[2] Agulhas rings may extend to the sea floor and in the Cape Basin, west of Cape Town, they may split, disperse or join with other rings.[52] They are believed to be the largest eddies of their kind in the world's oceans. Being warm at the surface in a windy and cold environment they rapidly lose their surface heat and thus their contrast with ambient waters.

The part of the Agulhas Current that is not involved in the shedding of rings flows eastwards after retroflection as the Agulhas Return Current.[53] At its start this current has much the same characteristics as its mother, the Agulhas Current. However, as it travels downstream, Agulhas water continuously leaks northward into the South Indian Ocean so that by about 70° E all Agulhas water has been dissipated. The remainder of the current along the Subtropical Convergence is called the South Indian Current. As it moves east the Agulhas Current traverses a number of ocean ridges and is forced northward in each case. Flowing along the Subtropical Convergence at the northern boundary of the Southern Ocean, the Agulhas Return Current also becomes part of a special habitat. Sporadically enhanced biological productivity at this front occurs in the form of short-lived events[54] over patches of a few hundred kilometres.

The Southern Ocean

The Southern Ocean is the name given to the oceanic region that surrounds the entire Antarctic continent (Fig. 4), and consists of the southern extents of the Atlantic, Indian and Pacific oceans. The continent of Antarctica forms its southern boundary, while the northern border is not physiographically fixed, but is usually considered to coincide with the location of the Subtropical Convergence.[55] The unique geography of the Southern Ocean makes it the only place where ocean currents run completely around the globe. Dominating the region is the strong, deep, eastward-flowing Antarctic Circumpolar Current (ACC) that lies between 40° S and 65° S.[56] This is often referred to as the 'West Wind Drift' after the strong westerly winds that prevail between the subtropical high pressure belt and the Antarctic trough.

The Southern Ocean plays a major role in global ocean circulation. Extending unbroken around Antarctica, it is the primary means whereby water, heat and salt are transferred between different ocean basins. As these exchanges play an important role in regulating the mean global climate, a better understanding of the physical and dynamic processes responsible for the variability of the ACC is very important.[57, 58]

As a large meridional gap between the African and Antarctic continents, the oceanic region south of Africa allows not only an eastward exchange of cold waters from the Pacific and the Atlantic basins into the Indian Ocean (conveyed mainly by the ACC), but also a direct exchange between waters of the subtropical Indian and South Atlantic gyres.[59, 60] This temperate inter-ocean exchange is supposed to affect the heat and freshwater budget of the Atlantic and be very important for global thermohaline circulation and stability.[61, 62]

Moreover, meridional heat transport is usually directed from the tropics to the polar regions, but in the Atlantic Ocean the oceanic heat transport is northwards in both hemispheres. This anomaly is related to the role that the Atlantic Ocean basin plays in the thermohaline circulation and in the exchange of water heat and mass with the Indian and Pacific via the Southern Ocean and around South Africa.[63]

The surface flow of the ACC (Fig. 4) is driven primarily by the frictional stress of the westerly wind. Combined with the Coriolis force, a northward component to the surface current is responsible for the formation of frontal bands, which act as boundaries separating zones of uniform water mass. These fronts are a distinct permanent feature in the Southern Ocean. They are circumpolar but show a strong temporal variability in latitude and structure in different oceanic sectors[55, 64] (Fig. 4). Their positions are mainly determined by topographic steering of the ACC[65] in combination with the global spatial pattern of wind fields, in particular the distribution of the wind stress curl.[66] Transects across the ACC show this current to consist of zones of several quasi-uniform water masses separated by fronts in which the horizontal gradients of properties are relatively large.[67] These broad zones are (from north to south) the Subantarctic Zone, the Polar Frontal Zone, the Antarctic Zone and the Continental Zone.[68] They are demarcated by the major circumpolar fronts of the ACC, namely the Subtropical Convergence, the Subantarctic Front, the Antarctic Polar Front and the Southern Antarctic Circumpolar Front respectively.[7, 55, 69–73]

The Subtropical Convergence

The Subtropical Convergence (STC) marks the boundary between warm, salty subtropical surface water to the north and cooler, fresher subantarctic surface water to the south. It is the most northerly front associated with the ACC and the most prominent surface thermal front, with surface temperature ranging from 10.3 °C to 16.9 °C.[7] In the South Atlantic its mean position lies at 41°40′ S.[55]

Two separate fronts associated with the northern and southern boundaries of the STC in a Subtropical Frontal Zone have been identified in the south-east Atlantic sector of the Southern Ocean.[7, 74]

In the south-west Indian sector of the Southern Ocean it is one of the strongest fronts in the world's oceans.[55] Here the Agulhas Return Current[53] strengthens the horizontal gradients of this front considerably, leading to the formation of intense mesoscale eddies. These heat anomalies have been shown to have a marked influence on the overlying atmosphere and on local biota. Satellite remote sensing and modelling of the Subtropical Convergence[54] have demonstrated the climatological importance of its primary production, which occurs as event-scale blooms and not as persistent enhancement as was previously thought.[75]

The Subantarctic Front

The Subantarctic Front (SAF) represents the northern boundary of the Polar Frontal Zone and marks the southern extent of subantarctic surface water. In comparison to the STC, which is clearly characterised by a sharp and consistent gradient in both surface and subsurface expressions, making identification extremely easy,[69] the SAF is less clear in its surface expression. Owing to the weak nature of this front, the exact boundaries of the Polar Frontal Zone can be difficult to identify. The SAF is predominantly a subsurface front and can be defined by the most vertically orientated isotherm within a temperature gradient lying between 3 °C and 5 °C, while its surface expression extends between 8 °C and 4 °C.[33] The SAF changes in structure west and east of the Crozet Plateau.[76] West of the Crozet Plateau it appears as a band typically 200–300 km wide consisting of a series of steps. As the surface expression is generally weak, the subsurface structure with the 6 °C isotherm as its core is usually the preferred indicator. Despite its ill-defined surface expression, recent investigations[58] have found that the SAF contributes over 30% to the volume flux of the ACC. Between 20° W and 80° E the subsurface SAF axial indices remain fairly stable, lying at a mean location of 47°25′ S.[7,69]

The Antarctic Polar Front

The Antarctic Polar Front (APF) or Antarctic Convergence marks the northern limit of the Antarctic zone and the area where Antarctic surface water moving northwards sinks below warmer subantarctic water masses. The APF is a region of elevated current speeds and strong horizontal gradients in density, temperature and salinity.[77,78] The subsurface expression of the APF is historically identified by the northern limit of the 2 °C temperature minimum at a depth of 200 m. However, the surface and subsurface expressions do not always coincide and are often approximately 50 km apart.[69] In these instances the surface expression can be identified by the maximum temperature gradient between 6 °C and 2 °C. The APF can be characterised further by a shallow temperature minimum associated with the remnants of winter water, which lie at depths of 50–150 m. It is seasonally variable: in winter it is nearly homogeneous, extending to 250 m, while in summer the mixed layer extends only to between 50 m and 100 m. Temperatures for this water mass range from −1.8 °C to 6 °C at the APF. South of Africa the APF has a mean location of 50°47′ S,[69] whereas the circumpolar average location of the APF lies between 52° S and 53° S.[79]

Current measurements at the Drake Passage have shown that the ACC includes an additional frontal feature south of the APF, called the Southern Antarctic Circumpolar Current Front.[80]

The Southern Antarctic Circumpolar Current Front

The Southern Antarctic Circumpolar Current Front (SACCF) has been described as a circumpolar, deep-reaching front lying south of the APF.[80] The position of this front corresponds to the position of the atmospheric low-pressure Antarctic trough, which separates the easterly and westerly wind belts at about 65° S. In contrast to the other fronts associated with the ACC, the SACCF does not separate distinct surface water masses; instead, it is defined by the temperature and salinity characteristics of the Upper Circumpolar Deep Water. South of Africa the SACCF has been identified at a mean position of 63°48′ S.[76]

In a narrow zone around most of the continent there is a westward flowing coastal current termed the East Wind Drift after the prevailing easterly winds near the coast.[77] Flowing westwards, this current is a counter-current to the ACC and can be found south of 66° S. Although the East Wind Drift is circumpolar, the Antarctic Peninsula partially impedes its flow, resulting in an increase in sea ice concentrations throughout the year.

Sea ice

The Antarctic sea ice zone undergoes one of the greatest seasonal surface changes on earth. It expands from a minimum extent of four million km^2 in February to a maximum of 19 million km^2 in September. At its greatest extent the area of sea ice is greater than that of Antarctica itself. The ice edge varies with location around Antarctica as a result of the ocean circulation, in particular the coastal current. Variability in sea ice influences the global heat budget, surface albedo, and consequently ocean and atmosphere circulation. Regionally, the advance and retreat of sea ice strongly affect the physical environment, dramatically changing ocean-atmosphere exchanges of momentum, heat, gases, and hence ocean mixed-layer properties (e.g. depth, light availability and nutrients). In turn, these seasonal changes in the ocean-atmosphere-ice environment significantly affect the marine ecosystem through changes in the vertical stability of the water column. Indeed, recent observations in the Ross Sea have shown that surface waters on the continental shelf are freshening[81,82] as a result of an increase in ice melt from the continent, a decrease in sea ice formation, and an increase in precipitation.

Anomalies in the extent of Antarctic sea ice, sea surface temperature, and pressure and winds at sea level propagate eastward around the Southern Ocean as a bimodal phenomenon called the Antarctic Circumpolar Wave (ACW).[83] The ACW has a periodicity of 4–5 years and takes approximately 8–10 years to encircle Antarctica. It is characterised by a persistent phase relationship between warm (cool) sea surface temperature anomalies and poleward (equatorward) meridional surface wind anomalies. It is not clear how these waves are triggered and maintained – likely factors are the strong westerly winds in the region, the bottom topography, and the meridional temperature gradient in the upper ocean.

Variability

Extensive investigations have been carried out in the past decades to understand the flow pattern of the ACC and the extent of its temporal and spatial variability throughout the Southern Ocean. A good correspondence between the increased surface speeds of the ACC and the locations of the three main fronts (the STC, the SAF and the APF) seems to exist.[84] A banded current structure clearly occurs within the ACC, with greater densities observed along the fronts. Meridional excursions from the otherwise zonal flow in the form of meanders of mesoscale eddies occur in regions of prominent bathymetric features, suggesting a close relationship between the level of mesoscale sea surface variability and bottom topography. These regions coincide with areas of prominent topographic relief, such as the Drake Passage, the South-West Indian Ridge, Crozet Plateau, Kerguelen Plateau, and the Macquarie Ridge.

Significant inter-annual variability occurs in ecosystems that are coupled tightly to local physical processes. Extreme events associated with the Benguela region – such as Benguela Niños, the intrusion of Agulhas rings into the south-east Atlantic, and the high bands of mesoscale variability associated with the Antarctic Circumpolar Current – are all examples of how the physical environment may have great implications for the distribution, growth, reproduction and mortality of living organisms.[85] Attention needs to be given to an increase in sustainable observations that provide, explore and monitor the role that large-scale ocean processes have in influencing the macro- and mesoscale oceanographic, climatic and background biological productivity between ocean basins.

References

1 Findlay *et al.* 1992
2 Cockcroft *et al.* 1990a
3 Lutjeharms 2001
4 Lutjeharms 2006a
5 Shannon 1985
6 Shannon and Nelson 1996
7 Belkin and Gordon 1996
8 Peterson and Stramma 1991
9 Shannon *et al.* 1987
10 Lutjeharms and Meeuwis 1987
11 Garzoli *et al.* 1996
12 Wedepohl *et al.* 2000
13 Parrish *et al.* 1983
14 Agenbag and Shannon 1988
15 Stander 1964
16 Taunton-Clark and Shannon 1988
17 Shannon and Agenbag 1990
18 Brundrit *et al.* 1987
19 Shannon *et al.* 1992
20 Agenbag 1996
21 Lutjeharms and Stockton 1987
22 Hart and Currie 1960
23 Moroshkin *et al.* 1970
24 Dias 1983a
25 Dias 1983b
26 Lass *et al.* 2000
27 Meeuwis and Lutjeharms 1990
28 O'Toole 1980
29 Mazeika 1967
30 Signorini *et al.* 1999
31 Boyd *et al.* 1987
32 Gammelsrød *et al.* 1998
33 Shannon *et al.* 1988
34 Shannon *et al.* 1986
35 Lutjeharms 2006b
36 Lutjeharms *et al.* 2000a
37 Lutjeharms and Jorge da Silva 1988
38 De Ruijter *et al.* 2002
39 De Ruijter *et al.* 2004
40 Gründlingh 1983
41 Gründlingh 1985
42 Lutjeharms and Machu 2000
43 Best *et al.* 2003a
44 Lutjeharms and Roberts 1988
45 Bryden *et al.* 2005
46 Beal and Bryden 1997
47 Lutjeharms *et al.* 1989
48 Lutjeharms *et al.* 2000b
49 Lutjeharms and Cooper 1996
50 Lutjeharms and Van Ballegooyen 1988
51 Shannon *et al.* 1989
52 Boebel *et al.* 2003
53 Lutjeharms and Ansorge 2001
54 Llido *et al.* 2005
55 Lutjeharms 1985
56 Taylor *et al.* 1978
57 Budillon and Rintoul 2003
58 Swart *et al.* (submitted)
59 De Ruijter 1982
60 Speich *et al.* 2001
61 Gordon *et al.* 1992
62 De Ruijter *et al.* 1999
63 Stramma and England 1999
64 Joyce and Patterson 1977
65 Gordon *et al.* 1978
66 Nowlin and Klinck 1986
67 Nowlin and Clifford 1982
68 Whitworth III 1980
69 Lutjeharms and Valentine 1984
70 Whitworth III and Nowlin 1987
71 Read and Pollard 1993
72 Park *et al.* 1991
73 Park *et al.* 1993
74 Smythe-Wright *et al.* 1998
75 Weeks and Shillington 1996
76 Holliday and Read 1998
77 Deacon 1937
78 Deacon 1983
79 Treshnikov and Baranov 1976
80 Orsi *et al.* 1995
81 Jacobs *et al.* 2002
82 Jacobs 2004
83 White and Peterson 1996
84 Hofmann 1985
85 Knox 1984

THE SOUTH AFRICAN SPECIES GATE

William F. Perrin

National Marine Fisheries Service, Southwest Fisheries Science Center, La Jolla, California, USA

DESPITE THEIR NEARNESS to each other, the tropical South Atlantic and the southern Indian oceans have very different dolphin faunas. Why? One possible part of an explanation is that the southern tip of Africa acts as a barrier for tropical marine animals, allowing them to evolve independently in the two regions. More importantly for our query about differential diversity, in the past when the oceans were warmer, subtropical water moving around this barrier may instead have created an open gate, allowing sporadic movement of tropical cetaceans between the two basins, perhaps more in one direction than the other.[1]

What are the faunal differences? The tropical Atlantic has five species of dolphins in the genus *Stenella*:[2–4] the striped dolphin (*S. coeruleoalba*), the spinner dolphin (*S. longirostris*), the Clymene dolphin (*S. clymene*), the pantropical spotted dolphin (*S. attenuata*) and the Atlantic spotted dolphin (*S. frontalis*). However, the Indian Ocean (and the rest of the world's tropical waters) have only three: *S. coeruleoalba, S. longirostris* and *S. attenuata*. The Indian Ocean has two bottlenose dolphins, the common bottlenose dolphin (*Tursiops truncatus*) and the Indo-Pacific bottlenose dolphin (*T. aduncus*), whereas the Atlantic has only the common bottlenose.[5] The humpback dolphins in the South Atlantic are considered to belong to a different species (*Sousa teuszii*) than those on the Indian Ocean coast (*S. chinensis* or *S. plumbea*, depending on whose classification you believe).[6]

Why would the tip of southern Africa act as a barrier for tropical dolphins at some times but not at others? During the hottest month of the year, subtropical to tropical water (20–25 °C) extends around the Cape of Good Hope and up into the Atlantic.[7] In the winter this connection is interrupted, with the 20-degree isotherm hitting the continent north of the Cape. Thus it is reasonable to suppose that during a Pleistocene or earlier cooling period the tropical Atlantic was completely cut off from the rest of the tropics.

The pattern of oceanic currents bordering southern South Africa suggests that the occasionally open gate may have been effectively a one-way filter, enabling greater movement from east to west than the reverse. When the sub-African corridor is open in the summertime, the fast Agulhas Current (up to 5 knots) runs from northeast to southwest, and the prevailing Benguela Current along the west coast of Africa flows from south to north, so a tropical dolphin moving from the Atlantic into the Indian Ocean would have to buck the currents for several thousand kilometres. One coming from the Indian Ocean into the Atlantic would have a more or less free ride in the eddies shed by the Agulhas Current that contribute to the origin of the Benguela.[8]

So, considering the opening and closing of the gate and its likely mostly one-way nature, an hypothesis explaining the existence of two spotted dolphins in the Atlantic and only one elsewhere goes like this: stage 1) one primordial spotted dolphin species or close ancestor existed in the world's oceans, stage 2) isolation of the tropical Atlantic occurred when the gate closed, stage 3) divergent evolution in the Atlantic and the Indian/Pacific oceans gave rise to two species, stage 4) re-opening of the gate, and stage 5) movement of the Indo-Pacific species westward through the gate into the Atlantic, resulted in the sympatric occurrence of a pantropical species and an Atlantic endemic.

The same vicariant scenario would fit the spinner/Clymene species pair: movement of the spinner west into the Atlantic after a period of separation and divergent evolution and speciation. In the case of the bottlenose dolphins, the movement would have had to have been in the opposite direction, as there are two species in the Indian and Pacific oceans but only one cosmopolitan one in the Atlantic.

These are simplistic scenarios, of course, and the real happenings may have involved a series of openings and closings of the gate and complex movements of dolphins back and forth. Submergence of the Isthmus of Panama during the Pliocene and sporadically in earlier periods may also have played a part in the picture, providing a connection between the tropical Atlantic and Pacific.[9] Alternative scenarios can certainly be imagined.

Another fly in the ointment is that the apparent similarities of species in pairs (spinner/Clymene dolphins, Atlantic and pantropical spotted dolphins, and common/Indo-Pacific bottlenose dolphins) don't accord well with current molecular phylogenies. For example, in a phylogeny based on cytochrome *b* gene sequences in the mitochondrial DNA, the Atlantic spotted dolphin is more closely related to the common bottlenose dolphin than it is to the pantropical spotted dolphin.[10] Similarly, the Clymene dolphin is more closely related to the striped dolphin than to the spinner dolphin. Indeed, the genus *Stenella* is indicated to be an artificial taxon, containing species that are more closely related to, or as closely related to, species in other genera (*Sousa*, *Tursiops* and *Lagenodelphis*) than they are to other species in *Stenella*. However, these conclusions, while suggestive, are based on a single marker in the mitochondrial genome, which is known to be subject to introgression (movement of sequences across populations and species independent of the nuclear genome) and homoplasy due to saturation (because of relatively high mutation rates). Data for more markers, including nuclear markers, will be needed to confirm relationships among the delphinid species.[11] In any case, evolution of five species of *Stenella* from three could have occurred in a more complex fashion than suggested by the species-pairs hypothesis. For example, the spinner dolphin could have been confined to the Indian Ocean (and other tropical waters to the east), and the spinner-like Clymene dolphin could have evolved in the Atlantic from a common ancestor with the striped dolphin through convergence. Similarly, the Atlantic spotted dolphin could have evolved in the Atlantic from a common ancestor with the common bottlenose dolphin by convergence, in the absence of the pantropical spotted dolphin.

A more fascinating idea is that the Clymene dolphin may have arisen in the Atlantic from hybridization between the striped dolphin and the spinner dolphin.[10, 12] This hypothesis arises from consideration of the pattern of similarities among the three species. The striped dolphin and Clymene dolphin are similar in mtDNA sequence and in skull morphology; this might be due to common descent. The Clymene dolphin is much more similar to the spinner dolphin in colour pattern. This could reasonably be ascribed to the retention of primitive characteristics possibly present in the ancestors of all the delphinids. However, both the Clymene dolphin and the spinner dolphin spin! It is hard to believe that this unique and wonderful behaviour exhibited only by these two species is ancestral in the delphinids. Perhaps the Clymene dolphin is descended from a striped mother and a spinner father? Research is underway to examine this possibility by using a series of mitochondrial and nuclear markers, including loci on the Y chromosome that are transmitted only through the paternal lineage.

Whatever the actual course of evolution has been, it would appear that the southern tip of Africa has played an important role in the evolution and present distribution of dolphin species in the tropical oceans.

References

1 Perrin *et al.* 1978
2 Perrin *et al.* 1987
3 Perrin *et al.* 1981
4 Archer 2002
5 Wells and Scott 2002
6 Ross 2002
7 Davies 1962
8 Longhurst 2007
9 Briggs 1974
10 LeDuc *et al.* 1999
11 Reeves *et al.* 2004
12 Leslie *et al.* 2005

A CLASSIFICATION OF LIVING CETACEA[1]

Order CETACEA / whales, dolphins and porpoises
- Suborder MYSTICETI / whalebone whales
  - Family BALAENIDAE / right whales (4 species)
  - Family NEOBALAENIDAE / pygmy right whale (1 species)
  - Family ESCHRICHTIIDAE / gray whale (1 species)
  - Family BALAENOPTERIDAE / humpback and rorquals (8 species)
- Suborder ODONTOCETI / toothed whales
  - Superfamily PHYSETEROIDEA
    - Family PHYSETERIDAE / sperm whale (1 species)
    - Family KOGIIDAE / short-headed sperm whales (2 species)
  - Superfamily ZIPHIOIDEA
    - Family ZIPHIIDAE / beaked and bottlenose whales (20 species)
  - Superfamily PLATANISTOIDEA
    - Family PLATANISTIDAE / Indian river dolphin (1 species)
    - Family INIIDAE / Amazon river dolphin (1 species)
    - Family LIPOTIDAE / baiji (1 species)
    - Family PONTOPORIIDAE / franciscana (1 species)
  - Superfamily DELPHINOIDEA
    - Family MONODONTIDAE / narwhal and white whale (2 species)
    - Family DELPHINIDAE / marine dolphins (36 species)
    - Family PHOCOENIDAE / porpoises (6 species)

Reference

1 After Rice 1998

KEY TO THE WHALES, DOLPHINS AND PORPOISE OF THE SUBREGION

THIS KEY HAS been designed primarily as a guide to the identification of animals 'in the flesh' and will be most useful in situations where one has a specimen to hand, such as at a stranding or in a fisheries by-catch. It will be less useful for identifying animals at sea, where the observer has to rely to a great extent on behavioural cues and brief glimpses of the animal, and for which a deal of experience is really necessary.

Even with a specimen to hand, identification can be problematic. Young or immature animals, whose teeth or baleen may not be fully erupted, can prove troublesome to key out, and decomposition can play havoc. The skin first turns black in the sun (masking most details of the colour pattern) and then falls off, the exposed blubber making the animal appear all 'white'. Eventually the animal loses much of its shape and the baleen or teeth can fall out (or be stolen).

Taking all these factors into account, it has been considered practical to take this key only as far as genus level, and to omit details of colour pattern.

If the observer is going to report the incident to an interested museum or fisheries official, it will also help if the following data are provided:

Location. Valuable time will be saved if the location is given as accurately as possible. GPS positions are ideal, but odometer readings (where you are allowed to drive on the beach!) or even walking time from an access point are useful.
Size. This is a really important piece of information, and most people's guesstimates are way off the mark. Pacing off a length (or using short steps with your shoe size as a measure) is quite acceptable if a tape measure is not available.
Photographs. Many people now have digital cameras and an image or two of the animal can save a thousand words and result in rapid identification. Views of the head are especially important, from all three sides if possible.
A skin sample. Species (and even gender) can be determined genetically from such a sample. Only a small piece (enough to cover half a postage stamp) of the outer, pigmented skin is necessary. A good place to find skin that has not become too dessicated is under the flipper. The sample should be collected with a clean implement, and then kept moist in a clean container with seawater until it can be refrigerated.

Key to the suborders

Baleen present in upper jaw; blowhole paired; ribs articulated only with transverse processes of vertebrae; no sternal ribs

More than one species

MYSTICETI
(Whalebone whales)

Teeth present in lower or both jaws; blowhole single; ribs articulated with centra of vertebrae; sternal ribs distinct, ossified or cartilaginous

More than one species

ODONTOCETI
(Toothed whales)

MYSTICETI – Key to families

No throat grooves present; wart-like callosities on head; lower jaw arched; baleen plates narrow and mainly black; no dorsal fin

Only one species in the subregion

BALAENIDAE
(Southern right whale)

No throat grooves or callosities on head; lower jaw arched, baleen plates narrow and either white or white with black outer edge; dorsal fin present

Only one species in the subregion

NEOBALAENIDAE
(Pygmy right whale)

At least 20 throat grooves; no callosities; lower jaw nearly straight; baleen plates short and broad; dorsal fin present

More than one species

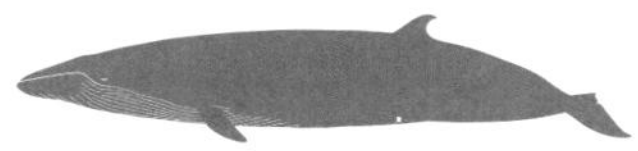

BALAENOPTERIDAE
(Rorquals, humpback whale)

BALAENOPTERIDAE – Key to genera

Flippers long, nearly one-third of body length; base of hair follicles on head in rounded protuberances; trailing edge of flukes highly scalloped

Only one species in the subregion

Megaptera
(Humpback whale)

Flippers short, about one-fifth of body length; no rounded protuberances on head; trailing edge of flukes unscalloped

More than one species

Balaenoptera
(Rorquals)

ODONTOCETI – Key to families

1 Upper jaw prolonged into a narrow beak; usually one or two pairs of teeth in lower jaw only (erupting in adult males); two grooves forming V-shape on throat; hind margin of tail flukes not notched in centre

More than one species

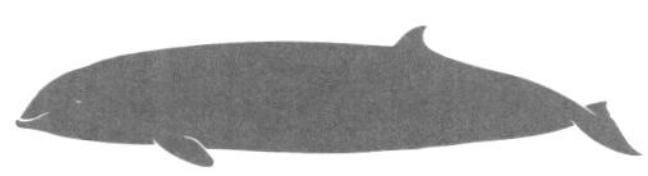

ZIPHIIDAE
(Beaked and bottlenose whales)

Several pairs of teeth present, usually in both upper and lower jaws; throat grooves absent or irregular in pattern; hind margin of flukes with central notch 2

2 From eight to 28 pairs of erupted teeth present, usually in lower jaw only; lower jaw narrow and much shorter than upper; no beak 3

Several pairs of teeth present in both jaws (except in *Grampus*, where 3–7 pairs in lower jaw only); beak may or may not be present 4

3 Large, up to 18 m in length; 18–28 pairs of teeth in lower jaw; blowhole S-shaped and at extreme tip of snout; body surface 'corrugated'

Only one species in the subregion

PHYSETERIDAE
(Sperm whale)

Small, up to 3.5 m in length; 8–16 pairs of teeth in lower jaw; blowhole straight or slightly bow-shaped and on top of head; surface of body smooth

More than one species

KOGIIDAE
(Short-headed sperm whales)

4 Beak moderately long (less than 10% of body length), short or absent; teeth conical; dorsal fin usually hooked, sometimes triangular or lobe shaped (or even absent)

More than one species

DELPHINIDAE
(Dolphins)

No beak; teeth spade-shaped at the tip; dorsal fin rounded

Only one species in the subregion

PHOCOENIDAE
(Porpoises)

ZIPHIIDAE – Key to genera

Two pairs of teeth at the tip of the lower jaw, which projects considerably beyond upper; blowhole is crescent-shaped, but the arms of the crescent point forward

Only one species in the subregion

Berardius
(Arnoux's beaked whale)

Single pair of large teeth at tip of lower jaw, but 17–22 pairs of smaller teeth (5–15 mm in diameter) in both jaws (not all of which may erupt); blowhole crescent-shaped but arms of the crescent point backwards

Only one species in the subregion

Tasmacetus
(Shepherd's beaked whale)

Single pair of teeth at tip of lower jaw, circular or oval in cross-section; snout broad and 'goose-beaked', with melon distinct but not bulbous; blowhole crescent-shaped but arms of the crescent point backwards

Only one species in the subregion

Ziphius
(Cuvier's beaked whale)

Single pair of teeth at tip of lower jaw, circular or oval in cross-section; snout narrow and pointed, melon bulbous and tending to overhang beak; blowhole crescent-shaped but arms of the crescent point backwards

Only one species in the subregion

Hyperoodon
(Southern bottlenose whale)

Single pair of teeth, usually near centre of lower jaw, but if at tip are noticeably flattened in cross-section; snout beak-like and obvious but melon not prominent and does not overhang beak; blowhole crescent-shaped but arms of the crescent point backwards

More than one species

Mesoplodon/Indopacetus
(Beaked whales)

DELPHINIDAE – Key to genera

1 Permanent teeth in lower jaw only; vertical furrow in front of head

Only one species in the subregion

Grampus
(Risso's dolphin)

Permanent teeth in both upper and lower jaws; no vertical furrow in front of head 2

2 No more than 13 pairs of teeth in upper and lower jaws 3

20 or more pairs of teeth in upper and lower jaws 6

3 Teeth confined to the front half of the rostrum; length of dorsal fin base tends to be far greater than its height

More than one species

Globicephala
(Pilot whales)

Teeth covering two-thirds or more of rostrum; length of dorsal fin base usually equivalent to, or less than, height of dorsal fin 4

4 Teeth flattened in front and behind; flippers large and almost oval in shape

Only one species in the subregion

Orcinus
(Killer whale)

Teeth more or less rounded in cross-section; flippers tapered 5

5 Small, up to 2.5 m; teeth occupy front two-thirds of rostrum; leading edge of flipper bow-shaped

Only one species in the subregion

Feresa
(Pygmy killer whale)

Large, up to 5.8 m; teeth occupy whole rostrum; leading edge of flipper S-shaped

Only one species in the subregion

Pseudorca
(False killer whale)

6 Teeth with roughened crowns; snout tapering into melon

Only one species in the subregion

Steno
(Rough-toothed dolphin)

Teeth with smooth crowns; snout absent or distinctly marked off from melon 7

7 Long-beaked dolphins with snout projecting clear of, and sharply demarcated from, remainder of head 8

Dolphins with beak very short or absent 11

8 Diameter of largest tooth 2–4 mm 9

Diameter of largest tooth 5–9 mm 10

9 Palate deeply grooved (has to be palpated or exposed with a knife)

More than one species

Delphinus
(Common dolphins)

Palate ungrooved or with only shallow grooves

More than one species

Stenella
(Spotted, spinner, striped dolphins)

10 Dorsal fin mounted on prominent fleshy ridge on back

More than one species

Sousa
(Humpback dolphins)

Dorsal fin not mounted on prominent ridge

More than one species

Tursiops
(Bottlenose dolphins)

11 More than 35 pairs of teeth in upper and lower jaws 12

Less than 35 pairs of teeth in upper and lower jaws 13

12 Dorsal fin absent

Only one species in the subregion

Lissodelphis
(Southern right whale dolphin)

Dorsal fin present

Only one species in the subregion

Lagenodelphis
(Fraser's dolphin)

13 Head melon-shaped with no trace of a beak

Only one species in the subregion

Peponocephala
(Melon-headed whale)

Head with beak which, even if indistinct, is marked off by lateral grooves from melon 14

14 Dorsal fin hooked

More than one species

Lagenorhynchus
(Dusky, hourglass dolphins)

Dorsal fin triangular or rounded at tip

More than one species

Cephalorhynchus
(Heaviside's, Commerson's dolphins)

ANATOMY OF CETACEAN SKULLS

WHEN THE KEY on pages 18–23 can no longer be applied because the animal is too decomposed, or only remains as a skeleton, the researcher is obliged to employ more traditional methods of identification, for which a familiarity with skull anatomy is of great use. The species accounts that follow contain several references to bones in the skull and how these may differ in size and shape between species. To assist the reader in such discussions, diagrams are provided below that illustrate dorsal and lateral views of 'typical' baleen and toothed whale skulls, with the names of the more important bones identified.

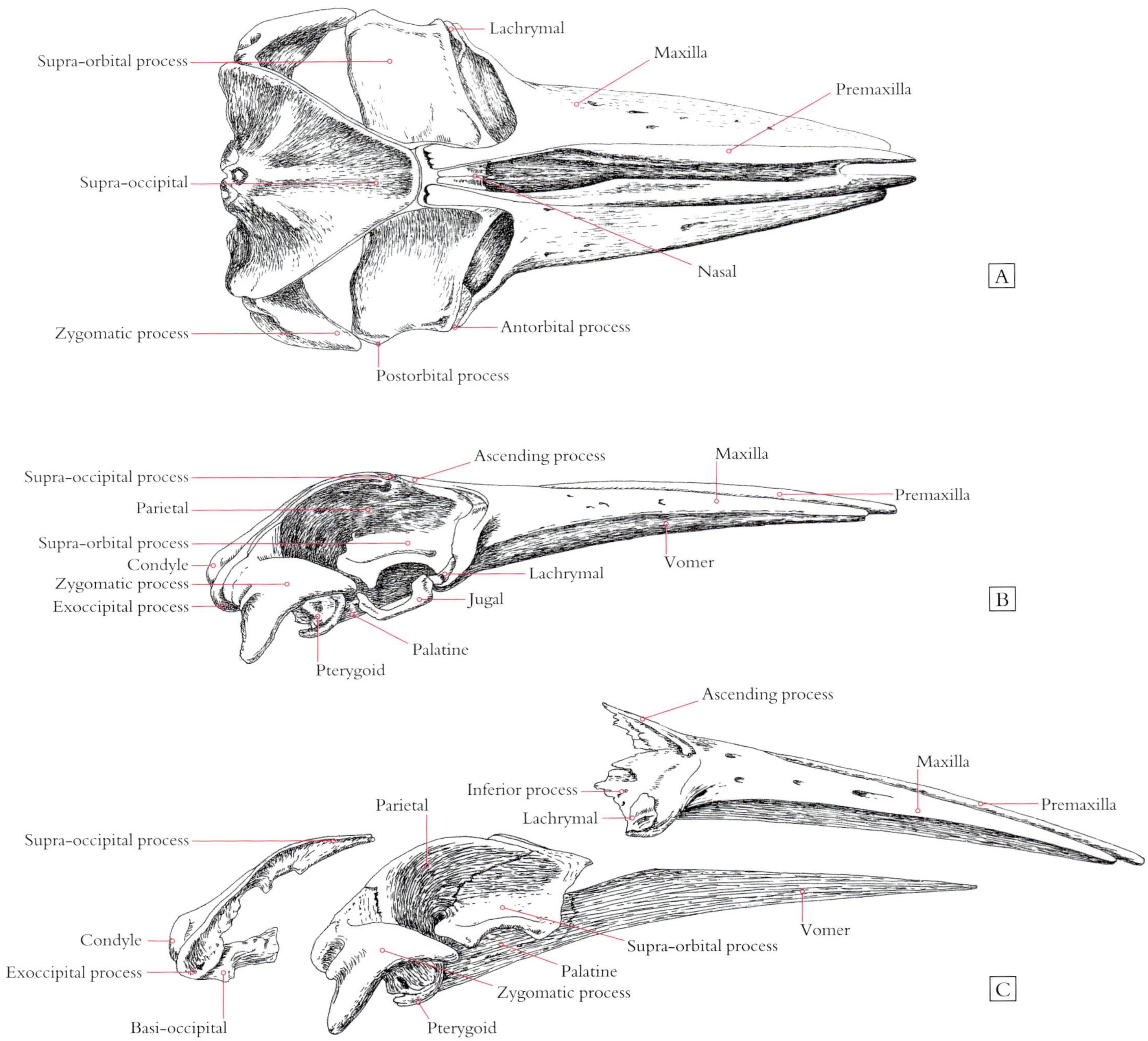

FIG. 5 (A) Dorsal and (B) right lateral view of whalebone whale skull. (C) Exploded right lateral view.

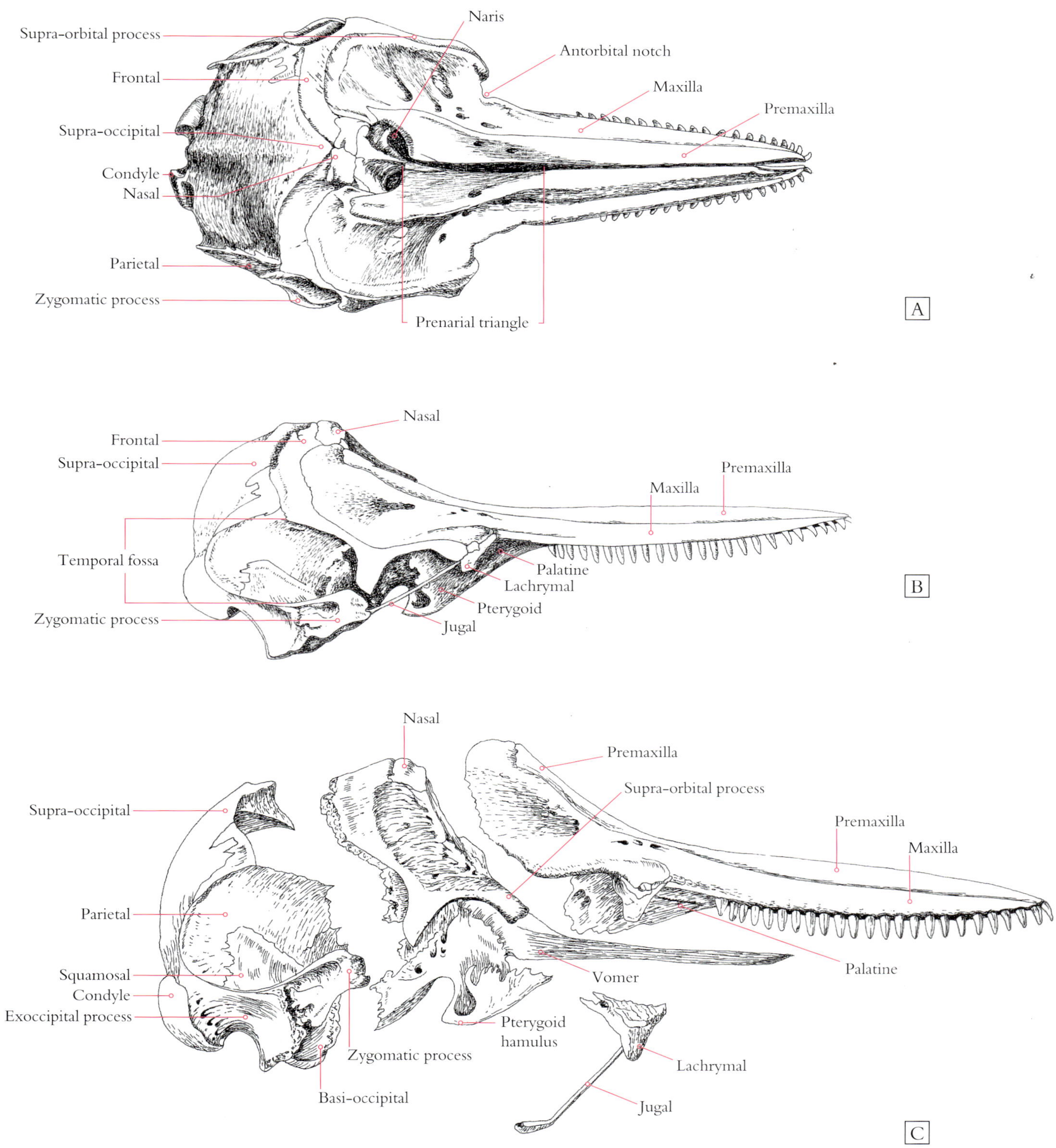

FIG. 6 (A) Dorsal and (B) right lateral view of toothed whale skull. (C) Exploded right lateral view.

Reference

Based on illustrations in Kellogg (1928)

Arguably the most widely distributed cetacean in the southern African subregion, a humpback whale spy hops off Gabon.

SPECIES ACCOUNTS

Family BALAENIDAE

Southern Right Whale

Eubalaena australis

Dorsal view of head and callosity pattern

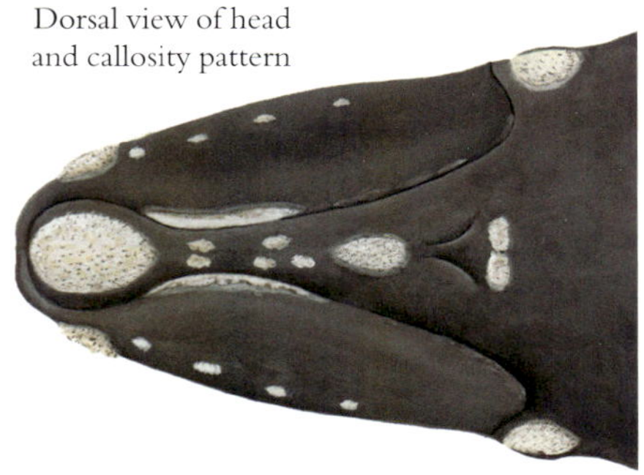

'RIGHT' OR 'TRUE' were epithets applied by early whalers to balaenid whales, apparently in recognition of their commercial value, especially the long tapering baleen that was superior to that of other whalebone whales. The fact that they were slow-moving, produced high oil yields, (usually) floated when dead, and showed fidelity to nearshore calving grounds, must also have helped. Right whales in the southern hemisphere are now regarded on phylogenetic grounds as a distinct species from those in the North Atlantic (*E. glacialis*) and North Pacific (*E. japonica*), although reliable morphological criteria for distinguishing between the three species are still lacking.[1,2]

Appearance

Right whales are hard to confuse with any other large whale. Firstly, they lack a dorsal fin, and, secondly, their head is covered with a pattern of yellowish-white wart-like callosities – sometimes referred to colloquially as the 'rock garden'. The blow is also characteristically V-shaped, although like all blows this can be affected by wind and whale behaviour, and humpback whales are also able to produce a similar blow. When lying at rest on the surface, which they tend to do a lot, right whales resemble overweight hippos.

Adult females average 13.9 m in length with a maximum of 15.5 m,[3] and are perhaps slightly larger than males.[4] The head is big, forming 27–31% of the length in adults, the upper jaw narrow and the lower lips strongly arched upwards. The flippers are large (16–21% of the body length) and broad (maximum width 50–58% of flipper length). Although they are sometimes described as paddle-shaped, in reality the flippers are more rectangular in outline, but with a pointed tip.[5] The flukes are pointed rather than hooked at the tip (as in humpback whales), and in adults their width averages 36% of body length.[3] Right whales are not shy to expose both their flippers and flukes above the surface.

The callosities are distributed over the head in a distinct pattern, but with sufficient variation to make them extremely useful in identifying individuals. The most prominent sits at the

Length:weight relationship

For 19 North Pacific right whales, $W = 0.0124L^{3.06}$, where W = weight in tonnes and L = total length in metres.[11] As these weights have been unadjusted for weight loss during cutting up, about 6% should be added.

Weights of only three southern right whales are available, all from South Africa. A 4.8 m female neonate weighed 1.1457 tonnes, a 9.85 m female 9.975 tonnes, and a 10.78 m female 16.34 tonnes: the first was weighed in pieces and the others whole.[8] If these are added to the weights for eight North Atlantic right whales (which are more like southern right whales in size than North Pacific right whales) already corrected for weight loss during dismemberment (6.8%),[12] then **$W = 0.0116L^{3.037}$**, where W = weight in tonnes and L = total length in metres. This implies that a 10 m whale would weigh 12.6 tonnes and a 15 m whale 43.3 tonnes.

Skeletal data

Vertebral formula: 7 cervical, 15 thoracic, 9–11 lumbar and 21–27 caudal; total 54–57.[13]

All cervical vertebrae are fused.

Phalangeal formula: I_{2-3}, II_5, III_6, IV_{5-6}, V_4.[13]

FIG. 7 Variations in pigmentation patterns of southern right whales: (A) White blazes stay permanently white and are found equally on both sexes; (B) Irregular markings such as these are white in the calf (C) but gradually darken with age, and are only found in females; (D) A few calves are born almost completely white (apart from a black collar and variable spotting), but also darken with age (E). Such 'brindled' animals are nearly always male.

tip of the snout and is known as the 'bonnet'. In front of the blowhole there is a 'coaming', and behind the blowhole a pair of 'post-blowhole' callosities. Between the bonnet and the coaming are a variable number of smaller, scattered 'rostral islands'. Above each eye is an 'eyebrow' callosity, on each side of the chin a 'chin callosity', and on the side of the lower jaw a number of small 'mandibular callosities'. Finally, and more typical of southern than northern right whales, there are often 'lip callosities' along the upper edge of each lower lip. These tend to be asymmetrical, being bigger on the right than on the left side of the head. Each callosity corresponds to the position of one or more rudimentary hairs that can be seen in stranded animals protruding bristle-like from the centre. Much of the contrast between the colour of the callosities and the surrounding black skin comes from the layer of external parasites (mostly cyamids) that cover the callosities (see page 32).[6]

Right whales were often referred to as 'black whales' by open-boat whalers, and indeed most individuals are black in colour, at least on the back. On the belly, however, most animals possess a white blaze in the region of the navel, varying in size from a small spot to a large irregularly-shaped blaze that may extend up the side and sometimes even onto the back. Separate white blazes on the back are found in 4.8% of adults and seem to be evenly distributed between the sexes. Some animals also carry grey streaks or flecking on the back, usually highly irregular in outline, and often in the shape of a V or partial V, pointing forwards. These marks are white at birth and sometimes difficult to distinguish from a white blaze, but unlike the blazes they gradually darken with age to become grey in adults. Such markings are only found in females, and 10.5% of adult cows carry them (including 1.4% with both white blazes and grey markings). Most interesting of all, perhaps, are the 3.5% of calves that are born almost completely white. They are never completely white, always having a collar of black around the neck and a variable amount of black spotting on the adjacent back, but they are invariably reported as 'white' calves. As in the grey markings of females, the white colour gradually fades to become grey in adults (but with the black collar and spotting still visible), giving a sort of 'brindled' appearance. About 94% of these brindled animals are male; females of this colour phase can only be produced by the union of a brindled male with a female with grey markings. In North Atlantic right whales neither white nor grey dorsal markings, nor white calves, have been recorded, and in the North Pacific only whales with a white dorsal blaze.[7]

Southern right whales carry an average of 222 (range 196–246, n = 11) baleen plates on each side of the upper jaw,[5, 8] with a conspicuous gap between the series at the front of the mouth. The plates are characteristically long and narrow, reaching a maximum length (including the gum) of 2.38 m and a width at the base of 0.20 m. They are a greyish-black and the fringe a yellowish-brown in colour, while the baleen gum (which is often exposed above water during skim-feeding) is a pale whitish-grey. Baleen plates are already erupted in newborn whales but only as a meagre 9 cm or so: by the time the whale is one year old the baleen may have reached a length of 70 cm (including the gum). In the second year the plates begin to wear away at the tip, but still continue to grow in overall length.[9] In young right whales the baleen bristles are among the finest in all baleen whales, averaging 0.13–0.19 mm in diameter and being spaced at about 45–55/cm.[10]

Distribution and migrations

With blubber up to 36 cm thick on the back, southern right whales are reluctant to venture into the tropics. Their distribution is often described as south of 20° S, but in regions where a cold current extends into the tropics (such as off the west coasts of southern Africa and South America) right whales will venture further north. On the west coast of southern Africa, for example, the northernmost known whaling ground for the species was at Baia dos Tigres in southern Angola (or c. 16° S), whereas on the east coast the most northern whaling ground was Maputo Bay (or c. 26° S). Obviously these are not absolute limits, since an individual was taken off Cap Lopez, Gabon (c. 1° S) in 1951[14] and a sighting was made in Antongil Bay, Madagascar (15°31′ S) in 1997.[15]

There has also been some controversy over the southern limit to right whale distribution, which has often been portrayed as the Antarctic Convergence, or about 50–55° S. There is little doubt, however, that a substantial number of whales penetrate further south than this in late summer – up to 64° S, judging by Soviet catches.[4] This may or may not be a recent development: the restriction of nineteenth century catches mainly to north of 50° S may have reflected operational avoidance of ice-filled waters.

The right whale migration paradigm is that the whales move into the coastal waters of southern continents and some island groups in winter, the females apparently being constrained to give birth and nurse their newborn in sheltered areas protected from swell and wind, and frequently in shallow water over gently sloping soft substrate.[16, 17] At other times of the year the whales are present on feeding grounds away from the coast, usually to the south of the nursery areas. The higher latitude grounds may only be utilised in late summer/early autumn.

Within the subregion, winter concentrations of right whales have been recorded from Tristan da Cunha in the South Atlantic,[18] and various localities along the coast of southern Africa, extending from Maputo Bay, Mozambique, in the east to Baia dos Tigres, Angola, in the west. Currently the most significant of these are located on the southern coast of South Africa, between Port Elizabeth and Cape Town,[19] but historically there were significant catches made at other localities to the north, such as Lüderitz and Walvis Bay in Namibia and Maputo Bay in Mozambique. While there are signs of right whales returning to Tristan da Cunha[18] and some of their historic haunts in Namibia,[20] there has so far been no such sign in Mozambique or Angola. Right whales arrive in coastal waters on the south coast of southern Africa in June, build up to a maximum in September/October and then depart again by December:[21] a similar pattern has been described for Tristan da Cunha.[18] Historically, however, a significant component of the southern African population moved up the west coast in spring and spent the summer in the area of St Helena Bay and Saldanha Bay. Re-occupation of this apparent feeding ground has been observed recently.[22]

Pelagic concentrations of right whales in the subregion have been recorded (A) in a band between 30° S and 40° S from Cape Town to Tristan da Cunha, (B) around the Crozet Islands in the southern Indian Ocean,[23] and (C) in an area between 50° S and 60° S from 0° E to 20° E.[4] Whales were caught in A mainly from October to December, and in C from December to April. At the Crozets, nineteenth century catches were largely made in February and March, but Soviet whalers also took substantial catches in December. It is likely that these were mainly feeding grounds.

Some long-distance movements between these grounds have been recorded. Three whales satellite-tagged on the South African coast in September 2001 subsequently moved down to area C,[24] and a whale photographed in C (at 58° S 2° E) in 1997 was a female that had (and has subsequently) calved on the South African coastline.[25] Three satellite-tagged animals (including two that had already visited C) moved from the South African coastline to area A,[24] and a whale photographed at Gough Island (south of Tristan da Cunha) was photographed with a calf on the South African coast five years later.[26] No links between the South African coastline and B have yet been demonstrated, but it is possible that the Crozet Islands are a destination for whales from the east coast of southern Africa.

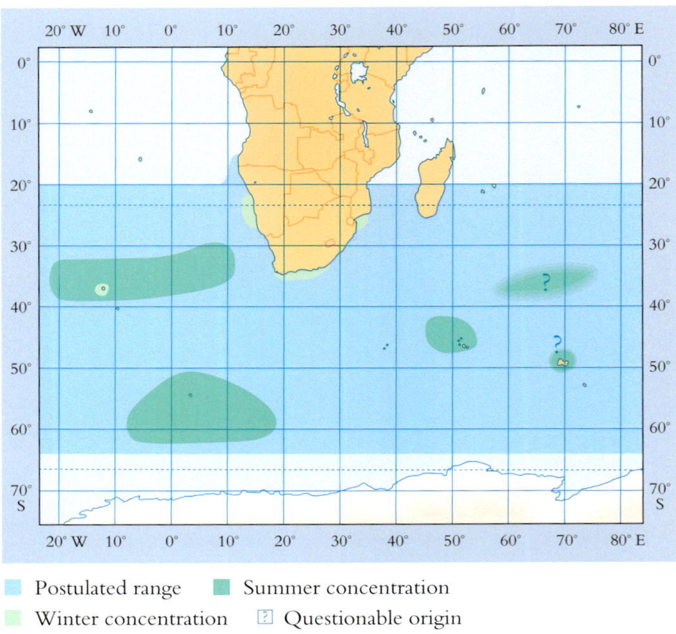

FIG. 8 Distribution map for southern right whale.

A genetic (mitochondrial DNA) comparison of right whales from wintering grounds in Argentina, South Africa, Western Australia and Campbell Island (New Zealand subantarctic) with feeding ground samples from South Georgia and south of Western Australia demonstrated differentiation between all four wintering grounds (although somewhat less between South Africa and Argentina). In general this indicated a strong maternal fidelity to a particular wintering ground. There appeared to be very little gene flow between the South Atlantic and Indo-Pacific ocean basins, and most gene flow occurred primarily between the adjacent populations in one ocean basin. The two feeding ground samples each contained representatives from both wintering grounds to the north of it, suggesting that mixing of breeding populations occurred on these grounds in summer.[27]

Right whale cow-calf pairs on the South African coastline tend to move westwards as the season progresses.

Right whales are not exactly the Ferraris of the whale world: animals tracked by theodolite from the shore in South Africa have been found to move at average speeds of 1.7 km/hr (maximum 3.6 km/hr).[19] This slow rate of movement was undoubtedly one reason why they were so easily taken by open-boat whalers, and currently may be a contributory factor to their vulnerability to ship strikes.

External commensals and parasites

Southern right whales are infested with cyamids or so-called whale lice virtually from the moment of birth. Three species are usually present, *Cyamus ovalis*, *C. gracilis* and *C. erraticus* (although *C. catodontis*, a species normally found on male sperm whales, was recorded from a stranded adult in South Africa[28]). These three species have different habitats on the whale. The first two are confined to the callosities, with *gracilis* predominating in the deeper pits of the callosity and *ovalis* in the more open areas. They reach such abundance that the whole callosity turns chalky white. *Cyamus erraticus* is normally elsewhere on the body surface, but particularly around the genital slit and anus or in fresh wounds. Shortly after birth, however, large patches of *erraticus* (of a reddish-orange colour) may be found on the head of the calf, but these disappear as the calf grows, to be replaced by *ovalis*. It has been speculated that the initial infestation of *erraticus* may be related to a high rate of skin production on the calf's head,[29] or to the dramatic post-natal ecdysis (moult) that occurs.[30]

The barnacle *Tubicinella* is only found on right whales, and then only in their callosities. It is unique in that it 'burrows' into the skin as far as its junction with the blubber, rather than sitting on the skin's surface.[31] *Tubicinella* seems to be universally present on adult southern right whales in southern African waters, though their incidence elsewhere in the southern hemisphere is either lower or under-reported. Historically, *Tubicinella* has been recorded from North Atlantic right whales, but not apparently in recent times, and there are no records from North Pacific right whales.[32] This seems to be another piece of evidence confirming the long separation of northern and southern right whales. There are few authentic records of coronuline barnacles on southern right whales.[33]

Films of the cold-water diatom *Bennettella* (=*Cocconeis*) *ceticola* either occur rarely on southern right whales or are difficult to detect macroscopically because of the dark colour of the skin.

Food and feeding behaviour

We know relatively little about the food of southern right whales. Until recently there were only incidental records of their feeding on krill, copepods or the post-larva of lobster-krill. More substantive data became available when the information on stomach contents of 249 whales taken illegally in the 1960s was published. South of 50° S virtually all (99.4%) of individuals examined were feeding on krill (unidentified), while north of 40° S nearly all (91.7%) were feeding on copepods (unidentified). Between 40° S and 50° S, whales could be feeding on either copepods (71.4%), euphausiids (24.3%) or 'small crustacea' (4.3%). The only items specifically identified were the copepods *Calanus propinquus* and *Pleuromamma robusta*, from whales south of Tasmania.[4] Faeces collected on the South African coast indicate feeding on copepods, probably *Calanoides carinatus* and (to a lesser extent) *Calanus agulhensis*.[8]

The results of stable isotope analysis of baleen plates from South Africa indicated a strong pattern of seasonality in feeding, falling off after May and only resuming again in August to December.[9] Between November and April the incidence of food in right whale stomachs taken pelagically also showed a seasonal trend, with a peak from January to March.[4] These observations seem to suggest that right whales follow the normal baleen whale paradigm in moving between summer feeding grounds in high

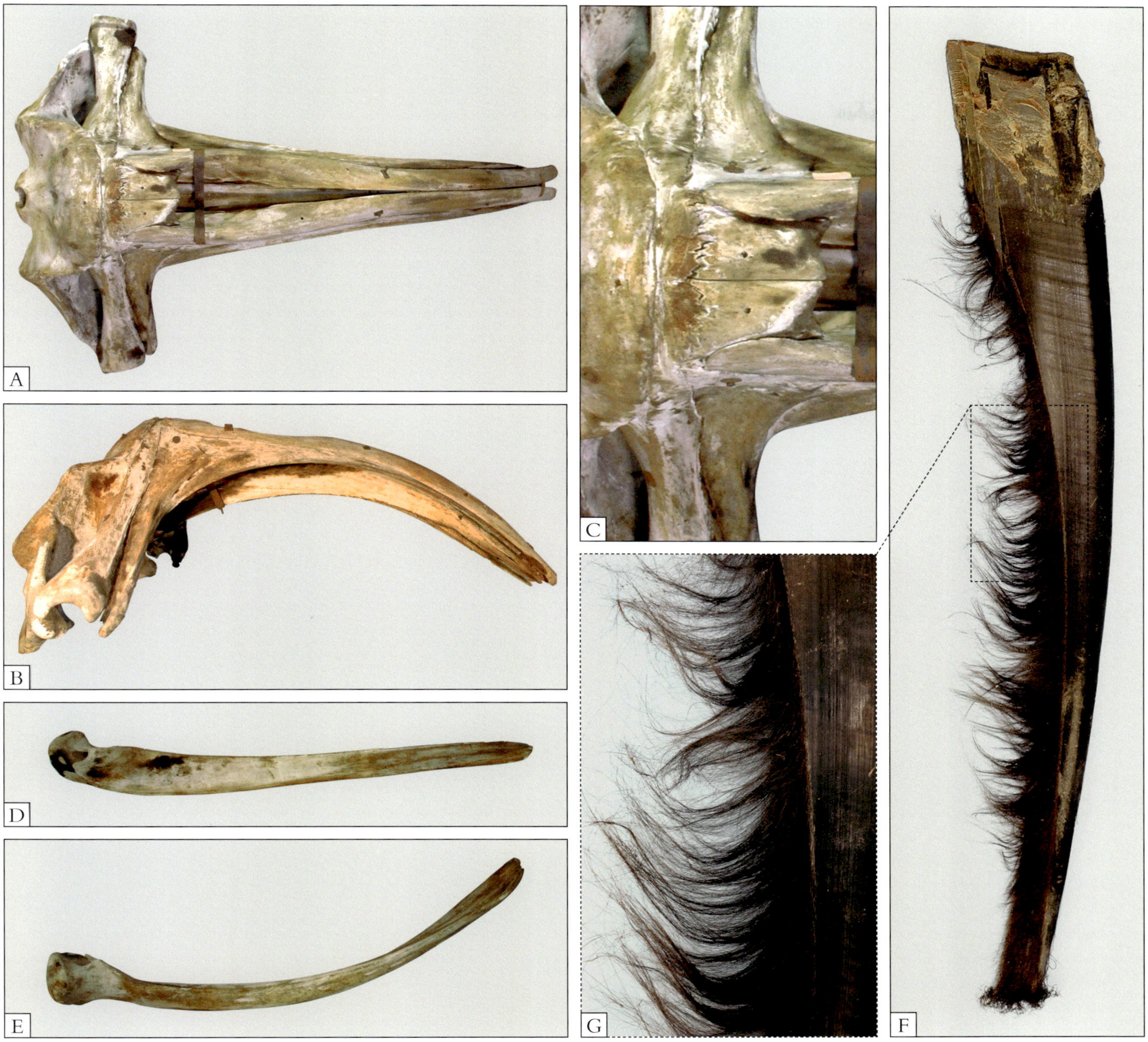

FIG. 9 Dorsal (A) and right lateral (B – retouched) views and vertex of skull (C); lateral (lingual) (D) and dorsal (reversed) (E) views of left mandible (USNM 267612); anterior view of left baleen (F) and baleen fringe detail (G) from a young right whale (ZM 05/15).

latitudes and winter breeding grounds in low latitudes, where feeding is minimal. However, right whales have recently been found feeding in spring and summer in regions of high productivity on South Africa's west coast, particularly St Helena Bay, with some individuals staying as long as 100 days in the vicinity.[22, 24] The importance of this ground has yet to be established, but it seems that not all intensive feeding takes place at high latitudes.

Right whales are designed as skim-feeders, that is to say they catch their prey by moving forward through the water column with their mouths open for protracted periods. This is in contrast to blue and fin whales, for example, which feed principally by engulfing their prey in one opening and closing of the mouth. Right whales have consequently developed a massive filtering apparatus (their baleen) rather than an expanding throat system. In order to open the mouth, the large lower lips have to be swung out away from the head, and they presumably then act as guides for funnelling the plankton into the mouth. When feeding at the surface the whale will swim up and down a plankton swarm, much like someone mowing a lawn, periodically closing its mouth and swallowing. At these times there is a distinct head nodding, which may be the whale attempting to get rid of plankton accumulated on the baleen fringes by backflushing. Such feeding bouts can continue for at least 1.75 hours. On the west coast of South Africa most feeding takes place at depth, where the behaviour cannot be seen, although headnodding is seen at the surface.[8]

Reproduction and growth

After a gestation period of 12–13 months, most southern right whale calves off South Africa are born over a 118-day period, with a peak around late August.[34] The size of the calf at birth varies from 4.5 m to 6 m, and is correlated with the size of its mother. For the next 3–4 months the calf grows in length at an average rate of 2.8 cm/day. A differential exodus of larger calves from the nursery area seems to occur when the calf reaches a critical size[3] (about 8 m), presumably when it is large enough to make the migration. Females usually give birth to their first calf when they are eight years old[35] and at least 12.4 m long.[3] It is not known for how long the calf is nursed, but calves have been seen attempting to skim-feed in February,[8] or at an estimated age of six months, and a few are still accompanying their mother at one year of age. Most subsequent calves are born at three-year intervals.[35]

Right whales are among the best endowed of all mammals. The male North Pacific right whale has been recorded as having a penis 2.7 m long and a combined testis weight of 972 kg[36] (the largest corresponding measurements from a southern right whale are 2.24 m and a single testis volume of 198 litres[5]). By comparison, a mature blue whale might have testes one-tenth the size of the North Pacific right whale. The reason for this difference probably lies in differing mating strategies. From analogy with primates it is thought that the right whale may be practising sperm competition, which means that instead of physically preventing another male from mating, the right whale may simply be attempting to dilute its sperm by sequential mating. Under this strategy, the larger the volume of sperm that can be produced, the better the chances of success, hence the evolutionary pressure for larger testes and a longer penis with which to contest access to the female.[37] Certainly the behaviour seen in surface-active groups (see below) would be consistent with such a strategy.

Behaviour and vocalisations

While in South African waters, right whales are found in groups of 1–10 animals, with cow-calf pairs predominating in nursery areas. From July to October the proportion of single animals drops, while more animals become involved in surface-active groups (known as SAGs). These are aggregations of 2–10 interacting animals that are conspicuous from the amount of splashing and exposure of heads, bodies and flippers at the surface. Characteristically, one individual is the focus of attention, usually a female, while the other members of the group are mostly males. These groups can persist for an hour or more, and membership can wax and wane. While mating seems to be what is taking place (and can be seen occasionally), closer study of some of the groups has shown that the females are immature (2–5 years old) and none has been seen with a calf in the following year. Furthermore, SAGs occur in coastal waters during at least seven months of the year, while we believe that conceptions take place over a relatively brief period (four months). One can only speculate that some sort of mating strategy is being practised by

In dense fog

The southern right whale whose skull photographs are shown in Fig. 9 is the sole representative of its species in the Smithsonian Institution's Natural History Museum. It was taken in the South Atlantic (at 62°38 S 20°48 W) on 23 February 1938 and processed on board the US floating factory *Ulysses*. This was of course some time after the species had been protected by the League of Nations agreement of 1931 (that came into force in 1935), to which the United States was a signatory. According to a cablegram from the whaling inspector on board the *Ulysses* (Lt Q. R. Walsh of the US Coastguard) to Dr Remington Kellogg of the Smithsonian, the animal was 'accidentally killed in dense fog'. This explanation was accepted and Scientific Permit No. 1, for the importation of the specimen for scientific purposes, was issued by the US Bureau of Fisheries on 7 April 1938. The plausibility of the explanation is somewhat compromised, however, by earlier wording in Walsh's cablegram—'For Doctor Kellogg Smithsonian Institution Stop Accordance your request if southern right whale sighted or accidentally killed …',[1] implying (to the suspicious mind at least) that there was some element of premeditation in this take. These suspicions are not allayed by the fact that the period 21–26 February 1938 represented the most productive six consecutive days of the season for the *Ulysses*, during which 20–27 whales were processed daily, including 27 on 23 February itself: if there was any fog around it must have been very localised. Furthermore, no right whale takes were reported to the Bureau of Whaling Statistics in Norway.[2]

This is not to deny that southern right whales have been taken by accident. In 1963, a 13.1 m male right whale was taken by a catcher attached to the Union Whaling Company at Durban, South Africa. According to a later interview with a crew member, a group of sperm whales had been sighted and during the chase a whale was seen coming up beneath the bow of the catcher. Although the sailor in the masthead barrel recognised that it was a right and not a sperm whale, his warning shout came too late to prevent the reflex firing of a harpoon by the gunner.[3] As it happened, science also benefited from this capture, as Dr G. Pilleri of the Brain Anatomy Institute, Berne, Switzerland, was collecting neurological tissue at the Durban whaling station at the time, and was able to secure and publish a paper on the brain of this individual.[4]

No excuses (other than pure ignorance) can be found for other post-protection takes in the area. At Donkergat in the 1937 season, for instance, no fewer than seven southern right whales were landed, five of which were pregnant.[5] Such takes of a protected species were termed 'deplorable' by the International Whaling Conference at its 1938 meeting,[6] and thereafter (until the depredations of the Soviets in the 1960s) illegal takes of right whales were few and far between.

References

1 Accession no. 148006, US National Museum, Smithsonian Institution
2 Allison, C., International Whaling Commission secretariat, pers. comm.
3 PBB pers. comm.
4 Pilleri 1964
5 Best and Ross 1986
6 Anon. 1947

juvenile animals, so that breeding with the fittest male will eventually occur. In the Antarctic, group sizes are smaller and SAGs are not so apparent.[38]

Southern right whales produce a variety of low frequency tonal and pulsive sounds. Most of these are indiscernible to an above-surface listener, but on occasion (and seemingly more often at night) a loud in-air moaning bellow, rather similar to a foghorn, can be heard. Their underwater vocabulary has been classified by starting frequency (low 55–110 Hz, medium 110–220 Hz, and high 220–440 Hz) and acoustic contour (up, down, flat and variable). Low up-calls, the commonest call-type, seem to serve as contact calls between individuals. Other calls, such as medium and high down-calls, seem to be produced mainly by surface-active groups, possibly by the focal female,[39] as a playback of the sounds can induce males to approach. A particularly distinctive call-type is the so-called 'gun shot', a very short, intense broadband cracking sound that can be painful to the listener. Nobody is sure how the whale can make such a sound, which appears to be produced by males as a type of threat.[40]

Conservation status

The southern right whale was the first species to be targeted by commercial whaling in the southern hemisphere. In 1603 Basque whalers set up a shore-based enterprise in Brazil that lasted for over a century, with right whales as the target.[41] In about 1770 pelagic whaling vessels from America began operating south of the equator, and these were soon followed by British, French and other whalers. The fishery for right whales rapidly spread to all southern oceans, and although reliable catch figures are hard to obtain, it has been estimated that between 1770 and 1850 some 125 000–151 000 southern right whales were killed from an original population of 55 000–70 000. This produced obvious depletion, but because the species was not given international protection until 1935, the southern right whale at its lowest point (in 1920) may have been reduced to as few as 300 animals.[42] Fortunately this protection, although by no means absolute,[4] was sufficient to ensure the onset of a recovery. From 1979 to 1998 the population calving on the south coast of South Africa increased steadily at 7.1% a year,[35] and similar rates of increase have been recorded for Argentina and Australia.[42] Recent sightings off Namibia,[20] Madagascar,[15] Réunion[42] and even Gabon[43] suggest that this recovery may be widespread. In 1997 the size of the South African population was estimated at 3 100 out of a total southern hemisphere population of about 7 500. The latter figure is 10–14% of the range of estimates for original population size,[44] indicating that the whales still have a long way to go.

Southern right whales are listed as Lower Risk: Conservation Dependent by the IUCN and as Least Concern in the South African Red Data Book. The species is also included on Appendix I of both CITES and CMS.

References

1 Gaines *et al.* 2005
2 Rosenbaum *et al.* 2000
3 Best and Rüther 1992
4 Tormosov *et al.* 1998
5 Matthews 1938c
6 Payne *et al.* 1983
7 Schaeff *et al.* 1999
8 MRI Whale Unit data
9 Best and Schell 1996
10 Nemoto 1959
11 Lockyer 1976
12 Moore *et al.* 2004
13 Van Beneden and Gervais 1880
14 Budker and Collignon 1952
15 Rosenbaum *et al.* 2001
16 Elwen and Best 2004a
17 Elwen and Best 2004b
18 Best 1988a
19 Best 2000
20 Le Roux *et al.* 2001
21 Best and Scott 1993
22 Best 2006
23 Townsend 1935
24 Mate, B. R., pers. comm.
25 Best 1997
26 Best *et al.* 1993
27 Patenaude *et al.* 2007
28 Best 1970a
29 Rowntree 1996
30 Reeb *et al.* 2005
31 Marloth 1900
32 Scarff 1986
33 Best 1991
34 Best 1994b
35 Best *et al.* 2001
36 Omura *et al.* 1969
37 Brownell and Ralls 1986
38 Best *et al.* 2003b
39 Hofmeyr-Juritz 2006
40 Parks 2003
41 Richards 1994
42 Le Journal de l'Ile de la Réunion, 26 Sept. 1991
43 Darling, J., pers. comm.
44 International Whaling Commission 2001a

Family **NEOBALAENIDAE**

Pygmy Right Whale

Caperea marginata

THIS, THE SMALLEST of the baleen whales, is in many ways the most interesting. Despite its common name, it bears only a superficial resemblance to a right whale (presumably because it has similar feeding strategies). Anatomically it is quite unique, and genetically appears to be closer to the rorquals and the gray whale than to right whales.[1]

Appearance

As landed specimens, pygmy right whales are readily recognised from their arched (but unscalloped) lower jaw, lack of throat grooves, slender flippers and small dorsal fin. At sea, however, they can easily be mistaken for minke whales, and can probably only be distinguished at very close range. Then it should be relatively easy to exclude the dwarf minke if a white shoulder blaze and flipper band are not evident. Differentiation from the Antarctic minke is more difficult: the bowed lower jaw of pygmy right whales should be apparent (at least in adults) but the narrow rostrum is probably more diagnostic. From the few available pictures of the living animal,[2–5] it can be seen that pygmy right whales have a crescent-shaped grey mark running up above the flipper onto the back, similar to that seen in the Antarctic minke, but do not have blowhole streaks or a flank patch. They also have a distinctive grey flare running up and forward from the eyes, more or less parallel with the jaw line. The blow has been described as thin (in adults) or small and hardly distinguishable (in a juvenile). The head and blow usually appear before the dorsal fin, in a movement described as 'throwing' the snout out of the water, and during some surfacings the fin may not be exposed at all.[3–5]

Pygmy right whales reach a maximum length of about 6.1 m in males and 6.34 m in females.[6] They are dark bluish-grey above, shading to a pale greyish-white below, with the crescent and flare markings described above, plus a third around the centre of the body that does not extend as far dorsally as the other two.[5] A white band along the upper jaw above the baleen mentioned in some descriptions[7] seems to refer to the colour of the baleen gum. Oval wounds and scars attributed to a small pelagic shark such as *Isistius* have been noted,[8] particularly on the flanks and belly of adults[9] (see box on page 40). Despite an earlier suggestion that the species might possess the rudiments of throat grooves as found in balaenopterids,[3] an examination of subsequent specimens suggests that this is not likely. The dorsal fin is hooked and curved backwards and is situated 64–69% (average 65.5%) of the body length back from the tip of the snout.

Length:weight relationship

The weights of seven pygmy right whales were available, five from South Africa and two from the literature.[8] One juvenile weighed in pieces was 94.9% of its weight when whole. Using 6% to account for weight loss, the length:weight relationship was $\mathbf{W = 0.00514L^{3.58}}$, where W = weight in tonnes and L = total length in metres. This implies that a 3.5 m whale would weigh 0.46 tonnes and a 6 m whale 3.1 tonnes, adjusted for weight loss.

Skeletal data

One of the more amazing anatomical features of this animal is the number and shape of the ribs. They are more numerous than in any other cetacean, extending almost as far back as the genital aperture, and are so greatly flattened in cross section that at their widest (around the centre of the series) they are in close apposition. Much speculation has surrounded the reason for this phenomenon, including the suggestion that they might spend time lying on the sea floor (!) and, more recently, that it may have something to do with the extraordinary development of the laryngeal sac.[11]

Vertebral formula (three South African specimens, seven from the literature[12]): 7 cervical, 16–18 thoracic, 1–4 lumbar and 14–17 caudal; total 40–44.

All the cervical vertebrae are fused.

Phalangeal formula (X-rays of two South African animals): I_3, II_{4-5}, III_{4-5}, IV_3.

It increases in height from 3.2–3.5% of the body length in juveniles to 4.1–5.5% in adults. The flippers increase in length from 10.7–11.7% of body length in juveniles to 12.5–13.6% in adults, and become more slender, their maximum width being 25.3–26.5% of their length in juveniles, declining to 22.8–23.7% in adults. The flukes increase in width from 20.5–25.9% of body length in juveniles to 28–30.8% in adults.[9]

Pygmy right whales carry an average of 239 (range 213–265, n = 10) baleen plates on each side of the upper jaw. The plates are long and very narrow, reaching a maximum length (including the gum) of 69 cm and a width of 12 cm. In juveniles (3.1–3.7 m) the longest plates are only 22–29 cm. They are a striking off-white colour with a black (or greyish) outer band on most of the plates. The front 15–40 and the rear 10–56 plates are usually completely white (or white with a greyish band), and the middle plates are white with a black outer band. There is no obvious asymmetry to this coloration. The baleen fringes are also off-white and the finest in texture (less than 0.1 mm in diameter) of all the baleen whales. They are spaced at 40–49/cm in adults, but are much denser (94–107/cm) in juveniles.[10]

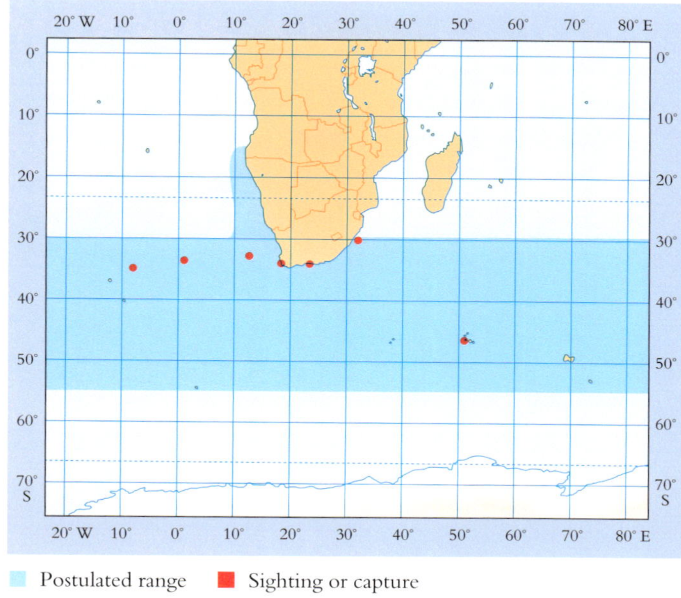

FIG. 10 Distribution map for pygmy right whale.

Distribution and migrations

Pygmy right whales are restricted to the southern hemisphere. Most records occur in a latitudinal band between 30° S and 55° S and in temperature regimes between 5 °C and 20 °C.[13] In the subregion they have been recorded from South Africa,[3] Namibia,[9] the pelagic South Atlantic (32°56′ S to 35° S, 0°56′ E to 12°40′ E)[8, 14] and to the south of the Crozet Islands at 46° S to 47° S.[8] The Namibian records include several strandings in the Walvis Bay region of Namibia (23° S), and one discovery of a vertebra alongside a Namibian coastal road at 19°28′ S 13°02′ E.[9] These records are easily the furthest north for the species, and are

FIG. 11 A narrow rostrum distinguishes the pygmy right whale from minke whales in the field.

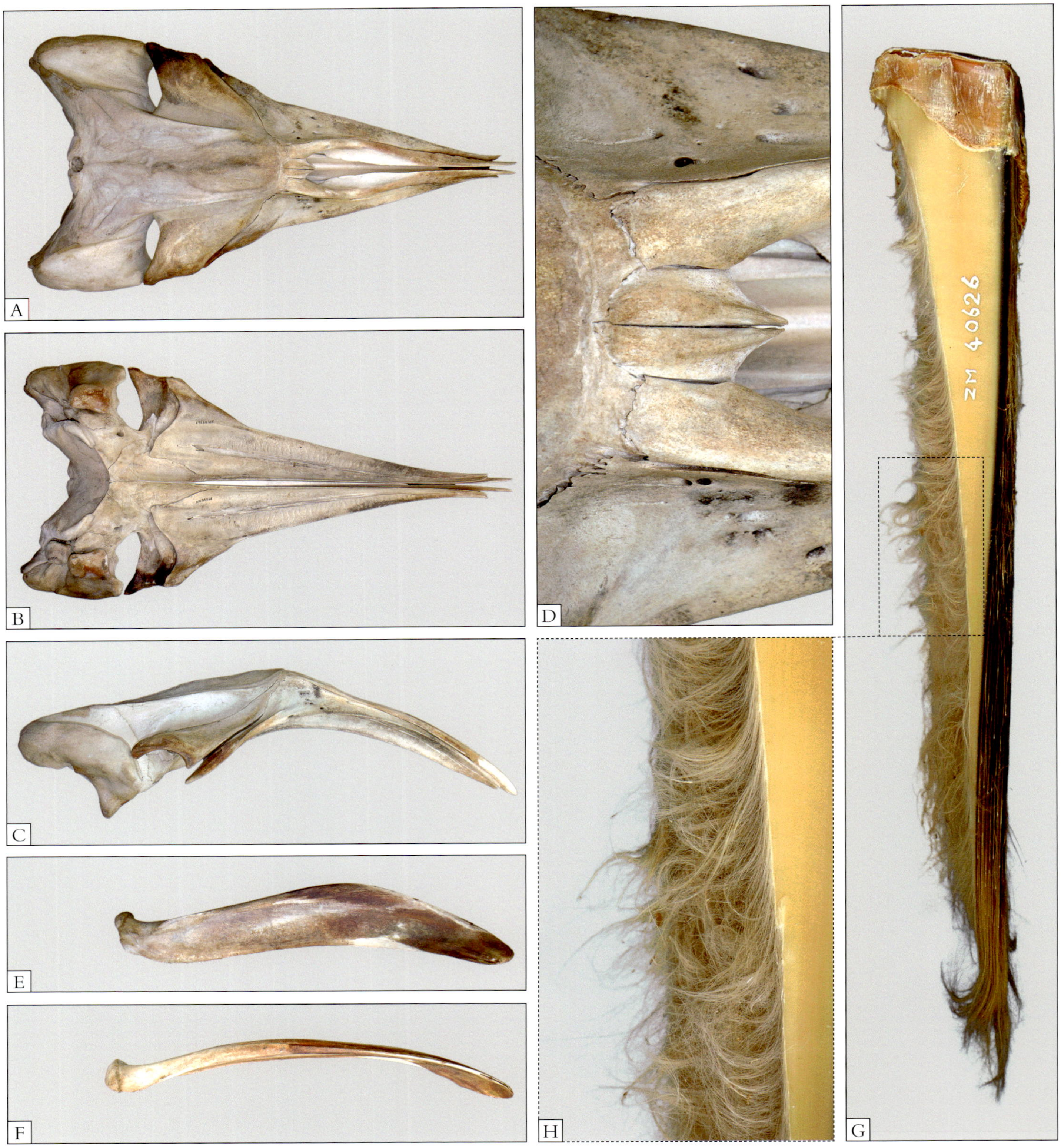

FIG. 12 Dorsal (A), ventral (B), and right lateral (C) views and vertex of skull (D); lateral (lingual) (E) and dorsal (F) view of left mandible (ZM 39768); anterior view of left baleen (G) and baleen fringe detail (H) (ZM 40626).

probably associated with the extension of the cold Benguela System northwards.

In Australia and New Zealand, pre-weaning juveniles made up 41% of the records north of 41° S, while subadults were rare (13%). South of 41° S the proportions were reversed (36% subadults, 17% juveniles).[15] Juveniles also predominate in southern African waters.[3] Pre-pubertal animals may therefore range further south than juveniles.

Feeding grounds are unknown, but from analogy with other baleen whales are likely to be in the highly productive regions of higher latitudes, particularly near the Subtropical Convergence. Coastal upwelling could be of local importance in both South African and Australian waters.[10, 15] Indirect evidence suggests that feeding in Australian waters may be relatively frequent.[15]

It is not known whether seasonal movements similar to those of the balaenopterids or the southern right whale occur, but there is a lack of seasonality to sighting and stranding records in Australasia.[15] Off southern Africa it has been suggested that there may be an inshore movement of juveniles in spring and summer, possibly as part of a general post-weaning dispersal.[3] The presence of only healed cookie-cutter shark bites on animals in the Antarctic suggests that they must move into warmer waters, and the apparent accumulation of such scars on larger animals implies that this movement may take place regularly.[16]

External commensals and parasites

No external parasites or commensals have been described for pygmy right whales. A statement that barnacles had been found on the lower lips seems to have been an error.[8]

Food and feeding behavior

Calanoid copepods have been described as the principal food items in the stomachs of five pygmy right whales, and euphausiids (or possibly euphausiids) in another two.[14] To these can be added amphipods *Themisto gaudichaudi* from a 3.11 m animal stranded in South Africa in 2003. A 3.67 m juvenile stranded in False Bay, South Africa had been feeding intensively on copepods, principally the species *Centropages brachiatus* and *Calanoides carinatus*.[10] Bright red faeces produced in the Antarctic are said to be consistent with a diet of copepods.[16]

Feathers found in the stranded False Bay juvenile and in other stomachs, together with some behavioural observations, suggest that pygmy right whales may use a surface skim-feeding technique similar to that of southern right whales.[14]

Reproduction and growth

The smallest recorded calf was 1.98 m and the largest foetus was 2.12 cm long,[14] indicating a length at birth of about 2.05 m. Available foetal records are really too few (n = 5) to assess the length of gestation or the seasons of calving/conception, and the data that are available do not suggest a very restricted breeding season. Nevertheless, by combining these records with the dates and sizes of juveniles, alternative estimates of 10 and 16 months have been obtained for the length of gestation. From analogy with other baleen whales, the former is probably the more likely.[17] Weaning is estimated to occur at a body length of 3–3.5 m[6] and at an estimated age of five months.[17] Sexual maturity in the female may occur at 4.4–5.65 m, and physical maturity at body lengths of at least 5.9 m.[6] There are no equivalent ages for these milestones.

Behaviour and vocalisations

Group sizes of sightings in South African and Australian waters have mostly been singles or pairs, with one group of five, and the pairs have usually been described as adults plus juveniles.[2, 3] Offshore and in the Southern Ocean, reported group sizes have been larger, ranging from five to 80.[5, 8, 16]

Swimming speeds have been reported as 3–5 knots.[3, 8] There are no reports of aerial behaviour, or exposure of flukes, and the animal's behaviour has usually been described as 'inconspicuous'.[12]

Whales taken in the South Atlantic had a high incidence of crescent-shaped scars as opposed to the completed bites usually attributed to a shark such as *Isistius*.[8] In an adult female stranded in South Africa this was also evident, and on the flanks the ratio was at least ten crescent-shaped scars to every successful bite.[9] In the author's experience this is most unusual for a baleen whale, where such a high incidence of crescent-shaped scars is usually only seen on the surface of the flukes themselves. It suggests that something about the pygmy right whale's behaviour makes it difficult for the sharks to attack successfully. In an analysis of an underwater film of a juvenile *Caperea*, attention was drawn to the extensive flexing of the animal's entire body during swimming,[3] and it is possible that such flexion forces the shark to let go before a bite can be completed.

Vocalisations have only been recorded once in the vicinity of pygmy right whales. A juvenile (estimated at 3 m) made a series of intense, thump-like pulses, usually in pairs but once as a trio, while swimming alone around a harbour. The sounds (mostly between 60 and 120 Hz) were produced at irregular intervals and only nine times over 23 hours of monitoring.[18] The adult male has an extraordinarily well-developed and distinct laryngeal sac that is only barely developed in juvenile males and females (the situation in adult females is currently unknown). It seems likely that this is somehow used in sound production,[11] but the vocalisations of adult pygmy right whales are as yet unrecorded.

Conservation status

The pygmy right whale has never been a target of commercial whaling. Although it is not a right whale, it has benefited from its nomenclatural association with right whales by being given international protection since 1935, first under the 1931 League of Nations Convention and latterly by the 1946 Convention, in both of which it was included in the definition of 'right whale'. This, plus its small size, means that the species has escaped the sort of exploitation suffered by most baleen whales. Individuals have been taken, mostly out of curiosity, and since 1957 at least four animals are known to have been caught by beach-seine fishermen in South Africa, some of which were utilised for meat. There are no estimates of population size or trend. The species is listed as Lower Risk: Least Concern by the IUCN and as Least Concern in the South African Red Data Book. It is included in Appendix I of CITES and Appendix II of CMS.

Demon whale-biter

FIG. 13 A cookie-cutter shark *Isistius brasiliensis*.

For a long time the presence of oval pits and scars on the skin of cetaceans puzzled scientists. The open pits are of a scooped-out nature, about 7 cm long, 4–5 cm wide and 3 cm deep, with the long axis usually parallel to the body of the whale. Some have flabby discs of greyish or brown tissue attached to their centre, obviously part of the process of healing, while others are only crescent-shaped wounds of varying depths of penetration but with no tissue removed. After healing, a similar-sized scar is left, either white (in the case of crescent bites) or grey with radiating dark lines (most oval pits). The bites and scars are normally concentrated in the rear half of the body, especially on the flanks and belly, and are so numerous in some individuals that the normal body pigmentation is completely obscured by scarring. The crescents often occur on or near the tail flukes.

Over the years numerous perpetrators were fingered, including the barnacle *Coronula*, the parasitic copepod *Pennella*, sucking or biting fishes, sharp rocks and even micro-organisms such as ciliated protozoa.[1] It was only in 1971 that the probable identity of the 'demon whale-biter' was revealed – the cookie-cutter shark *Isistius*.[2,3] These little sharks (up to 50 cm long) are unusual in that while the upper teeth are quite small and fine, the lower teeth are large and juxtaposed in a continuous semicircular band, just like a cookie-cutter. The lower teeth are not shed individually, as in most sharks, but as a unit. The shark's method of attack seems to be to hit the whale with open mouth, apply suction to latch on, and then to bring its lower teeth down in one continuous bite before dropping off. The crescent-shaped wounds presumably represent incomplete bites, and this would explain why they are often found on the tail flukes where it must be difficult to make a successful attack. The crescents are almost always open to the front, indicating that the shark is facing in the same direction as the whale when attacking (the original suggestion that the shark attacked from the opposite direction and allowed the force of the water flow to swing it round in a circle seems unlikely, as this would produce a round rather than oval wound). *Isistius* is roughly the same size and conformation as a remora: could this deception allow it to get close to the whale? These sharks are tropical in distribution, which accounts for the higher incidence of fresh scars on whales in warmer waters.

FIG. 14 The business end of a cookie-cutter shark *(I. plutodus)* and a typical plug it has cut.

References

1 Mackintosh and Wheeler 1929
2 Jones 1971
3 Shevchenko 1970

References

1 Árnason *et al.* 2004
2 Kemper *et al.* 1997
3 Ross *et al.* 1975
4 Arnold 1987
5 Matsuoka *et al.* 1996
6 Kemper and Leppard 1999
7 Hale 1964
8 Ivashin *et al.* 1972
9 MRI Whale Unit data
10 Sekiguchi *et al.* 1992a
11 Reeb and Best 1999
12 Baker 1985
13 Kemper 2002b
14 Budylenko *et al.* 1973
15 Kemper 2002a
16 Matsuoka *et al.* 2005
17 Pavey 1992
18 Dawbin and Cato 1992

Family **BALAENOPTERIDAE**

Humpback Whale

Megaptera novaeangliae

Appearance

Humpback whales are relatively easily recognised at sea from the characteristic shape of their dorsal fin, which has an extended leading edge with a distinct 'interruption' to its outline. If the animal is sporting itself at the surface, it may also expose its long, wing-like flippers, and when diving it tends to 'hunch' the back and often exposes its tail flukes before descending. You will then see that the trailing edge of the flukes is unlike that of any other whale, being extensively scalloped rather than exhibiting a gentle curve. The humpback's blow can appear V-shaped, as in a right whale, but is usually straight though somewhat more bushy than in other balaenopterids. At close range the pattern of knob-like protuberances on the surface of the rostrum will be obvious.

Humpback whales are not among the largest of the baleen whales. Females grow bigger than males, the largest female measured in a large sample from Australia being 50 ft (15.2 m, n = 1 605) and the largest male 47 ft (14.3 m, n = 2 031) long.[1]

The head comprises about 30% of the body's length, and the upper surface of the snout, the chin and the sides of the mandibles are covered with raised tubercles, each of which can carry a single hair up to 2.5 cm long. The numbers and positions of the tubercles vary individually, but there is usually a single median row and two lateral rows on each side of the top of the head, with as many as 36 tubercles in total. On the chin there may be as many as 40 tubercles, arranged in two groups, one at the tip and the other paired a short distance below, while there may be up to 11 on each mandible. The dorsal fin is situated about two-thirds of the way back from the tip of the upper jaw. Along the back, between the fin and the tail stock, there are often a number of small triangular peaks or 'knuckles' that vary in number and spacing. They are permanent features present from birth. The flippers are extremely long and tapered, equal to 30–33% of the body length, but they are only about a quarter of that at their widest. Their leading edge carries about five major knobs that protrude at gradually decreasing intervals towards the tip, while the trailing edge is much straighter. These protuberances each correspond to the position of a cartilaginous joint in the underlying flipper bones, and may help the whale to execute tight banking and turning manoeuvres when catching prey.[2,3]

Humpback whales are dark above with variable amounts of white pigmentation on the belly, the outer side of the flipper and the underside of the flukes. In the southern hemisphere there is a trend for the proportion of white to increase from the South

Atlantic eastwards to at least the New Zealand/Ross Sea area.[2, 4–6] Thus, although animals from South Africa and South Georgia included individuals of all colour patterns, the majority (60%) were of the darkest category, where white is confined to patches on the throat and belly with little or no extension up the side; the flippers are dark on their outer side and there may only be 'windows' of white on the underside of the flukes.[2] In 75 whales examined at South Georgia in 1904/05, none had white on the outer side of the flippers.[7] A similar situation was described for 93 humpback whales examined off Porto Alexandre, Angola, in 1911. Of these, 61% had dark bellies, 21.5% white-spotted or striped bellies, and 17.2% a completely white belly. Only six out of 115 humpbacks from the same locality had completely white flippers.[8]

The throat grooves extend as far back as the umbilicus in the midline, and number 21–36 (average 29, n = 8).[2] Because these are much fewer than in blue, fin, sei, Bryde's or minke whales, they tend to be further apart. At the front the grooves join together in the midline just short of the chin, and a median ridge, sometimes referred to as a second chin or cutwater, is present.

There are 300–361 (average 325, n = 4) baleen plates in each series. The longest plates are about 80–100 cm long in the largest whales and are narrow, their maximum width being about 29–45% of their length.[2, 4] They are olive grey to black in colour, although some anterior plates (up to 38 in number) may be white. The baleen bristles are white to greyish-white,[2] coarse (about 0.6–0.8 mm in diameter) and spaced at about 18–28/cm along the inner edge of the plate.[9]

Distribution and migrations

Humpback whales occur worldwide in all oceans, including the tropics, where some trans-equatorial movement during their winter migration has been recorded for southern hemisphere animals on the western seaboard of central America.[14] However, a six-month difference in the breeding season between northern and southern hemispheres probably means there is little genetic interchange between them. The species is assumed to occur throughout the subregion. Besides the Southern Ocean, it has been recorded in whaling statistics from Pagalu,[15] Gabon, Angola, Namibia, South Africa, Mozambique,[16] Madagascar[17] and Kerguelen,[18] and has been sighted at Ascension,[19] St Helena,[20] Tristan da Cunha,[21] Tanzania (including Zanzibar),[22] Kenya,[23] the Comoro Islands,[24] the Seychelles,[25] Mauritius[26] and Réunion.[27]

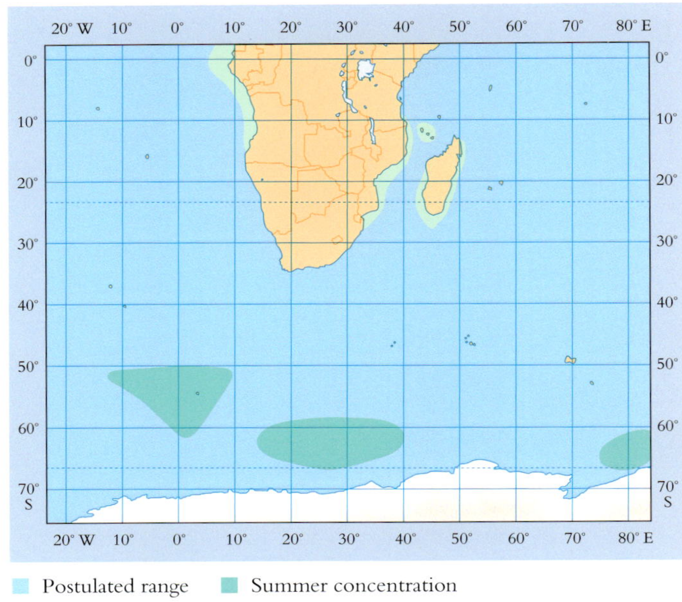

FIG. 15 Distribution map for humpback whale.

In general, the migrations of southern humpback whales follow the usual baleen whale paradigm of moving between summer feeding grounds in higher latitudes and winter breeding grounds in lower latitudes. In the subregion there are considered to be two main winter concentration areas, one on the west coast of Africa, involving Angola, Congo, Gabon and offshore islands such as São Thomé and Pagalu, and one on the east coast of Africa, including Mozambique, Tanzania, the Comoro Islands and Madagascar. These are believed to be linked principally to feeding grounds around Bouvet Island (10° W to 10° E) and off Queen Maud Land (10° E to 60° E) respectively.[28]

Direct links with feeding grounds are confined so far to two recoveries of whales marked with 'Discovery' tags. One was tagged in January 1935 at 64° S 54°03′ E and recovered in August 1938 at 25°03′ S 47°10′ E, and the other was tagged in March 1937 at 62°18′ S 10°41′ E and recovered in August 1938 at 25°36′ S 44°52′ E.[29] These show that whales wintering around Madagascar are feeding over a considerable portion of the south-west Indian Ocean sector of the Antarctic, roughly equivalent to Queen Maud Land. There are as yet no direct links established between west coast humpback whales and the Antarctic.

From a combination of shore- and boat-based sightings, acoustic monitoring, historical catch data and abundance indices from spotter aircraft and catcher boats attached to the Durban whaling station, three principal migratory routes in the south-west Indian Ocean were proposed. The first strikes and leaves

Length:weight relationship

A length:weight regression is available for humpback whales from the Antarctic, based on 17 cooker fillings, corresponding to 781 humpback whales.[10, 11] The length:weight relationship was **$W = 0.0158L^{2.95}$**, where W = weight in tonnes and L = length in metres.

As all these whales were cut up, a 6% correction factor for weight loss must be applied. This means a 5 m whale would weigh 1.9 tonnes and a 10 m whale 14.9 tonnes, adjusted for weight loss (but see the box on page 46!).

Skeletal data

Vertebral formula (North Atlantic, n = 10[12]): 7 cervical, 14 thoracic, 10–11 lumbar and 19–22 caudal; total 51–54.

The cervical vertebrae are usually all free, but some anterior ones may fuse in older animals.[13]

Phalangeal formula (North Atlantic, n = 4[12]): II_2, III_{7-8}, IV_{6-7}, V_{2-3} (excludes metacarpals).

The scapula is unique in the rudimentary nature of the acromion and coracoid processes, creating an almost perfect fan shape.

FIG. 16 Rounded and raised tubercles on the head of the humpback whale correspond to the positions of rudimentary hairs.

the coast as far west as Knysna (23° E), and seems to continue as far as central Mozambique. Movement along the east coast has been shown by the recoveries of lost harpoons in 1912 from Durban (30° S) to Linga Linga (23°30′ S), and from Angoche (16°20′ S) to Durban.[8] The second route supplies the coast of Madagascar directly from the south, probably via the Mozambique Ridge, and may proceed as far as northern Madagascar. Walters Shoal (400 nautical miles south of Madagascar) seems to be an important staging post for whales on this migratory path. The third route (the existence of which is not as well established) may travel up the centre of the Mozambique Channel, its destination being the Comoro Islands and Aldabra.[30]

On the first route, the northern migration seems to reach the coast in the vicinity of Knysna about April, and continues through to August/September, when the southern migration begins and continues through to December. Further up the east coast at Cape Vidal in northern KwaZulu-Natal, the peak of northerly moving animals occurs in mid–late July, but some are still moving north in October. Southward-moving animals were first seen in early July, reached a peak in late August, but were also still moving south in late October when the surveys stopped. Historically there were unimodal peaks in catches in late July/early August at Linga Linga in Mozambique, indicating that this might be close to the limit of the winter migration on this coast.[30] At Cape Vidal, the modal distance offshore for the northern migration (2–3 km) tended to be less than for the southern migration (3–4 km), while the modal speed (3–4 km/hr) was slower than in the southern migration (4–5 km/hr): these differences may reflect the use of the southward-flowing Agulhas Current offshore by southward-moving animals.[31, 32]

Far less is known about migratory routes in the south-east Atlantic. Coastwise movements have been demonstrated by two same-season genetic matches between whales in Gabon and the Saldanha/St Helena Bay region in South Africa, one being a cow-calf pair.[33] Because early catches affected the availability of humpback whales at Saldanha Bay so drastically, while catching to the north continued, it has been speculated that only a small proportion of the main migration chooses to come close inshore that far south, possibly only those whales that summer in the Antarctic due south of Africa. All the others would choose the shortest route to and from the west coast wintering ground, and thus pass far offshore of the Western Cape.[8]

Genetic comparisons of humpback whales using mitochondrial DNA have confirmed the separation of the wintering grounds on the west coast of South America, and on the east and west coasts of southern Africa, but also revealed some heterogeneity within the last two regions. On the east coast, whales from the Comoro Islands differed from those off Mozambique or Madagascar, while on the west coast there were significant differences between whales from the Saldanha/St Helena Bay region and those off Gabon or Angola.[34] An analysis of nuclear microsatellite DNA (inherited from both parents rather than just the mother) generally supported the differentiation between wintering grounds but with less resolution, and did not find the same degree of heterogeneity on either west or east coast of southern Africa. Some degree of gene flow was occurring across all boundaries,[33] and a juvenile male had moved from Madagascar in one year to Gabon two years later.[35]

The migrations of southern humpback whales between feeding and breeding grounds tend to be highly structured. Females in late lactation leave the Antarctic first, followed by immatures of both sexes, then mature males, resting females and lastly females in late pregnancy, with the mean dates between the first and last categories being about 31 days. The first to depart from the breeding grounds are the newly pregnant and resting females, followed by immatures, mature males, and lastly females

FIG. 17 Dorsal (A – retouched), ventral (B), and left lateral (C) views and vertex of skull (D) (USNM 269982); lateral (lingual, reversed) (E) and dorsal (F) views of left mandible (USNM 504216); anterior view of right baleen (G) and baleen fringe detail (H) (USNM 504216).

in early lactation: the difference in timing between the first and last categories is again about a month.[36] This is an overall pattern, and not rigidly adhered to by all members of the population. Several studies on the breeding grounds and on migration corridors have found that there is a disparate number of males, suggesting either that some females fail to make the winter migration, or that males are somehow more available to be sampled than females.[37] Catches of humpback whales at South Georgia between 1910 and 1918 (when whaling was carried out for much of the year) did occur between April and September, but very few indeed were taken between July and September,[38] so although some humpback whales may remain in the Antarctic in winter, their numbers appear low. Studies on the West Indies breeding ground in the North Atlantic have indicated that males have a longer residence time than females, which may explain their higher representation in samples.[39] On the west coast of southern Africa, humpbacks may still be present in some numbers during summer months, even as far north as Walvis Bay, Namibia, presumably induced to stay by the high productivity of the Benguela System at that time of year,[40] and monitoring of movements of humpback whales from the shore at Cape Columbine, South Africa, in October–November indicated that most animals were meandering in the area rather than passing through.[41]

External commensals and parasites

Possibly because they are relatively slow-moving, humpback whales tend to become a favoured substrate for a number of organisms, most of which do no real harm to their host and so are not technically parasites. The barnacle *Coronula* is universally present on all humpbacks other than newborn calves. It tends to settle in greatest numbers on the 'cutwater' under the chin (which may well then become a useful tool with which to rough up an opponent), around the genital and anal apertures, and on the knobs of the flippers and tips of the flukes. It occurs less often on the ventral grooves, the surface of the head, and scattered over the sides, and rarely on the upper surface of the body. White circular rings can often be seen on the skin where a barnacle has fallen off.[2] At Porto Alexandre, Angola, whalers recorded that humpbacks in June–July were densely covered with barnacles, whereas in October they were almost clean or had only very small barnacles. Their explanation was that the adults fell off in the relatively fresh water at the mouth of the Congo River to the north.[8] A similar pattern was described for humpback whales in Madagascar, where the barnacle population changed from individuals 2.5–5 cm in diameter in mid-June to mid-August, to very small (0.5–1 cm) ones from mid-September to early October.[4] The natural life of the barnacle may therefore be short, with larval settlement and loss of the adults occurring close together. The stalked barnacle *Conchoderma* is found on about 85% of whales, almost invariably attached in bunches to the shells of the *Coronula*. Cyamids seem to occur on all whales except, presumably, on neonates: they are found in greatest numbers around the genitalia, among the barnacles, and between the ventral grooves, with lesser numbers over the rest of the body.[2] Examples from three whales landed at Donkergat in 1962 and 1963 were all identified as *Cyamus* (= *Paracyamus*) *boopis*.[21] The ciliate protozoan *Haematophagus* has also been recorded on the baleen plates of humpback whales at South Georgia, as have patches of diatom film (*Cocconeis wheeleri*) on the skin, but neither has been recorded from South Africa.[2]

The oval wounds and scars usually attributed to the shark *Isistius* are less frequently seen on humpback whales than on most other rorquals, possibly because of their more coastal distribution at low latitudes.[2] In four individuals examined at Donkergat in 1962 or 1963, 'no recent scars' were recorded on one, 'no recent scars, few old' on another, with no comments on the other two.[21]

Food and feeding behaviour

Two of four stomachs examined at Saldanha Bay in 1926 contained fish remains, identified in one case as possibly clupeoids,[2] and a stomach full of 'herrings' was illustrated from a whale presumably taken in 1912 or 1913.[8] No food was found in four stomachs examined there in 1962 and 1963.[42] During observations in October–November 1993 at Cape Columbine north of Saldanha Bay, humpback whales were observed feeding on 10 occasions, and defecating on seven occasions. Faecal samples indicated that they had been feeding on euphausiids (possibly *Euphausia lucens*) and the amphipod *Themisto gaudichaudi*.[41] A juvenile 6.15 m long (and estimated to be 4.5 months old), ensnared in a rock lobster trap line off Saldanha Bay, was found to have been feeding on stomatopods (*Pterygosquilla armata*).[40] A whale taken off Gabon in July 1952 had unidentified small crustaceans ('krill') in its stomach, although humpbacks in general were recorded as having empty stomachs.[43] At Durban, all 13 whales examined in 1926 or 1930 had empty stomachs,[2] while 16% of 25 whales examined in 1962 or 1963 contained some food remains, identified in two cases as the euphausiid *Thysanoessa gregaria* and in one case as a mixture of *T. gregaria* and *E. recurva*.[44]

At South Georgia, 82% of 33 stomachs contained food, exclusively krill (*E. superba*), of which 52% were half full or full. The dominant size classes of krill eaten by four whales ranged from 2.5 cm to 5 cm.[2] In the Antarctic, euphausiids (unidentified) were found in the stomachs of 77.8% of nine humpback whales examined.[45]

Feeding behaviour observed at Cape Columbine consisted exclusively of 'lunge-feeding', where the whale surfaced vertically or horizontally with its mouth open and the throat grooves 'bagging'. This may have been because this was the only feeding strategy recognisable from the distances at which the observations were being made.[41] Although 'bubble-netting' – in which a screen of bubbles is formed around a prey school to restrain it while a number of whales lunge from below – has not been formally described for humpback whales in the subregion, some incidental observations suggest that it is used to catch fish.[46] The net may not only act as a physical barrier but may also trap sound produced by the humpbacks within its walls and so discourage fish from escaping.[47]

Reproduction and growth

Births take place mainly between July and October, with a peak in August.[2] Some calves are born on the northern migration (such as at the latitude of Durban), but cow and calf continue northwards.[8] Length at birth is 3.96–4.57 m[48] and twins occur in about 0.28% of pregnancies. Lactation lasts 10.5–11 months, with weaning occurring in temperate regions[49] and at a length of 8–10 m[48] (calves may start to take solid food well before final weaning, as shown by the entangled 6.15 m juvenile mentioned above). Cows can fall pregnant again shortly after giving birth, resulting in an inter-calf interval of about a year, but the usual calving interval is two years and the average birth rate about 37%.[49] Puberty is reached at a length of 11.2 m in males[50] and 11.7 m in females.[51] Ages at puberty have been given as 4–5 years in both sexes,[1] and although some longitudinal studies of known-age animals in the North Atlantic have tended to support these estimates,[52] other similar studies in Alaskan waters have suggested a mean age of 11.5–11.8 years for animals seen with their first calf.[53] The average composition of humpback whale milk consists of 42.9% water, 43.8% fat, 9.1% protein and 2.1% ash.[54]

Genetic sampling of known females and three to five of their offspring has shown multiple paternity in each case, indicating a promiscuous mating system.[55]

Physical maturity is reached at a length of 13.1 m in males and 13.5 m in females.[51] Equivalent ages seem to be in excess of 20 years (or 40 growth layers in the ear plug), while the oldest aged individual had 95 growth layers (equivalent to 48 years by the author's interpretation).[1] The recovery of a marked humpback whale 17 years after tagging[56] provides direct evidence for longevity of about 20 years (given that the age of the whale at marking was unknown).

Behaviour and vocalisations

Group sizes of humpback whales encountered at sea off Cape Columbine between October and November 1993 ranged from one to five (mean 2.3, n =12). None of these contained calves. Off Cabinda in northern Angola in September 1998, groups encountered by boat ranged from one to seven in size (mean 2.6, n = 19). This may have been biased upwards by selection. Marked social interactions were seen in 21.1% of groups, including 25% of cow-calf pairs.[57] At Cape Vidal in 1990, the size of observed groups covered a similar range: 1–6+ on the northern migration (mean = 2.3, n = 251) and 1–5 (mean 2.2, n = 55) on the southern migration.[58] Off Mozambique in August–September 1991, confirmed group sizes ranged from one to seven (mean 2, n = 69), with escorts accompanying 31% of cow-calf pairs.[59] In southern Madagascar in August–September 1994, confirmed group sizes ranged from one to six with outliers at eight and 10 (mean 2.4, n = 102). Escorts accompanied 80% of cow-calf pairs.[60] In the Antarctic in summer, group sizes tended to be smaller than at any of these low latitude areas: 1–5 with a mean of 1.9 (n = 128).[61]

As shown above, humpback whales form distinct social (or competitive) groups on their breeding ground. These consist of

The heaviest humpback ever?

FIG. 18 The whale in question, with the propeller damage clearly visible as parallel diagonal cuts.

On 27 July 1995, a male humpback whale 14.2 m long washed ashore dead at Mouille Point, Cape Town. The reason for its death seemed pretty obvious, as there were four diagonal slashes 1.1–1.3 m apart, up to 2.9 m in length and 35–44 cm deep, along the animal's back, suggesting marks from the propeller of a large ship. Heroic attempts were made over the ensuing days to remove the carcass from the rocks and out to sea, using a helicopter and a tug, all to no avail. Finally the carcass obligingly drifted to an accessible beach, where it was cut up over 31 July and 1 August and removed to a municipal dump site in 13 separate truckloads. The City Engineer, Mr Bruce Black, arranged for each truck to be weighed on arrival at the dump site, and after subtracting the tared weights the total consignment amounted to 56 060 kg. This is about 37% more than the heaviest humpback whale previously weighed, and compares with a predicted weight from its length of 39.6 tonnes.[1] Given that the carcass was several days old and partly decomposed when cut up, it seems reasonable to add the 6% loss factor (although a case could be made that some beach sand might inadvertently have been weighed as well). This translates to a total weight of 59.4 tonnes. How come the difference from expectation? Although no blubber measurements were taken, the whale did appear particularly fat, with pink visceral fat being visible in the propeller cuts. It has been calculated that individual humpbacks may put on as much as 1 ton in weight every week of the Antarctic summer,[2] or double their body weight during one Antarctic feeding season,[3] so it is possible that this animal was a recent arrival after a prolonged and very successful spell of foraging.

The episode had a bizarre sequel: part of the load of one of the trucks fell onto the highway during transit to the dump, and was struck by a car and motorcycle, causing considerable damage to both. The subsequent reaction of the insurance assessors is not recorded.

References
1 Lockyer 1976
2 Ash 1953
3 Lockyer 1981a

3–15 animals (average 4.7), composed of a nuclear animal (centrally placed, generally passive), a principal escort (spatially closest to the nuclear animal, fending off challenges from other whales), and one or more challengers or secondary escorts (not seen to act as a challenger). Nuclear animals are usually female, principal escorts and challengers almost invariably male, and secondary escorts usually male. Calves may or may not be present.[62] Interactions between the principal escort and challengers may be particularly vigorous, including tail-swiping and more serious bodily impact that draws blood and even in one case caused death.[63] Paternity analyses suggest that the principal escorts and challengers have a higher reproductive success than the secondary escorts, although this was not statistically significant.[64]

Humpback whales are famous for their 'songs'. These are constructed of three to nine themes that tend to be sung in a particular order. Each theme is made up of repetitive phrases containing a series of sounds. A phrase may last about 15 sec., and a full song (after all the themes have been repeated) may last 10–15 min.[65] Singing is heard mainly (but not exclusively) on the wintering grounds.[66] All singers tend to be singing the same song on a wintering ground at the same time, although the nature of the song evolves within a season.[67] Almost all singers are lone adult males,[68] and bouts of singing by an individual may last for several hours.[65] The purpose of the song is still a matter of debate. Despite suggestions that it might be a form of long-range sonar for locating other whales,[69] most evidence indicates that the song is some form of reproductive display, either to attract females,[70] or to establish dominance ranking among males.[68] Although the balance of observations suggests that the song may play a role in mediating male-male interactions, this does not preclude the possibility that females also use the song in choosing a mate.[71]

Conservation status

Although humpback whales were a target for open-boat whalers in the nineteenth century, with the US fishery estimated to have landed between 14 000 and 18 000 worldwide,[72] there has been no analysis of the likely effect of such catches on the stocks. But when modern whaling started in the southern hemisphere, humpback whales immediately became the principal target, being closest to shore and therefore the most accessible of the large baleen whales before the onset of pelagic whaling with factory ships. Between 1908 and 1915, an estimated 24 700 humpback whales were taken on the African coast from the Congo to Mozambique.[16] This was clearly too much for the stocks on either coast to withstand and their collapse led to many operations closing down or switching to other, more offshore species. Similar collapses were recorded in other southern populations, but humpback whales only received some sort of protection in 1949, when the IWC set a quota for the following Antarctic season. Thereafter a number of seasonal limitations and closed areas were applied, but full protection was only afforded from the 1963/64 season.

Unfortunately all these measures did not stop some IWC member nations from continuing to take humpbacks illegally: between 1948 and 1972, Soviet fleets reported a legal catch of 2 707 humpbacks, but actually killed 48 646, including 4 963 taken after full protection was afforded.[73] These infractions only ceased when the International Observer Scheme came into effect in the 1972/73 season. Fortunately it seems that as a species the humpback is fairly resilient to depletion, and all indications are that stocks on both the east and west coasts of South Africa are now recovering. In 1990 the number migrating past Cape Vidal was estimated at 1 711,[32] and in August/September 1991 those in the coastal waters of southern and central Mozambique (presumably the same component of the east coast population) as 1 954.[59] In August/September 2003, a survey of Mozambique's coastal waters between 14°26′ S and 26° S resulted in a population estimate of 5 811 humpback whales. A simple comparison with the estimate for 1991 over comparable areas suggested that the population had increased at 7.9% a year in the interim.[74] A comparison of relative abundance indices at Cape Vidal between 1988–1991 and 2002 indicated a rate of increase of 9.9% a year.[75] In August–September 1994 the population in southern Madagascar was estimated as 2 532,[60] and a mark-recapture estimate for Antongil Bay in northern Madagascar produced a figure of 1 746 for the period 1996–1999.[76] These probably represent overlapping components of the same population. If it is assumed that Mozambique and Madagascar are separate migratory destinations, and allow for a rate of increase of 9.9% per annum for the intervening years, the total East African population in 2003 might have been almost 12 000 animals. Population modelling suggests that this would be around 79% of the original population size.[75] There is much less information available for the west coast of South Africa. An aerial survey of the Gabonese coast in August 2002 produced an estimate, corrected for a range of possible dive times, of 1 594–2 798 humpback whales. It was thought that this was likely to be an under-estimate.[77] The results of population modelling, which indicate that the population is less than 10% of initial stock size,[75] are therefore likely to be too pessimistic. The species is listed as Vulnerable by the IUCN but Near Threatened in the South African Red Data Book. It is included on Appendix I of both CITES and CMS.

References

1 Chittleborough 1965
2 Matthews 1938a
3 Miklosovic *et al.* 2004
4 Angot 1951
5 Lillie 1915
6 Omura 1953
7 Lönnberg 1906
8 Olsen 1914
9 Nemoto 1959
10 Lockyer 1976
11 Ash 1953
12 True 1904
13 Tomilin 1957
14 Flórez-González *et al.* 1998
15 Aguilar 1985
16 Best 1994a
17 Mackintosh 1942
18 Tønnessen and Johnsen 1982
19 Pelembe, T., pers. comm.
20 MacLeod and Bennett 2007
21 MRI Whale Unit data
22 Amir and Berggren 2001
23 Weru 2001
24 Avolio *et al.* 2002
25 Marine Conservation Society Seychelles 2004
26 Corbett 1994
27 Boucaud and Rota 2006
28 Anon. 1998
29 Rayner 1940
30 Best *et al.* 1998
31 Findlay and Best 1996a
32 Findlay and Best 1996b
33 Pomilla *et al.* 2005
34 Rosenbaum *et al.* 2004a
35 Pomilla and Rosenbaum 2005
36 Dawbin 1966
37 Brown *et al.* 1995
38 Anon. 1920
39 Mattila *et al.* 2001
40 Findlay and Best 1995
41 Best *et al.* 1995
42 Best 1967
43 Budker 1953
44 Bannister and Baker 1967
45 Nemoto 1973
46 Craig 2001
47 Leighton *et al.* 2004
48 Clapham *et al.* 1999
49 Chittleborough 1958
50 Chittleborough 1955a
51 Chittleborough 1955b
52 Clapham 1992
53 Gabriele *et al.* 2007
54 Oftedal 1997
55 Clapham and Palsbøll 1997
56 Brown 1977
57 Best *et al.* 1999
58 Findlay 1994
59 Findlay *et al.* 1994
60 Best *et al.* 1996
61 Kasamatsu *et al.* 1988
62 Clapham *et al.* 1992
63 Pack *et al.* 1998
64 Nielsen *et al.* 2001
65 Payne and McVay 1971
66 Clark and Clapham 2004
67 Payne *et al.* 1983
68 Darling and Bérubé 2001
69 Mercado III *et al.* 1996
70 Tyack 1981
71 Tyack 2000
72 Best 1987
73 Golovlev 2000
74 Findlay *et al.* 2004
75 Johnston and Butterworth 2005
76 Rosenbaum *et al.* 2002a
77 Rosenbaum *et al.* 2004b

Antarctic Minke Whale

Balaenoptera bonaerensis

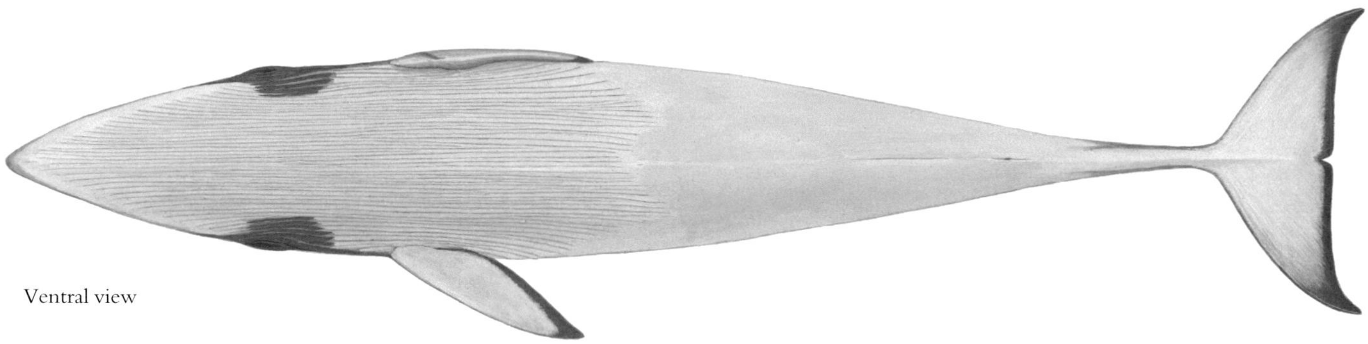

Ventral view

MINKE WHALES ARE the second smallest (and in the northern hemisphere the smallest) of all the baleen whales. As the story goes, a sailor named Meincke on the boat of Svend Foyn, the Norwegian pioneer of modern whaling, mistook a school of minke whales for blue whales.[1] His mistake was commemorated by having his name incorporated into whaling slang, as 'minke' whales. They are also called 'little piked whales', from the very pointed shape of the upper jaw.

For a long time confusion reigned over the identity and number of minke whale species in the southern hemisphere. Northern minke whales invariably carry a striking white flipper band, but there were records of southern minkes both with and without flipper bands. It was only when minke whales became the focus of commercial whaling (from 1968 at Durban and 1971/72 in the Antarctic) that the situation was clarified.[2, 3] There are two forms of minke whale in the southern hemisphere, a larger, more abundant Antarctic minke and a smaller, more tropical dwarf minke whale: it is the latter that carries the distinctive white flipper blaze. The Antarctic minke whale is morphologically quite distinct from northern minke whales (*B. acutorostrata*), and the genetic difference between the two is greater than between some other species of rorquals.[4, 5] As there is also a wide divergence genetically between Antarctic and dwarf minke whales,[6] the former is now recognised as a distinct species, *B. bonaerensis*, originally described in 1867 from a whale found floating dead in the River Plate in Argentina.[7] The identity of the dwarf form remains a taxonomic problem (see species account of *B. acutorostrata*).

Appearance

At sea, Antarctic minke whales can easily be confused with dwarf minke whales. Their main distinguishing feature is a light grey flank patch extending high up the back in front of the dorsal fin, higher than a similarly-coloured patch on the side of the chest. Such a flank patch is not so prominent in dwarf minke whales, where the grey crescent-shaped shoulder blaze is much more prominent than in Antarctic minke whales. Although the more striking white flipper and shoulder patch of the dwarf minke might seem an obvious distinguishing field characteristic, some Antarctic minke whales have strongly two-toned flippers that can appear misleadingly bright when viewed through the water. The dorsal fin of the Antarctic minke whale may carry a grey flare on its trailing half (not seen in the dwarf minke).[2] Blows are not a particularly reliable identification character, but adult Antarctic minke whales seem to produce visible blows, whereas those of the adult dwarf minke are much fainter or invisible. Both species (but more especially dwarf minke) are often attracted to drifting vessels.

When on land, it can be seen that just about the whole of the throat and belly of the Antarctic minke whale is completely white, except for dark pigment in the folds of the throat grooves (which are only visible when the grooves are relaxed), and a pair of curved chevrons extending down from the dark back pigment at the base of the tail. The outer surface of the flipper is either a plain steel-grey, or two tones of grey separated by a distinct line running across the flipper near its base, with the lighter tone covering most of the flipper. The left flipper may be more brightly coloured than the right.[2] The dorsal fin is situated further back in the Antarctic minke whale than the dwarf minke whale, the tip being 73–76% of the body back from the snout, versus 70–72% in the dwarf minke, and is somewhat shorter (its height being 3–4.2% of body length, versus 4.8–5.6% in the dwarf minke).[3] The throat grooves number 44–76 with a mean of about 65,[2] and in the midline extend from the tip of the lower jaw to just short of the umbilicus.

There are 215–310 plates in each baleen series, the biggest being 25–27 cm long with a length:breadth ratio of 1:0.4–0.8. Dwarf minke whales have smaller plates but with similar dimensions. In one of the major features distinguishing Antarctic from dwarf minke whales, less than half the plates in each series are all white, the majority on each side having an outer band of black. The relative width of this band varies according to the position of the plate in the series, getting wider towards the back of the series. In the largest plates (around mid-series) it forms 25–35% of the maximum width of the plate. There is marked asymmetry in the coloration of the baleen, with all-white plates (always the most anterior) usually forming 20–25% of the length of the series on the left and 35–40% on the right.[2] The bristles are white and relatively fine. Although measurements have not been published for Antarctic minke whales, in northern minke the bristles are 0.2–0.35 mm in diameter and widely spaced (18–21/cm along the edge of the plate).[8]

Antarctic minke whales reach physical maturity at a length of 8.9–9 m in females and 8.5–8.6 m in males,[9] although there may have been some temporal trends in this size due to density-dependent changes in growth rate.[10]

Length:weight relationship

For 22 Antarctic minke, **$W = 0.0076L^{3.23}$**, where W is weight in tonnes and L is body length in metres.[11] This means that a 3 m whale would weigh 260 kg and a 9 m whale 9.2 tonnes (about 6% should be added to these figures to account for the fact that most weights were obtained by weighing animals in parts, incurring fluid loss).

Skeletal data

The following skull characteristics distinguish Antarctic minke whales from dwarf minke whales. In the vertex of the skull, the anterior face of the nasal bones is straight or concave, and extends forward laterally, and the parietal and interparietal bones are not in contact. The anterior border of the frontal bone is concave, and the anterior margin of the supra-occipital is convex anteriorly. On the ventral surface of the skull, the posterior edge of the palatine bones is smoothly curved, and the hamular process of the pterygoid bone is short and stout.[3, 12]

Vertebral formula: 7 cervical, 11–12 thoracic, 10–12 lumbar, 18–20 caudal; total 49–50.[7, 13, 14]

Phalangeal formula: I_{4-5}, II_{7-8}, III_{6-7}, IV_4.[13, 14]

Distribution and migrations

Antarctic minke whales probably occur throughout the subregion, having been recorded in the South Atlantic from South Africa, Namibia and Angola (as far north as Porto Alexandre,

FIG. 19 Grey streaks from the blowhole, a thin grey shoulder crescent and a grey shoulder region reveal that this is an Antarctic minke whale.

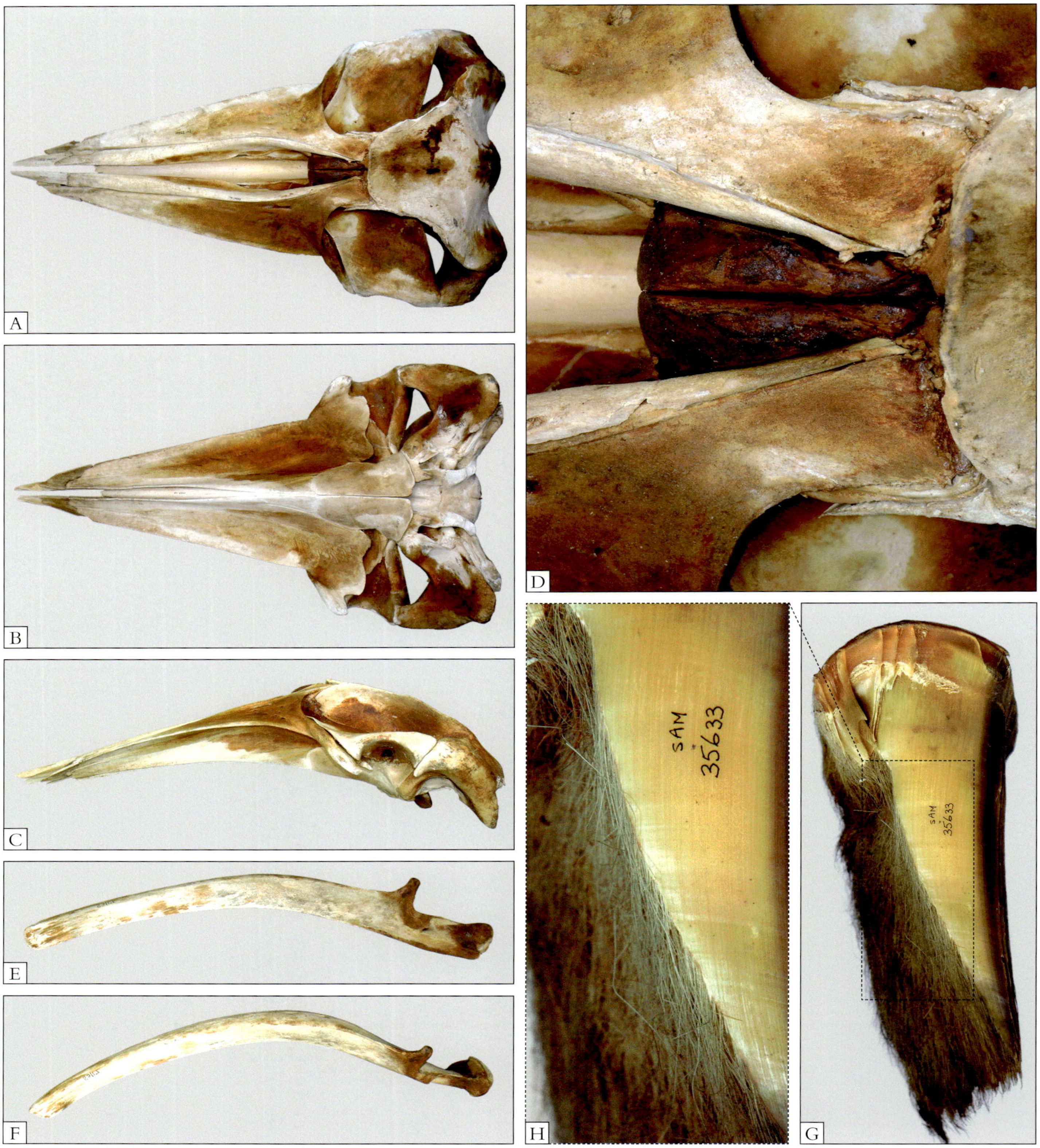

FIG. 20 Dorsal (A), ventral (B), and left lateral (C) views and vertex of skull (D); lateral (lingual) (E) and dorsal (F) views of left mandible (reversed) (ZM 39861); anterior view of left baleen (G) and baleen fringe detail (H) (ZM 35633).

c. 15° S[15]), and in the Indian Ocean from South Africa and pelagically as far north as 21° S.[16] In summer they occur as far south as (and often within) the pack ice, with adult females ranging further south than adult males. Off Durban they were mainly caught beyond 30 nautical miles from the coast,[2] suggesting that they are more pelagic in distribution when in warmer waters. Breeding populations may indeed be relatively dispersed in open waters compared to those of some other baleen whales.[17]

Direct evidence of migration between the Antarctic and warmer waters is confined to two whales that were tagged with 'Discovery' marks in the Antarctic and killed in the Brazilian whaling grounds 2.5 and 3.5 years later. One of these whales was tagged almost directly south of South Africa (69° S, 19° E) and the other south of South Georgia, at 62° S and 35° W.[18] Indirect evidence of movements obtained from sighting data suggest a southward migration from 10° S to 20° S in October, with most whales being in the Antarctic by January. Overall migration speed was estimated at 20 nautical miles per day north of the Subtropical Convergence and 40–50 nautical miles per day south of the Convergence.[17]

This migration may not be undertaken by all individuals. Antarctic minke whales seem to be present in the temperate waters of the south-western Indian Ocean throughout the year. During summer and autumn, a large proportion of this population consists of small, sexually immature animals, and the densities north of 40° S are substantially lower than those recorded for the Antarctic. In April and May the abundance of minke whales off Durban begins to increase, and remains at a high level in June, July, August and especially September. Separate northward and southward migration waves could not be identified, and the incidence of a cold-water skin-film diatom was high throughout the June and September period.[9] It seems that the migrations of minke whales are more protracted and less well defined than those of fin and sei whales. Minke whales have indeed been seen within the Antarctic pack ice in winter,[19] when the majority of the population is believed to be in warmer waters.

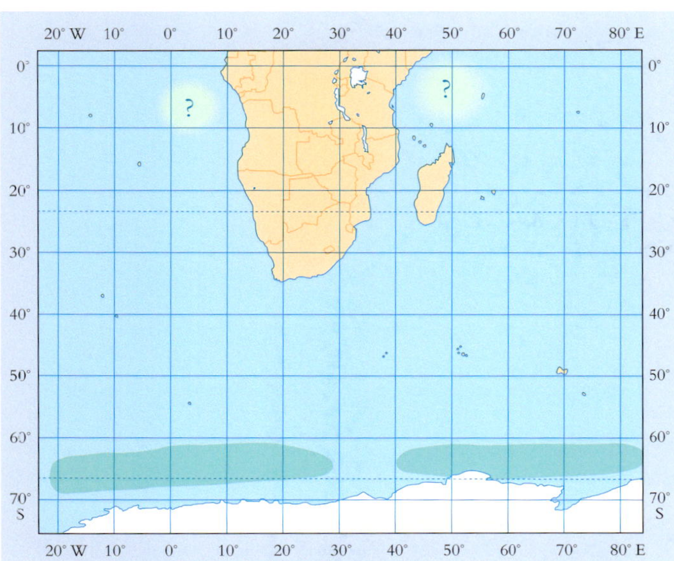

FIG. 21 Distribution map for Antarctic minke whale.

The two mark recoveries in the Brazilian whaling ground suggest that whales from a particular breeding ground may disperse widely on the feeding grounds. Recoveries of marked whales within the Antarctic confirm that individuals may range over as much as 100° of longitude between years on the feeding ground, but that most (90%) remained within about 40° of longitude.[20]

External commensals and parasites

About 20% of minke whales examined in the Antarctic carried whale lice (*Cyamus balaenopterae*), mostly at the posterior end of the ventral grooves or around the navel, with counts on one side of the body ranging from one to 405 with a mean of 55. The incidence at Durban seemed much lower but may have been affected by the longer post-mortem times before examination. Skin-film diatoms *Bennettella ceticola* were present as orange-brown or yellowish patches on the sides and belly of the animal in 67.5% of whales in the Antarctic and 35.7% at Durban. The bootlace-like parasitic copepod *Pennella* sp. was also recorded from one individual in the Antarctic.[9]

Food and feeding behaviour

Antarctic minke whales feed predominantly on euphausiids, both on their summer feeding grounds and winter migrations, and, unlike dwarf minke whales, there is no substantial fish component to the diet. In the Antarctic, krill (*Euphausia superba*) formed 100% of the diet at the ice edge and 94% of that offshore of the ice-edge zone, despite another euphausiid species (*Thysanoessa macrura*) dominating plankton samples in the latter zone. This suggests that adult Antarctic minke whales may preferentially feed on *E. superba*.[21] At Durban, *T. gregaria* and *E. recurva* were the principal species eaten, while *E. lucens* and *E. diomedeae* were also occasionally taken.[9] As might be expected, the incidence of stomachs containing food was much higher in the Antarctic (55–82%)[21, 22] than at Durban (15.2%),[9] and amounts consumed were much greater. Young, recently weaned Antarctic minke may feed on copepods such as *Calanus tonsus*,[22] and juveniles on *Thysanoessa* sp.[21]

At the ice edge, minke whale feeding activity appeared to be highest in the morning (06h00–09h00) and lowest at dusk (15h00–sunset), but there were no observations at night. Away from the ice edge no particular trend was evident during the day.[21]

Krill consumption rates for an average male (8.4 m) and female (8.9 m) minke whale were estimated by a number of methods as 359–491 kg/day and 423–512 kg/day for an assumed 90-day and 120-day summer feeding period for males and females respectively.[23] The total food for the completion of a 14-month reproductive cycle has been estimated at 13.8 tonnes (or an average of 32.3 kg/day),[24] the relatively modest amount reflecting an averaging of the high summer inputs estimated above and the much lower food consumption in winter.

North Atlantic minke whales employed a variety of feeding techniques while foraging at the surface. Entrapment manoeuvres involved circles, spirals, ellipses, figures-of-eight and hyperbolas, with the animal on its side with its white belly towards the prey. At the conclusion of the entrapment, the whale performed an engulfing manoeuvre. This could be an oblique, lateral, vertical or ventral lunge (either at or below the surface), or a plunge. In the latter, the whale was upright and approached

Whaling for science

Article VIII of the International Convention for the Regulation of Whaling (1946) states: 'Notwithstanding anything contained in this Convention any Contracting Government may grant to any of its nationals a special permit authorising that national to kill, take and treat whales for purposes of scientific research subject to such restrictions as to number and subject to such other conditions as the Contracting Government thinks fit, and the killing, taking, and treating of whales in accordance with the provisions of this Article shall be exempt from the operation of the Convention.'

Before 1982 over 100 such permits had been issued by nations such as Canada, the USA, the USSR, South Africa and Japan, to catch species or classes (lactating, undersized) of whale that were otherwise protected. Most of these were for bona fide research purposes, although the motives for South Africa's permit for up to 250 undersized sperm whales at Durban and Donkergat in 1963 deserve closer inspection. But in nearly every other case the numbers involved represented relatively minor proportions of the catch, and it is only since the 1986 moratorium on commercial whaling that some nations (Iceland, Norway and Japan) have taken advantage of this Article to run whole whaling operations for scientific reasons. This has naturally led to a great deal of suspicion about the real motivation for such permits. Although some findings of considerable scientific interest have resulted (notably regarding trends in recruitment of Antarctic minke whales[1]), there is little doubt that the origins of the permits lie more in a desire to keep whaling options open during the moratorium than in a thirst for knowledge,[2] and Japan's expansion of its scientific programme in the Antarctic is a matter of concern.[3]

References

1 Aron *et al.* 2002
2 Clapham *et al.* 2003
3 Morell 2007

the surface at an angle of less than 30°. A patch fishing strategy was applied when prey was dense, and a line fishing strategy when it appeared to be scattered.[25]

Growth and reproduction

Antarctic minke whales are born at a length of 2.7–2.9 m.[26] The calf is suckled for about six months (by which time its length has grown to about 4.6 m). Puberty is reached at 7.2–7.6 m in males and 8.0–8.1 m in females, and at an age of 5–9 years in both sexes.[9] Some observations suggest that there may have been a decline in the age at puberty between the 1940s and the late 1960s/early 1970s, possibly in response to the decline in the abundance of other baleen whale species.[10] According to age readings in the ear plug, individuals may live for more than 40 years.[27] The oldest 'known age' individual is a marked minke whale that was 24 years at large before recapture, but its age at marking was unknown.[28] Different sexes and reproductive classes of minke whale tend to be segregated in time and space, making it difficult to piece together the reproductive cycle. For example, males outnumbered females by two to one in the Durban catch, where only about 1% of mature females were pregnant,[9] while lactating females were almost completely absent from the catch in the high latitudes of the Antarctic, where apparent pregnancy rates were 82–92%.[27] A variety of evidence therefore has to be used to reconstruct the cycle. Males show a marked seasonality in testis size and spermatogenic activity: testes are some 40% larger off Durban than in the Antarctic,[9] with 83% of testes containing spermatozoa compared to 8%.[29] This indicates that conception is more likely to occur in winter in warmer waters. Indeed, from August to September most females with calves off Durban were accompanied by another 1–3 adults, usually males, whereas by December–February, such pairs were mostly alone.[9] From foetal size distributions, it appears that the peak of conceptions lies between August and November.[29] Available evidence suggests that pregnancy lasts *c.* 10 months, and that the majority of females ovulate about four months after giving birth (hence the interest of mature males in cow-calf pairs). The resultant calving interval of 14 months means that a female will conceive progressively later and later, producing the very wide range of foetal lengths seen at any one time. Eventually the female will get out of phase with the breeding season, and pass into a resting condition once lactation has ended. The 'real' pregnancy rate is estimated to be around 78%, equivalent to a calving rate of about four calves every five years.[9]

The average composition of Antarctic minke whale milk from July to September is 51.9–60.8% water, 20.3–30.2% fat, 13.6–14.6% protein and 1.67–1.69% ash.[30]

Behaviour and vocalisations

Group sizes in the subregion sector of the Antarctic ranged from one to 400 with a mean of 3.6 (n = 3 608), but very few (2.5%) sightings were of more than 15 animals.[15] A positive correlation between group size and density has been shown,[31] suggesting some temporary associations around locally abundant food. Off Durban, group sizes were smaller, ranging from one to 15 with a mean of 2.1 (n = 283).[9] In both areas single animals were the most frequently encountered group (36.8% and 45.9% of all sightings respectively).

Apart from the cow-calf pair during lactation (which, as indicated above, may be accompanied seasonally by one or more escort males), there does not seem to be any long-term association between individuals.

While Antarctic minke whales usually produce visible blows, their strength and duration are dependent on (among other things) environmental conditions, the angle with the sun and whale behaviour. For example, a small change in orientation (or whale behaviour) may result in the blows changing from visible to invisible, even in very calm, windless conditions. The intervals between blows measured in the Antarctic depended on group size, averaging 1.5 min. for single animals (with a maximum of 6.15 min.) and 0.11 min. for groups of over 20 animals. The maximum interval (from a group of three animals) was 7.73 min. Blow durations ranged from 1.4 to 16.1 sec., with an average of 6.2 sec.[32]

Very little monitoring of acoustic activity by Antarctic minke whales has been undertaken, and none in the winter (breeding) grounds. Sounds recorded in the vicinity of minke whales in the Ross Sea included stereotypical intense downsweeps in frequency, a series of whistles, a clanging bell series, a click screech and a low-frequency grunt (although the presence of other species during some of these recordings made species attribution problematical).[33, 34]

Conservation status

Antarctic minke whales were hardly taken at all by Antarctic whaling fleets before the 1971/72 season, being too small to be of commercial interest until all the larger species had been depleted. Catching at Durban had begun a little earlier (in 1968), for similar reasons. For the first time in whaling history, catches were limited by an individual species quota from the outset of exploitation in the Antarctic, and later there was even subdivision by longitudinal area. Compliance was also ensured by the implementation of an International Observer Scheme in 1972/73. These measures were undoubtedly part of the reason why, when the moratorium on commercial whaling came into effect in 1986, there were no signs of over-exploitation of Antarctic minke whales. Since that date, Japan has circumvented the moratorium and continued taking a limited number of minke whales in the Antarctic for 'scientific purposes', as permitted in terms of the International Whaling Commission's Convention (see box on page 52). Takes elsewhere have ceased. Since 1978/79, the population of Antarctic minke whales south of 60° S has been assessed in a series of research cruises organised by the International Whaling Commission. These cruises have now been around Antarctica three times, producing circumpolar estimates of 608 000, 766 000 and 268 000 whales respectively.[35] The third set of estimates is significantly lower than the first two, but it is a controversial point how much of this trend represents a real decline in whale numbers and how much can be attributed to changes in survey methodologies. What is clear is that the 'scientific permit' catches (which have been a few hundred a year and confined to the South Pacific and Eastern Indian Ocean sectors) cannot have contributed significantly to the decline. Some modelling of the recruitment rate for Antarctic minke whales indicates a decline between 1970 and the mid-1980s,[36] which, if real, could have contributed to the apparent recent population trend.

Antarctic minke whales are listed as Lower Risk: Conservation Dependent by the IUCN and as Least Concern in the South African Red Data Book. However, they are included on Appendix I of CITES and on Appendix II of CMS.

References

1 Jonsgård 1951
2 Best 1985
3 Arnold *et al.* 1987
4 Wada and Numachi 1991
5 Arnason *et al.* 1993
6 Pastene *et al.* 1994
7 Burmeister 1867
8 Nemoto 1959
9 Best 1982
10 Kato 1987
11 Lockyer 1976
12 Zerbini and Simões-Lopes 2000
13 Omura 1975
14 Omura and Kasuya 1976
15 MRI Whale Unit data
16 Gambell *et al.* 1975
17 Kasamatsu *et al.* 1995
18 Buckland and Duff 1989
19 Taylor 1957
20 Kato *et al.* 1993
21 Ichii and Kato 1991
22 Ohsumi *et al.* 1970
23 Armstrong and Siegfried 1991
24 Lockyer 1981c
25 Lynas and Sylvestre 1988
26 Ohsumi and Masaki 1975
27 Kato 1982
28 Kato and Miyashita 2005
29 Kato 1986
30 Oftedal 1997
31 Kasamatsu *et al.* 1998
32 Joyce 1982
33 Leatherwood *et al.* 1981
34 Schevill and Watkins 1972
35 Branch and Butterworth 2001b
36 Butterworth *et al.* 1999

Dwarf Minke Whale

Balaenoptera acutorostrata subsp.

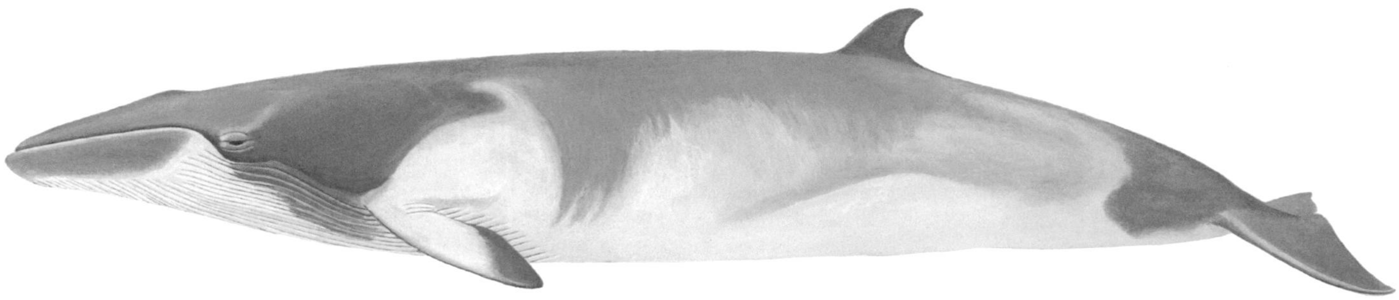

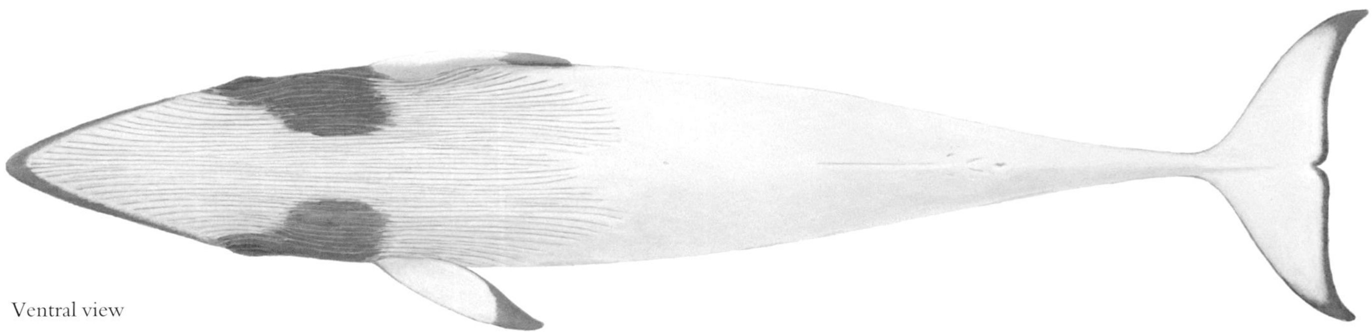

ALTHOUGH THIS IS clearly a very different creature from the Antarctic minke whale *Balaenoptera bonaerensis*, both morphologically[1, 2] and genetically,[3, 4] the identity of the dwarf form remains a taxonomic problem. It is much closer in appearance, habits and genetics to the northern minke whale *B. acutorostrata*, but a comprehensive morphological and genetic comparison of minke whales from all ocean basins is needed before the taxonomic status of the dwarf minke whale can be established. For the moment it is referred to here as an unnamed subspecies of *acutorostrata*, which is a somewhat undignified situation and not very useful in management or conservation terms.

Appearance

Dwarf minke whales are most easily distinguished at sea from Antarctic minke whales by the presence of a conspicuous sabre-like grey crescent extending up above the flipper and onto the back, and by the lack of a grey flank patch. When particularly close to the observer, the grey rostrum (the upper jaw in front of the blowhole) may be seen.

They also seem attracted to stationary or drifting vessels, which they may circle for minutes on end, although ship-attraction tends to be a characteristic behaviour of minke whales in general.

The length of the largest female measured was 7.77 m and of the largest male 7.62 m[1] (the latter specimen is annotated as being measured to the nearest foot; the next largest male is 7.01 m[5]). Although the available measured sample is relatively small (some 15 males and 32 females[1, 5, 6]), it seems likely that there is a similar sexual dimorphism in size to other balaenopterids. The dwarf minke has the same overall body shape and dimensions as the Antarctic minke whale, but with the dorsal fin situated further forward (69.6–71.5% vs. 72.9–76.1% of body length from the tip of the snout). The dorsal fin is also taller (4.8–5.6% vs. 3.0–4.2% of body length) and longer at the base (8.3–8.4% vs. 3.7–7.7% of body length).[2]

In coloration, the dwarf minke whale is characterised by a flipper with a black tip but a pure white base. This base is connected to a prominent white blaze on the shoulder region, variable in size and shape but usually with a dark region in the 'flipper pit'. Above the white blaze is a roughly triangular grey area, which extends upwards onto the back as the sabre-like crescent described above. Between the eye and the leading edge of the flipper, a patch of dark pigment extends from the back onto the throat grooves. Behind the dorsal fin there is an extension of light grey up the side with some white near its trailing edge. The rostrum forward of the eye and blowhole appears light grey in colour.

As in Antarctic minke whales, the throat grooves usually do not extend as far back as the umbilicus, falling short by 2.8–5.4% of the body length. They number 44–66 between the flippers[5] and 54–76 between the eye-axilla axes.[1]

In two specimens there were at least 225–235[2] and 244–251[1] baleen plates in each series; a third juvenile had 200–205 plates but some may not have been fully erupted.[1] The longest plates reach a length of 18–20 cm in the largest whales, and seem to be 2–3 cm shorter than the longest plates in Antarctic minke whales of a similar size.[1] Their width is about 43–45% of their length.[2] The plates are either all creamy white in colour, or have a variable number (up to 70%) with a thin outer band of grey or black, with no consistent asymmetry between the left and right sides.[5] In the longest plates this dark band never exceeds 10% of the maximum width of the plate.[2] The bristles are all creamy white in colour.

Distribution and migrations

Dwarf minke whales per se are only known from the southern hemisphere, although their uncertain taxonomic relationship with northern minke whales means this may not be technically correct. In the subregion they have been recorded from South Africa,[1] Mozambique,[8] Crozet Islands[9] and the Southern Ocean.[10] Within the Antarctic they may range as far south as 65° S, but are generally considered to be commoner north of 60° S.[5] Their summer distribution is therefore somewhat to the north of the main concentration of Antarctic minke whales. In the Durban whaling grounds, dwarf minke whales tended to be caught closer inshore than Antarctic minke whales, most (77%) being within 30 nautical miles of the coast,[1] and a similar pattern of sightings has been described off Brazil, with dwarf minke mainly occurring in water 4–200 m deep and Antarctic minke in water over 600 m deep.[11] Dwarf minke whales have been seen close inshore in Table Bay harbour, Cape Town, in St Helena Bay on the west coast of South Africa, and in the Durban roadstead.[10]

Length:weight relationship

A length:weight regression is available for 15 dwarf minke whales from the Antarctic[5] and one from South Africa. $W = 0.01036L^{3.095}$, where W = weight in tonnes and L = length in metres. All these whales were weighed whole, so no correction factor is needed. This means a 3 m whale would weigh 0.3 tonnes and a 7 m whale 4.3 tonnes.

Skeletal data

The following skull characteristics distinguish dwarf minke from Antarctic minke whales. In the vertex of the skull, the anterior face of the nasal bones is straight or convex, and the parietal and interparietal bones are in contact. The anterior border of the frontal bone is straight, and the anterior margin of the supra-occipital is curved posteriorly in the midline. On the ventral surface of the skull, the hamular process of the pterygoid bone is elongated and gracile.[2, 5]

Vertebral formula (Antarctic, n = 16[5]): 7 cervical, 10–11 thoracic, 11–14 lumbar and 18–20 caudal; total 47–50.

Cervical vertebrae are presumably all free, as in the Antarctic minke whale.[7]

Phalangeal formula, including metacarpals (n = 1[2]): II_4, III_7, IV_7, V_4.

Patterns of migration have not been described for dwarf minke whales. Off Durban they were taken mainly in the months of April to June;[1] in the northern Great Barrier Reef off Queensland, Australia, about 80% of sightings were in June and July.[12] Off Brazil, sightings and strandings occurred between June and February but with the highest numbers in August.[13] In the Southern Ocean, all sightings have been between December and March,[5] but this may largely be a reflection of the seasonality of effort. While these data would suggest that dwarf minke whales follow the normal baleen whale paradigm of summering in high latitudes and wintering in low latitudes, sightings attributed to dwarf minke whales have been made in summer (January) off Brazil at 23° S.[14] The migration may therefore not be comprehensive, with some individuals staying in middle and low latitudes throughout the summer.[11]

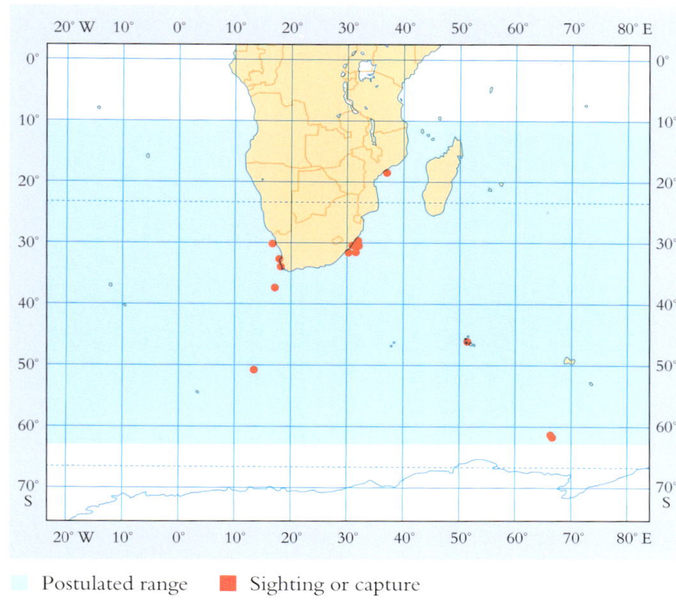

FIG. 22 Distribution map for dwarf minke whale.

External commensals and parasites

The stalked barnacle *Xenobalanus* attaches itself to the trailing edge of the flukes and occasionally the dorsal fin and flippers of the dwarf minke whale, where several dozen may attach themselves at one time. The parasitic copepod *Pennella* has also been recorded, and remoras *Echeneis* are occasionally seen on dwarf minke whales, although the whales make attempts to avoid them. Oval scars attributable to the cookie-cutter shark *Isistius* have also been reported.[15]

Food and feeding behaviour

In the Antarctic, 81.3% of stomachs (n = 16) contained food, in all but one case myctophid fish, identified as *Krefftichthys anderssoni* in one stomach. The exception was an animal that had fed on euphausiids (unidentified), this individual being taken furthest south. Amounts found in the stomach were generally small, 2.6–47.3 kg, with an average of 21.2 kg.[5] Off Durban, the stomachs of all the 13 dwarf minke known to have been landed were recorded as 'empty',[10] but this was standard practice for baleen whales not examined by biologists. Dwarf minke whales have

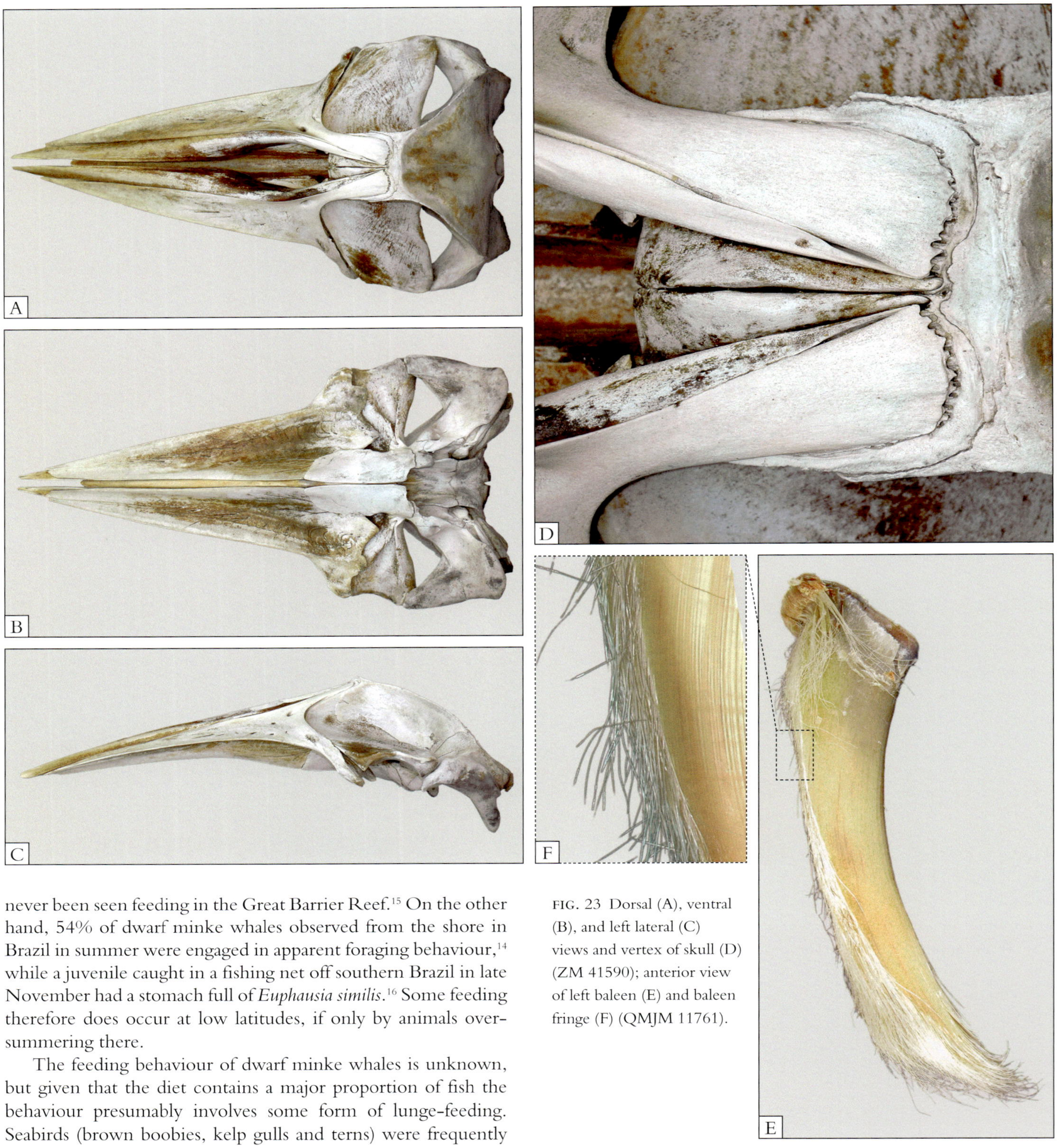

FIG. 23 Dorsal (A), ventral (B), and left lateral (C) views and vertex of skull (D) (ZM 41590); anterior view of left baleen (E) and baleen fringe (F) (QMJM 11761).

never been seen feeding in the Great Barrier Reef.[15] On the other hand, 54% of dwarf minke whales observed from the shore in Brazil in summer were engaged in apparent foraging behaviour,[14] while a juvenile caught in a fishing net off southern Brazil in late November had a stomach full of *Euphausia similis*.[16] Some feeding therefore does occur at low latitudes, if only by animals over-summering there.

The feeding behaviour of dwarf minke whales is unknown, but given that the diet contains a major proportion of fish the behaviour presumably involves some form of lunge-feeding. Seabirds (brown boobies, kelp gulls and terns) were frequently seen in association with foraging whales in Brazilian waters.[14]

Reproduction and growth

The largest foetus known for a dwarf minke whale was 169.6 cm,[5] and the smallest calf was 1.92 m.[1] The calf had a raw umbilicus and air-filled lungs, indicating recent birth. The next largest calf recorded was 2.2 m long, but with no record of umbilical condition.[6] A 2.54 m male had a fully healed umbilicus and heavy lungworm infestation, suggesting it had been alive for some time.[1] Birth may thus take place around 2 m. A male of 3.43 m was already feeding independently.[16] From the evidence of seven foetuses, conception has been deduced to occur in a relatively narrow time frame around mid-July to early August,[5] with birth presumably about a year later. The frequency of reproduction is unknown, although a female off Queensland, Australia, was seen with two different calves in successive years.[17] Age and size at maturity are poorly known: in females the largest known immature animal is 5.94 m[5] and the smallest mature is 6.40 m.[1]

FIG. 24 A dwarf minke off Mozambique exposes its broad grey shoulder crescent, and hints of the white shoulder blaze can be seen subsurface.

Maturity may thus occur at a length of just over 6 m. An equivalent age is 7–10 years.[5] The data are too few to make similar determinations for males. Physical maturity may be achieved at about 7 m in females, and somewhat less in males.[5]

The oldest aged individual in a very small sample (n = 13) was 26 years old.[5]

Behaviour and vocalisations

In the Antarctic 98% of dwarf minke whale sightings were single animals, the exception being a pair (n = 49).[5] All minke whales landed at Durban were taken as single animals.[1] Of the 22 sightings off Brazil in summer, 18 were of single animals and four of pairs.[14] On the Great Barrier Reef, group sizes were somewhat larger, 53.3% being singles, 28.9% pairs, and the rest groups of 3–8 animals (n = 180).[12] Group sizes seemed to be larger in open water away from the reef.[15] These observations were made mostly from drifting vessels that may act as an attractant to whales in the area, which might account for some of the larger sightings.

A variety of behaviours have been reported from whale-watching vessels on the Great Barrier Reef that are in a unique position to make underwater observations. Apart from spyhops and breaches, these include jaw gapes, jaw claps, belly presentations and bubble blasts.[15]

The acoustic behaviour of dwarf minke whales at low latitudes includes one of the most spectacular of all whale sounds. A complex and stereotyped sound sequence is produced, which from its unusual and synthetic-sounding characteristics has been termed the 'Star Wars' vocalisation. This sound spans a wide frequency range (50 Hz–9.4 kHz) and is composed of distinct and stereotypically repeated units with both amplitude- and frequency-modulated components. Typically, three identical pulses are followed by a broadband sound and then a tonal sound with a harmonic series. The whole vocalisation lasts about 3 sec. and is repeated at consistent intervals (from a few sec. to 3–4 min.).[18] Whether this sound is produced at high latitudes is unknown, but it has not been reported as part of the repertoire of minke whales recorded in the Antarctic, although these are more likely to have been *B. bonaerensis*.

Conservation status

Although dwarf minke whales have only been recognised as a distinct taxon since about 1985, or after the closure of most commercial whaling, it seems unlikely that they were ever taken widely in the fishery. Antarctic whaling for minke whales took place mainly close to the ice edge, where weather conditions were better and concentrations of Antarctic minke whales were highest in summer; dwarf minke whales are believed to penetrate only rarely to such latitudes. Being smaller, it is also unlikely that dwarf minke whales would have been selected in the commercial minke whale fishery. In 1978/79, personnel aboard the Japanese factory ship and scientists with experience in the fishery claimed never to have seen a whale resembling sketches/photographs of a dwarf minke.[1] Since the start of whaling under special scientific permit in 1987/88, a total of 16 dwarf minke whales have been taken, at an average of 2–3 a season, but sampling of the subspecies stopped in 1993/94.[5] At Durban the incidence of dwarf minke whales in the minke whale catch was estimated at 3–4%,[1] which translates to a total catch of about 33–45 animals between 1969 and 1975, or an average of 5–7 a year. The only other operation known to have taken dwarf minke whales is the land station at Costinha in Brazil, where about 0.2% of the catch may have comprised this form,[1] converting to a total of some 29 individuals between 1964 and 1985, or about one a year on average.

Although there are no population estimates for dwarf minke whales, the generally low level of catches historically suggests that it is unlikely that the subspecies has been substantially depleted. The dwarf minke whale is not specifically recognised by the IUCN and seems to be effectively unlisted, as only northern populations of *B. acutorostrata* are included. It is listed as Data Deficient in the South African Red Data Book. All *B. acutorostrata* populations except that off West Greenland are on Appendix I of CITES. No minke whales are listed by CMS.

References

1 Best 1985
2 Arnold *et al.* 1987
3 Pastene *et al.* 1994
4 Pastene *et al.* 2007
5 Kato and Fujise 2000
6 Zerbini and Simões-Lopes 2000
7 Omura 1975
8 Findlay *et al.* 1994
9 Saint Thomas Productions 2000
10 MRI Whale Unit data
11 Zerbini *et al.* 1997
12 Arnold 1997
13 Zerbini *et al.* 1996
14 Hassel *et al.* 2003
15 Birtles and Arnold 2002
16 Secchi *et al.* 2003
17 Anon. 2006
18 Gedamke *et al.* 2001

Bryde's Whale

Balaenoptera brydei

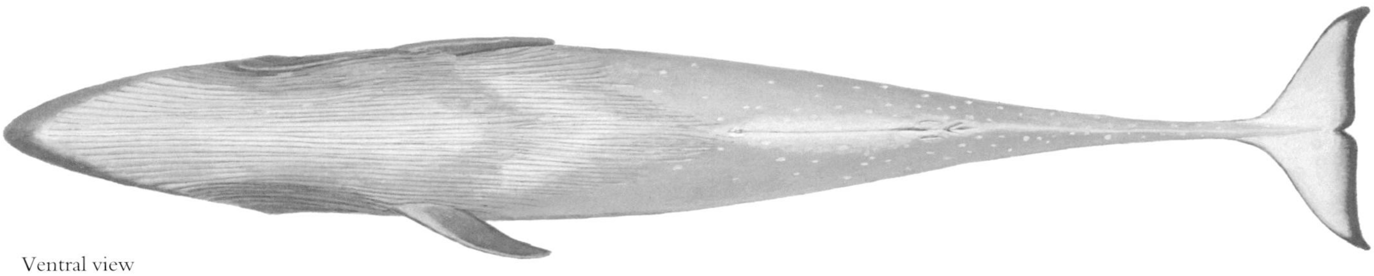

Ventral view

BRYDE'S WHALE IS named after Johan Bryde, a Norwegian entrepreneur involved with the start of modern whaling in southern Africa (who eventually fell foul of the authorities through trading with the enemy[1]). The name should be pronounced 'Brewder's whale', but 'Breeder's whale' seems to have become the accepted anglicised version.

The common name is the easy part, whereas the scientific nomenclature is confused. In several parts of its range (including southern Africa) there appear to be two forms or ecotypes of Bryde's whale, one smaller (usually inshore) and one larger (usually offshore).[2] Genetic analyses suggest that at least in the western North Pacific, these two forms may be differentiated at higher than the population (possibly at subspecific) level.[3] The original description of Bryde's whale (as *B. brydei*) from Saldanha Bay was based on external characters only, and clearly included elements of both the inshore and offshore forms.[4] Later, *brydei* was considered (by most authors) to be the same as *B. edeni*, a whale that had been described earlier from a skeleton from Burma,[5, 6] but although the external appearance of *edeni* is almost completely unknown, it is clearly much smaller than even the inshore form in southern Africa. More recently, a small form of Bryde's whale has been described from the tropical eastern Indian Ocean and the western Pacific Ocean. This appears to be at least as genetically distinct from other Bryde's whales (including inshore and offshore ecotypes) as Bryde's whale is from the sei whale.[3, 7] This has now been described as a new species, *B. omurai*, with specimens from the Sea of Japan, the Solomon Sea, and the eastern Indian Ocean near the Cocos Islands, and is considered different from both *edeni* and *brydei*.[8] So now we have three species instead of one: *brydei* that occurs pelagically in several oceans, a sister taxon *edeni* that occurs in at least Hong Kong, Australia and the coastal waters of south-western Japan, and *omurai* from the Philippines and the Solomon Sea. The last-mentioned is considered genetically to be an ancient lineage that diverged much earlier than sei whales or either *brydei* or *edeni*.[9]

Here I have chosen to retain the name *brydei* for all animals in the subregion, recognising that earlier synonymisation with *edeni* may have been incorrect, and that there are no records yet from the subregion of a small form referable to either *edeni* or *omurai* (although the taxonomic status of the smaller inshore form off South Africa is still in doubt – see the box on page 62).

Appearance

At sea Bryde's whales are most likely to be confused with sei whales, being of similar size, dorsal coloration and fin shape (even experienced whalers have failed to distinguish between them). Although adult sei whales in the region are often covered with oval white scars, so too are Bryde's whales from some populations. The physical characteristic that most readily distinguishes the two species at sea (but can be seen only at close quarters) is the presence of a ridge running along either side of the rostrum

FIG. 25 In an unusual posture, a Bryde's whale lifts its rostrum almost clear of the water, revealing the accessory ridges running either side of the central ridge.

in Bryde's whale that is absent in sei whales. The typical surfacing behaviour of Bryde's whales is also said to be more similar to that of blue and fin whales, with four or five breaths being taken between dives, whereas sei whales tend to blow only once or twice. On land, the two species are easier to distinguish. Unlike the sei whale, the throat grooves in Bryde's whale extend as far back as the navel, and a single groove usually runs in the midline between the navel and the genital aperture.

Bryde's whales tend to differ in size between populations. In the subregion, whales belonging to the inshore population on the south coast of South Africa reached adult lengths of 12.8–13.1 m in males and 13.7–14.0 m in females, compared to 13.7 m and 14.3–14.6 m respectively in whales from the offshore population on the west coast.[2] Animals from south of Madagascar seemed to be even smaller than those from the South African inshore stock.[10] The body shape is highly tapered, the girth at the umbilicus being only 30.9–46.6% of body length. The rostrum is pointed and flat and carries three ridges, one central and one on each side. The flippers are narrow and pointed and their length is 9.8–12.9% of body length. The dorsal fin is erect and hooked, 1.8–3.6% of body length in height, and situated 66.5–74.9% of the way back down the body.[2] The major ventral grooves number 42–54 and extend back from the chin as far back as the umbilicus,[4] and 96.4% of individuals also possessed a median groove running from the umbilicus to the genital aperture.[2]

Bryde's whales are generally dark greyish-black above and white below without any sharply-defined border between the two. In most individuals in South African waters the sides of the throat are bluish-grey, as is an area of the body about 70 cm below the flippers and running back to about level with its tips. This grey area is interrupted by irregular white streaks and areas. The centre of the throat and belly is yellowish-white, and this extends back onto the caudal peduncle behind the genital

Length:weight relationship

Mass data from 27 Bryde's whales weighed in pieces in Japan have been used to produce the following relationship: $W = 0.0122L^{2.74}$, where W = weight in tonnes and L = total length in metres. This implies that a 10 m whale would weigh 6.7 tonnes and a 15 m whale 20.4 tonnes, but 6% should be added to these values to account for weight loss during dismemberment.[12]

Skeletal data

Vertebral formula (n = 2–15): 7 cervical, 13 thoracic, 12–13 lumbar and 21–22 caudal; total 54–55.

All seven cervicals are free, or occasionally the third and fourth are fused.

Phalangeal formula (n = 1): I_5, II_7, IV_7, V_4.

The following skull characteristics distinguish Bryde's whales from sei whales:

1 The rostrum is flat in side view, rather than down-curved.
2 Seen from above, the outer margin of the rostrum is convex, but straighter in sei whales.
3 The front margin of the nasal bones is concave or straight and extends forward at its outer edge, but in sei whales it is convex and does not extend forward at its outer edge.
4 The front margin of the nasal bones is markedly behind the front margin of the rear maxillary concavity, whereas in sei whales the two are at about the same level.
5 On the underside of the skull, the palatine bones do not extend as far back as in sei whales, so that the portion of skull exposed behind them is far longer than broad (whereas in sei whales this area is square or broader than long).[13]

aperture. At the rear end of the throat grooves this white may be interrupted by a bluish-grey band running across the stomach. In some individuals, however, this more or less symmetrical coloration is replaced by a strong asymmetry of pigmentation on the head and throat, with the lower jaw being white on the right side and black on the left, and the right side of the baleen having a greater number of white baleen plates at the front than the left side.[4]

Although not strictly part of the species' colour pattern, two aspects of scarring on Bryde's whales in the subregion are important. Adult whales from the offshore population on the west coast of southern Africa accumulate so many oval scars from bites by what is believed to be the cookie-cutter shark *Isistius* that their original skin pigmentation (particularly on the rear half of their body) disappears under a welter of scarring that resembles galvanised iron. Such scars are rare, even on adults, from the inshore population. On the other hand, about 82% of individuals from this population carry long, thin scars on the underside of their tail flukes and caudal peduncle, suggesting that the animals had come into contact with the sea floor.[2] Such scarring is not seen in offshore Bryde's whales.

Each series of baleen contains 276–289 plates, the largest of which grow to a length of nearly 40 cm. The plates differ in shape between populations, with those from the inshore population off South Africa being narrower (length/breadth ratio 2.22–2.43) than those from the offshore population (length/breadth ratio 1.83–2.24).[2] Most of the plates are greyish-black in colour, although some individuals have up to 0.70 m of the front plates white, and this may be asymmetrically placed as in the fin whale. The baleen bristles are whitish-grey or straw coloured[4] and coarse, ranging in thickness from 0.19–1.11 mm with a mean of 0.48 mm,[2] and spaced usually at around 22–30/cm.[11]

Distribution and migrations

Bryde's whales are found worldwide in tropical, subtropical and temperate waters, often portrayed as between latitudes 40° N and 40° S, or in sea surface temperatures of 20 °C or higher.[5]

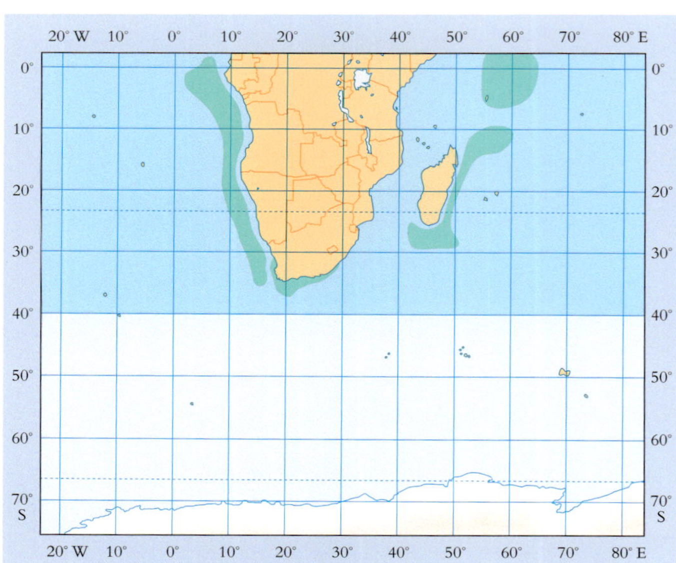

Postulated range General concentration area

FIG. 26 Distribution map for Bryde's whale.

Three distinct populations of Bryde's whales have been described in the subregion. One occurs inshore over the continental shelf of South Africa south of *c.* 30° S, primarily based on the Agulhas Bank. A second population occurs on the west coast of southern Africa, ranging from equatorial regions to around 34° S and within 300 nautical miles of the coast. A third population occurs south of Madagascar, which may or may not be linked to whales seen off the Seychelles.[10]

Although Bryde's whales are unique among all southern baleen whales in not undertaking feeding migrations to the Antarctic, it is not entirely correct to conclude that the species is non-migratory. The inshore population is essentially resident over the Agulhas Bank, and while it may make small seasonal excursions in winter up both west and east coasts, following movements of pelagic fish, these can hardly be rated as migrations.[10] On the other hand, evidence from the pattern of catching by a catcher/factory ship on the west coast of southern Africa revealed distinct seasonal movements by (presumably) the offshore population, with whales being furthest north (*c.* 5° S) in winter and furthest south (25–28° S) in summer. A whale tagged with a 'Discovery' mark off Saldanha Bay in summer was recovered 10 years later in winter off Gabon almost on the equator (or 3 500 km away), confirming the extent and seasonality of this pattern.[14] Possible seasonal movements of Bryde's whales south of Madagascar are unclear.

The swimming speed of relatively undisturbed inshore Bryde's whales was estimated in four experiments as 6.5–8.3 km/hr.[15]

External commensals and parasites

The copepod parasite *Pennella* was found on 9.5% of inshore Bryde's whales (n = 21) and on 11.1% of offshore Bryde's whales (n = 45) in South African waters. The numbers on each whale were either one (in five whales) or two (in two whales). One of these copepods itself carried a stalked barnacle *Conchoderma virgatum*.[16] Nauplii of the harpacticoid copepod *Balaenophilus unisetus* were found on the baleen of one of 67 Bryde's whales examined at the same land stations.[17] No cyamids or skin-film diatoms seem to have been recorded from Bryde's whales.

Food and feeding behaviour

Whales from the inshore population on the west coast of South Africa fed almost exclusively on pelagic fish, principally anchovy *Engraulis japonicus* (61%) but also juvenile maasbanker *Trachurus trachurus* (14.7%) and pilchard *Sardinops ocellata* (1.6%). Feeding was intensive, with 92.3% of the stomachs examined containing food[10] and fills reaching 92 kg.[2] On the south coast of South Africa, at-sea observations confirmed the predominance of pelagic fish and anchovies in particular in the diet, while the shoaling squid *Lycoteuthis diadema* was identified from faecal samples collected at one sighting.[15]

Whales from the offshore population on the west coast of South Africa fed on both euphausiids (80%) and fish (18.5%). Feeding was intensive, with 83.3% of stomachs containing food[10] and fills reaching 120 kg.[2] Euphausiids eaten were mainly *Euphausia lucens* (24.3%), but *E. recurva*, *Thysanoessa gregaria*, and *Nyctiphanes capensis* were also consumed, while the fish –

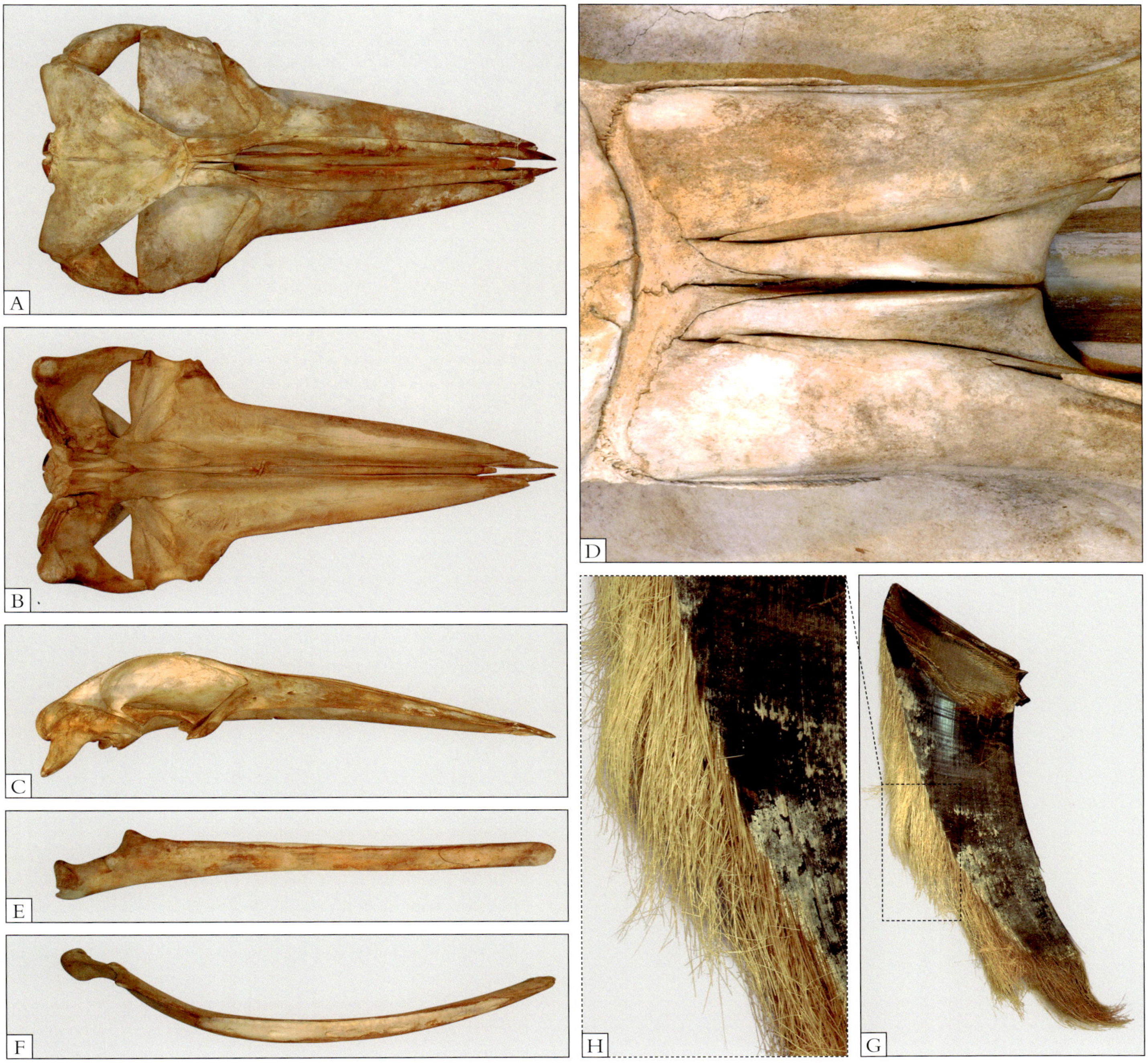

FIG. 27 Dorsal (A), ventral (B), and right lateral (C) views and vertex of skull (D); lateral (lingual-inverted) (E) and dorsal (F) views of right mandible (USNM 572922); anterior view of a left baleen plate (G) and baleen fringe detail (H) (USNM 572922).

lightfish *Maurolicus muelleri* (4.6%), barracudina *Lestidium* sp. (2.3%) and saury *Scomberesox saurus* (2.3%) – were more oceanic and deep-dwelling in nature than those taken by whales from the inshore population. Observations from a pelagic whaling operation over the entire west coast of southern Africa indicated a diet of fish (68.6%) and 'shrimps' (31.4%) for whales from the same population, with higher incidences of feeding in spring/summer than autumn/winter. There was also a marked change in food-type, 'shrimps' being relatively unimportant in autumn (6%) and winter (10%), but becoming more important in spring (21%) and predominating in summer (75% of stomachs containing food).[10]

On the east coast at Durban, the incidence of food in the few stomachs examined was low (23.8%), and individuals were feeding on fish ('sardines', carangid sp.), squid (*L. diadema*) and 'shrimp remains'. Bryde's whales south of Madagascar in autumn were feeding more intensively (46.7%) and exclusively on euphausiids (*E. recurva* and *E. diomedeae*).[10]

Two types of feeding behaviour have been observed in inshore whales off South Africa. In the first, whales were seen lying on their sides at the surface with their throat grooves distended, possibly representing horizontal pursuit of prey at the surface ('horizontal lunges'). In the second, whales came up from below and engulfed a shoal pinned at the surface by other predators ('vertical lunges').[15] Such behaviour could explain early accounts of the presence of small sharks, penguins and gannets in the stomach of Bryde's whales at Saldanha Bay.[4] In New Zealand waters, a Bryde's whale seemed to use the underwater release of a bubble cloud as part of its feeding strategy.[18]

Reproduction and growth

Size at birth (without distinction among populations) is estimated as 3.81–3.96 m. Females in the South African inshore population and those south of Madagascar reach sexual maturity at a smaller length (11.9–12.5 m) than those in the offshore population off the west coast of South Africa (12.8–13.1 m). A similar pattern is shown for males (11.6–12.5 m compared to 12.8–13.7 m). Bryde's whales in general do not show such a marked seasonality in breeding as other baleen whales. Conceptions can occur in any month of the year in all populations in the subregion, but in the offshore population conceptions peak in autumn, while there is no such seasonality in the inshore population. Insufficient information is available for populations on the east coast. There are other apparent differences in the reproductive behaviour of the populations. While whales from the offshore population (like most other baleen whales) seem to ovulate once at each seasonal cycle, those from the inshore population ovulate several times during a reproductive cycle, including during lactation. This unusually high frequency of ovulation may be the result of the year-round abundance of food available to this non-migratory population.[10]

Details of the reproductive cycle are not well known. Observed pregnancy rates ranged from 31.4% in offshore animals to 42.9% in inshore animals on the west coast of South Africa.[10] The overall proportion of calves in the inshore population was estimated as 11.2%,[15] which, given a 50:50 sex ratio and approximately 50% mature in the population, would be equivalent to a pregnancy rate of *c.* 45%. An inshore whale 8.5 m long (and estimated to be two years old) had milk in the stomach, but lactation on average may be much shorter (*c.* six months).[2]

A sample of milk from an inshore whale in South Africa contained 52.1% water, 29.6% fat, 14.7% protein and 1.2% ash.[19]

Pygmy Bryde's whales

Although neither Eden's whale *Balaenoptera edeni* nor Omura's whale *B. omurai* have been recorded from the subregion, the latter at least has been recorded from 10–11° S, 94–97° E,[1] or close enough for it to have a reasonable chance of occurring in the tropical western Indian Ocean.

Apart from genetic markers, the diagnostic characters for *B. omurai* appear to be the following:

Externally

1) Maximum length of around 12 m; 2) about 200 baleen plates in each series, substantially less than in any other balaenopterid known; 3) baleen plates short and broad, the ratio of length to breadth being 1.22:1; 4) asymmetry of baleen coloration, with a greater number of white plates at the front of the right series than the left; 5) asymmetry of pigmentation of head and throat, with a greater extent of dark pigmentation on the left side; 6) lateral ridges on rostrum absent.[1]

Because of our almost total ignorance of the external appearance of *edeni*, these may not all be appropriate for distinguishing *omurai* from that species. In addition, some of these features – 4) and 5) – can occur in *brydei*, and contradictory statements about 6) exist.[2]

Internally

1) The posterior extension of the maxilla on the skull vertex expands sufficiently towards the midline to cover the posterior portion of the premaxilla adjacent to the nasal bone (whereas in *edeni* it is slender throughout, leaving space for the premaxillae); 2) the frontal bone is only exposed as a narrow belt between the ends of the maxillae, the nasals and the supra-occipital (whereas in *edeni* it is broadly exposed and forms a low but noticeable protuberance); 3) the alisphenoid bone is separate from the squamosal, whereas in *edeni* they are in contact.[1]

Obviously to make these distinctions one needs to have a cleaned skull at one's disposal, and a reasonable knowledge of cranial anatomy, both of which may be rare occurrences.

References

1 Wada *et al.* 2003
2 Wada and Numachi 1991

Behaviour and vocalisations

Inshore Bryde's whales off South Africa were encountered in groups of 1–8 animals with a mean of 1.4 (n = 100), but all but five schools were of two animals or less. Aggregations of individuals sometimes occurred, presumably in response to high prey densities, in the form of loose associations of individuals 0.9–1.8 km apart (in one case numbering 23 animals spread out over 18 km).[15] In the western Indian Ocean, the average school size encountered was 1.1 animals (n = 156).[20]

Feeding associations between inshore Bryde's whales and a number of other vertebrate species (gannets, dolphins, fur seals, penguins, gulls, Cape cormorants, predatory fish, etc.) were seen in 38% of encounters. Some of the predatory fish seemed to be hunting with the Bryde's whale, perhaps using its bulk to ambush prey.[15]

When Bryde's whales surface between shorter dives, the dorsal fin only just clears the water completely, and is the last part of the body to submerge. In longer dives, the back is arched more steeply, the fin lifted clear of the water and the peduncle is the last part of the body to disappear. The normal surfacing pattern of the species is for a series of 1–6 short dives (averaging 0.58 min.) to be followed by a longer dive (averaging 4.87 min., with a maximum of 9 min. 5 sec.). Alternatively, the whale may surface regularly, with an average dive time of 1 min. 44 sec.[15] Surfacing behaviour may change with time of day: dives of radio-tagged whales off Venezuela shifted at dusk from 6–8 min. (followed by 4–7 blows) to 0.5–2.0 min. (followed by 1–2 blows).[21]

Bryde's whale vocalisations seem to show great geographical variation. In the eastern tropical Pacific three call-types were only recorded north of 9° N and three call-types south of 5° N. The commonest call-type to the north had an average frequency of 41.7 Hz with two to four harmonics up to 160 Hz, while the commonest call to the south was a complex, two-part frequency-modulated call with an upper component around 37 Hz and a lower component around 23 Hz. These geographical differences seemed to reflect a division between stocks. Calls from the southern Caribbean were of only one type, with a fundamental frequency around 44 Hz plus 2–3 harmonics and lasting 0.8–2.5 sec. In the north-west Pacific two different but similar call-types were recorded, the most common being a pulsed moan with a fundamental frequency about 45 Hz and an average pulse rate of 51.6 pulses/sec. In the eastern tropical Pacific some of the different call-types seem to have been related to group size and behaviour.[22] Two call-types have also been noted around northern New Zealand, one seasonal and the other year-round, possibly reflecting two different populations.[23] The possibility of age-related changes in vocalisations has also been raised by a comparison of calls from free-ranging cow-calf pairs in the Gulf of California, and a captive juvenile from the Gulf of Mexico.[24]

Conservation status

The inability of the whaling industry to distinguish Bryde's from sei whales consistently until very late in the history of the fishery means that there is no reliable set of catch figures for Bryde's whales in the subregion before about 1962. Nevertheless, Bryde's whales rarely range south of 40° S, so most Antarctic whaling fleets (apart from those of the Soviet Union[25]) never seriously exploited them. Furthermore, the two main South African land stations (Durban and Donkergat) were situated at the periphery of the ranges of the inshore and offshore populations, so although Bryde's whales did feature in their catches (especially on the west coast), they may not have been heavily impacted. An exception is the operation of the catcher/factory ship *Run/Sierra*, which between 1971 and 1978 killed at least 1 645 whales off the coasts of South Africa, Namibia, Angola and Gabon, ostensibly 'sei whales' but almost certainly all Bryde's whales from the offshore population. The only population numbers available are for the inshore population, which in 1983 was estimated as 582±184 whales, although this might be an underestimate owing to the way secondary sightings were handled.[15] A scientific catch of 105 Bryde's whales was taken from the population south of Madagascar in 1977.

The limited gene flow between oceans and populations of this species dictates that they should be managed as separate entities.[26]

Bryde's whales are listed as Data Deficient by the IUCN, but as Vulnerable in the South African Red Data Book. They are included on Appendix I of CITES and Appendix II of CMS.

References

1 Best and Ross 1989
2 Best 1977
3 Yoshida and Kato 1999
4 Olsen 1913
5 Omura 1959
6 Junge 1950
7 Wada and Numachi 1991
8 Wada *et al.* 2003
9 Sasaki *et al.* 2006
10 Best 2001
11 Nemoto 1959
12 Lockyer 1976
13 Omura *et al.* 1981
14 Best 1996
15 Best *et al.* 1984a
16 MRI Whale Unit data
17 Bannister and Grindley 1966
18 Neumann and Orams 2003
19 Best 1960
20 Ohsumi 1978
21 Watkins *et al.* 1979
22 Oleson *et al.* 2003
23 McDonald 2006
24 Edds *et al.* 1993
25 Mikhalev 2000
26 Kanda *et al.* 2007

Sei Whale

Balaenoptera borealis

Ventral view

THE SEI (PRONOUNCED 'say') whale is named after a small fish, the 'seje' (coal-fish or saithe *Pollachius virens*) that appears in northern Norwegian waters at about the same time as sei whales each year.

Appearance

Sei whales can be confused at sea with fin or Bryde's whales, but here we will only consider the comparison with fin whales. The principal difference lies in the shape of the dorsal fin. In fin whales the leading edge of the fin is straighter and makes a more acute angle (27–40°) with the body than in sei whales, where the fin's leading edge is more convex and forms a more obtuse angle with the body (38–51°). The effect is for the sei whale's fin to appear more erect than the fin whale's.[1] In addition, the pigmentation of the head is not as asymmetrical in sei whales, so that the right side of the lower jaw is dark. Differences in surfacing behaviour are also apparent (see page 68).

Sei whales from Area III of the Antarctic (or between 0° E and 70° E) reached physical maturity at lengths of about 14.6 m in males and 15.4 m in females (very similar to data from animals landed at the Durban and Donkergat whaling stations in South Africa).[2–4] Their body shape is slender and similar to that of Bryde's whales. The rostrum is pointed and slightly arched in side view, with a single median ridge running forward from the blowhole. The flippers are similar in shape to those of a fin whale, and in length they average 11.7–12.4% of body length. The dorsal fin is relatively large compared to that of a fin whale, its height averaging 3.4–3.8% of body length, and it is more erect, as described above. In shape it has a gently convex leading

Length:weight relationship

Mass data from 16 North Pacific sei whales weighed in pieces have been used to produce the following relationship: **$W = 0.0242L^{2.43}$**, where W = weight in tonnes and L = total length in metres. This implies that a 10 m whale would weigh 6.5 tonnes and a 15 m whale 17.4 tonnes, but 6% should be added to these values to account for weight loss during dismemberment.[8]

Skeletal data

Vertebral formula: 7 cervical, 13–14 thoracic, 13–14 lumbar and 21–23 caudal, for a total count of 55–57.

All seven cervical vertebrae are normally unfused. The first rib characteristically has a bifurcated head.[9]

Phalangeal formula (including metacarpals): II_{4-5}, III_{6-8}, IV_{5-7}, V_{3-4}.[9]

(See Bryde's whale account for differences in skull characteristics between sei and Bryde's whale.)

FIG. 28 The down-curved rostrum of the sei whale distinguishes it from all of its congeners.

FIG. 29 The dorsal fin of the sei whale inserts at a much steeper angle than it does on fin whales. Compare these photos with Fig. 33.

edge but a strongly concave trailing edge. It is situated about two-thirds of the way back from the tip of the snout. The throat grooves number 40–62 (average 47), and the longest ends well in front of the umbilicus (the gap being about 6–8% of the body length). Only rarely is there a central groove joining the umbilicus to the genital aperture.[5]

The pigmentation pattern of sei whales is very variable. The animal is dark greyish-black above, shading to white below, with no marked asymmetry of coloration in 90% of cases. On the throat and chest there is a white area that is highly variable in extent and configuration. It runs from the chin backwards, and at its maximum extent covers the whole of the throat groove area but does not extend onto the lower jaw, while at its minimum it may end only 1.5 m from the chin. Usually it stops short of the end of the grooves, where it fades gradually into the coloration of the rear half of the body. The side margins of this white area are highly irregular through the invasion of tongues of dark pigment. Behind the white area there may be a light grey chevron marking in the midline, with its apex close to the umbilicus, and isolated white spots may also be scattered within the dark pigmentation at the rear of the throat grooves. Behind the throat grooves the body is a light grey in the midline shading gradually to darker grey on the sides and back. The upper surface of the flukes and flippers is the same colour as the back, while the undersurface is usually lighter in colour (sometimes even white on the flipper). Other markings that may occur are a light grey V on the shoulder region, a white steak running back from the ear, and a dark streak from the eye, all features that are similar to those seen in fin whales.[5] Finally, individuals in some populations may carry a variable number of oval white scars, attributable to the bites of the cookie-cutter shark *Isistius*. These are usually located on the flanks and belly towards the rear of the animal, and are especially numerous in whales that migrate to low latitudes in the south-east Atlantic in winter.[6]

There are 296–402 baleen plates (average 343) in each series. The plates reach a maximum length of *c.* 80 cm, and are quite narrow, their width at the base being 25–50.9% (average 36.9%) of plate length. They are mostly dark bluish-black in colour, but

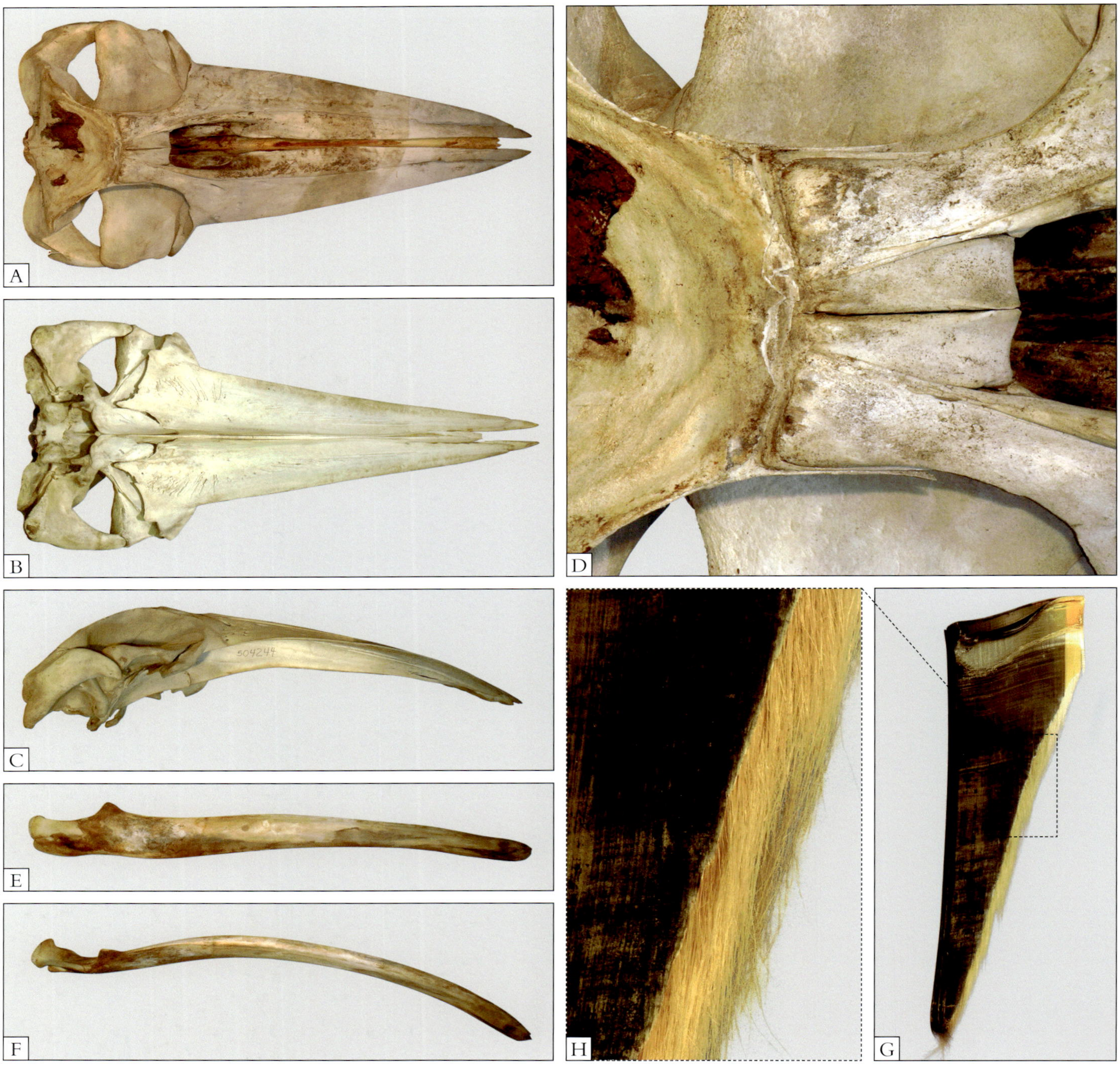

FIG. 30 Dorsal (A), ventral (B), and right lateral (C) views and vertex of skull (D) (USNM 504244); lateral (lingual) (E) and dorsal (F) views of left mandible (USNM 593415); anterior view of left baleen (G) and baleen fringe detail (H) (USNM 504244).

up to 45 plates at the front of either series may be white (and some other plates may be streaked with white). This may be more prevalent on the right side of the head. The baleen bristles are greyish-white[5] and characteristically very fine (0.1–0.25 mm in diameter), silky in texture and relatively dense (42–50/cm).[7]

Distribution and migrations

Sei whales are distributed worldwide in oceanic waters but avoid the coldest regions and marginal seas. Within the subregion there is the likelihood of some confusion between sightings (and even catches) of sei and Bryde's whales in warmer waters, complicating the determination of distribution.[10]

In summer, most sei whales are found between the Subtropical Convergence and the Antarctic Convergence (or between about 40° S and 50° S), with the latter convergence seeming to act as a barrier to the southern migration of younger, smaller sei whales. However, even the largest sei whales are rarely found south of 60° S.[11] In winter they migrate north to temperate and tropical waters, but, as with blue and fin whales, their ultimate destination is poorly known. The movements of five sei whales either tagged or recovered (with 'Discovery' tags – see box on page 69) in South African waters showed that sei whales could range widely. Whales from south of 40° S and as far east as 50–85° E in summer were linked to the east coast of South Africa in winter, while a whale from the region of Tristan da Cunha (12° W) moved to the South African west coast.[12]

Historically, the annual migrations of sei whales were deemed capricious because of the erratic appearance of the species in catches. However, their low value in the early years of whaling meant that they were only taken when other species were unavailable, and in summer their availability in high latitudes may have been affected by annual variability in oceanographic conditions. Off southern Africa in the 1960s their seasonal appearance was consistent and predictable. On the west coast of South Africa there were two peaks in availability, a smaller one in May or June and a much larger one from August to October, believed to correspond to northern and southern migrations respectively. Most animals were in water deeper than 200 m, and the northward migration (especially of mature animals) was further offshore than the southward migration.[4] On the east coast of South Africa there was a similar pattern of availability, with a smaller peak in June (when most animals were moving northwards) and a larger peak in September (when most animals were moving southwards).[13] These observations imply that sei whales migrated further north than the latitudes of either Saldanha Bay or Durban in winter, and the species was certainly declared in early catches off both Angola and the Congo, although it is unclear how many of these may have been Bryde's whales.

Sei whale migrations off the South African coast were highly structured. Catches from the northern migration consisted mainly of immature whales of both sexes plus a few mature males – presumably most mature whales passed further offshore. In the southward migration, pregnant females and immature whales were in the vanguard, with females in their first pregnancy travelling a little later than older pregnant females. The wave of mature males arrived about a month later than the peak of pregnant females, while lactating females and calves only started to arrive at the end of September and peaked at the end of October.[4]

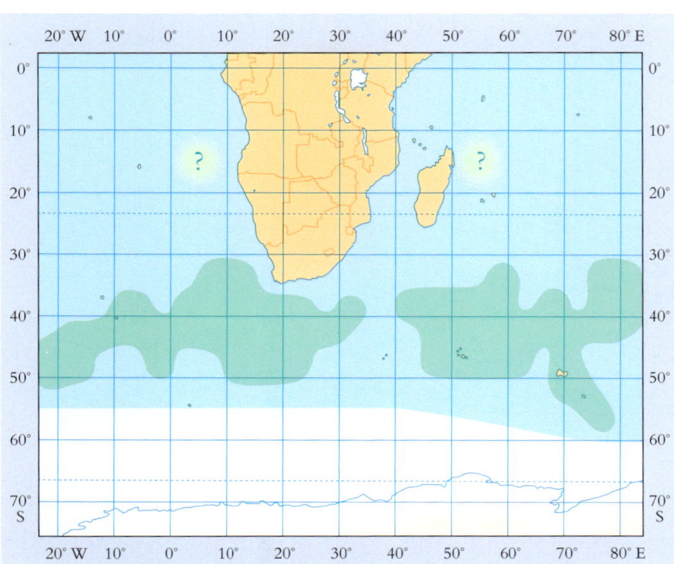

FIG. 31 Distribution map for sei whale.

Swimming speeds of 2–6.5 km/hr have been noted for feeding sei whales, increasing to 5–30 km/hr during migration.[14]

External commensals and parasites

The most characteristic external parasite of the sei whale is one that is least apparent, the harpacticoid copepod *Balaenophilus unisetus*. Its usual habitat is on the flat surface of the baleen plates, so that to spot it one has to prise the baleen plates apart. As many as 75.6% of 942 sei whales examined at the Donkergat whaling station on the west coast of South Africa carried this parasite, and the numbers present could reach several hundred. The parasitic copepod *Pennella* was also present on 5.4% of whales (n = 826). From one to 15 (mean = 2.2) individuals were carried on a whale, and these tended to occur on the side or flank (36.6%), shoulder (25.6%), and belly, especially the throat grooves (20%). Ten were lodged in healing wounds probably caused by the cookie-cutter shark *Isistius*. Stalked barnacles *Conchoderma virgatum* (six whales) or *C. auritum* (one whale) were sometimes attached to the *Pennella*, and *Xenobalanus* occurred on three whales, being attached to the rostrum, throat grooves, flank and caudal peduncle. Acorn barnacles *Coronula* sp. were found embedded in the skin of two whales.[15] Cyamids do not seem to have been recorded from sei whales,[16] and no skin-film diatoms such as *Bennettella ceticola* were recorded from 300 sei whales examined at Donkergat.[15] The latter may reflect the fact that sei whales in general do not penetrate far enough south in summer to become as heavily covered with diatoms as blue, fin or minke whales.

Food and feeding behaviour

Sei whales are like most other mysticetes in that they feed extensively in summer in higher latitudes, but less intensively in winter in low latitudes. In summer, the nature of their prey tends to change with latitude. Between about 40° S and 47° S, copepods (especially *Calanus tonsus* and *C. simillimus*) predominated in stomachs, to be replaced further south by small euphausiids such as *Euphausia vallentini*. South of the Antarctic Convergence amphipods (*Themisto gaudichaudii*) predominated initially but were gradually replaced in higher latitudes by krill, *E. superba*. These changes took place over about 20° of latitude. There were also changes by ocean, so that copepods seemed to be more important in the Atlantic than in the Indian Ocean sector. Feeding intensity is high, with on average 39.5–61.2% of stomachs containing food. 'Full' stomachs contain an average of 177.5 litres, with a maximum of 312 litres (equivalent to 228 kg).[17] Surprisingly, in winter off the west coast of South Africa the proportion of stomachs containing food was similar to that in the Antarctic (47–55%), but the quantities present were much less, averaging only 4–5 litres per whale. Euphausiids (mainly *E. lucens*, *E. recurva* and *E. spinifera*) comprised 64–73% of the food eaten and copepods (mainly *Calanus tonsus*, *Clausocalanus arcuicornis* and *Calanoides carinatus*) 24–31%.[18] At Durban on the east coast of South Africa, only 24% of sei whales were feeding when killed, and euphausiids (*E. recurva*, *E. spinifera* and *Thysanoessa gregaria*) were the principal prey.[19]

Sei whale feeding methods have been classified as 'swallowing' and 'skimming'. In the first, whales feed by lunging at prey aggregations and taking discrete mouthfuls of food, while in the second, prey is collected by the whale swimming through aggregations with its mouth open for protracted periods.[20] The ability of sei whales to skim-feed may be assisted by the size of the baleen filter area, which is much greater relative to body size than in Bryde's, humpback or fin whales of the same size.[7] When sei whales have been seen feeding at the surface their behaviour has been described as skimming[21] but seems more like a compromise between the two techniques. The mouth is opened quite wide at irregular intervals and then closed slowly (but not necessarily completely) over the next 20 sec. to 1 min. before opening it again. The throat is slightly distended each time the mouth is almost closed, and then contracted to its usual shape before the next mouth opening.[22] Turning on their side to engulf prey is not commonly seen.[9]

Within the Antarctic there is a diel pattern of feeding activity, with the proportion of whales having food in the stomach being highest at 05h00 and gradually declining thereafter (up to 21h00–22h00 at night).[17] The annual food requirement for a mature female to successfully complete a reproductive cycle has been estimated at 19 tonnes, or an average of 52 kg/day over the whole year.[23]

Reproduction and growth

The size of the sei whale at birth is not well established but is about 4.5 m. Although single calves are the rule, a few pregnancies (0.7%) involve twins (and larger multiplets). Sexual maturity is reached on average at a length of 14.1 m in females and 13.8 m in males, at an age of 8–9 years in both sexes. The typical reproductive cycle is believed to take two years, with pregnancy occupying 12 months[24] and lactation about five months,[5] and the female resting for the remaining seven months. However, estimated pregnancy rates are somewhat less than 50%, and annual ovulation rates 0.47, suggesting that the average cycle must occupy more than two years. Breeding is strongly seasonal, with about 90% of conceptions occurring over 70 days around a peak in June, and males on the northward migration in autumn on the west coast of South Africa had testes 89% heavier than those on the southward migration in spring.[4] An exponential rise in serum testosterone from late summer into autumn has been shown for North Atlantic sei whales, with levels increasing three- or fourfold every 30 days, indicating an imminent breeding season.[25] Females in their first pregnancy may conceive (or their conceptus may implant) about a week later than older females (which may explain the similar difference in the timing of their southern migration).[4] There was little evidence of a decline in pregnancy rate with age,[11] and both sexes can live for an estimated 53–60 years.[4]

The average values for sei whale milk are 58.3% water, 26.8% fat, 10% protein and 1% ash.[26]

Behaviour and vocalisations

Group sizes in the subregion (Antarctic) ranged from one to five, with an average of 1.8 (n = 73).[15]

When undisturbed, the sei whale has a distinctive way of surfacing, which it may only share with Bryde's whale. It comes to the surface very obliquely, with the snout, top of the head and anterior portion of the back appearing in quick succession before it blows. Thereafter the animal continues forward and downward, so that the body sinks lower and lower until the dorsal fin is submerged.

The sei whale does not arch its body when it dives, as do most of the other large rorquals, so that a normal surfacing and one preceding a dive are barely distinguishable from one another.[9] It has been described as 'dragging its fin at the surface'.[22]

Surfacing bouts by sei whales tend to be less well defined than in other rorquals, with only one or two blows being made at each surfacing followed by a prolonged submergence, while other species tend to produce a set of 5–6 blows and then submerge.[27]

Sei whales recorded in the Antarctic produced low-frequency tonal and 'sweep' calls, as well as broadband growls and whooshes. Characteristically, many of the tonal and sweep calls consisted of multiple parts with a frequency step between them. On average each of the tones within a tonal call lasted 0.45 sec. with an average frequency of 433 Hz. Sweep calls had an average duration of 1.1 sec. and covered a frequency range of 178 Hz with a midpoint around 432 Hz. Only a few were multi-part in nature. All calls were generally quieter than those produced by other baleen whales, which may account for the infrequency of their documentation.[28]

Conservation status

Although sei whales featured in catches almost from the start of modern whaling in the subregion, their relatively low oil yield (and somewhat separate distribution) compared to blue, fin and humpback whales meant that the numbers taken annually over the first five decades were low.

This all changed in the 1960s with the depletion of the stocks of the larger species, and the increasing focus of the Japanese expeditions, in particular, on meat rather than oil production. Catches increased from 3 629 in 1959 to a maximum of 22 207 in 1965, and as quotas were not allocated by species or by area, the resulting depletion of sei whale stocks could not be avoided. In Area III (or the sector of the Antarctic between 0° E and 70° E), sei whales dropped from an estimated 21 000 animals in 1960 to about 5% of that number in 1979, with corresponding declines in relative abundance off South Africa.[29]

The species was finally protected by the International Whaling Commission in 1979. Current trends are unknown, as there has been no concerted attempt to monitor the status of sei whale stocks since that date.

Sei whales are listed as Endangered by the IUCN and as Data Deficient in the South African Red Data Book. Sei whales are included on Appendix I of CITES, and are listed on both Appendix I and II (presumably different populations) of CMS.

References

1 Gambell 1966
2 Lockyer 1977
3 Ohsumi and Masaki 1978
4 Best and Lockyer 2002
5 Matthews 1938d
6 Shevchenko 1977
7 Sekiguchi *et al.* 1992a
8 Lockyer 1976
9 Andrews 1916
10 Best 1994a
11 Horwood 1987
12 Best and Ross 1989
13 Bannister and Gambell 1965
14 Lockyer 1981a
15 MRI Whale Unit data
16 Martin and Heyning 1999
17 Kawamura 1974
18 Best 1967
19 Bannister and Baker 1967
20 Nemoto 1959
21 Ingebrigtsen 1929
22 Watkins and Schevill 1979
23 Lockyer 1978
24 Gambell 1968
25 Kjeld *et al.* 2003
26 Oftedal 1997
27 Collett 1886
28 McDonald *et al.* 2005
29 Best and Gambell 1968b

'Discovery' marking

Although experimentation with tagging whales began as early as 1910, it was only in 1932/33 that widespread marking of whales commenced, using the so-called 'Discovery' mark, named after its designers, the Discovery Committee in England. Supposedly based on the fact that lost harpoons were frequently found in the bodies of whales killed in nineteenth-century whaling, the final design involved a fully implantable device. The standard 'Discovery' mark was a stainless steel tube 1.7 cm in diameter and 23 cm long with a lead tip, the shaft being engraved with a unique serial number and the words 'Reward for return to Discovery British Museum (Nat. History) London' (usually the princely sum of £5). The mark was fired from a 12-bore shotgun, and was supposed to pass through the blubber layer and lodge in the dorsal muscle, from where it might be recovered when the whale was killed and processed (protruding marks were believed to be shed rapidly and so were not counted as 'effective hits'). These marks were really designed for blue, fin and humpback whales, and a .410 mark identical in design but smaller (*c.* 14 cm long by 0.6 cm) was developed in the 1960s and used widely on minke whales.

Between 1932 and 1975 at least 20 573 whales were marked worldwide,[1] and subsequently at least another 2 804 minke whales.[2] Recovery rates varied widely, but in the post-war period for all species combined were 5.7% in the North Atlantic and 14.8% in the Antarctic.[1] Although the system had many drawbacks, including uncertainty over how many whales were effectively marked, varying and often unknown efficiencies of detection/reporting of marks upon recapture, and delayed recoveries (from cookers, etc.) for which positional information was inexact and from which associated biological data could not be gathered, some invaluable (and in some cases still the only) information on whale movements and age were obtained. Marks were retained for up to 37 years in fin whales, 22 years in sperm whales and 19 years in Bryde's whales.[1] The introduction of the moratorium on commercial whaling in 1986 effectively brought to an end this form of tagging, as in the absence of an industry few if any recoveries could be expected. 'Discovery' marking did, however, contribute to the thinking that went into the initial design of radio and later satellite tags.[3]

References

1 Brown 1977 2 Buckland and Duff 1989 3 Watkins 1983

FIG. 32 Three Discovery-type whale marks: the upper two 12 gauge (for large whales) and the lower one .410 gauge (for small whales); the lower two still have their cartridges attached. The uppermost mark is from the Soviet scheme and the lower two from the International scheme.

Fin Whale

Balaenoptera physalus

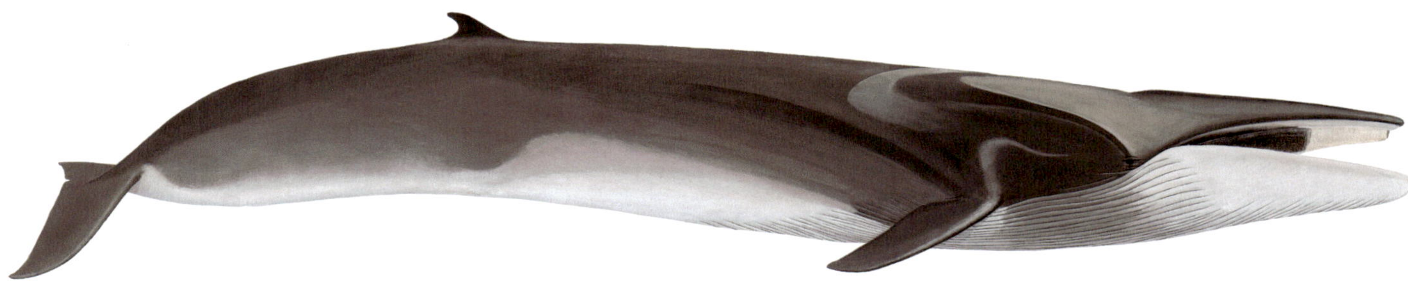

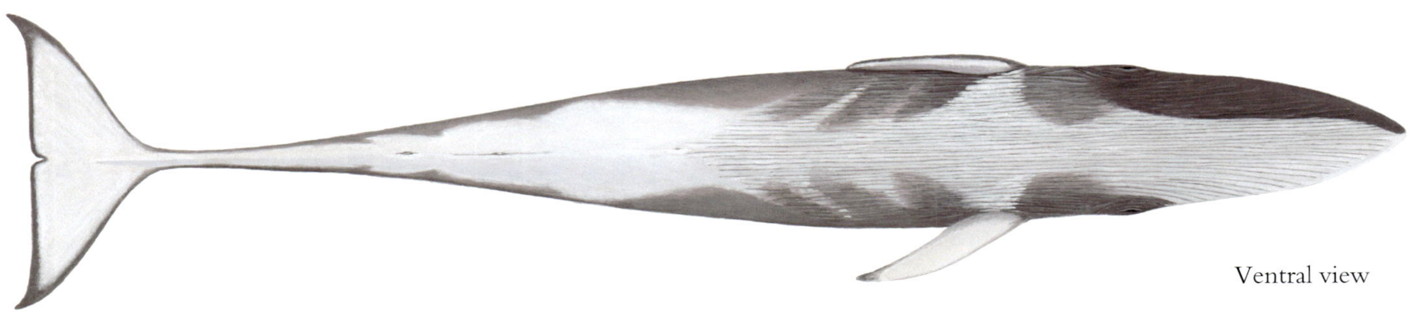

Ventral view

A SUBSPECIES, THE pygmy fin whale *B. physalus patachonica*, has recently been proposed from the southern hemisphere.[1] The basis for this proposal is not well founded, so although the nominated type specimen is from Argentina and the putative distribution may well include the subregion, it has not been included here (the proposed type specimen also apparently no longer exists in the Buenos Aires Museum[2]).

Appearance

Appropriately enough, fin whales are most easily recognised at sea from the size and distinctive profile of their dorsal fin. In most other balaenopterids, the leading edge of the dorsal fin is quite erect – in the case of the sei whale making an angle of 38–51° (average 46.5°) with the back – while the fin whale's fin is more backward sloping, making an angle of 27–40° (average 32.6°) with the back.[3] Only the blue whale has a fin with a similar slope to it, but is diminutive in size (average height 1.2–1.3% of body length) when compared with that of the fin whale (average height 2.4–2.5% of body length).[4] At closer range, the complex streaking around the head region may be seen, and the marked asymmetry of coloration (especially on the lower jaw) detected.

Fin whales are the second largest of the whales in the subregion. Females grow larger than males, the length of the largest female measured by the Discovery Investigations staff being 24.53 m and the largest male 22.4 m.[4] The head comprises about 25% of the body's length, and the sides of the rostrum are straighter than in the blue whale, making the snout appear more pointed (there may also be sexual dimorphism in the shape of the snout, with the tip becoming swollen in mature males[5]). The dorsal fin is situated about three-quarters of the way back from

Length:weight relationship

A length:weight regression is available for 34 fin whales from the Antarctic:[8] **$W = 0.0238L^{2.53}$**, where W = weight in tonnes and L = length in metres. As all these whales were cut up to be weighed, a 6% correction factor for weight loss must be applied. This means a 10 m whale would weigh 8.6 tonnes and a 20 m whale 49.6 tonnes, adjusted for weight loss.[8]

Skeletal data

Vertebral formula (North Atlantic, n = 16[6]): 7 cervical, 14–16 thoracic, 13–15 lumbar and 24–27 caudal; total 60–63.

The cervical vertebrae are usually all free, but the first two or three may be fused.[9]

Phalangeal formula (North Atlantic, n = 17[6]): II_{2-4}, III_{5-7}, IV_{4-7}, V_{2-4} (excludes metacarpals).

FIG. 33 A fin whale surfacing in rough seas in the Antarctic reveals, firstly, the white mandible and complicated shoulder pattern on its right side, and then the distinctive dorsal fin.

the tip of the upper jaw. Flipper length is about 11–12% of body length,[4] and its shape is different to that of the blue whale, having a straighter leading edge and a less concave trailing edge (giving a less 'hooked' appearance).[6] The throat grooves extend as far back as the umbilicus, and number 68–114 (average 87, n = 80).[4]

Fin whales are a bluish slate-grey above and white below. The borderline between the two colours is generally diffuse and variable in position. In the chest region the pigment reaches down as irregular projections over the throat grooves, which in places may be pigmented between the grooves but are white on the ridges. The white extends onto the underside of the flukes, but behind the anus may be constricted or cut off by extensions of pigment from above. But the most interesting aspect of the fin whale's pigmentation is its asymmetry on the front half of the body. Externally, both upper and lower jaws on the left side are pigmented, while on the right side the entire lower jaw and the front of the upper jaw are white. On the side of the head there may be several distinct streak marks: a pale streak running back from the external ear opening, a thicker black band extending back and up from the eye, and a pale streak running up and back from the flipper. Because of the asymmetry of the head coloration these marks (and particularly the eye streak) are more obvious on the right than the left side of the head. On the back at the level of the flippers there is a V-shaped pale mark with its apex pointing towards the snout.[4]

There are 262–473 (average 363, n = 70) baleen plates in each series. The longest plates reach a length of about 70–80 cm in the largest whales, and their width is about 60–74% of their length. Reflecting the asymmetry of head coloration, on the left side of the head only the first few baleen plates at the front have white outer edges, while on the right side the first 100–200 (average 153, n = 103) plates may be so coloured.[4] The remaining plates are a dark bluish-black with vertical light and dark streaks of varying width and number, with the dark streaks tending to prevail towards the back of the mouth.[6]

The bristles are all whitish-yellow and coarse, most having a diameter of 0.35–0.5 mm in young and 0.6–0.9 mm in adult whales, with densities of 20–28/cm.[7]

Distribution and migrations

Fin whales occur worldwide in all oceans but less commonly in the tropics, and a six-month difference in the breeding season between northern and southern hemispheres probably means there is little interchange between them. The species is assumed to occur throughout the subregion, at least in waters away from the continental shelf. Besides the Southern Ocean, it has been recorded in commercial catches from South Africa, Namibia, Angola, Gabon, Mozambique,[10] and Madagascar[11] (some of these identifications from early catches may be suspect, and involve other balaenopterids).

The migration of southern fin whales is supposed to follow that of the baleen whale paradigm – summer feeding in higher latitudes and winter breeding in lower latitudes. Sightings south of South Africa during cruises in the Antarctic between 1933 and 1939 fitted this pattern, with fin whales first appearing in October, rising to a peak in February and disappearing by July.[12] The seasonality of some of the lower latitude catches also suggests migration. On the west coast of southern Africa in the 1920s there were two peaks in fin whale catches (May–July and October–November), suggesting that the migratory destination was to the north of the whaling grounds.[13] By the 1960s, only the first peak was apparent, but nearly all the fin whales seen from the spotter aircraft were moving northwards, so their destination may still have been to the north.[14] In the summer (December–March) fin whales were seldom seen off the west coast.[15] On the east coast (at Durban) there was only a single peak in catches (June–July) in both the 1920s and the 1950s to 1960s, suggesting that fin whales did not travel much further north.[12, 13] The proportion of northward-moving animals (as seen from the spotter aircraft attached to the whaling station) declined from May to August, and yet by August–September over half the whales were still moving north. It is possible that the southward migration takes place further offshore.[12]

Direct evidence of migration comes from 12 instances where fin whales tagged with 'Discovery' marks have moved from the Antarctic to South African waters or vice versa.[16] Intervals between marking and recovery ranged from seven months to 10 years and 10 months. Ten of these animals were marked in summer (November–February) in latitudes 52–64° S and recovered from May to September at either Donkergat on the west coast (two) or Durban on the east coast (eight). One animal was marked in the subantarctic just north of the Crozet Islands in November and recovered at Durban in June, while one animal was marked off Durban in August and recovered at 43° S in April. These recoveries show that fin whales migrating through South African coastal waters in winter spend the summer in the Antarctic at longitudes between 7° E and 50° E (with an outlier at 81° E), and that there is a definite overlap of the feeding grounds utilised by whales from the east and west coasts of South Africa.

The actual locations of the winter breeding ground(s) for fin whales, and whether they are dispersed or concentrated, remain a mystery.[17] Catches made at South African land stations tended to be dominated by immature animals (79% of males, 92% of females at Donkergat in 1926, for instance[4]), a feature that was consistent throughout the catch history.[12] One possible explanation for this might be that mature individuals moved past further offshore, which would be consistent with an offshore breeding ground. Sightings made on a transatlantic voyage in August 1943 from South America to southern Africa, mostly along 20° S, are of particular interest. In the central south Atlantic, between about 5° W and 25° W, 40 whales were seen over six days, and some of these were identified as fin whales or finners. The sighting rate was similar to contemporary rates achieved in the Antarctic in summer.[18]

There is some evidence that fin whales may use the earth's geomagnetic field to navigate during migration.[19]

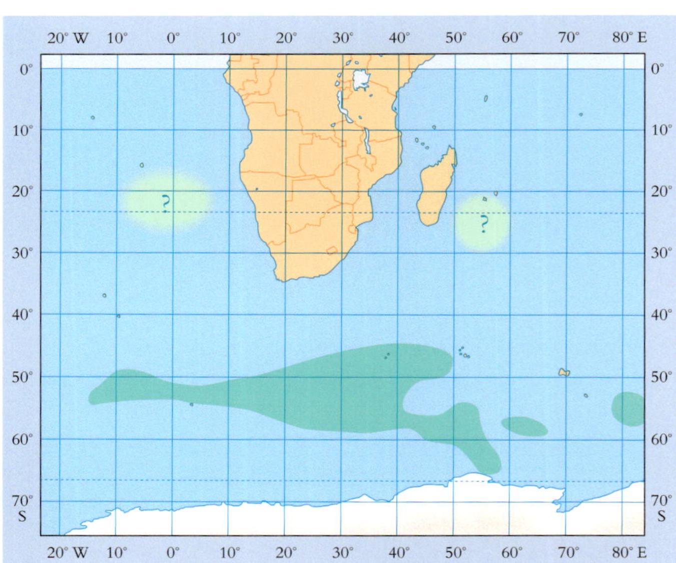

Postulated range
Summer concentration
Winter concentration
Questionable origin

FIG. 34 Distribution map for fin whale.

External commensals and parasites

At Donkergat, the copepod *Pennella* was the commonest external parasite, occurring in 11.5% of 61 whales examined. The numbers present ranged from one to six (average two), and were mostly found on the flanks. Five cyamids (*Cyamus balaenopterae*) were found at the edges of the blowhole in one whale. No diatoms were recorded in 42 whales examined, and no *Balaenophilus unisetus* on the baleen of 39 whales.[20] In an earlier study, no diatoms were recorded from adults, but small spots were seen on a few immature whales.[4] At Saldanha Bay, sessile barnacles *Coronula reginae* have been reported on the tail of one whale (one specimen) and two small individuals embedded in the skin of another, and many specimens of the stalked barnacle *Xenobalanus globicipitis* on the tail, fluke, etc. of a third fin whale.[21]

In the West Pacific sector of the Antarctic in summer, the incidence of *Pennella* on fin whales averaged 0.8% (n = 605), while cyamids occurred on 7.1%, diatoms on 34.3%, sessile barnacles *Coronula* sp. on 8.9% and stalked barnacles *Conchoderma* sp. on 0.8%. A tendency for the incidence of *Pennella* to decline and that of diatoms and cyamids to increase with time during the season was noted.[22] Together with the relative incidences seen in South African waters, this would suggest that *Pennella* tends to be contracted in warmer waters, and diatoms and cyamids in colder waters.

FIG. 35 Dorsal (A – retouched), ventral (B), and right lateral (C) views and vertex of skull (D); lateral (lingual) (E) and dorsal (F) views of left mandible (USNM 237566); anterior view of left baleen (G) and baleen fringe detail (H) (USNM 504258).

Small oval wounds and scars attributed mostly to the shark *Isistius* are commonly seen on fin whales. At Donkergat, 86% of males (n = 36) and 94% of females (n = 36) carried unhealed wounds, which ranged from one to 43 in number (mean 10). Of these, 8.9% were still pink and thus recently inflicted, indicating that the biting agent must occur relatively close to the whaling ground. There was no obvious relationship between the incidence of unhealed wounds and the size or sex of the whale (although very few mature whales were examined). Old healed wounds were more numerous and difficult to quantify.[20]

Food and feeding behaviour

On both the east and west coasts of South Africa, fin whales fed mainly on euphausiids. On the west coast, species identified included *Euphausia recurva*, *E. lucens*, *E. spinifera*, *Thysanoessa gregaria* and *Nyctiphanes* (*africanus*) *capensis*,[4, 14] with the principal species being *E. recurva* and *E. lucens* (together comprising 87.1% by weight).[14] On the east coast *E. recurva* and *T. gregaria* predominated (either one or both in 32 of 38 stomachs with food). Other minor food items on either coast included copepods, amphipods and megalopa larvae.[14, 23] One fin whale examined at Saldanha Bay in 1926 contained fish (unidentified).[4] In the Antarctic proper the diet seems to consist largely of krill (*E. superba*), but in the subantarctic other species, especially *E. vallentini* and *E. frigida*, are more important.[24]

Feeding is supposedly more intense in the Antarctic than on the wintering grounds. At South Georgia, the percentage of fin whales with food in the stomach in 1925–27 averaged 88.4% (n = 510), compared to 42% (n = 205) at Saldanha Bay in 1926,[4] and 35.8% (n = 67) at the same locality in 1962–63.[14] At Durban, two indices of feeding activity – the incidence of food in the stomach and the colour of the faeces – gave essentially identical results, 22.1% (n = 299) and 21.3% (n = 310).[23] In 1930 only 4.1% of 49 stomachs examined there were said to contain food.[11] Quantities consumed were also much less in South African waters than in the Antarctic: whereas 300–900 kg is given as the amount of food usually found in a fin whale stomach in the Antarctic,[25] an equivalent figure for the west coast of South Africa was estimated at less than 2 kg.[14]

Theoretical food requirements for fin whales at puberty have been estimated as 1 500–1 660 kg/day while in the Antarctic for 120 days, and 150–166 kg/day for the rest of the year. For fully mature females these amounts increase to 2 258–2 286 kg/day and 226–232 kg/day respectively.[26]

Fin whales normally use lung-feeding to ingest prey, in which they swim rapidly at a prey patch and engulf a large quantity of water containing aggregated prey in one mouthful, expelling the water through their baleen as they close their mouth (see box on page 75). One to eight such lunges may be made during one feeding dive, usually from below the swarm, and for average vertical distances of 21 m.[27] While the whales are at or near the surface, most (81–97%) of these lunges take place with the whale on its right side, which would be consistent with the hypothesis that fin whales use their asymmetric coloration to maintain counter-shading (and thus avoid detection). However, other symmetrically-pigmented whales (such as blue and Bryde's) show a similar tendency to turn on their right sides when lunge-feeding. There is equally little support for another hypothesis, namely that fin whales use the white side of the head to herd prey into tighter formations before lunging. It is possible that the asymmetric coloration has some ecological or social function unrelated to feeding.[28]

Fin whales fitted with telemetry devices in the Mediterranean frequently dived below 150 m to a maximum of at least 470 m, with the deeper dives tending to occur in the late afternoon, coinciding with the upward vertical migration of krill.[29]

Reproduction and growth

The length of southern fin whales at birth is about 6.5 m,[4] after a pregnancy lasting 11 months. The calf is said to be weaned after six or seven months, when it has reached a length of 12 m,[26] but there is at least one instance of a calf still suckling when 12–13 months old and 15.2 m long.[30] Sexual maturity is reached at an average length of 20 m in females and 19 m in males.[26] An apparent decline in the age at maturity from 10 years around 1930 to five or six years in the 1950s has been attributed to the effects of exploitation,[31] although other interpretations of such data are possible.[32] The average calving interval is generally taken as two years, and although there are signs of an increase in the pregnancy rate from the 1930s to the 1960s, which have also been attributed to the effects of exploitation,[31] these results have been challenged.[33] Breeding is quite seasonal, with 86% of conceptions occurring over five months (April–August). There is only a slight reduction in fertility with age.[34]

Average values for the composition of fin whale milk are 53.2% water, 33.4% fat, 10.6% protein and 1.2% ash.[35]

Physical maturity is reached at a length of 21 m in males and 22.25 m in females.[26] Individuals with an estimated age of 100–101 years (based on counting the layers in the ear plug) have been recorded in the North Pacific.[36] The recovery of a marked fin whale 37 years after tagging[37] provides direct evidence for longevity in the region of 40 years (given that the age of the whale at marking was unknown).

Behaviour and vocalisations

Group sizes of fin whales on the east coast whaling grounds off South Africa in 1972–75 ranged from one to three with a mean of 1.6 (n = 52).[38] On the west coast whaling ground off South Africa between 1961 and 1965, fin whales were seen from the spotter aircraft in groups of 1–7 animals, with a mean of 1.5 (n = 77).[20] In the Antarctic group sizes tended to be larger: 1–16 with a mean of 2.8 (n = 60).[39]

Swimming speeds have been reported as 2.0–6.5 km/hr when feeding and 5–30 km/hr when migrating, with a maximum of 41 km/hr when alarmed.[26] Aerial behaviour, or exposure of flukes, is rare.

Fin whales produce very characteristic low-frequency (20 Hz) sounds, primarily as downsweeps of about 1 sec. in duration, and usually in long, regularly-spaced patterns. These sounds are very loud, and (under the right conditions) might be able to propagate over hundreds of kilometres.[40] In the western North Atlantic there is a definite seasonal pattern to the vocalisations, with a distinct increase in late summer rising to a peak in winter and diminishing by spring. This coincides with a move into lower latitudes during the autumn and a reverse movement in spring.[41] The sounds seem to be only produced by males, in what has been proposed is a display to attract females to prey aggregations,[42] and the seasonality noted above would be consistent with the peak of the breeding season in winter. In the Antarctic Peninsula region, fin whale calls were strongly seasonal, with a peak in May and an absence of calls from July to January.[43] Fin whales from different ocean basins may also have different vocal patterns.[41]

Conservation status

Fin whales were for a long time the mainstay of the whaling industry in the southern hemisphere. Some 716 000 southern fin whales were taken in the twentieth century, mostly in the Antarctic.[44] At an average body weight of 40 tonnes, this represents the removal of about 28.6 million tonnes of biomass. The initial population size has been estimated at 405 000 whales, with Area III (south of the African continent) comprising 152 000 or 37.5% of the total. By 1970 it was estimated that the total population had been reduced to about 65 000 whales, and the Area III population to 30 200.[45, 46]

There is a need to review these assessments, particularly in the light of previously falsified Soviet catches. Indices of fin whale availability at Durban from 1954 to 1975 also indicated a much greater decline (by as much as 89–97%) than suggested by the earlier assessments.[47] The current status of southern fin whales is unclear. Three sets of circumpolar sighting surveys in the Antarctic between 1978/79 and 1997/98 produced estimates of 2 100, 2 100 and 5 500 fin whales respectively. As the surveys were confined to south of 60° S, they are unlikely to have covered the whole latitudinal range of the species' summer distribution.[48] No significant population trend could be detected from these estimates.

Fin whales are listed as Endangered by the IUCN but Data Deficient in the South African Red Data Book. They are included on Appendix I of CITES but (as different populations) on Appendices I and II of CMS.

Lunging for lunch

Blue, fin, sei, Bryde's, minke and humpback whales are all referred to as rorquals, a word that originates from the Norwegian *röyrkval*, or 'furrow whale', the furrows of course being the throat grooves that are a distinctive feature of these species. These grooves, along with other unique anatomical adaptations, play a fundamental role in the feeding mechanism that these whales most often employ, referred to as engulfment, gulping, swallowing or lunge-feeding. As a whale closes in on a prey shoal, it lowers its mandible, positive inertial pressure opens the mouth, and the water plus prey is passively enveloped in one large gulp. This process is aided by the nature of the tongue which is flaccid and deformable, the accordion-like throat grooves, and the presence of a space between the belly blubber and the overlying body, known as the *cavum ventrale*, or ventral cave.[1] The whale is also able to open its mouth to an extraordinary degree, up to 80° and possibly 90°, thanks to the nature of the hinge connecting the lower jaw to the skull, which is fibrous and very different to that of a normal mammal. As the water plus prey enters this impressive mouth, the throat grooves expand and the tongue everts through the floor of the mouth into the *cavum ventrale*, forming a huge sac-like structure inside the front half of the animal.[2] As an indication, it has been calculated that a 150-tonne blue whale engulfs 50 tonnes in one of these bite-sized pieces of the ocean. Elastic recoil of the pouch and reformation of the tongue helps the whale to close its mouth and expel water through the baleen. This whole process takes place as the whale continues to move forward in its lunge, and a fibrous stay running between a process on the lower jaw and the skull not only provides a limit to the extent the mouth can open but could also be using kinetic energy generated during mouth-opening to drive the jaw closed.[3]

Having got its prey bagged and trapped against the inside of the baleen, the whale is then faced with the problem of removing this compacted mass of plankton from the inner surface of the baleen rack so that it can be swallowed. Three possible mechanisms have been proposed. The tongue could be used to dislodge prey from the baleen into the centre of the mouth, either by physically scraping it off the baleen or by agitating the baleen plates to release it. Alternatively, the head could be shaken from side to side to achieve the same effect. Or the whale might use a form of back-washing, opening the mouth slightly and using the tongue to generate suction pressure so that water reverses through the baleen and dislodges the prey. Observations of head-nodding and rapid opening and closing of the mouth in right whales support the second and third suggestions, but different techniques may well be employed by different species or for different prey types.[4]

References

1 Pivorunas 1979
2 Lambertsen 1983
3 Lambertsen *et al.* 1995
4 Werth 2001

FIG. 36 A humpback whale lunges on its right side at a swarm of zooplankton in St Helena Bay, South Africa.

References

1 Clarke 2004
2 Cappozzo, L., Estación Hidrobiológica de Puerto Quequén, Buenos Aires, pers. comm.
3 Gambell 1966
4 Mackintosh and Wheeler 1929
5 Nemoto 1962a
6 True 1904
7 Nemoto 1959
8 Lockyer 1976
9 Tomilin 1957
10 Best 1994a
11 Mackintosh 1942
12 Bannister and Gambell 1965
13 Harmer 1931
14 Best 1967
15 Olsen 1914
16 Best and Ross 1989
17 Mackintosh 1966
18 Wheeler 1946
19 Walker *et al.* 1992
20 MRI Whale Unit data
21 Nilsson-Cantell 1930
22 Nishiwaki and Hayashi 1950
23 Bannister and Baker 1967
24 Kawamura 1980
25 Kawamura 1974
26 Lockyer 1981a
27 Croll *et al.* 2001
28 Tershy and Wiley 1992
29 Panigada *et al.* 1999
30 Best 1966
31 Gambell 1973
32 Cooke and de la Mare 1984
33 Mizroch and York 1984
34 Laws 1961
35 Oftedal 1997
36 Ohsumi *et al.* 1958
37 Brown 1977
38 Findlay 1989
39 Kasamatsu *et al.* 1988
40 Payne and Webb 1971
41 Clark 1995
42 Croll *et al.* 2002
43 Širović *et al.* 2004
44 Allison and Smith 2004 (catches prior to 1930 obtained from BIWS data)
45 Breiwick 1977
46 Allen 1977
47 Best 2003
48 Branch and Butterworth 2001a

Blue Whale

Balaenoptera musculus

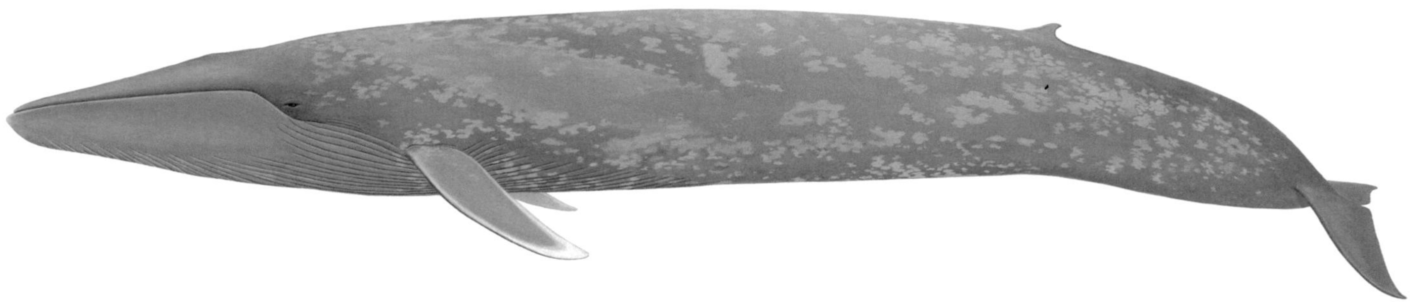

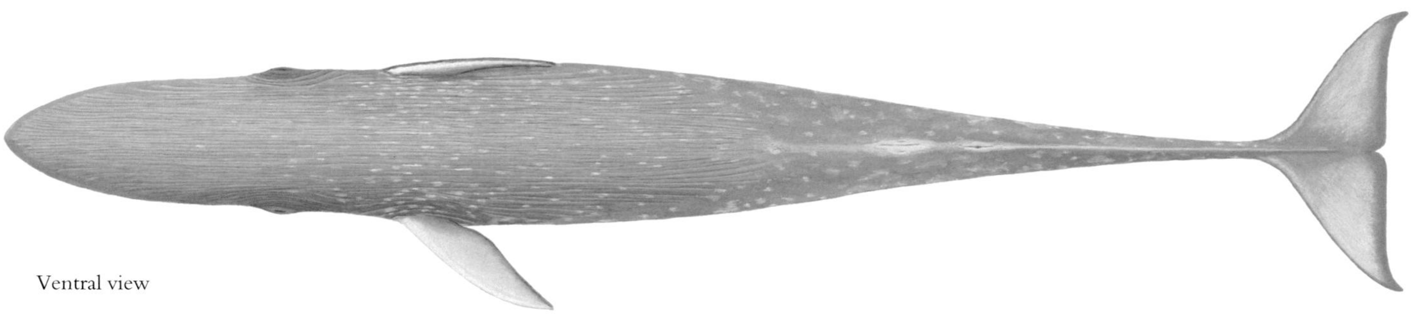

Ventral view

TWO SUBSPECIES OF blue whale have been described in the southern hemisphere, but they are considered together here because it is difficult to tell them apart at sea, or even when one is looking at a specimen on land. Physically they differ mainly in size and body proportions, with the smaller form being universally known as the 'pygmy' blue whale *B. m. brevicauda* and the larger form usually as the 'true' blue whale *B. m. intermedia* (I prefer the name 'Antarctic' blue whale, as it better reflects the difference in distribution between the two). A third, unnamed subspecies may occur in the south-east Pacific[1] but is not considered here because it does not occur in the subregion.

Appearance

These whales only really appear blue when seen through the water: in air their bodies are more aptly described as a gunmetal colour, with a variable mottling of lighter grey spots scattered over the back and flanks, which may only be seen clearly when the sun is behind the observer. The dorsal fin (which in a big individual seems to take an age to appear during a surfacing) seems absurdly small and is often sloped back with a rounded tip. The head of a blue whale, if seen from above, is much blunter in outline than that of other rorquals, with a distinctively rounded tip to the snout. When the animal dives, the flukes are often lifted clear of the water, although (unlike the flukes of right and humpback whales) they are not usually hoisted vertically.

Features said to distinguish the pygmy blue whale at sea are 1) its distinctive tadpole shape (large head, small body), 2) an extension of the median groove between the blowholes beyond the leading edge of the blowholes themselves, and 3) the presence of knobs on the caudal peduncle.[2]

Antarctic blue whales reach physical maturity at a length of 25.9 m in females and 24.1 m in males,[3] and are the largest animals known to have existed (see box on page 81). The length of pygmy blue whales at physical maturity is not well established but growth is estimated to cease at about 21.6–21.9 m in females and 20.4–20.7 m in males, with the longest whale taken by Japanese expeditions being a female of 24.1 m.[4] The head is large, the distance from the tip of the snout to the eye averaging 19.5–20.5% of body length in Antarctic blue whales, and the lower jaw projects some 1–1.7% of the body length beyond the tip of the snout.[5] Pygmy blue whale proportions are similar.[4] The rostrum is flat with a single central ridge extending forward from the blowhole, and is characteristically rounded at the tip compared to other members of the genus. The flippers have a markedly convex leading edge and a somewhat S-shaped trailing edge. Flipper length (from anterior insertion to tip) averages 12.9–13.2% of body length. The dorsal fin is small, its height averaging only 1.2–1.4% of the body length, but is quite variable in size and shape: usually it is small and quite sloped back with a distinctly concave trailing edge, but sometimes it is larger, more erect and less falcate.[6] It is situated far back on the body (74.7–75.5% of the way back from the tip of the snout in Antarctic blue whales),[5] although the proportion is possibly different in pygmy blue whales, given the differences in 'tail length'. The distance from the anus to the notch in the flukes is given as a diagnostic character between the two subspecies, ranging from 27–31% (with an outlier of 35%) of body length in

FIG. 37 The rostrum of a blue whale is characteristically broad and U-shaped, and some of its typical grey speckling can be seen on the back.

FIG. 38 An Antarctic blue whale surfacing amongst broken ice shows just how blue the body appears when just submerged.

Antarctic blue whales and 22–30% in pygmy blue whales, the means being 29.5% and 26% respectively.[7] These values are calculated over a limited size range (21–23 m), and may not be applicable to individuals of a different size. The ventral grooves extend from just behind the tip of the lower jaw back to a point just behind the navel. In females a variable number of separate grooves may be found on either side of the genital slit. The number of main ventral grooves in Antarctic blue whales ranges from 70 to 118 (average 91),[5] and in adult pygmy blue whales from 76 to 94 (average 84).[4]

As previously mentioned, the basic coloration of blue whales is dark blue-grey, although much of the body is covered with a pale mottling of roughly oval marks (averaging 10 x 7.5 cm in size), which are slightly lighter in tone than the ground colour. Areas not normally covered by such mottling include the head and lower jaw, flippers and tail flukes. The mottling is most intense on the shoulder and flanks, where the marks may coalesce in some individuals. It also occurs on the ventral grooves but is less obvious, and may be accompanied by a number of white flecks of similar size and shape, particularly under the flipper and at the rear end of the grooves. The flippers are darkly pigmented on their outside, usually with a white tip, and are white (sometimes streaked with grey) on their inside. The underside of the flukes is usually paler than the upperside and sometimes carries fine whitish striations.[5] Pygmy blue whale coloration is similar, though the basic coloration is said to be silvery grey rather than the steel blue of Antarctic blue whales.[4]

In Antarctic blue whales there are 270 to 395 (average 322) baleen plates in each series, the longest of which reaches nearly 1 m in the largest whales.[5] In pygmy blue whales the number of baleen plates in a series ranges from 280 to 350 with a mean of 317.[4] The shape of the baleen plates is said to differ between Antarctic and pygmy blue whales: those of the former are narrower, the ratio of breadth to length ranging from 1.3 to 2.2 with a mean of 1.79, while in the pygmy blue this ratio ranges from 1.1 to 1.8 with a mean of 1.48, but the overlap is considerable.[4] The baleen plates in both subspecies are blue-black in colour with black bristles. There are 10–30 (mostly 15–23) bristles per cm and the bristles are relatively coarse in texture, most of them being 0.8–1.2 mm in diameter.[8]

Length:weight relationship

Mass data from 43 Antarctic blue whales weighed in pieces have been used to produce the following relationship: **$W = 0.0046L^{3.09}$**, where W = weight in tonnes and L = total length in metres. This implies that a 20 m whale would weigh 48.2 tonnes and a 25 m whale 96 tonnes, but 6% should be added to these values to account for weight loss during dismemberment.[9]

Mass data from five pygmy blue whales weighed in pieces have been used to produce the following relationship: **$W = 0.0004L^{3.97}$**, where W = weight in tonnes and L = total length in metres. This implies that a 20 m whale would weigh 58.5 tonnes, or somewhat greater than an Antarctic blue whale of the same size, although 6% should also be added to this to account for weighing in pieces.[9]

Skeletal data

Vertebral formula for five pygmy blue whales (one adult, four foetuses): 7 cervical, 14–15 thoracic, 14–15 lumbar and 27–29 caudal; total 63–66.[4, 10] Published data for documented Antarctic blue whales do not seem to exist, but in North Atlantic and North Pacific blue whales the formula is similar to that of pygmy blue whales: 7 cervical, 15–16 thoracic, 14–16 lumbar and 27–28 caudal; total 63–65.[4, 6]

All seven cervical vertebrae are unfused.[10]

The phalangeal formula is: I_{4-5}, II_{5-8}, IV_{5-7}, V_{3-4}.[11] The pygmy blue whale is similar: I_5, II_6, IV_5, V_3.[10]

The skull characteristics that distinguish pygmy blue whales from northern hemisphere blue whales are:

1 The nasal bones are convex anteriorly, and their inner borders project further forward than the outer borders, whereas in northern blue whales they are concave anteriorly, and their inner and outer borders are more or less on the same level.
2 The rostrum is more tapered, so that its width at its mid-length is about 4% of skull length less than its width at the base, whereas in northern blue whales the two widths are about comparable.
3 The length of the mandible around the curve is shorter, being about the same measurement as the condylo-premaxillary length (CPL) of the skull, whereas in northern blue whales this length is 5–10% greater than the CPL.[10]

Distribution and migrations

Blue whales occur worldwide in most oceanic waters but are rarer in the tropics.

Antarctic blue whales are found in summer mainly south of 60° S, especially close to the edge of (and even within) the pack ice. Exactly where they go in winter is a mystery that remains to be fully solved, but there are some clues from the seasonal patterns of catches in the 1920s (before blue whales were severely depleted). Off Saldanha Bay, on the west coast of South Africa, blue whales were most abundant from May to July, with a secondary peak sometimes in August to October (corresponding possibly with the northward and southward migrations respectively). Further north, at Walvis Bay in Namibia and Mossamedes (Namibe) in Angola, catches showed a single peak, in July or August, suggesting that these latitudes were close to the northern limit of migration. Off Durban on the east coast of South Africa there was a single peak in catches, usually in June or July, suggesting that they did not travel much further north in this locality.[12] Not all individuals necessarily make these migrations: the monitoring of underwater vocal activity at high latitudes in the Antarctic detected blue whale calls year round, although the lowest calling rates occurred between June and September (winter).[13] On the other hand, the detection of Antarctic blue whale vocalisations (see page 82) only occurred at low latitudes (6–8° S) in the eastern tropical Pacific and central Indian Ocean, and at middle latitudes (35° S) of the eastern Indian Ocean, from May to September, or during winter.[14] This confirms that at least some Antarctic blue whales seasonally visit tropical waters. Immature animals formed about 80% of southern African catches of blue whales, and many of them were exceedingly thin:

this led to the suggestion that the west coast of South Africa in particular might be a 'nursery area',[5] and that such animals might fail to make the normal migration to higher latitudes in summer.[15]

While in high latitudes in summer, Antarctic blue whales tend to range widely. Recoveries of tagged individuals show that after two or more years at large they could have dispersed by up to 130–180° of longitude from the tagging site, or nearly halfway around Antarctica, although most animals moved 90° or less.[16]

Pygmy blue whales in summer do not migrate as far south as Antarctic blue whales, most remaining north of 55° S, especially (in the subregion) in the vicinity of the Prince Edward and Crozet islands. It is believed that these animals move north in winter past Madagascar and may reach the vicinity of the Seychelles and the Amirante Islands.[17] Pygmy blue whales were

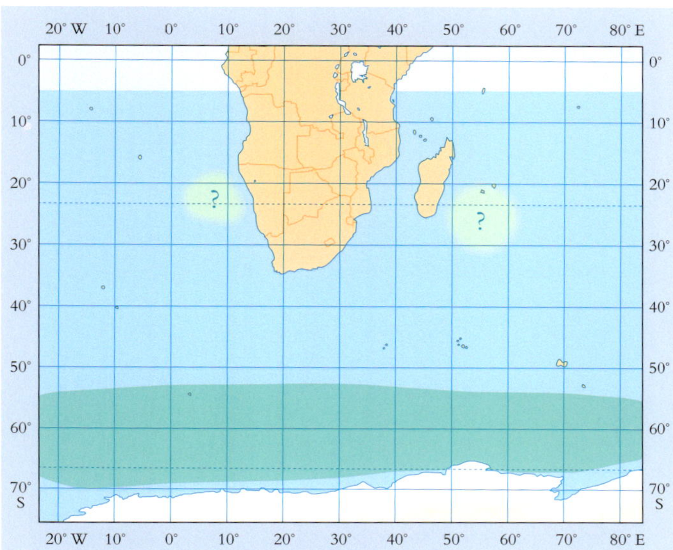

Postulated range ■ Summer concentration
Winter concentration ? Questionable origin

FIG. 39 Distribution map for Antarctic blue whale.

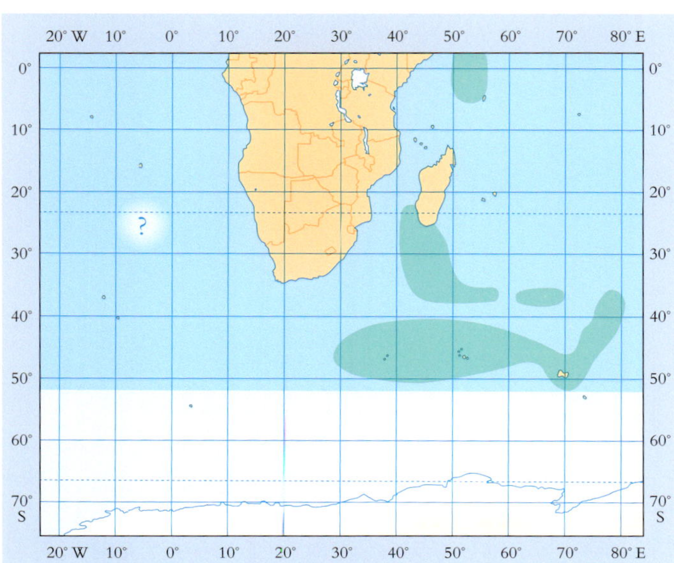

Postulated range ■ Summer concentration
? Questionable origin

FIG. 40 Distribution map for pygmy blue whale.

recorded from both the Durban[18] and Donkergat whaling stations,[19] but are generally regarded as being rarer in the Atlantic Ocean than in the Indian Ocean.

Blue whales off the South African west coast were said generally to stay far from land, usually 40–60 nautical miles offshore, but occasionally (especially in midwinter) they would come quite close to land and stay there for some time. Large aggregations (sometimes several hundred) could appear off Saldanha Bay for a week and then disappear.[20] It may have been on one such occasion that a 28.7 m 'sulphurbottom' was taken by open-boat whalers in St Helena Bay, South Africa, in June 1855.[21] Recent sightings have been made as close as 1.5 nautical miles off North Head, Saldanha Bay.[19] Off Durban, on the east coast of South Africa, early reports were that blue whales could approach within a few kilometres of the coast, but that most stayed far out to sea.[20] The last catches made on both coasts of South Africa before protection were all in water over 500 m deep.[22]

During migration blue whales may travel at average speeds (over 3–7 days) of 5–7 km/hr.[23] Maximum short-term speeds of 37 km/hr, and longer-term speeds of 27 km/hr, have been claimed.[11]

External commensals and parasites

The remora (or whalesucker) *Remilegia australis* seems to be carried more often by blue whales than by any other baleen whale in warmer waters, up to 12 occurring on a whale.[24] One of four blue whales landed at Donkergat, South Africa, carried 'plenty' of remoras. The parasitic copepod *Pennella* (resembling a black bootlace protruding from a small wound in the blubber) also occurs on blue whales (two copepods on one of four whales examined at Donkergat, South Africa),[19] apparently more frequently in lower than higher latitudes.[5] Cyamids (*Cyamus balaenopterae*) also occur on blue whales,[25] but not on four whales examined at Donkergat.[19] The small harpacticoid copepod *Balaenophilus unisetus* may be found attached to the inner surfaces of the baleen plates, especially near their base and even spreading onto the gum. Heavy infestations resemble a greyish scum. Blue whales in the Antarctic have very low infestation rates (0.3%), but those taken in lower latitudes have a much higher rate (25%). Some of the latter were definitely pygmy blue whales.[26] Skin-film diatoms *Bennettella* (= *Cocconeis*) *ceticola* occur frequently on blue whales in the Antarctic, with between 25% and 85% of individuals in most summer months being infested, and 10–40% carrying thick films.[27] It is massive infestations of this diatom that caused the whalers to dub the species 'sulphurbottom', as such films can give the skin a yellowish hue. The only infestations detected at the Saldanha Bay whaling station were some small spots of diatoms on immature whales in late winter/early spring.[5] The barnacles *Coronula reginae* and *C. diadema* were found attached to the skin of blue whales from South Africa and in the Antarctic. They were located on the mandible, flank, tip of the flipper and tail flukes. The stalked barnacle *Conchoderma auritum* was found attached to the shells of *Coronula* and directly to baleen plates, *C. virgatum* was attached to a *Pennella*, and *Xenobalanus globicipitis* was attached to the tip of a flipper and to the tail flukes.[28]

Food and feeding behaviour

Blue whales are compulsive euphausiid eaters. In summer Antarctic blue whales mainly consume krill *Euphausia superba*,[5]

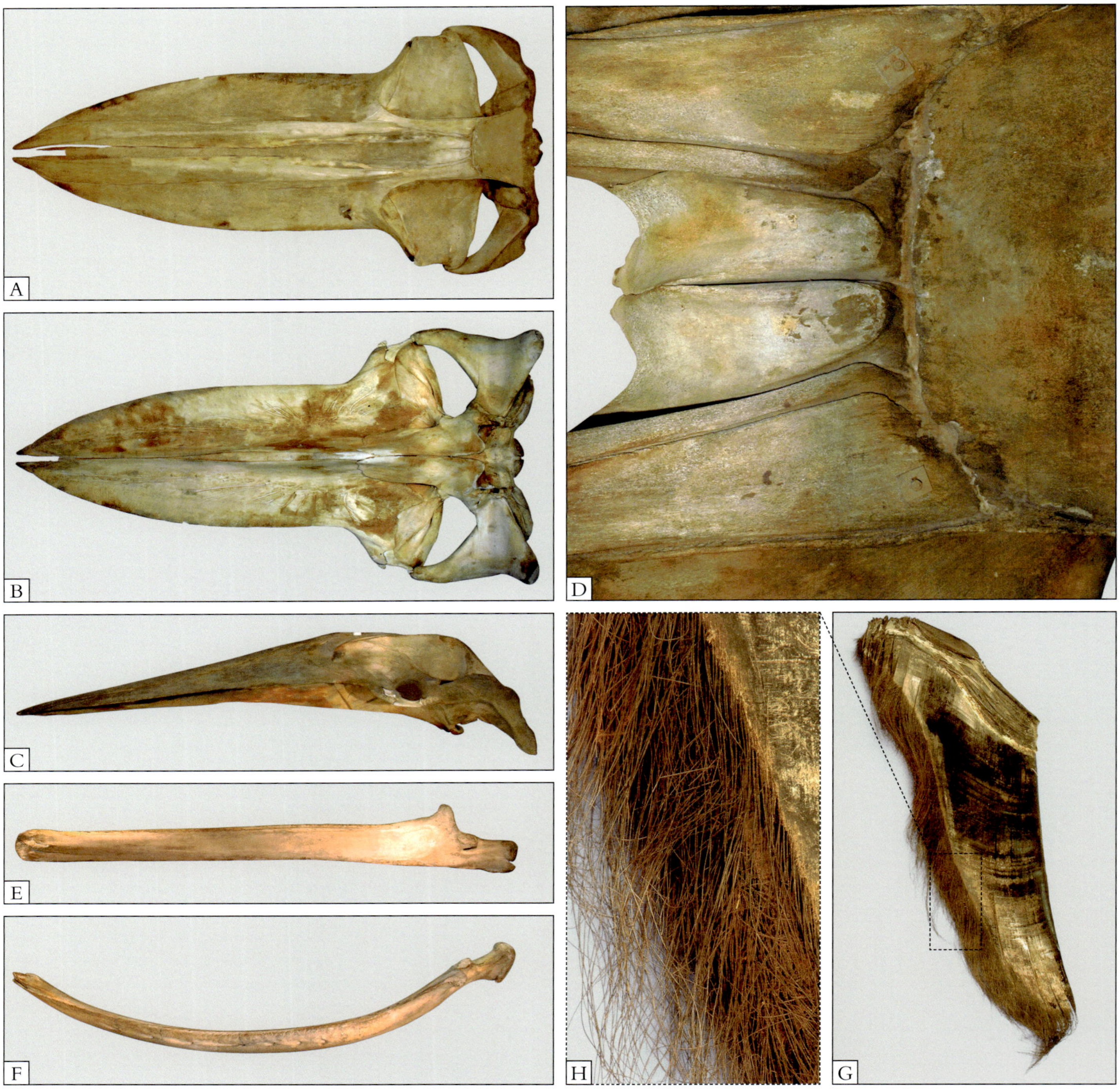

FIG. 41 Dorsal (A – retouched), ventral (B), and left lateral (C – retouched) views and vertex of skull (D) (USNM 124326); lingual (E) and dorsal (F) view of left and right mandibles, both reversed (ZM 15304); anterior view of left baleen (G) and baleen fringe detail (H) (USNM 504996).

and pygmy blue whales smaller species, *E. vallentini*[29] or *E. frigida*.[30] Little is known about the composition of food in the subregion in winter: a pygmy blue whale at Durban, South Africa had been feeding on a mixture of *Euphausia recurva* and *E. diomedeae*;[31] faeces from probable pygmy blue whales south of Madagascar indicated feeding on euphausiids;[32] and a blue whale at Donkergat had been feeding on crab megalopa larvae.[33] There is great seasonal variation in the intensity of feeding: in the Antarctic, between 80% and 100% of whales had food in the stomach, compared to 0–15% of mature and 25–75% of immature whales at Donkergat, South Africa,[34] and the quantities consumed were also much larger in the Antarctic.[25]

While as much as 400 kg of krill has been found in the stomach of a blue whale, the size of a 'meal' is not known precisely but has been estimated to be about 1 tonne. Food requirements for mature blue whales have been estimated at 4 130–4 170 kg/day during a 120-day stay in the Antarctic and at 413–420 kg/day in lower latitudes for the remainder of the year.[34]

Blue whales feed using a 'swallowing' rather than 'skimming' technique, i.e. they engulf their prey through taking in a mouthful of water rather than attempting to filter them out of the water column by swimming forward with their mouth held open.[8] At the surface, Antarctic blue whales make sorties through a plankton patch on their sides with mouths open and ventral grooves distended, closing their mouths and rolling upright to breathe.[35]

Foraging blue whales in the North Pacific dived below their target swarm and then made up to six successive lunges upwards from roughly the same depth, each lunge covering about 30 m, before surfacing again. Such foraging dives were deeper and longer than non-foraging dives, being to an average depth of 140 m and for an average time of 7.8 min., with maxima of 204 m and 14.7 min.[36]

Most of the lunges seem to be made with the animal on its back, and the energetic costs of making repeated lunges without surfacing to breathe in between are believed to limit the amount of time a whale can stay at depth and enforce a longer recovery period at the surface.[37]

Reproduction and growth

Antarctic blue whales are born at a length of about 7 m between late March and early July.[5] Some calves may be born en route to and from the Antarctic: on 14 September 1912, a 29 m female and its 7.03 m calf with an umbilical stalk still attached were caught off Saldanha Bay. This female was exceptionally fat, producing 40 tons of oil.[20] Similarly fat females (presumably representing individuals in late pregnancy or early lactation) were caught at Walvis Bay, Namibia, in the 1920s, where the simultaneous occurrence of large adult females and juveniles of probable nursing size in the catch also suggested that this was a nursery area.[38] Weaning takes place at a length of between 14 and 16 m, when calves are probably seven months old.[5] By puberty, male Antarctic blue whales have reached a length of 22.3–22.6 m and females 23.5–24.4 m.[4] The ages at puberty are not accurately known, as most blue whale catching had ceased before a reliable method of age determination was discovered, but is believed to

The biggest blues

FIG. 42 A 27 m blue whale at the Donkergat whaling station, South Africa, *c.* 1930.

Methods of measuring whales were not standardised until the 1937 Convention for the Regulation of Whaling came into force, by which time most of the biggest blue whales had already been taken. Some of the earlier lengths were probably taken from the tip of the lower jaw to the tip of the tail flukes, and so would be intrinsically longer than those measured using the currently recognised protocol of measuring from the tip of the snout to the notch in the flukes. Although individuals of 100 ft (30.5 m) and up to 106 ft (32.3 m) were recorded during this early period, in reality very few specimens over 30 m long have been reliably measured.

Mainly owing to the post-war efforts of Japanese whalers, 42 blue whales from the Antarctic have been weighed.[1] Even with the use of steam-driven winches and bone saws, this must have been a Herculean task. As one author states: 'Indeed, what a hard job the whale body weighing was. It was carried out in the frigid Antarctic Ocean with storming snow and piercing winds for some hours.'[2] The heaviest blue whale ever weighed is listed as a 27.6 m female, weighing 190 tonnes, but the sum of the component parts given is a much smaller proportion of the total than for other blue whales, suggesting an error somewhere. The next heaviest was a 29.5 m female weighing 163 tonnes. These animals were, of course, cut up to be weighed, and the associated loss of blood and other body fluids is usually estimated at about 6%.[1] An adjusted weight for the 29.5 m whale would therefore be 173.4 tonnes. This must be considered the largest blue whale specimen ever documented.

A blue whale landed at Walvis Bay, Namibia, in July 1924 yielded 305 barrels (or about 52 tonnes) of oil – this seems to have been the highest oil yield recorded for this species.[3]

References

1 Lockyer 1976
2 Nishiwaki 1950
3 Risting 1928

be about 10 years in females.[39] The reproductive cycle is thought to last two or possibly three years, about one year of which is spent pregnant, seven months lactating, and the remainder resting.[5] Observed pregnancy rates were variable but averaged 43%, although these data may be biased by several factors, including selection against lactating females.[40] An upper limit is set by the estimated ovulation rate of 0.4 per year,[39] indicating that pregnancy must occur sometimes at intervals of more than two years.

Ranges of published mean values for the composition of blue whale milk are 44.3–60.5% water, 20.0–40.8% fat, 11.3–13.5% protein and 1.4–1.6% ash.[41]

Information for pygmy blue whale reproduction is much less. Birth seems to take place at 5.5–5.6 m, after a gestation period of 10.5–11.5 months. The peak month of births, throughout most of the Indian Ocean south of the equator, appears to be in April. Whales as large as 12–15 m were said to be still accompanying lactating females, and lactation is said to last 7–8 months. Females reach sexual maturity at about 19 m and males at 18.2–18.5 m.[42] Recorded pregnancy rates are low, around 36%[4] or 41.3%,[42] and these may be inflated by selection against lactating females: the proportion of calves in sightings south of Madagascar was only 8.4%.[32] Age frequency data would suggest that maximum life expectancy for both sexes might be in the region of 65–70 years.[4] The annual survival rate of adult blue whales in the North Atlantic has been estimated as 0.975.[43]

At least five suspected cases of hybridisation between blue and fin whales have been described, all in the northern hemisphere. One of these hybrids was pregnant.[44]

Behaviour and vocalisations

Sightings of Antarctic blue whales in the subregion have varied in size from one to 20, with a mean of 2.2 whales (n = 75), but excluding one school of 20 the next largest group was only eight whales and the mean was two whales. Four of these were mixed schools with Antarctic minke whales.[19]

Sightings of blue whales south of Madagascar (and believed to be pygmy blue whales) ranged in size from one to 10 animals with a mean of 1.2 (n = 95). Most (86.5%) were singles, and none occurred in association with other species. A somewhat stereotyped respiratory cycle consists of a series of 4–13 blows produced at intervals of about 15 sec., followed by a more prolonged submergence of about 2–15 min. (average 7.1 min.). The last blow of the series may be followed by exposure of the tail flukes above the surface as the animal dives.[32]

Most dives in a satellite-tagged North Pacific blue whale were either to depths of less than 16 m, or between 97 m and 152 m (and up to 200 m). The former are considered to represent blow intervals or surface feeding, while the latter may represent exploratory foraging dives, possibly into the deep scattering layer.[45]

Blue whales in the Antarctic produce a single, circumpolar call-type. It lasts about 19 sec. on average and has three components. The first is a tone around 28 Hz that on average lasts 9.5 sec. This is followed by a short (1.1 sec.) downsweep from 27.7 to 19.5 Hz, while the third component is a slightly downsweeping tone around 18.9 Hz that lasts on average 8 sec. The calls are produced in sequences with inter-call intervals of about 62 sec., with longer intervals (about 3 min.) presumably representing pauses for breathing.[13] Outside the Southern Ocean, four other geographically distinct (and different from Antarctic) call-types have been detected: in the south-west Indian Ocean, the south-east Indian Ocean, off New Zealand and in the south-east Pacific.[46] These presumably represent different populations of pygmy blue or other non-Antarctic blue whales. Those south of Madagascar produce a two-part call. The first part is a 38 Hz tone that lasts 15–20 sec., followed after a 22–27 sec. pause by the second part, which starts with a 1–2 sec. downsweep from 35 to 28 Hz and continues as a 15–20 sec. slightly modulated tone. This call is repeated at intervals of 90–100 sec.[47]

Calling rates in North Pacific blue whales vary with time of day, being higher at night than during the day, with peaks around sunrise and sunset. This pattern is probably related in some way to the vertical migration of their prey.[48]

Conservation status

Antarctic blue whales are virtual exemplars of the effects of whaling on whale stocks. From an estimated original population size of 239 000 in 1905, stocks had been reduced by more than 99%, possibly to as few as 360 animals, when legal protection finally became effective in 1973. Recently there have been definite signs that blue whales in the higher latitudes of the Antarctic are increasing, but their population size is still small (estimated as 1 700 animals in 1996).[40] There are no published population estimates for the subregion, but 71.6% of all Antarctic blue whale catches were made in the South Atlantic and western Indian Ocean sectors (60° W–70° E), suggesting that the subregion must have included a very significant part of the original Antarctic blue whale population. Pygmy blue whales, exploited intensively only from 1959/60, may not have been so heavily depleted, but they were never as abundant as Antarctic blue whales. They, too, were effectively protected from 1973. One preliminary assessment placed the original south-west Indian Ocean stock as 7 600 animals, reduced to 4 000 by 1971/72.[17] In December 1996 a partial assessment of blue whales south of Madagascar (and so probably of pygmy blue whales) revealed 424–472 animals.[32]

Blue whales (*B. musculus*) are listed as Endangered by the IUCN (although pygmy blue whales are listed as Data Deficient), and in the South African Red Data Book Antarctic blue whales are also listed as Endangered and pygmy blue whales as Data Deficient. Blue whales (*B. musculus*) are included on Appendix I of both CITES and CMS.

References

1 Branch *et al.* In press
2 Kato *et al.* 2002
3 Nishiwaki and Hayashi 1950
4 Ichihara 1966
5 Mackintosh and Wheeler 1929
6 True 1904
7 Ichihara 1963
8 Nemoto 1959
9 Lockyer 1976
10 Omura *et al.* 1970
11 Tomilin 1957
12 Harmer 1931
13 Širović *et al.* 2004
14 Stafford *et al.* 2004
15 Mackintosh 1966
16 Branch *et al.* 2007
17 Zemsky and Sazhinov 1982
18 Gambell 1964
19 MRI Whale Unit data
20 Olsen 1914
21 Best and Ross 1986
22 Findlay 1989
23 Mate *et al.* 1999
24 Rice and Caldwell 1961
25 Mackintosh 1942
26 Bannister and Grindley 1966
27 Hart 1935
28 Nilsson-Cantell 1930
29 Nemoto 1962b
30 Kawamura 1994
31 Bannister and Baker 1967
32 Best *et al.* 2003a
33 Best 1967
34 Lockyer 1981a
35 Corkeron *et al.* 1999
36 Croll *et al.* 2001
37 Acevedo-Gutiérrez *et al.* 2002
38 Best 1998
39 Ohsumi 1979
40 Branch *et al.* 2004
41 Oftedal 1997
42 Mikhalev 2000
43 Ramp *et al.* 2006
44 Bérubé and Aguilar 1998
45 Lagerquist *et al.* 2000
46 McDonald *et al.* 2006
47 Ljungblad *et al.* 1998
48 Wiggins *et al.* 2005

Family PHYSETERIDAE

Sperm Whale

Physeter macrocephalus

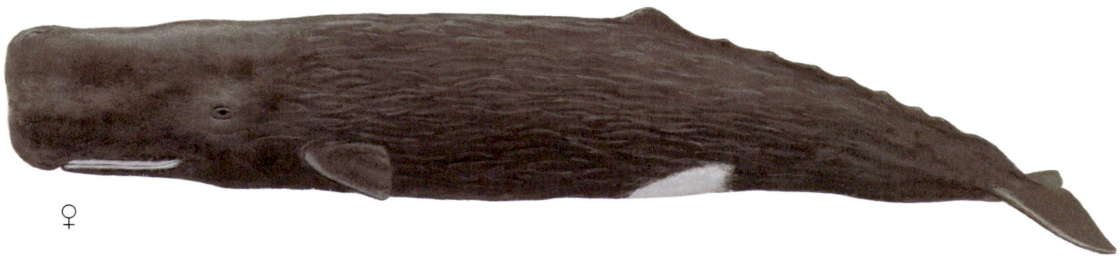

♀

♂

THE SPERM WHALE gets its name from the sixteenth and seventeenth century belief that the spermaceti contained in the vast reservoir in its head was the animal's semen.

Although only one species of sperm whale is recognised worldwide, there has been a great deal of confusion over what to call it scientifically. The principal candidates, *P. catodon* and *P. macrocephalus*, were two of four *Physeter* species listed by Carl Linnaeus in 1758. Although these were all considered one species as early as 1823, for which the name *macrocephalus* was adopted in 1836, the first time that the other names (including *catodon*) were explicitly listed as synonyms was in 1839. This was technically the 'first revision', and so from that date, *macrocephalus* has been the correct scientific name. Confusion has arisen because subsequent authors have preferred to use *catodon* on the grounds that it had 'line priority' (it was mentioned on the same page as *macrocephalus*, but on an earlier line), an argument that is not recognised officially. Others objected to *macrocephalus* on the grounds that Linnaeus's description was inaccurate, placing the blowhole in the neck rather than at the tip of the snout. However, it is clear from the rest of the description (and some of the figures on which the description was based) that *macrocephalus* was indeed a sperm whale.[1]

Appearance

It is difficult to confuse the sperm whale with any other cetacean: its box-like head and narrow underslung lower jaw make it virtually the iconic whale. This status may have been achieved through the immortality of the 1851 novel *Moby Dick*, which featured a large male sperm whale. The species is the largest of the toothed whales, and the most sexually dimorphic in size. In modern times males have been reliably recorded up to 17.3 m long, although individuals of 20 m or more have been recorded in earlier literature.[2] In contrast, females rarely exceed 11.7 m. This dimorphism in size is accompanied by differences in adult body shape: the head is proportionately larger in males and the rear part of the body longer in females. A roughened callus near the tip of the dorsal fin has been used to distinguish adult females

Length:weight relationship

For 44 sperm whales from the Antarctic and the North Pacific, **$W = 0.0196L^{2.74}$**, where W = weight in tonnes and L = total length in metres. This implies that a 10 m whale would weigh 10.8 tonnes and a 15 m whale 32.7 tonnes. These weights have not been adjusted for weight loss during cutting up, and so about 11% should be added on. (A note of warning, however: two sperm whales weighed whole and then in pieces at Durban gave a greater weight loss factor, averaging about 13%[7]).

Skeletal data

Vertebral formula: 7 cervical, 11 thoracic, 8–9 lumbar and 20–25 caudal; total 47–51.[2]

The first cervical vertebra (the atlas) is free, while the remaining six are fused – a unique situation for a cetacean.[1]

Phalangeal formula: I_{0-2}, II_{5-6}, III_{4-6}, IV_{4-5}, V_{2-3}.[2]

FIG. 43 Resting sperm whales tend to lie awash at the surface, exposing just their back between the triangular dorsal fin and bulbous blowhole.

from juveniles of both sexes, but is not entirely reliable.[3] Two other specific characteristics are the elongated S-shaped blowhole placed on the left side of the tip of the snout, and the peculiarly 'rippled' surface of most of the body except the head. The throat region also carries a highly variable pattern of longitudinal (and some transverse) grooves. The flippers are small (7–10% of body length), rounded and set low down on the sides of the body. The dorsal fin is roughly triangular with a wide base and is only about 1–3% of the body length in height. There may be up to six (usually 3–5) smaller protuberances or knuckles on the back between the fin and the flukes.[4] The flukes grow to almost one-third of the whale's length in width, and tend to have a straight or slightly convex trailing edge. They are commonly lifted above water when the whale begins a long dive.

Sperm whales are a fairly uniform dark bluish-grey to brown in colour, somewhat lighter underneath and on the sides of the head. Off South Africa about 20–40% of whales have a white blaze in the region of the navel, associated with a pattern of grey streaks/flecking converging anteriorly, both of which are highly variable in nature. There seems to be a higher incidence of grey flecking on the west coast than on the east coast. Most sperm whales also have a white lining to the mouth and white edges to the upper and lower lips. The front of the head may carry a whorl of light streaks: fewer females and small males carry such a head whorl than medium-sized or large males, and overall the head whorls are more prevalent on west coast than east coast whales.[4] Individuals, particularly old males, can become heavily scarred in the head region: such scars are often linear and parallel in nature, and although they have been attributed to the last despairing grasp of the tentacles of squid sliding to oblivion down the whale's throat, they are more likely to represent tooth scars from intra-specific fighting.[5] Apart from the legendary Moby Dick, at least nine instances of albinistic sperm whales have been recorded.[6]

The narrow lower jaw carries 18–30 (usually 22–25) pairs of teeth that in adult males may reach 20–25 cm in length and 6–8 cm in diameter, while in females and young males the teeth may not exceed 15 cm in length and 5 cm in diameter. However, in both sexes only a quarter to a third of the tooth is exposed above the gum. The teeth are roughly oval in cross-section, with curved tips that point backwards in young animals but may become rounded and worn in older individuals. No lower teeth are erupted at birth: eruption only takes place around puberty, commencing at the centre of the toothrow and progressing faster towards the back of the mouth. Although the majority of adult sperm whales have no erupted teeth in the upper jaw, a few possess up to 9–10 pairs. These are much smaller than the lower teeth, with a maximum length of 8–11 cm, and are strongly curved: they often erupt close to or within sockets in the palate for the lower teeth. The incidence of maxillary teeth increases with body length in both sexes, with a secondary increase at a body length of 14.0–14.9 m in males (coinciding with the attainment of social maturity – see below). In 0.5–1% of individuals the lower jaw becomes bent, curved or curled up ('scrolled'), or even abbreviated, to such an extent that some or all of the dentition becomes functionless in grasping food. Such individuals do not seem to suffer any nutritional deficiencies.[2, 4]

Distribution and migrations

Sperm whales occur worldwide in oceanic waters (including those of the subregion) but occasionally extend onto the continental shelf when feeding conditions are particularly favourable. In general, females and immature males are found in warmer waters than larger males, the rule of thumb being that they are restricted to the latitudes between approximately 40° N and 40° S. Males become increasingly migratory with age. In the southern hemisphere, they first seem to move south of the Subtropical Convergence in summer when they reach an average length of 12 m, and the larger (and presumably older) males tend to range further south than smaller ones.[8]

Within the west coast whaling ground off South Africa, some groups of male sperm whales, particularly medium-sized and large individuals, tended to visit the upper reaches of the continental slope, in water mostly 200–500 m deep. Females, on the other hand, were rarely found in water less 1 000 m deep.[9]

On both the east and west coasts of South Africa, where the species formed the target of a shore-based whaling industry, sperm whales showed distinct patterns of seasonal availability. The most comprehensive data are for the Durban whaling ground, where during some seasons in the 1940s and 1950s sperm whaling continued year-round. In summer (December–February) large males (more than 13.7 m) averaged 3% of the catch and were at their lowest relative abundance for the year, whereas at their peak in winter (June–August) they comprised 36% of the catch. Medium-sized males (12.2–13.7 m) showed a similar trend but were slightly more abundant than large males in summer and had a later winter peak (around August). By contrast, females were at their lowest abundance in winter (about 20%) and their highest in summer (70–75% of the catch).[10] Trends in absolute abundance during the whaling season were similar to these, except that medium-sized male abundance had two peaks (May and August–September) instead of one.[11] On the whaling ground on the west coast of South Africa, females and small males were most abundant in autumn (March/April) and declined thereafter, with no sign of an increase before the end of October. Both medium-sized and large males had two peaks in abundance, in April and October, and in May and August–October, respectively.[12] These patterns of availability, plus the incidence of certain external parasites (see below), and the recoveries of whales marked in higher latitudes,[13] indicate that sperm whales off southern Africa at least are migratory, with medium-sized and large male sperm whales being largely winter visitors from higher latitudes.

The different scales of movement undertaken by males and females are illustrated by the distances between positions of marking and recovery of sperm whales tagged in the southern hemisphere (and at large for more than one month) – an average of 1 570 km for 17 males and 690 km for 22 females.[8]

Speeds of movement of relatively undisturbed whales have been estimated at 2–5 km/hr with averages of 3–4 km/hr at the surface. Maximum speeds (maintained over a minute or two) might be about 18–22 km/hr.[14]

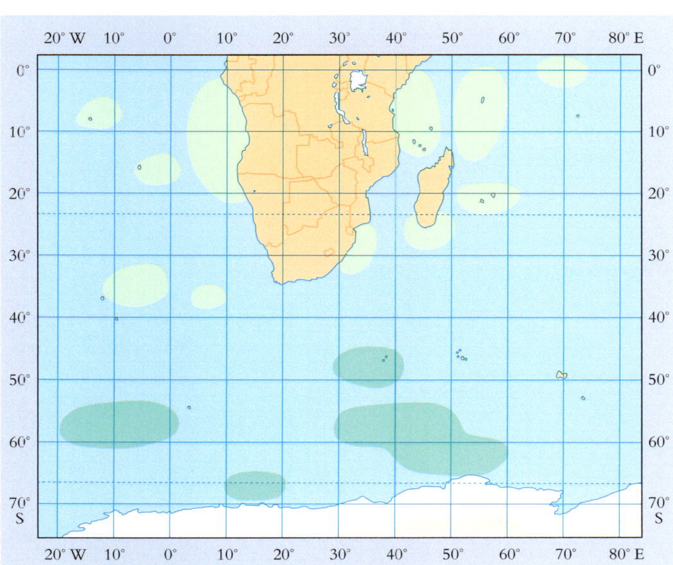

Postulated range   Male concentration areas in summer
Historic whaling grounds

FIG. 44 Distribution map for sperm whale.

External commensals and parasites

The majority (98%) of sperm whales with whale-lice (cyamids) in the whaling ground off the west coast of South Africa carried either *Neocyamus physeteris* or *Cyamus catodontis* or both, but two males were found with *C. erraticus*, a species more usually associated with right whales. Female and small male sperm whales were infested almost exclusively with *N. physeteris*, but above a body length of 11.3 m the proportion of males carrying *N. physeteris* relative to *C. catodontis* fell rapidly, so that by a length of 12.8 m males were infested exclusively with *C. catodontis*. This is assumed to be evidence of the stage at which males leave their maternal school and start to migrate to higher latitudes than females.[8, 15] When adult males over 13.7 m are found with females during the breeding season, they have rarely picked up *N. physeteris*, suggesting that such rendezvous are brief in duration.[8]

Films of the cold-water diatom *Bennettella* (= *Cocconeis*) *ceticola* occurred rarely on female or small male sperm whales examined on the South African west coast whaling ground, but were more frequent on medium-sized and large males, especially in autumn.[12] As these films seem to be contracted in the Antarctic,[16] this confirms that these males must have recently arrived from higher latitudes.

As is its wont, the stalked barnacle *Conchoderma auritum* is found on hard substrates, such as the sperm whale's teeth if its lower jaw is damaged or deformed. Sessile barnacles are rarer, although *Coronula* does sometimes occur on the body surface. The bootlace-like copepod *Pennella* is also found infrequently.[2]

In warmer waters, remoras (probably *Remilegia* sp.) can be seen on sperm whales, but usually on the calves and juveniles as it is possible that they cannot tolerate the deeper dives of older whales. They feed largely on the dead outer layers of the whale's skin, but can forage independently on zooplankton.[17]

Tablespoon-sized scoop marks on the back and flanks of sperm whales, which heal to form brown and then white oval scars, are evidence of attacks by small sharks such as the cookie-cutter shark *Isistius*.

Food and feeding behaviour

Although frequently portrayed as battling to the death with the giant squid *Architeuthis*, in reality the principal food items of sperm whales in the subregion are more modestly-sized squid averaging 1–3 kg in weight. Most of these live in midwater or on or near the sea floor, and include a wide variety of species. In South African waters the principal squid eaten on the Durban whaling grounds were *Taningia*, *Ancistrocheirus*, *Octopoteuthis* and *Todarodes* (in total making up *c.* 60% by weight of 27 species eaten), and on the west coast whaling grounds *Pholidoteuthis*, *Moroteuthis robsoni*, *Taningia*, and *Ancistrocheirus* (*c.* 56% by weight of 26 species eaten). These calculations exclude species such as *Kondakovia* and *Mesonychoteuthis*, which are known to be Antarctic squid and which were only found in the stomachs as trace remains (beaks, or horny mouthparts). As they occurred mainly in medium-sized or large males, it is assumed that they had been imported into the area by whales migrating from higher latitudes.[18] Medium-sized and large males also tended to eat larger species or larger individuals of squid species than females.[9, 18]

FIG. 45 Dorsal (A), ventral (B), and right lateral (C) views of skull; right lateral (labial) (D) and dorsal (E) views of mandible (ZM 15266).

Apart from squid, sperm whales in South African waters are known to eat a variety of other items. West coast males that moved onto the continental slope fed on benthic organisms such as rays (*Bathyraja*), monkfish (*Lophius vomerinus*) and crabs (mainly *Geryon* sp.). Offshore, both sexes fed in midwater on oilfish (*Ruvettus pretiosus*), ceratid anglerfish (*Ceratias holboelli* and *Cryptosaras couesii*), mysids (*Gnathophausia ingens*) and prawns (*Sergestes*). Tunicates (*Pyrosoma atlanticum*) were eaten by both sexes but predominantly by males.[9]

In the Antarctic, the diversity of cephalopods eaten was much less, only nine or 10 species. Of these, *Mesonychoteuthis*, *Kondakovia* and *Moroteuthis knipovitchi* were by far the most important (98% by weight).[18] Fish were recorded from 5.2% of sperm whale stomachs in the region and included *Ceratias holboelli*, *Dissostichus eleginoides*, *Micromesistius australis*, and *Raja griseocaudata*.[19]

The marked seasonality in the intensity of feeding shown by most baleen whales is not obvious in this, the largest of the toothed whales.[9] Similarly little evidence of a diurnal cycle in feeding has been found except in the Antarctic, where stomachs seemed to be fuller in the early morning (03h00–08h00) and late at night (22h00–24h00) compared to the intervening hours.[20] Elsewhere, female sperm whales have been recorded as spending as much as 75% of daylight hours in foraging behaviour, with a peak in socialising in the late afternoon ('tea-time').[14] The typical accumulation of hundreds if not thousands of squid beaks in a sperm whale's stomach may be indicative of an animal that nibbles rather than indulges in discrete meals.[9]

Daily food intake has been estimated at about 190 kg for a female at puberty and 420 kg for an adult female. For males, daily intake is estimated to increase from 270 kg at puberty to 1 505 kg in a physically mature individual.[21] The stomach of an 11 m male from the west coast of South Africa contained 130 kg of food,[9] and it is believed that the stomach must be filled several times a day to meet the daily food requirement.[21]

There has been a great deal of speculation about the methods used by sperm whales to catch their prey. One of the most intriguing is the suggestion that they use very loud acoustic pulses to stun their prey.[22] Such sounds seem to be produced too infrequently to be widely used in food capture,[14] and experimentally similar pulses have been found to have little or no effect on cephalopods (but are more effective on fish).[23] It has been suggested that the white lining of the mouth, perhaps accentuated by bioluminescence created by movements of the whale through the water, would attract squid, which might then be swallowed using suction.[24]

Reproduction and growth

Calves are born between January and April, after a pregnancy lasting 15–16 months. There is no difference in the size of the two sexes at birth (4 m). Although on average lactation is believed to last two years, with the next calf being born after an interval of five years, there is substantial individual variation. Younger females tend to become pregnant more frequently, and older cows to lactate for longer, and it has been speculated that the older cows nurse calves other than their own, so sharing the costs of reproduction. Although calves may start to take solid food sometime in their first year, clues to milk ingestion (lactose) have been found in the stomachs of females as old as 7.5 years and males as old as 13 years.[17] Average values for milk composition at

Durban were 63.8–65.4% water, 23–25.7% fat, 8.5–9.8% protein and 0.6–0.8% ash.[25]

Females reach sexual maturity when they are nine years old and 8.5 m long, and cease growing when they are about 11 m long and 28–29 years old.[26,27] Males undergo an extensive period of maturation (and its diagnosis depends to an extent on the site in the testis that is examined). When sampling the centre of the testis, physiological maturity (the production of spermatozoa) is attained at a similar age to sexual maturity in females, and at a length of 9.2 m, but it seems unlikely that males of this size normally play any role in reproduction. A secondary spurt in male growth takes place around 11.9–12.2 m in length (and an age of 19–20 years), and this corresponds to the stage at which 50% of males are physiologically mature when sampling the periphery of the testis. It also marks the stage at which males begin to migrate to higher latitudes than females in summer. At a length of around 13.7–14.0 m, and an age of 25–27 years, males undergo a rapid increase in testis growth, and it is believed that this corresponds to social maturity, or the stage at which males begin to compete for access to schools of females. It is around this size that they begin to exhibit the head scarring often associated with aggressive interactions (coincidental with a marked rise in the density of spermatozoa in the seminal fluid). The timing of the cessation of growth in males is less well established, but seems likely to commence around 35 years and when individuals are about 15.9 m long.[15,17,27]

Maximum longevity (obtained from dentinal growth layer counts) is of the order of 60–70 years,[1] although the very oldest individuals may have been removed by whaling before age determination commenced.[14] Females may live somewhat longer than males.[28]

Behaviour and vocalisations

The basic social unit of sperm whales consists of 10–15 mature females and their associated offspring. These units probably

Ambergris

Ambergris was a substance of mysterious origin to the Ancient Greeks and Romans, as it was only found floating in the sea or washed up on the beach. Those who believed it was the droppings of a giant bird feeding on aromatic herbs had it half-right: in the early 1700s it was established that ambergris originated in sperm whale intestines. Although explanations for its formation are speculative, ambergris appears to be nothing more than hardened whale faeces[1], characteristically containing the parrot-like mouth parts of squid that the whale has eaten. It occurs as solid, irregularly shaped lumps, almost black in colour and (as one might imagine) quite offensive in odour when fresh but gradually becoming greyish and more sweet-smelling (like dried manure) over time when stored dry. Generally, the lighter the colour the higher the value.[2] The lumps vary in size but are usually less than 10 kg in weight, with the largest recorded being boulder-sized and weighing up to 450 kg. Although such pieces are unlikely to be passed naturally by the whale, they do not seem to be injurious to its health. Ambergris occurs in about 1–5% of individuals of both sexes[3] (although more often in males[2]), with sperm whales from the Arabian Sea having a substantially higher incidence than in other areas.[4] There have been many recorded uses for the substance, but its main commercial value has been as a base for perfumes, where it has the properties of adsorbing, intensifying and stabilising the fragrance. At one stage it was said to be 'worth its weight in gold', but recently it has at least been partly replaced by synthetics. Getting an estimate of its current value seems to be almost as difficult as finding it in the first place, but figures equivalent to $14–$65/kg are being quoted (much of the variation being related to quality).

FIG. 46 A freshly-discovered lump of ambergris at the Donkergat whaling station, South Africa.

How do you know you have got it? Heat a piece of wire in a flame for about 15 seconds and then press it into the lump to a depth of about 3 mm. If it is genuine ambergris, a dark brown to black, opaque, resinous liquid will form around the wire and appear to boil. Withdraw the wire and before the melted material on the wire cools, touch it with your finger. True ambergris will leave tacky, pitch-like 'strings' sticking to the skin. If placed in a flame it will emit a white smoke that has the same characteristic odour as the solid, and then burn with a luminous blue flame. Ambergris should also dissolve in ether or warmed methyl alcohol, and (the clincher) it sticks to the teeth when chewed.

Ambergris was widely known throughout the Muslim world as early as the ninth century (possibly because of its high incidence in Arabian Sea whales). This awareness still exists even among the Cape Town Muslim community, where the author was assured it had a reputation as a cure for erectile dysfunction.

References

1 Clarke 2006
2 Berzin 1972
3 Rice 2002
4 Mikhalev 2000

maintain stable relationships over several years, although transfers between units are not uncommon. It was originally thought that these females were all matrilineally related, but genetic sampling to date suggests that this is not normally the case, and that most such groups consist of more than one matriline, some of which can be maternally unrelated. Units are frequently associated together into groups of 20–30 animals, possibly because this conveys some benefit in foraging.[14]

Juvenile males leave their natal school at around puberty, when they tend to associate together in so-called bachelor schools, ranging from one to 50 in number. Such aggregations do not seem to involve the same long-term relationships as the schools of females. The size of the associations decreases as the males get older, so that socially mature males are often found alone or in groups of up to six in number.[8]

Although the scarring evident on the heads of adult male sperm whales would seem to testify to their being involved in aggressive interactions, actual observations of such fights are rare in modern times.[14] Based on historical accounts of sperm whales ramming and even sinking ships, the use of the head as a battering ram in intraspecific disputes has been proposed.[29] However, this is only one of several proposed functions for the extraordinarily developed spermaceti organ that gives the sperm whale its characteristic head shape. It has also been suggested that it is used for buoyancy control,[30] and as a sound generating and reverberating mechanism that has resulted in the extreme sexual dimorphism of the organ through sexual selection.[31]

Although newborn calves do not seem capable of prolonged or deep dives, as a species the sperm whale is a deep diver par excellence. On average, schools of females spend about 62.5% of each day on prolonged dives, submerging for about 37.5 min. and then resting at the surface for 7.5 min. between dives.[14] An adolescent male estimated at 12 m long and tracked for 4.6 days spent 63.3% of its time on deep dives, of which 13.2% exceeded an hour (and up to 65.3 min.) in duration.[32] Studies of whales in three different localities (the eastern seaboard of the United States, the Gulf of Mexico and the Ligurian Sea) showed very consistent dive cycles, consisting of an average dive time of 45 min. followed by a 9 min. recovery period at or near the surface. The maximum depth of dive differed between regions, the averages being 985 m, 644 m and 827 m respectively.[33] It has been speculated that longer (and deeper) dives are made by males.[9] The longest recorded dive (monitored acoustically) lasted at least 2 hr 18 min.,[34] while the deepest (monitored with a transponder) was to 2 035 m.[35] A large male harpooned off Durban in a water depth of 3 193 m was found to have the remains of a bottom-dwelling shark in its stomach, suggesting that greater depths may be possible.[36] During ascents (which are faster than descents), whales employ a stroke-and-glide swimming style, whereas they usually stroke steadily throughout descents.[37] Pathological changes (osteonecrosis) in the skeleton of sperm whales have been attributed (although not exclusively) to the chronic effects of deep prolonged diving.[38]

When at the surface between foraging dives, sperm whales appear somewhat lethargic, and can lie log-like with the body almost horizontal and just awash. At such times the blows are produced at regular intervals (about 12.5 sec. for females and immatures and 17.5 sec. for large males). The flukes are almost invariably exposed above the water as the animal descends on its next dive. During periods of socialising, activities such as breaching, lob-tailing and spy-hopping are much more prevalent, and there is much less fluking.[14]

The usual vocalisations of sperm whales are long series of regularly spaced clicks, often lasting for minutes. Each click is composed of a short burst of pulses, and the inter-pulse interval may be correlated with the size of the spermaceti organ and hence (indirectly) of the whale itself.[31] Inter-click intervals are about 0.5 sec. for females and young males and 1 sec. for large males. Occasionally clicks are much more closely spaced, forming creaks or buzzes that can last for up to 45 sec. Both the click trains and creaks are highly directional in nature, and may be used principally for echolocation and homing in on prey respectively. (Nearly all creaks are produced in the deepest half of the dive, indicating foraging at depth.[39] On average 18 creaks are produced per dive, and if these are all taken to be prey capture attempts, then about 53% of each dive cycle is spent in pursuit of prey.[33]) Appearing at the end of some click series, but more often independently, are patterned series of clicks known as codas. These seem to be principally used in social contexts: they fall into a limited number of types, and groups of females may have regionally distinctive coda repertoires. Maturing or male sperm whales produce slow clicks or clangs: these are loud ringing clicks produced about 5–8 sec. apart that other whales might hear at a distance of 60 km (compared to about 16 km for normal clicks). The function of these is unknown but may include some kind of sexual display (in which bigger whales flaunt their bigger noses at every click[31]), although they have been recorded in contexts where reproduction seems very unlikely.[14]

Conservation status

Sperm whales were the subject of a protracted period of commercial exploitation, commencing in the early eighteenth century and only ceasing in the late twentieth century. It is possible that over a million sperm whales worldwide were killed during this period, of which at least 300 000 were victims of the open-boat fishery. However, the absence of official catch records for much of the first two centuries of exploitation mean that this figure contains a great deal of uncertainty. Apart from this long and poorly documented catch history, assessing the current status of sperm whales is made more difficult by the fact that, at least during the twentieth-century fishery, catches were dominated by males, owing to their larger size and higher product yield. Although in a species with a social organisation such as that of the sperm whale it might be expected that some 'bachelor' males could be removed without any adverse effect on the population, if the resulting drop in the ratio of males (and especially the large, socially mature males) to females is too great, there could be detrimental effects on calf production. Unfortunately, too little is known about the sperm whale mating system to be able to predict the effect of such relative male depletion. The most recent assessment of sperm whale stocks worldwide ignores the question of differential catches by sex.[40] So while its conclusion (that there are currently some 300 000 sperm whales worldwide) confirms subjective impressions that the species is far from rare, ignorance of the relative levels of depletion of each sex means that it is difficult to speculate on the species' status. The wisdom of attempting a global assessment (i.e. considering all stocks as one population) must also be questioned. There are no reliable estimates of population size for the subregion.

In some areas of the world, interactions between sperm whales and fisheries are of concern. In the South Atlantic, for instance, sperm whales have developed an association with long-lining vessels catching principally the Patagonian toothfish *Dissostichus eleginoides*. Although purloining of fish is rarely witnessed (because it must take place at depth), the circumstantial evidence is convincing that sperm whales remove fish from the hooks.[41, 42] Resolution of such conflicts without harming the whales does not always occur.

The sperm whale is listed as Vulnerable by the IUCN and in the South African Red Data Book. It is included on Appendix I of CITES, and on both Appendices I and II (presumably different populations) of CMS.

References

1 Rice 1989
2 Berzin 1972
3 Kasuya and Ohsumi 1966
4 Best and Gambell 1968a
5 Kato 1984
6 Fertl *et al.* 1999
7 Lockyer 1976
8 Best 1979
9 Best 1999
10 Best 1974
11 Gambell 1967
12 Best 1969c
13 Best and Ross 1989
14 Whitehead 2003
15 Best 1969b
16 Matthews 1938b
17 Best *et al.* 1984b
18 Clarke 1980
19 Korabelnikov 1959
20 Matsushita 1955
21 Lockyer 1981b
22 Norris and Mohl 1983
23 Clarke, M. R., pers. comm.
24 Fristrup and Harbison 2002
25 Oftedal 1997
26 Best 1968
27 Best 1970c
28 Ralls *et al.* 1980
29 Carrier *et al.* 2002
30 Clarke 1978
31 Cranford 1999
32 Watkins *et al.* 2002
33 Watwood *et al.* 2006
34 Watkins *et al.* 1985
35 Watkins *et al.* 1993
36 Clarke 1976
37 Miller *et al.* 2004b
38 Moore and Early 2004
39 Miller *et al.* 2004a
40 Whitehead 2002
41 Ashford *et al.* 1996
42 Nolan *et al.* 2000

Family **KOGIIDAE**

Pygmy Sperm Whale

Kogia breviceps

TWO SPECIES OF *Kogia* are recognised, *K. breviceps* and *K. sima* (but see the account of the latter for a possible complication). For many years only *breviceps* was widely used, and was commonly referred to as the pygmy sperm whale. This was despite the two species being distinguished as early as 1871. The 'rediscovery' of *sima* and the adoption of the common name dwarf sperm whale means that members of the genus *Kogia* should perhaps now be called 'short-headed' sperm whales, rather than the somewhat confusing pygmy sperm whales.[1]

Appearance

Because of their small size, low profile in the water and low gregariousness, both pygmy and dwarf sperm whales are difficult to spot at sea, and are usually only seen in conditions close to dead flat calm. Sightings tend to be rare events, so telling them apart at sea may be a little optimistic, but reportedly they can be distinguished from the relative size and position of the dorsal fin (see below).[2]

Pygmy sperm whales attain a maximum length of about 3.3 m, with some evidence that males may not grow as large as females[3] (see the box on page 93 about distinguishing the sexes in kogiids). There are distinct changes in the head shape with age. In young animals the profile of the head slopes down towards the tip of the snout, but when they are more than about 2 m long this slope decreases and the head becomes squarer and more like that of a sperm whale in outline. In adult pygmy sperm whales the snout is more elongated than that of dwarf sperm whales, with an almost bulbous appearance in some animals. The head is then larger, the distance from the tip of the snout to the centre of the blowhole, for instance, being 10.4–12.8% of body length (compared to 7.5–10.1% in dwarf sperm whales). The dorsal fin is relatively smaller in pygmy than dwarf sperm whales (less than 5% cf. more than 5% of body length in height), and is less erect in shape. It also tends to be situated nearer the tail than the head, its anterior insertion usually being more than 50% of body length from the tip of the snout (cf. less than 50% in dwarf sperm whales).[4]

Pygmy sperm whales are dark bluish-grey above fading to a pinkish-white on the belly, with a transition zone running from eye to fluke insertion. The flipper is dark above and connected by a light grey isthmus to the dorsal pigmentation. The eye is surrounded by a dark ring, backed by a light bracket-shaped mark (or 'false gill') running up in front of the grey flipper isthmus. No significant differences have been recorded between the colour pattern of pygmy and dwarf sperm whales.[4]

There are 11–17 pairs of erupted teeth in the narrow lower jaw (n = 36), and usually (94.7%) no erupted teeth in the upper jaw. When teeth are present in the upper jaw, they number only 1–3 on each side (n = 38).[5] The teeth in adults are almost 40 mm in length (including the unerupted root), are strongly curved

Length:weight relationship

Mass data from 18 pygmy sperm whales[4-6] and one near-term foetus[4] weighed whole have been combined to produce the following relationship: **$W = 14.86L^{3.0681}$**, where W = weight in kilograms and L = total length in metres. This implies that a 2 m whale would weigh 125 kg and a 3 m whale 432 kg.

Skeletal data

Vertebral formula (n = 18): 7 cervical, 11–13 thoracic, 8–12 lumbar and 22–26 caudal; total 50–55.[5]

All seven cervical vertebrae are fused.[7]

Phalangeal formula: I_2, II_{5-8}, III_{4-8}, IV_{4-8}, V_{2-7}.[8]

The cranial characteristics that separate pygmy from dwarf sperm whales include the following:

1 The dorsal cranial fossae are not cupped posteriorly.
2 The dorsal sagittal septum is 20–38 mm thick at its narrowest.
3 The mandibular symphysis is long (more than 86 mm) and keeled ventrally.
4 The foramen magnum is near the midpoint of skull height.[9]

backwards, and reach a maximum diameter of 8–9 mm. They are more numerous, longer and stouter than the teeth of adult dwarf sperm whales.[4]

Distribution and migrations

Pygmy sperm whales apparently occur worldwide in tropical and warm-temperate seas.[10] Because of the rarity of sightings, distribution in the subregion has to be largely deduced from the incidence of strandings. Pygmy sperm whales have been recorded stranded around the southern African coast from Cape Cross in Namibia (22° S) on the west coast to the Durban area (29°50′ S) on the east coast.[11] A nineteenth century record of the species being hunted in the Aldabra region[12] is discounted as the original source is considered unreliable, although a subsequent sighting north of the Seychelles (and just north of the equator)[13] suggests that pygmy sperm whales may well occur in that region.

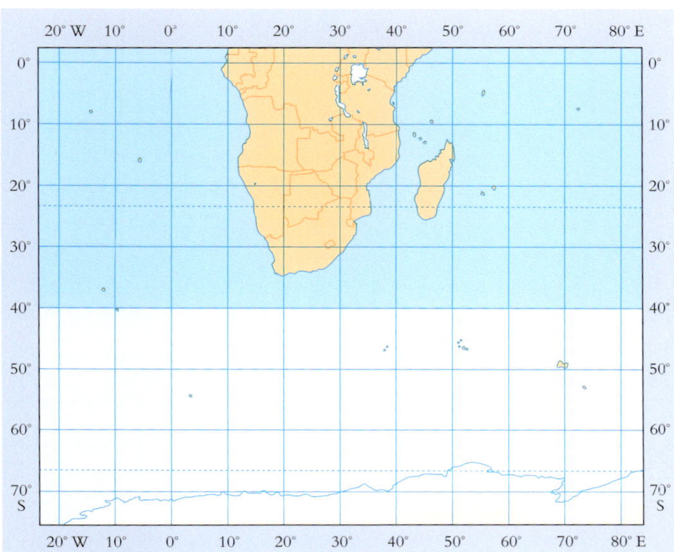

FIG. 47 Distribution map for pygmy sperm whale.

Strandings have occurred throughout the year on the coasts of the Eastern and Western Cape provinces of South Africa, with indications of a peak in July to October,[3] whereas in South Australia and New Caledonia, strandings were reported to occur mainly between April and October[14] and June and December[15] respectively. Along the west coast of the USA a lack of strandings in June, July and August prompted a suggestion that there was a northward movement between autumn and spring.[16] Whether these stranding patterns really indicate differences in migratory behaviour, or indeed reflect seasonal occurrence at all, is unknown.

External commensals and parasites

There are no records of external parasites on pygmy sperm whales in the subregion (n = 21).[5] Elsewhere, the cyamid *Isocyamus kogiae*[17] and the parasitic copepod *Pennella*[18] have been recorded.

Wounds, both open and healed, that are attributable to attacks by small sharks such as *Isistius* have been found on about 13% of whales examined in the subregion (n = 23).[5]

Food and feeding behaviour

The pygmy sperm whale has the most diverse diet of 20 small odontocete species studied in the subregion.[19] Its recorded diet consists of 50 different cephalopod, 12 fish, five crustacean, and one salp species. Overall, the most important species by mass are all cephalopods: *Histioteuthis* sp. (29.3%), *Loligo vulgaris* (13.9%), *Lycoteuthis diadema* (9.6%), *Octopoteuthis* sp. (6.9%) and *Anchistrocheirus lesueuri* (5.2%). Non-cephalopod items frequently eaten include light fish (*Photichthys argeneteus*), hake (*Merluccius* sp.), a deepwater mysid (*Gnathophausia ingens*) and the stomatopod *Squilla*. Differences between sexes and ages are not clear, but males tend to feed on a wider variety of items, and smaller animals on smaller individual cephalopods.[3] It has been suggested that the lower incidence of oceanic species in immatures, calves and accompanying adult females (55%) compared to other adults (85%) implies that adults (apart from cows with calves) tend to be found in deeper water than immatures.[20] Seasonal variations in the intensity of feeding have not been recorded.

Reproduction and growth

Length at birth is estimated to be 1.2 m,[3,4] and a male foetus as large as 1.262 m has been recorded from the subregion.[5] Although the birth of a 1.53 m male calf has been reported, the circumstances are not entirely convincing.[21] Gestation lasts 11–12 months. Births are said to take place between March and August,[3] but the plot of foetal/calf length against date is more consistent with a peak season of birth around January/February. Calves are said to nurse for between one and two years, although the latter estimate would be inconsistent with annual breeding, which has been suggested as the norm for the species.[3] Certainly 86% of lactating females were also pregnant (n = 7), but only 50% of pregnant females (n = 12) were also lactating,[3,5] which suggests that the reproductive cycle may not be annual in a number of individuals. Weaning takes place when the calf is 1.8–2.11 m long. Sexual maturity is reached at a length of 2.41 m in males and 2.62 m in females, the equivalent ages being about five years for both sexes. Growth ceases at an average of 3.06 m in females and 2.86 m in males at around 15 years of age. These results indicate reversed sexual dimorphism. No indications of reproductive senescence have been found.[3]

Behaviour and vocalisations

Pygmy sperm whales are reported to occur in small groups of 1–6 animals.[2,13,22,23] They tend to be relatively inconspicuous, and when breathing may rise slowly to the surface, blow faintly and then dive without rolling sharply at the surface or exposing the flukes. Others may be found lying still at the surface with just the top of the head and back exposed, and the tail hanging loosely in the water. When startled under such circumstances, they may release a cloud of reddish-brown faeces,[23] possibly as a distracting device. The lower intestine of both *Kogia* species is greatly expanded and filled with dark sepia liquid with the consistency of chocolate syrup, affectionately known to some as 'intestinal glop': in a large individual pygmy sperm whale there might be as much as 12 litres of it.[24]

A juvenile female that was released after rehabilitation in captivity spent most of the next four days in water more than

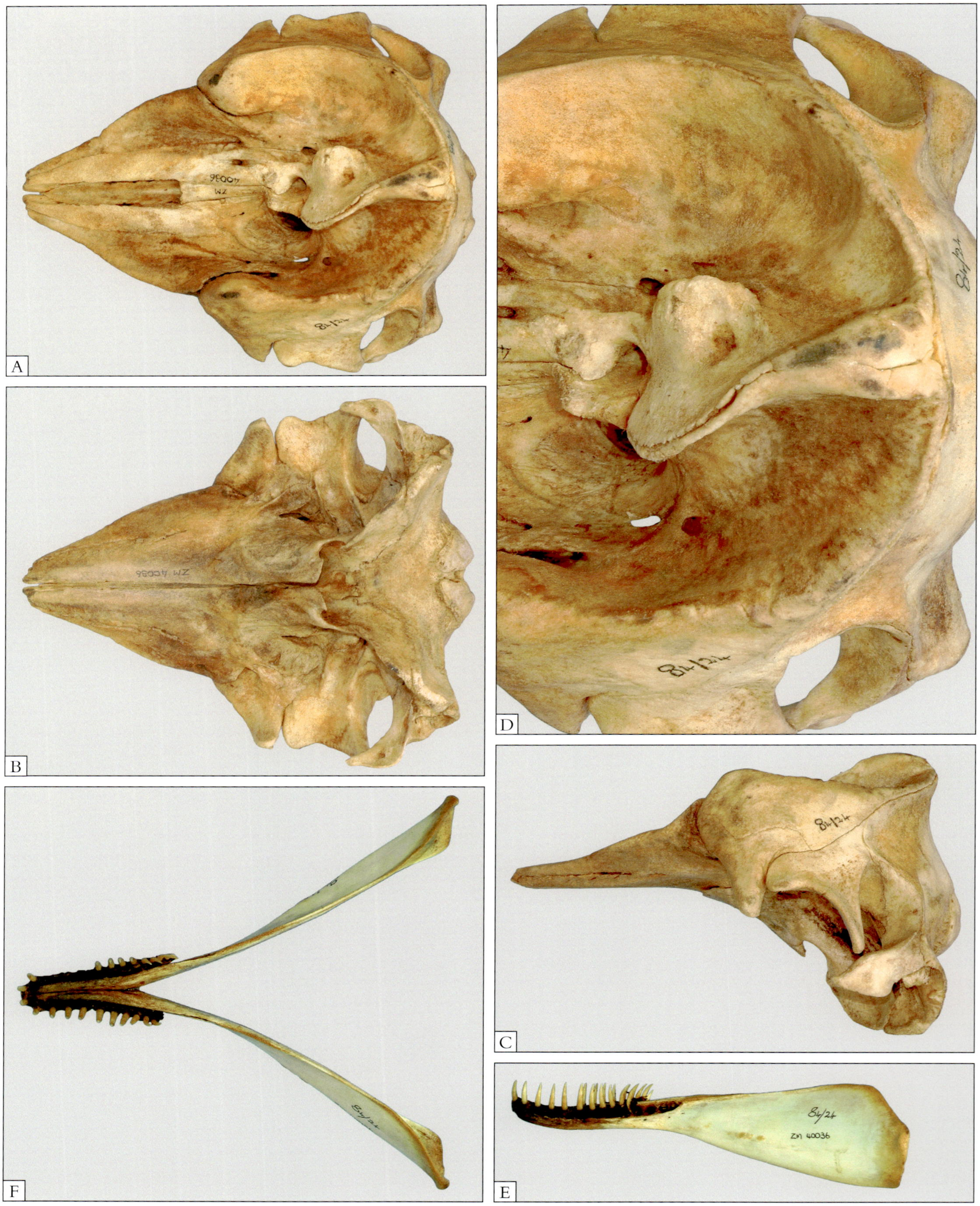

FIG. 48 Dorsal (A), ventral (B) and left lateral (C) views and vertex of skull (D); left lateral (labial) (E) and dorsal (F) views of mandible (ZM 40036).

FIG. 49 In this picture of a cow-calf pair lying at the surface, the dorsal fin of a pygmy sperm whale can be seen to be relatively smaller and situated further aft than that of a dwarf sperm whale (see Fig. 50).

200 m and mostly between 500 m and 800 m deep, and travelled over 255 nautical miles at the same time. Its surfacing behaviour seemed to be linked to ambient light conditions and time of day. At night and in overcast conditions it undertook more long (greater than 8 min.) dives than in bright daylight, while it spent more time at the surface between noon and 18h00 than during the rest of the day. Its longest dive lasted 18 min.[25]

Pygmy sperm whales have been recorded as producing ultrasonic echolocation pulses with peak frequencies at 125–130 kHz,[26, 27] as well as a low-frequency, low-amplitude 'cry', 0.42 sec. in duration with typical starting and ending frequencies of 1.36 kHz and 1.48 kHz respectively.[28] Hearing ability, as tested physiologically rather than behaviourally, indicated maximum sensitivity to sounds of 90–150 kHz.[26] Some of the sounds may be produced by structures associated with the nasal passages in the head.[29]

Sex determination in short-headed sperm whales

The textbooks tell us that male cetaceans can be recognised by the wider separation of the genital and anal apertures compared to females. This is true, but the statement will not have prepared you for the first occasion you are confronted with a male kogiid. The two apertures are so far apart that initially you can be forgiven for thinking that either the thing has no sex or no anus; in either case it is in trouble. In reality, while the anus is in its 'normal' position (about 70% back from the snout), the genital slit is right up there around the middle of the belly, posing as an umbilicus (which is actually just ahead it). The two apertures are fully one-fifth to one-quarter of the body length apart, a situation that seems to be found in no other cetaceans apart from phocoenids. An explanation for this is not forthcoming.

Conservation status

Apart from the odd specimen taken incidentally, pygmy sperm whales do not seem ever to have been the target of a directed fishery. Although it has been claimed that both species of short-headed sperm whales were taken in the Japanese coastal small cetacean fishery,[30] with about 10 a year being taken in one port, this seems rather to refer to the dwarf sperm whale.[31] Substantial numbers (more than 80 a year) were reported taken as a by-catch in gill net fisheries off Sri Lanka,[32] although the origin of this estimate is obscure. The species does appear susceptible to the ingestion of marine debris, with the stomachs of 18% of 32 stranded specimens examined on the eastern seaboard of the USA containing substantial amounts of plastic.[33] There are no estimates of abundance in the subregion, and estimates for other oceans are usually assigned to *Kogia* spp., given the difficulty of separating the species at sea.

The pygmy sperm whale is listed as Lower Risk: Least Concern by the IUCN and as Least Concern in the South African Red Data Book. It is included on Appendix II of CITES, but not listed by CMS.

References

1 Rice 1998
2 Baird 2005
3 Plön 2004
4 Ross 1979a
5 MRI Whale Unit data
6 Ross 1984
7 Nishiwaki 1963
8 Tomilin 1957
9 Handley 1966
10 Baird *et al.* 1996
11 Findlay *et al.* 1992
12 Palmer 1948
13 Ballance and Pitman 1998
14 Kemper and Ling 1991
15 Sylvestre 1988
16 Gunter *et al.* 1955
17 Martin and Heyning 1999
18 McAlpine and Murison 1997
19 Sekiguchi *et al.* 1992b
20 Ross 1979b
21 Hückstädt and Antezana 2001
22 Vidal *et al.* 1987
23 Leatherwood *et al.* 1982
24 Caldwell and Caldwell 1989
25 Scott *et al.* 2001
26 Ridgway and Carder 2001
27 Marten 2000
28 Thomas *et al.* 1990
29 Clarke 2003
30 Mitchell 1975
31 Yamada 1954a
32 International Whaling Commission 1994
33 Laist *et al.* 1999

Dwarf Sperm Whale

Kogia sima

ALTHOUGH IT IS now generally recognised that there are two species in the genus *Kogia*, the pygmy sperm whale *K. breviceps* and the dwarf sperm whale *K. sima*, recent genetic work has suggested that the populations of dwarf sperm whale in the Atlantic and Indo-Pacific regions may be different. Whether this difference represents two distinct species is currently unresolved, but further specimens from the subregion will be critical in addressing the question.[1]

Appearance

Remarks about the detectability of the pygmy sperm whale at sea apply equally to this species. Apparently the dwarf sperm whale can be distinguished from its larger congener by its prominent dorsal fin situated at the centre of the body rather than nearer the tail.[2]

Dwarf sperm whales attain a maximum length of about 2.74 m. It has been suggested that males grow slightly larger than females,[3] but this is not supported by the relative sizes of the two sexes at sexual maturity (see below). As with pygmy sperm whales, there is a distinct change in head shape with age: the profile of the head in younger animals slopes down towards the tip of the snout, but when body length exceeds about 2 m this slope decreases and the head becomes squarer and more like that of a sperm whale in outline. In dwarf sperm whales longer than about 2.3 m, the snout is distinctly squarish and blunt in side view (as opposed to the more elongated and even bulbous appearance in adult pygmy sperm whales). In general, dwarf sperm whales have a more pug-like appearance, due to their shorter head, the distance from the tip of the snout to the centre of the blowhole being 7.5–10.1% of body length (compared to 10.4–12.8% in pygmy sperm whales). The dorsal fin is relatively larger in dwarf than in pygmy sperm whales (more than 5% cf. less than 5% of body length in height), and is more erect in shape. It also tends to be situated more or less in the middle of the body, as opposed to being nearer the tail as in the pygmy sperm whale, with its anterior insertion usually being less than 50% of body length from the tip of the snout (cf. more than 50% in pygmy sperm whales).[4]

Dwarf sperm whales are dark bluish-grey above fading to a pinkish-white on the belly, with a transition zone running from eye to fluke insertion. The flipper is dark above and connected by a light grey isthmus to the dorsal pigmentation. The eye is surrounded by a dark ring, backed by a light bracket-shaped mark (or 'false gill') running up in front of the grey flipper isthmus.[4] Some animals may have a second light mark running almost vertically up in front of the false gill, and joining it at the tip to enclose an 'island' of dark pigment. Individuals may also have small dark spots or freckles in the lighter pigmented areas, and the pale underside may only extend as far back as the anus.[5]

Length:weight relationship

Mass data from 25 dwarf sperm whales[4,7,8] weighed whole have been used to produce the following relationship: **$W = 22.617L^{2.5368}$**, where W = weight in kilograms and L = total length in metres. This implies that a 2 m whale would weigh 131 kg and a 2.5 m whale 231 kg.

Skeletal data

Vertebral formula (n = 12): 7 cervical, 12–13 thoracic, 8–12 lumbar and 22–26 caudal; total 52–57.[7]

All seven cervical vertebrae are fused.[5]

Phalangeal formula (n = 2): I_{1-2}, II_{5-7}, III_{4-7}, IV_{4-5}, V_{2-4}.[7,9] From this small sample it seems that the number of phalanges may be extremely variable as in pygmy sperm whales.

The cranial characteristics that separate pygmy from dwarf sperm whales include the following:

1 The dorsal cranial fossae are cupped posteriorly.
2 The dorsal sagittal septum is 6–14 mm thick at its narrowest.
3 The mandibular symphysis is short (less than 50 mm) and rounded ventrally.
4 The foramen magnum is well below the midpoint of skull height.[10]

FIG. 50 A dwarf sperm whale in the North Pacific shows little of itself above water.

No significant differences have been recorded between the colour pattern of pygmy and dwarf sperm whales.[4] There may be several short longitudinal grooves on the throat.[6]

There are 8–11 pairs of erupted teeth in the narrow lower jaw (n = 15), and usually (in 80% of cases) 1–2 erupted teeth on one or both sides of the upper jaw (n = 15).[7] This contrasts with pygmy sperm whales, where erupted upper teeth are definitely the exception rather than the rule. The teeth in adults reach a length of 26 mm (including the unerupted root), are curved backwards, and have a maximum diameter of 4–5 mm. They are less numerous, shorter and more slender than the teeth of adult pygmy sperm whales.[4]

Distribution and migrations

Dwarf sperm whales apparently occur worldwide in tropical and warm-temperate seas, possibly preferring slightly warmer waters to pygmy sperm whales.[11] In the eastern tropical Pacific, sightings attributable to this species all occurred south of 24° N, while pygmy sperm whales were found to the north of this latitude.[12] In the northern Gulf of Mexico, both species (but principally the dwarf) seemed to be found largely over the upper continental slope.[13] Because of the rarity of sightings, the distribution in the subregion has to be deduced largely from the incidence of strandings. Dwarf sperm whales have been recorded stranded around the southern African coast from the vicinity of Rocher Pan, South Africa (32°36′ S) on the west coast to about 28° E on the east coast,[3] but with a record from northern KwaZulu-Natal at 27°38′ S 32°39′ E.[14] This record was originally considered to be linked to a warm-water event,[15] but in view of subsequent sightings in the Indian Ocean just north of the Seychelles[16] it is likely that the species is found throughout the tropical western Indian Ocean. However, a mysterious record to the north of Mauritius shown on some maps[11, 17] remains just that, a mystery. The lack of records from Namibia is perhaps confirmation of the species' preference for warmer waters, although it might well occur offshore outside the influence of the cold Benguela Current. Several sightings have been made off northern Angola (6–8° S).[18]

Strandings have occurred on the coasts of the Eastern and Western Cape provinces of South Africa in all months except June, and there is a slight suggestion of an increased incidence between March and May.[3] Dwarf sperm whales are reported to be summer visitors to the waters off Taiji, Japan,[5] and have been taken as a by-catch in gill nets off Sri Lanka in all months except November/December, with some suggestion of peaks during summer (although these data were unadjusted for fishing effort).[17] Elsewhere there have been few signs of a seasonality in strandings.[11] All 14 sightings off Angola have been in August or September.[18]

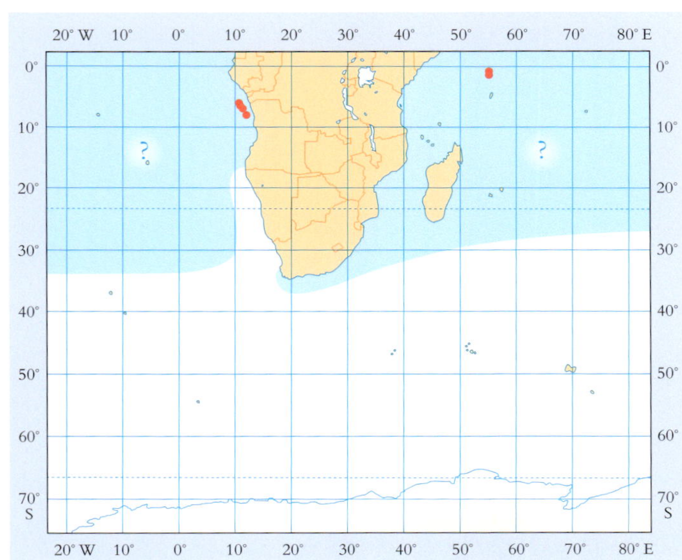

FIG. 51 Distribution map for dwarf sperm whale.

External commensals and parasites

There are no records of external parasites on dwarf sperm whales in the subregion (n = 17).[7] Elsewhere, the parasitic copepod *Pennella* has been recorded.[19]

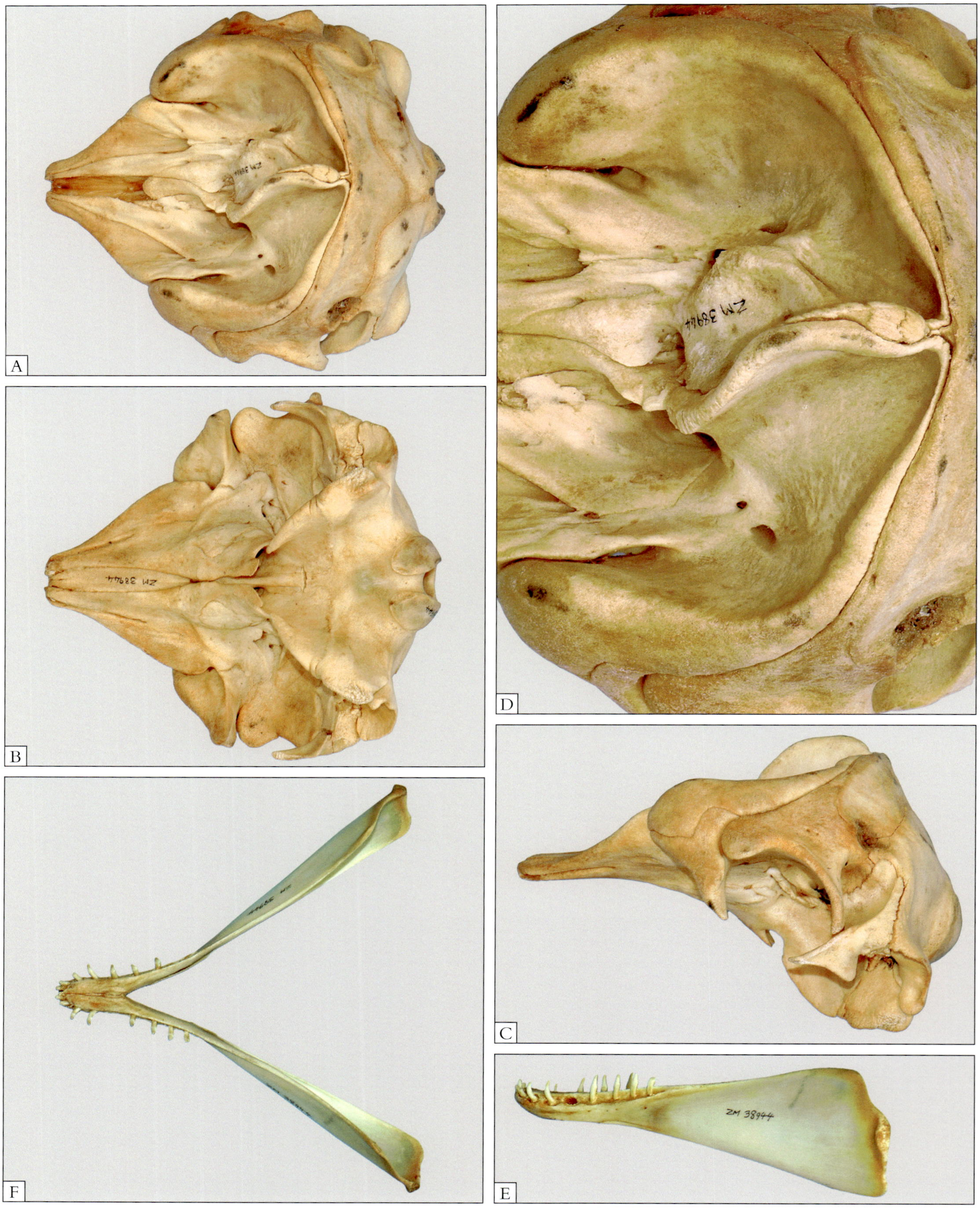

FIG. 52 Dorsal (A), ventral (B) and left lateral (C) views and vertex of skull (D); left lateral (labial) (E) and dorsal (F) views of mandible (ZM 38944).

Wounds, both open and healed (up to seven in number), that are attributable to attacks by small sharks such as *Isistius* have been found on about 18% of whales examined in the subregion (n = 17).[7]

Food and feeding behaviour

Dwarf sperm whales had the third most diverse diet of 20 smaller odontocete species studied in the subregion.[20] A total of 32 species of cephalopods, three fish species and three crustacean species have been recorded from 33 stomachs. The most important cephalopods by mass were *Histioteuthis* sp. (45.3%), *Lycoteuthis diadema* (16.6%) and *Loligo vulgaris* (5.2%), while hake (*Merluccius capensis*) and the deepwater mysid *Gnathophausia ingens* were the most important fish and crustaceans consumed respectively.[3, 20] There was considerable overlap between the diet of dwarf and pygmy sperm whales, although the dwarf sperm whale fed on a smaller variety of species and less often on fish and crustaceans. As in pygmy sperm whales, immature individuals tended to feed on smaller-sized cephalopods than mature individuals, and females and young plus immature individuals seemed to feed closer inshore than mature males and non-reproducing mature females.[3, 8] Seasonal variations in the intensity of feeding have not been recorded.

Reproduction and growth

Length at birth is estimated to be 1.0 m,[3, 4] after a gestation period of about 12 months.[3] Births seem to take place between December and March. Lactation can last for a year (at which stage the calf reaches a length of about 1.5 m), but animals can be weaned at 1.3 m. The ovulation rate (although based on relatively few individuals) suggests an average calving interval of more than a year, with some animals conceiving at a post-partum oestrus and others after an interval of two years.[3] Only 12.5% of lactating females (n = 8) were pregnant, and 14.3% of pregnant females were lactating (n = 7), suggesting that annual reproduction is the exception rather than the rule,[3, 7] and that the dwarf sperm whale may not be as fecund as the pygmy sperm whale. No indications of reproductive senescence were found.[3]

Sexual maturity is reached at a length of 1.97 m in males and 2.15 m in females, and at the age of about three and five years respectively. Although fitted growth curves predict growth to cease at 2.64 m in males and 2.49 m in females, the two largest individuals were female and the smallest known physically mature individual was male.[3] Coupled with the apparently smaller size of the male at sexual maturity (even if this is reached at a younger age than the female), this suggests that there may actually be little difference in size between the sexes.

Behaviour and vocalisations

Reported school sizes for dwarf sperm whales range from one to six, with averages of 1.6 (n = 20) for the western tropical Indian Ocean[16] and Maldives (n = 17),[21] and from one to three with an average of 1.6 (n = 14) for northern Angola.[18]

Typically, their behaviour has been described as logging at the surface for a few minutes, then slowly sinking or rolling out of sight, not to be seen again.[2] When they are frightened at the surface they may utilise the same technique as the pygmy sperm whale, and release a cloud of opaque reddish-brown faeces. This has earned them the Japanese name *tsunabi*, or rocketing firework, and the somewhat less attractive alternative, skunk of the sea.[5] A cow with a calf encircled in a tuna purse seine was seen to repeat this behaviour 6–8 times. She would attempt to hide herself and the calf within the subsequent cloud (which could cover an area as large as 100 m^2), suggesting an anti-predator strategy. Aggressive tactics (charging at or breaching near boats) were also described under the same circumstances.[22]

There are no published descriptions of vocalisations made by dwarf sperm whales, though it can be assumed that they probably resemble those made by pygmy sperm whales. Similar anatomical structures in the head that could be responsible for sound-generation and control have been described.[23]

Conservation status

The only places where dwarf sperm whales seem to have been the target of a directed take are in the Japanese coastal small cetacean fishery, with about 10 a year being taken in one port,[5] and in the pilot whale fishery at St Vincent in the West Indies.[6] Significant numbers (more than 230 a year) were reported taken as a by-catch in gill net fisheries off Sri Lanka,[24] although the origin of this estimate is obscure. Like its congener, the dwarf sperm whale is susceptible to the ingestion of marine debris, with plastic bags and even a pair of latex examination gloves being recorded from stomachs.[4, 25] There are no estimates of abundance in the subregion, and estimates for other oceans are usually assigned to *Kogia* spp., given the difficulty of separating the species at sea.

The dwarf sperm whale is listed as Lower Risk: Least Concern by the IUCN and as Least Concern in the South African Red Data Book. It is included on Appendix II of CITES, but not listed by CMS.

References

1 Chivers *et al.* 2005
2 Baird 2005
3 Plön 2004
4 Ross 1979a
5 Yamada 1954a
6 Caldwell and Caldwell 1989
7 MRI Whale Unit data
8 Ross 1984
9 Nagorsen 1985
10 Handley 1966
11 Willis and Baird 1998
12 Wade and Gerrodette 1993
13 Baumgartner *et al.* 2001
14 Findlay *et al.* 1992
15 Ross *et al.* 1985
16 Ballance and Pitman 1998
17 Chantrapornsyl *et al.* 1991
18 Weir, C., pers. comm.
19 Allen 1941
20 Sekiguchi *et al.* 1992b
21 Ballance *et al.* 2001
22 Scott and Cordaro 1987
23 Clarke 2003
24 International Whaling Commission 1994
25 Waring *et al.* 1999

Family **ZIPHIIDAE**

Cuvier's Beaked Whale

Ziphius cavirostris

Adult ♂

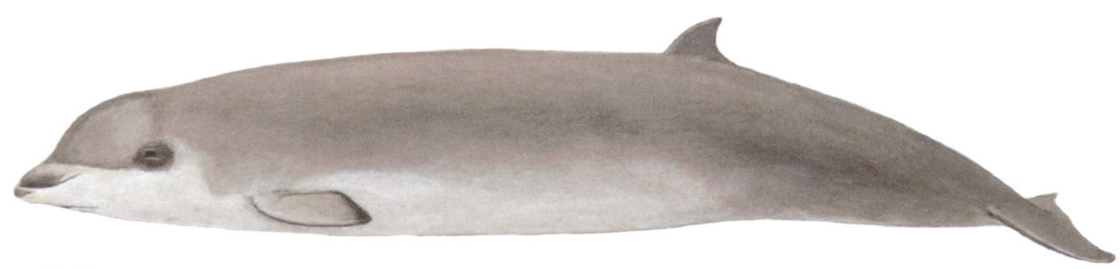

Calf

BECAUSE OF ITS characteristic head shape, this animal is sometimes referred to as the goose-beaked whale. Although Cuvier's beaked whale is considered to occur as a single species worldwide, there is a great deal of morphological variation[1] and an analysis of the genetic data indicates considerable differentiation between ocean basins.[2]

Appearance

At sea the species may be recognised by its characteristic head shape – which has a melon that is not very prominent and a short beak – and by the adult colour pattern.

The largest accurately measured specimen is a 6.93 m male. Other larger figures in the literature are believed to be either misidentifications or estimated lengths. Judging by the size composition of the commercial catch from Japan, there is little or no difference in length between the sexes.[1] The beak (such as it is) is short and not well differentiated from the melon, and is therefore difficult to measure precisely: in profile the head really does look a bit goose-like. The V-shaped throat grooves typical of ziphiids are present. The flippers are small (9.5–12.1% of body length, n = 8), oval in outline and placed low down on the side of the animal with the associated depression or 'pocket' typical of ziphiids. The dorsal fin is erect and strongly hooked, and in height is between 3.8% and 5.5% of body length (n = 8). It is situated 61.7–66.8% of the body back from the snout (n = 5).[3] There is usually no central notch in the tail flukes.

From a comparison of calves, subadults and adults, it appears that most of the variability in colour pattern seen in this species is the result of age-related change. Calves are dark greyish above fading to whitish-grey below, with the border between the two running from the region of the anus along the side above the flipper to the side of the head, where it extends upwards as a pale patch behind the eye, and in front of the eye onto the side of the head, leaving a narrow band mid-dorsally from the blowhole forward onto the beak. The tip of the lower jaw and outer

Length:weight relationship

Combining nine published weights[1] with three from South Africa,[3] **$W = 0.0147L^{2.8768}$**, where W = weight in tonnes and L = total length in metres. Of these 12, certainly the three from South Africa were weighed in pieces and it is not clear whether the other published weights were of whole animals. One animal that was weighed whole and then in parts in South Africa had lost 3.9% of body weight during cutting up.[3] The equation above indicates that a 3 m animal would weigh about 350 kg and a 6 m animal about 2.5 tonnes.

Skeletal data

Vertebral formula (n = 21): 7 cervical, 9–10 thoracic, 9–11 lumbar and 19–22 caudal; total 46–49.[1]

Usually the first four cervical vertebrae are fused and the remainder are free, but fusion can range from the first three to the first six vertebrae.[5]

Phalangeal formula, including metacarpals (n = 2): $I_2, II_7, III_6, IV_5, V_3$.[3]

FIG. 53 Two different Cuvier's beaked whales, illustrating some of the age-related changes in coloration of the head and back. In the top photograph, the erupted apical teeth and paired scars on the body indicate that this is an adult male.

margins of the flippers are dark brown. In subadults the belly colour darkens to become closer to that of the sides and back, while dark pigmentation extends onto the head, forming a cheek patch. In adults the ground colour of the sides and belly darkens further to become indistinguishable from that of the back, the head whitens, and a greyish to white blaze develops on the back behind the head: at its greatest development this blaze can extend back past the dorsal fin, and can include the entire shoulder and outer surface of the flipper. Although present in both sexes, the blaze seems to become whiter and more extensive in the oldest males than in any females. The animals become heavily scarred: single or paired white scratches appear on the back and all surfaces behind the dorsal fin, especially on males, while white oval spotting develops in both sexes, particularly in the anogenital region.[4]

A single pair (usually) of mandibular teeth erupts in adult males only: these are situated at the tip of the lower jaw. Small rudimentary teeth sometimes occur in the gum (one South African male did have a single erupted maxillary tooth), and occasionally there is a second pair of erupted teeth behind the principal mandibular teeth, similar to those in Arnoux's beaked whale.[1] The principal mandibular teeth are conical in shape and reach a length of 5.5 cm in adults. In females they are more or less circular in cross-section, measuring 0.9–1.2 cm in diameter, while in adult males they become more robust and less symmetrical in cross-section, measuring 1.5–2.2 cm in diameter.[5]

Distribution and migrations

Although mostly recorded as strandings, Cuvier's beaked whale is believed to have a worldwide distribution in all seas apart from polar waters,[1] although sightings south of the Antarctic Convergence have been reported.[6] In the subregion the species is known from Angola,[7] Namibia,[8] South Africa,[9] the Seychelles,[10] the Comoro Islands[11] and Aldabra.[9] Sightings attributed to the species (or specimens collected at sea) have been recorded off the South African coast,[8] Angola[7] and in the south-west Indian Ocean.[3, 12]

The Japanese coastal catch of Cuvier's beaked whale normally occurred in water that was 1 000 m deep or greater,[13] and all the sightings in the subregion have been in water of such depths.[3, 8] Off Hawaii the species also seemed to show a preference for water depths greater than 1 000 m,[14] while in the eastern tropical Pacific the species was encountered in average water depths of 3 400 m (and up to 5 100 m and more), but with a tendency to be sighted most often in water that was 2 000 m deep (or at the offshore edge of the continental slope).[15]

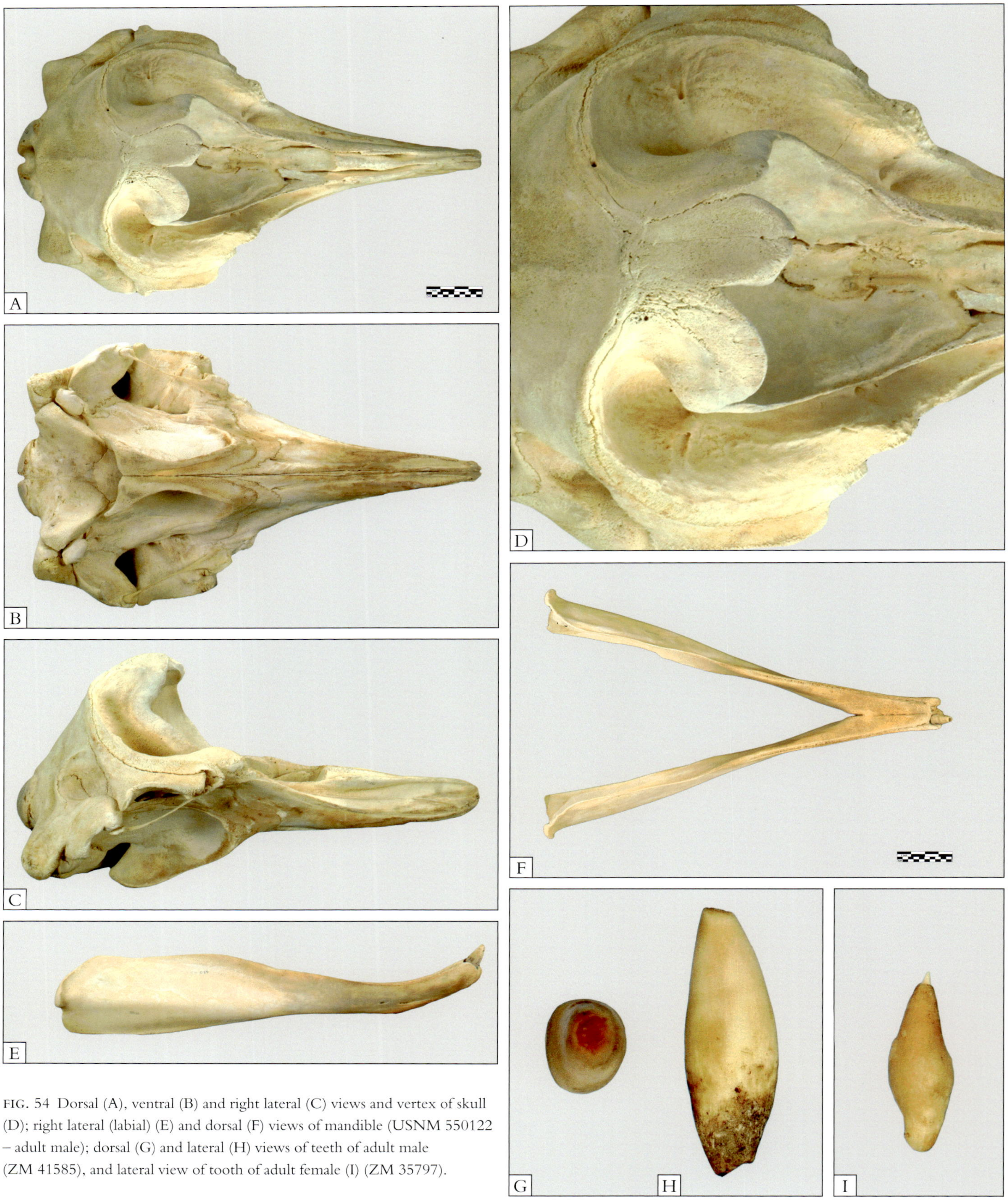

FIG. 54 Dorsal (A), ventral (B) and right lateral (C) views and vertex of skull (D); right lateral (labial) (E) and dorsal (F) views of mandible (USNM 550122 – adult male); dorsal (G) and lateral (H) views of teeth of adult male (ZM 41585), and lateral view of tooth of adult female (I) (ZM 35797).

Catches occurred throughout the year off Japan,[13] although there seemed to be a peak in summer,[16] and strandings have taken place in all seasons on the South African coast.[3] It is not known if the species makes seasonal migrations, but parasite and prey data suggest that this is a possibility.

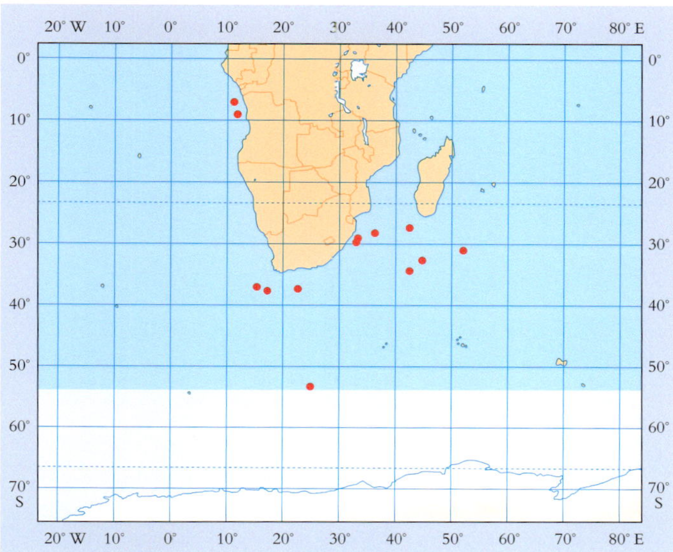

Postulated range ■ Sighting or capture

FIG. 55 Distribution map for Cuvier's beaked whale.

External commensals and parasites

Stalked barnacles (*Conchoderma* sp.) occur on the apical teeth of males (35 were found on the teeth of an adult male in South Africa[3]), and *Xenobalanus* sp. on the trailing edge of flukes and the dorsal fin[1] (69 specimens were found on the flukes of a young female stranded on the South African coast[3]). A single specimen of the parasitic copepod *Pennella* occurred behind the dorsal fin of another young female stranded on the South African coastline. Wounds and scars attributable to the cookie-cutter shark *Isistius* are commonly seen, 14 being counted on one side of a 4.11 m female.[3] Greenish or brown patches on the skin have been seen on at least three stranded individuals in South Africa, and a sample from one proved to contain the diatom species *Bennettella* (=*Cocconeis*) *ceticola* f. *berardiusii* and *Stauroneis olympica*. The presence of the former species suggests that Cuvier's beaked whales may move from higher latitudes to South African waters.[17]

Food and feeding behaviour

The species is usually considered primarily a squid-eater, with 87% of stomachs from 38 (mainly) stranded animals containing cephalopod remains, 13% crustacean remains and 8% fish remains.[18] In the fishery off Japan, however, whales taken in water deeper than 1 000 m had been eating mainly fish, while those caught in shallower water were eating mainly cephalopods.[13] It has been suggested that stranded animals may give a biased picture of the diet, either because any fish eaten may have digested more quickly than the squid, or because the whales were presumably in shallower water before stranding.[18]

The stomachs of four Cuvier's beaked whales from South Africa contained 18 species of squid, of which eight species (*Moroteuthis knipovitchi*, *M. robsoni*, *Pholidoteuthis boschmai*, *Teuthowenia* sp., *Onychoteuthis banksi*, *Octopoteuthis* sp., *Taningia danae* and *Moroteuthis* sp.) formed 83.4% by mass. *Moroteuthis knipovitchi* is a subantarctic species, suggesting that some Cuvier's beaked whales move from south of 40° S to South African waters. One individual contained an unidentified fish otolith and another 10 examples of the midwater mysid prawn *Gnathophausia ingens*. A fifth whale only contained tunicates. The sizes of the squid eaten were generally large, species with a mean weight of over 1 kg forming 64% of the mass of stomach contents.[19] It has been suggested that this represents a form of niche separation from *Mesoplodon* beaked whales that generally eat much smaller prey items.[18]

There is no information on seasonal variation in feeding.

Reproduction and growth

Length at birth has been estimated at 2.7 m.[20] The length of gestation is unknown, and from the little foetal data available there does not seem to be a strong seasonality to reproduction. The smallest known mature female was 5.13 m long (North Atlantic) and the largest immature 5.49 m (North Pacific),[1] although an immature female 5.77 m long has been recorded from South Africa.[3] Inter-ocean differences may account for this disparity. Reproductive data for males are equally scarce and confusing: two North Pacific males 5.49 m and 5.85 m long had combined testis weights of 8 300 g and 7 500 g, and so were obviously mature,[13] while three physically (and therefore presumably sexually) mature males from South Africa that were 5.69 m, 5.87 m and 5.75 m long had combined testes weights of 310 g, 395 g and 590 g respectively.[3] It has been proposed that the North Pacific animals were sexually active, and that the others were senescent animals with atrophied testes.[20] Although the dates of death of the North Pacific animals are not given, the three South African males stranded in March, July and September, making the possibility of a seasonal cycle in spermatogenic activity seem unlikely. Another possibility is that there is a misplaced decimal point in the North Pacific data, as the weights are so much greater than in most other beaked whales.[21]

Readings of growth layer groups in the dentine and cementum resulted in several males with 45 or more combined layers, but females only in the high twenties. This could mean that the rate of layer accumulation differs between the sexes, or that females cease to accumulate layers before males, or that (as in *Berardius*) males live longer than females.[21]

Behaviour and vocalisations

In the eastern tropical Pacific, Cuvier's beaked whale is found in groups of 1–7 animals, with most sightings (39%) being of singles (n = 135).[1] Similar-sized groups were encountered in the North Atlantic: from one to seven with a mean of 2.4, with 31% of the sightings involving single animals (n = 59).[22] In the subregion, sightings have ranged from one to five animals with a mean of 2.3 (n = 12).[3,7]

The composition of groups in the North Atlantic has been assessed on the basis that only males are heavily scarred, and that immatures have a different colour pattern to adults. Males comprised 83% of single animals, pairs were composed of 67% females and 25% immatures (the highest proportion in any grouping), groups of 3–6 were 76% female, while groups

Ping, ping – you're dead

In March 2000, 17 cetaceans (nine Cuvier's beaked whales, three Blainville's beaked whales, two unidentified beaked whales, two minke whales and one spotted dolphin) stranded or were found dead over 36 hours along a 240 km arc of the Bahama Islands. Post mortems were carried out on six animals. The spotted dolphin had a debilitating disease and its demise was considered unrelated to the mass stranding. In four of the beaked whales, some type of auditory structural damage was found, specifically bloody effusions or haemorrhaging in the ear region. The most likely cause of this damage was felt to be either acoustic or impulse (shock wave) injuries. By themselves the injuries would not have been fatal but could have compromised the animals' hearing or navigational abilities causing them to strand, after which overheating, cardiovascular collapse and physiological shock caused their death. As it happened, these events coincided in time and space with a United States naval exercise including 4–5 ships using active sonar. A joint US Navy-Department of Commerce report subsequently concluded that the ships' mid-range frequency sonars were 'the most plausible source' of the acoustic or impulse trauma.[1] This was the first time that a possible causal link between such mass strandings of beaked whales and naval operations had been established, although fingers had been pointed previously.[2,3]

Autopsies conducted after a mass stranding event in the Canary Islands in 2002 involving 14 beaked whales (of which eight Cuvier's beaked whales, a Blainville's beaked whale and a Gervais' beaked whale were examined) revealed similar congestion and haemorrhaging in the jaw fat, ears, brain and kidneys. In addition, gas bubble-associated lesions and fat embolisms were found in vital organs, consistent with acute trauma resulting from rapid decompression (as in 'the bends' of human divers). This event was also associated in time and space with a naval exercise using mid-frequency sonar. The post-mortem findings suggested either that the whales surfaced too rapidly as a fright reaction, or that the sonar itself had some direct effect on bubble formation in the tissues.[4] Other suggested scenarios include a behavioural response in which the whales made repeated dives shallower than the depth at which their lungs collapse, thus facilitating tissue supersaturation with nitrogen.[5] Beaked whales may also be predisposed to haemorrhaging, which might be provoked by the stress of exposure to sound.[6] Why Cuvier's beaked whales should be particularly susceptible to these sound effects is unknown, but it could simply be that there are more of them or that they occur further offshore than other beaked whales. On 13 July 1973 groups of three and two Cuvier's beaked whales were found floating dead 10 nautical miles apart and about 100 nautical miles offshore on the Durban whaling grounds off the east coast of South Africa, and one specimen was brought in by a whale catcher for research purposes. Could these have been early victims of some military exercise? Or were their deaths a consequence of the use of asdic by Durban whalers in the hunt for sperm whales?

References

1 United States Department of Commerce and U.S. Navy 2001
2 Simmonds and Lopez-Jurado 1991
3 Frantzis 1998
4 Jepson *et al.* 2003
5 Tyack *et al.* 2006
6 Cox *et al.* 2006

containing immatures consisted solely of mother-calf pairs. This suggests a polygynous society with males contesting access to females but females with young avoiding adult males.[22]

The blow is low, diffuse and directed forwards, which is supposed to make sighting them more difficult, and they may avoid vessels by diving.[1]

The species is believed to be one of the most proficient divers among air-breathing animals. Telemetered individuals made foraging dives to an average depth of 1 070 m for an average duration of 58 min. The maximum depth attained was 1 888 m (in water depths of up to 2 000 m) and the maximum duration was 84.9 min. Rates of descent (averaging 1.5 m/sec.) were always greater than rates of ascent (averaging 0.7 m/sec.).[23] Elsewhere an adult female monitored visually made 12 long dives over 7 hours ranging from 19 to 87 min. in duration with an average of 31.5 min.[14] A series of increasingly shallower dives (extending on average for 63 min.) normally followed the deep foraging dives, during which the whales presumably recovered from their oxygen debt.[23]

Cuvier's beaked whale emits two types of clicks. Regular clicks have an inter-click interval of 0.4–0.43 sec. and a principal frequency range of 20–60 kHz, while buzzes represent clicks produced at a much shorter interval (less than 0.15 sec.) and at a lower source level.[24] Buzzes may be associated with prey capture attempts. Whales only start clicking below 200 m,[25] indicating that feeding takes place at depth, as would be suggested by the prey items recorded. Whistles lasting about 1 sec. and sweeping up from 8 kHz to 12 kHz have also been recorded.[26]

Conservation status

Cuvier's beaked whales formed part of a small-whale type fishery in the coastal waters of Japan until the 1970s, but do not seem to have featured in the catch recently. Elsewhere it has only been targeted occasionally. It is generally considered the most widespread of the beaked whales, and not as rare as originally thought.[1] A genetics-based estimate of 456 000–916 000 breeding adults worldwide[2] needs to be interpreted with some caution, given the unlikely estimates obtained for large whales elsewhere using the same approach. The only identifiable threat to the species seems to be from loud sounds associated with seismic exploration or military activity, which in several instances have been implicated in mass strandings of Cuvier's and other beaked whales (see box above). Cuvier's beaked whale is listed as Data Deficient by the IUCN and in the South African Red Data Book. It is included on Appendix II of CITES, but not listed by CMS.

References

1 Heyning 1989
2 Dalebout *et al.* 2005
3 MRI Whale Unit data
4 Ross *et al.* 1988
5 Omura 1972
6 Kasamatsu *et al.* 1988
7 Weir 2006b
8 Findlay *et al.* 1992
9 Ross 1984
10 Keller *et al.* 1982
11 Robineau 1975
12 Gambell *et al.* 1975
13 Nishiwaki and Oguro 1972
14 Baird *et al.* 2004
15 Ferguson *et al.* 2006
16 Omura *et al.* 1955
17 Nemoto *et al.* 1980
18 MacLeod *et al.* 2003
19 Sekiguchi 1994
20 Mead 1984
21 Heyning 1988a
22 Cresswell and Walker 2002
23 Tyack *et al.* 2006
24 Zimmer *et al.* 2005
25 Johnson *et al.* 2004
26 MacLeod and D'Amico 2006

Arnoux's Beaked Whale

Berardius arnuxii

THE ONLY MORPHOLOGICAL characteristic differentiating Arnoux's beaked whale from the southern hemisphere and Baird's beaked whale (*B. bairdii*) from the North Pacific appears to be that of size,[1] while the genetic distance between the two species is less than that between most other species of beaked whales (but greater than most intraspecific differences).[2] Further analyses are required to decide whether *bairdii* is really a subspecies of *arnuxii* or whether the two should be considered distinct species. Because of this similarity, I feel justified in borrowing heavily from published information for *bairdii* to supplement our meagre knowledge about many aspects of the biology of *arnuxii*.

Appearance

At sea the species can be recognised by its long beak and prominent bulging forehead, which are often exposed when it surfaces to breathe. When rafting at the surface after a long dive, Arnoux's beaked whales could briefly be mistaken for sperm whales as they may retain their dorsal fin below the surface. Once exposed, the shape of the dorsal fin (not strongly hooked or pointed) should permit the species to be distinguished from southern bottlenose whales, whose fin is more erect and pointed at the tip.[3]

The largest known specimen appears to be a male from Western Australia with a length of 10.06 m[4] (or over 2 m shorter than the largest known Baird's beaked whale[5]). Members of the genus are considered the least dimorphic members of the ziphiid family,[6] with female Baird's on average being only 35 cm longer than males.[7] The beak is long (5–6% of body length) and robust, with a lower jaw that protrudes noticeably beyond the snout and in adults carries two pairs of prominent teeth. The head may be slightly larger in Arnoux's beaked whale, the distance from snout to blowhole being 11.7–13% of body length (compared to 10–12.1% in Baird's). The blowhole itself is unusual, because although it is crescent-shaped the arms of the crescent point backwards (rather than forwards as in most other odontocetes). The V-shaped throat grooves typical of ziphiids may be accompanied (in Baird's at least) by several smaller grooves behind them[8] (see box on page 105 for the possible function of the grooves). The flippers are small (7.9–11.1% of body length) and somewhat smaller than those of Baird's (10.5–13.4% of body length). The dorsal fin is low (1.6–3% of body length in height), positioned about two-thirds of the way down the body,[9] with a characteristically long leading edge and a very rounded tip.

The coloration of juvenile Arnoux's beaked whales is not well documented – in Baird's beaked whales it is described as slate grey with little scarring.[8] In adult Arnoux's beaked whales the basal colour has darkened to a brownish-black, but this is almost totally obscured from the front of the melon to the level of the dorsal fin by light grey blotching and a number of linear white scars running irregularly over the back. This patterning largely avoids the dorsal fin area, which then forms a dark 'island' dividing the grey on each side. In contrast to the light grey melon, most of the beak remains dark, although the tip of the snout and the lower jaw are lighter. The rear third of the animal is dark except for light grey scarring on the tail flukes, especially near their leading edge and tips.[10] Baird's beaked whale also has white patches ventrally, either on the throat, between the flippers or at the umbilicus.[11]

Unlike any other ziphiid, two pairs of mandibular teeth erupt in adults of both sexes around sexual maturity,[11] the anterior pair (which are apical and protrude in front of the upper jaw) being

Length:weight relationship

For one foetus and three post-natal Baird's beaked whales, $W = 0.0092L^{3.0812}$, where W = weight in tonnes and L = total length in metres.[7] This implies that a 5 m whale would weigh 1.3 tonnes and a 10 m whale 11.1 tonnes.

Skeletal data

Vertebral formula (n = 4): 7 cervical, 10–11 thoracic, 12–13 lumbar and 17–19 caudal; total 47–49.[9]

The first three cervical vertebrae are fused, the remaining four are free.[12]

Phalangeal formula (including metacarpals): I_2, II_5, III_4, IV_4, V_3.[12]

FIG. 56 Arnoux's beaked whales photographed outside the harbour entance, Table Bay docks, South Africa.

the larger. The teeth are strongly compressed laterally and are triangular in side view, the front pair measuring (in *bairdii*) up to 8–9 cm in height and 7–10 cm in antero-posterior width, and the rear pair 5–6 cm in height and 3–4 cm in antero-posterior width.[5]

Distribution and migrations

Arnoux's beaked whale is confined to the southern hemisphere and is generally portrayed as having a contiguous circumpolar distribution in temperate and polar waters, with the furthest north record being a stranding from Brazil at 24° S.[6] Records in the subregion are few. Apart from three single strandings on the South African coastline,[13] a live stranding of four animals occurred during the night of 6/7 January 1998 in Sandy Point harbour, St Helena Bay, on the west coast of South Africa (all the animals were successfully refloated), and a group of seven animals appeared at the entrance to Table Bay harbour, Cape Town on 7 December 2005.[10] The furthest north sighting in the subregion is two animals seen on 28 February 1997 at 27°52′ S 36°09′ E.[14] In the Antarctic numerous sightings have been reported off Kemp Land at about 62° E.[15]

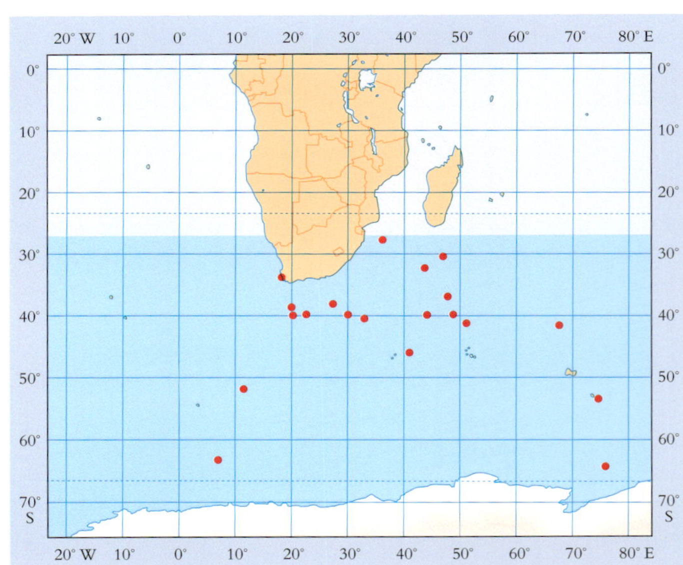

Postulated range ■ Sighting or capture

FIG. 57 Distribution map for Arnoux's beaked whale.

The few records of Arnoux's beaked whale in lower latitudes of the subregion have all been in the months of October to February (or spring/summer), and a similar seasonality of sightings and strandings (December to March) has been noticed in Australasia.[4] Sightings of the species are frequently made in the Antarctic in summer,[16] and it has been observed in the pack ice in August.[17] Because of the sparse nature of the data (and the general lack of associated effort data), it is difficult to determine whether there is any definite migratory pattern.

For Baird's beaked whale, a summer movement onto the continental slope of the Japanese coast has been proposed,[18] and if a similar seasonal slopewards movement occurred in Arnoux's beaked whale it would be consistent with the incidence of strandings and coastal sightings in the subregion. The wintering ground for Baird's beaked whales is still unknown. However, there are believed to be three separate stocks in the north-west Pacific, including one at high latitude (in the Okhotsk Sea).[18] The simultaneous presence of Arnoux's beaked whales in the Antarctic and off the coasts of southern continents in summer might suggest a similar stock separation.

External commensals and parasites

No external commensals or parasites seem to have been described for Arnoux's beaked whale. Two species of cyamid have been recorded from Baird's beaked whale: *Platycyamus flaviscutatus*, which occurs singly on the head, back, flanks and flukes, and *Cyamus orubraedon*, which occurs in clusters adjacent to the teeth at the tip of the lower jaw.[19] Stalked barnacles (*Conchoderma auritum*) are sometimes found attached to the teeth,[8] and films of the diatoms *Cocconeis* and *Navicula* may occur in patches on the skin.[20] These organisms could presumably all occur on Arnoux's beaked whale, although the specific identity of any cyamids present would be of special interest given the wide geographical separation of the two host populations. Wounds or scars attributable to the cookie-cutter shark *Isistius* have also been described for Baird's beaked whale.[8]

Food and feeding behaviour

Cephalopod beaks (unidentified) have been recorded from the stomach of an Arnoux's beaked whale from New Zealand.[5] Otherwise there is no information on feeding habits. In the western North Pacific, Baird's beaked whales feed mainly on fish (rat-tails and hakes), while in the Sea of Okhotsk, pollock and squid are also important. A comparison with the depths at which some of these species are caught commercially suggests that the whales can dive to 1 000 m or more.[21] There is no information on seasonality of feeding.

Reproduction and growth

There is no information for Arnoux's beaked whale, but the following data for Baird's beaked whale (given an appropriate downward adjustment in any length criteria) may well apply. Calves are born at a length of about 4.56 m[11] after a gestation that could last either 10 months or 17 months. The average calving interval is estimated to be 3.4 years, of which lactation might occupy as little as six months. Sexual maturity is reached at a length of 9.1–9.8 m in males and 9.8–10.7 m in females, and at ages of 6–11 and 10–15 years respectively. Physical maturity is reached within, at most, five years of sexual maturity, and at body lengths of 10.1 m in males and 10.45 m in females.[7]

In a most unusual demographic profile, males are estimated to live for up to 84 years, while the oldest known female was 54 years old, indicating a sex ratio among mature animals of 3.3 males per female. There is no sign of reproductive senescence in either sex, while testis size is correlated with body size even in physically mature males. The cause of the higher apparent mortality in females is unknown and the evolutionary benefit of greater male longevity is a matter for speculation. One suggestion to explain this longevity is that males of this deep- and long-diving species may play some role in protecting or guiding the young while the mothers feed.[7]

Behaviour and vocalisations

The 26 sightings off Australia attributed to Arnoux's beaked whale consisted of groups of 1–16 animals, with a modal value of four. Only two calves were seen, in two cow-calf pairs.[4] The group that stranded in St Helena Bay in 1998 was said to consist of two males and one female about 9 m long and a calf about 5 m long.[10] Large aggregations of up to 80 animals have been reported from the Antarctic,[8] but with an average number per pod of 6.7 (n = 15).[15]

Within groups, whales generally surface and blow together for 15 or more breaths, during which time they swim characteristically close to the surface. The raising of flukes on diving is rarely seen,[4, 8] but the species can indulge in energetic aerial displays.[15] A swimming speed of about 4 knots has been reported.[8]

Suckers for squid

One thing most beaked whales don't have plenty of is teeth. With the notable exception of Shepherd's beaked whale, the best that most of them can boast is a single pair in the lower jaw, and then only in mature males. As they are usually squid-eaters, one might well ask how they manage to catch and consume such slippery customers without any meaningful dentition. The clue seems to lie in three special adaptations common to all beaked whales: a restricted gape, a pair of V-shaped throat grooves, and relatively large hyoid bones (to which the tongue muscles are attached). By simultaneously retracting the tongue and distending the throat grooves, the whale is able create a sudden drop in pressure inside the mouth, thereby sucking prey in. The restricted gape assists this process by concentrating the suction to a relatively small area (like sucking on a straw). Once the prey is secured in the mouth against the roughened palate, the water can be expelled by contracting the throat grooves and moving the tongue forward. Suction may be used by other toothed whales (even those with a full set of teeth, such as sperm whales) to ingest their prey, explaining why such prey often don't carry tooth marks.[1]

Reference

1 Heyning and Mead 1996

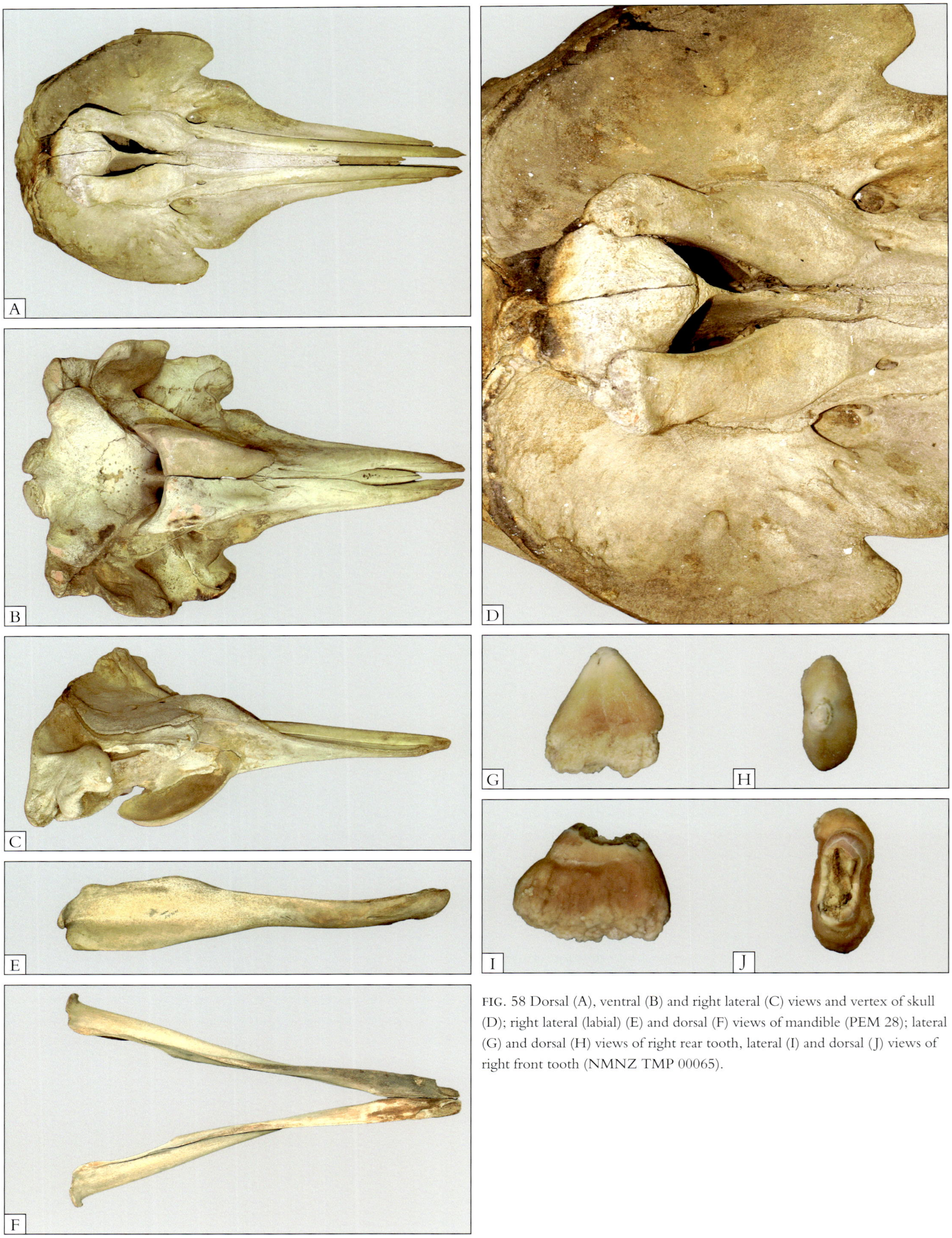

FIG. 58 Dorsal (A), ventral (B) and right lateral (C) views and vertex of skull (D); right lateral (labial) (E) and dorsal (F) views of mandible (PEM 28); lateral (G) and dorsal (H) views of right rear tooth, lateral (I) and dorsal (J) views of right front tooth (NMNZ TMP 00065).

Arnoux's beaked whale seems to be one of the more accomplished mammalian divers. Dive durations of 17 individual animals in leads in Antarctic sea ice had a mode of 35–65 min. and a maximum of at least 70 min. (n = 70). Periods of inter-dive respiration varied from 1.2 to 6.8 min., during which time an average of 9.6 blows per minute were made (n = 8).[22]

Recordings made at the ice edge showed the animal to be highly vociferous, with a characteristic warbling sound to the ear. The commonest vocalisation was an amplitude-modulated call of between 1 000 Hz and 8 500 Hz, with a 'wavering' frequency modulation. Broadband clicks were also produced as trains of about 25 clicks at 34 clicks/sec., and as burst pulses lasting about 0.53 sec. Whistles with a wavering frequency modulation between 4 869 Hz and 5 222 Hz were the second commonest call recorded.[15]

Conservation status

Arnoux's beaked whales have never formed the target of a directed hunt, and seem to have little direct involvement with fisheries. Although there are no published population estimates, the species is generally considered uncommon. The only identifiable threat to it would seem to be from loud sounds associated with seismic exploration or military activity, which have been implicated elsewhere in mass strandings of other beaked whales.[23]

Arnoux's beaked whale is listed as Data Deficient in the South African Red Data Book, and as Lower Risk: Conservation Dependent by the IUCN. It is included on Appendix I of CITES, but not listed by CMS.

References

1 Rice 1998
2 Dalebout *et al.* 1998
3 Leatherwood *et al.* 1982
4 Paterson and Parker 1994
5 McCann 1975
6 Kasuya 2002
7 Kasuya *et al.* 1997a
8 Balcomb 1989
9 Ross 1984
10 MRI Whale Unit data
11 Omura *et al.* 1955
12 Flower 1872b
13 Findlay *et al.* 1992
14 Leslie, R., Marine and Coastal Mgt, pers. comm.
15 Rogers and Brown 1999
16 Ponganis *et al.* 1995
17 Taylor 1957
18 Kasuya and Miyashita 1997
19 Waller 1989
20 Pike 1953
21 Ohizumi *et al.* 2003
22 Hobson and Martin 1996
23 Fernández *et al.* 2005

Shepherd's Beaked Whale

Tasmacetus shepherdi

Adult ♂

Calf

Appearance

At sea the species should be distinguishable from other large beaked whales by its long dark beak, pale melon, and pale shoulder patch, and the fact that all animals in the group will have a similar coloration irrespective of age. From the air (or high up on a ship), the pale grey areas on the flanks behind the dorsal fin should be obvious. The species does produce a clearly visible blow on occasion.[1]

The largest known specimen was 6.8 m, although longer specimens (7 m and 7.35 m) have been measured by non-specialists.[1] It is not clear whether there is any sexual dimorphism in size. The beak is long (5.6–7.5% of body length) and decidedly pointed at the tip. The melon is more prominent than in most *Mesoplodon* species but not as swollen as in *Hyperoodon* or *Berardius*. There is a single pair of anteriorly converging throat grooves. The flippers are about 10–11% of the body in length and situated well down on the side of the whale as in other ziphiids.

Length:weight relationship

There are no published weights.

Skeletal data

Vertebral formula (type specimen): 7 cervical, 9 thoracic, 11 lumbar and 18 caudal; total 45.[3]

The first six cervical vertebrae are fused, and the seventh is free.[3]

Phalangeal formula (including metacarpals): I_1, II_5, III_4, IV_3, V_2.[3]

The dorsal fin is placed 65–72.5% of the body back from the snout, is about 5% of the body length in height[2] and distinctly dolphin-like in shape.

The colour pattern of Shepherd's beaked whale is quite distinctive. The beak is dark but the melon as far back as the blowhole is a pale blue-grey. The rest of the body is dark bluish-grey brown above and pale grey-white below, but with extensions of grey up onto the flanks behind the dorsal fin, and diagonally up from just behind the flipper onto the shoulder. Between these two extensions the dark dorsal colour dips down the side almost to the midline of the belly. There is a dark stripe running from the flipper towards the eye, which continues forward towards the angle of the gape, effectively separating the pale melon from the white throat. Unlike many other ziphiids, this colour pattern seems to be essentially the same in neonates as in adults.[1]

The dentition of Shepherd's beaked whale is unique among ziphiids in that there are multiple teeth in both upper and lower jaws. The pair at the tip of the mandible is distinctly larger than the others, measuring 41–43 mm in height, with an anterior-posterior diameter of 22–24 mm and a transverse diameter of 16 mm. These teeth have a bulbous base that suddenly contracts to steeply conical crowns, and only erupt in adult males. The remaining teeth (17–21 pairs above, 22–28 pairs below) are much smaller: 15–28 mm in height, and with an anterior-posterior diameter of 5–15 mm. They seem to erupt in both sexes.[2, 3] The rows of these 'secondary' teeth only start several centimetres back from the tip of the snout and lower jaw, so that there is a gap of some 7 cm between the apical mandibular teeth and the first secondary teeth. It is not clear if all of the secondary teeth erupt.

FIG. 59 This aerial photograph from Tristan da Cunha is among the few available of living Shepherd's beaked whales.

Distribution and migrations

Shepherd's beaked whale is only known from the southern hemisphere. All records to date have been between the latitudes of 30° S and 55° S,[1] and the species has been described as likely to have a circumpolar distribution in cold-temperate waters.[4] The only records within the subregion are six strandings on the island of Tristan da Cunha between 1983 and 1995, and sightings from the vicinity of Tristan da Cunha and Gough Island.[1] Statements that it has been recorded from South Africa[5, 6] are incorrect, and a sighting from the Seychelles region attributed to this species[7] is now considered unlikely to have been Shepherd's beaked whale.[1]

From the few sightings available, the species seems to inhabit a wide range of water depths, from 350–500 to 3 600 m.[1] Strandings at Tristan da Cunha have occurred six months apart, so although most strandings elsewhere are reported to have occurred in summer,[4] it would be premature to infer any seasonal movement from these data.

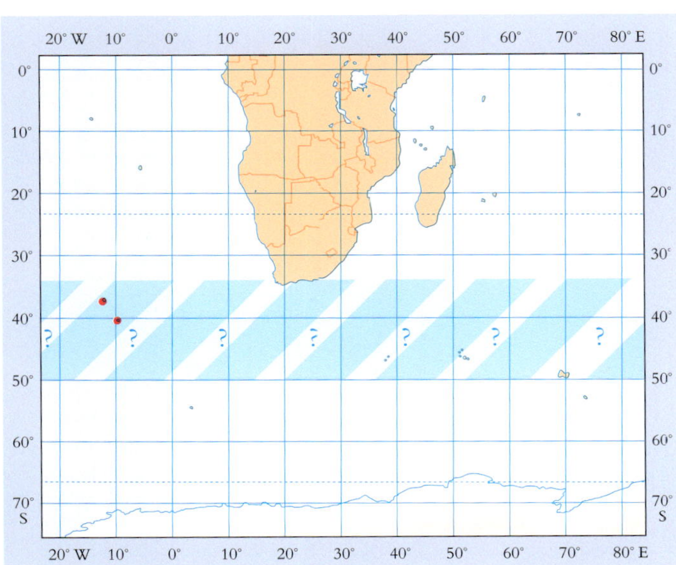

FIG. 60 Distribution map for Shepherd's beaked whale.

External commensals and parasites

No external commensals or parasites have been described for Shepherd's beaked whale.

Food and feeding behaviour

Two stomachs have been examined. In the first (from Argentina), fish predominated, particularly an unidentified brotulid, but hake (*Merluccius hubbsi*) and an unidentified serranid were also present. Remains of a crab (*Peltarion spinulosum*) and an unidentified squid beak were considered likely to have come from the stomachs of the fish.[2]

Only cephalopod remains were found in the second stomach (from Tristan da Cunha), consisting of *Todarodes filippovae*, *Teuthowenia pellucida*, *Ancistrocheirus lesueuri*, *Histioteuthis* (possibly *corpuscula*) and one unidentified beak.[1]

The claim that Shepherd's beaked whale is predominantly a fish-eater[4] may therefore be premature.

Reproduction and growth

There is almost no information. A stranded 3.4 m male still bore foetal fold marks on its flank.[1] In bottlenose dolphins these may remain visible for up to two months after birth,[8] so this individual must have been very young but not necessarily newborn. A 6.6 m female was determined to be sexually mature.[2] A 6.6 m male had 23 growth layers visible in the cementum of one of the mandibular teeth.[9]

Behaviour and vocalisations

The four known sightings of this species involved groups of 3–6 animals. A calf was reported in the largest group. Some individuals were reported to lift their beak clear of the water on surfacing. No exposure of flukes and no aerial activity were recorded. Shipboard observers recorded no visible blows, whereas an aerial observer noted and photographed visible blows.[1]

No vocalisations have yet been recorded from Shepherd's beaked whale.

Conservation status

Shepherd's beaked whales have never formed the target of a directed hunt, and seem to have little direct involvement with fisheries (although if they really are mainly fish-feeders this might create at least the potential for some indirect interaction with commercial fishing).

The species is generally considered uncommon simply because of the rarity of sightings, but this may partly be because up to now it has not been recognisable at sea.[1]

The only identifiable threat to it would seem to be from loud sounds associated with seismic exploration or military activity, which have been implicated elsewhere in mass strandings of other beaked whales.[10]

Shepherd's beaked whale is classified as Data Deficient by the IUCN and unlisted in the South African Red Data Book. It is included on Appendix II of CITES, but unlisted by CMS.

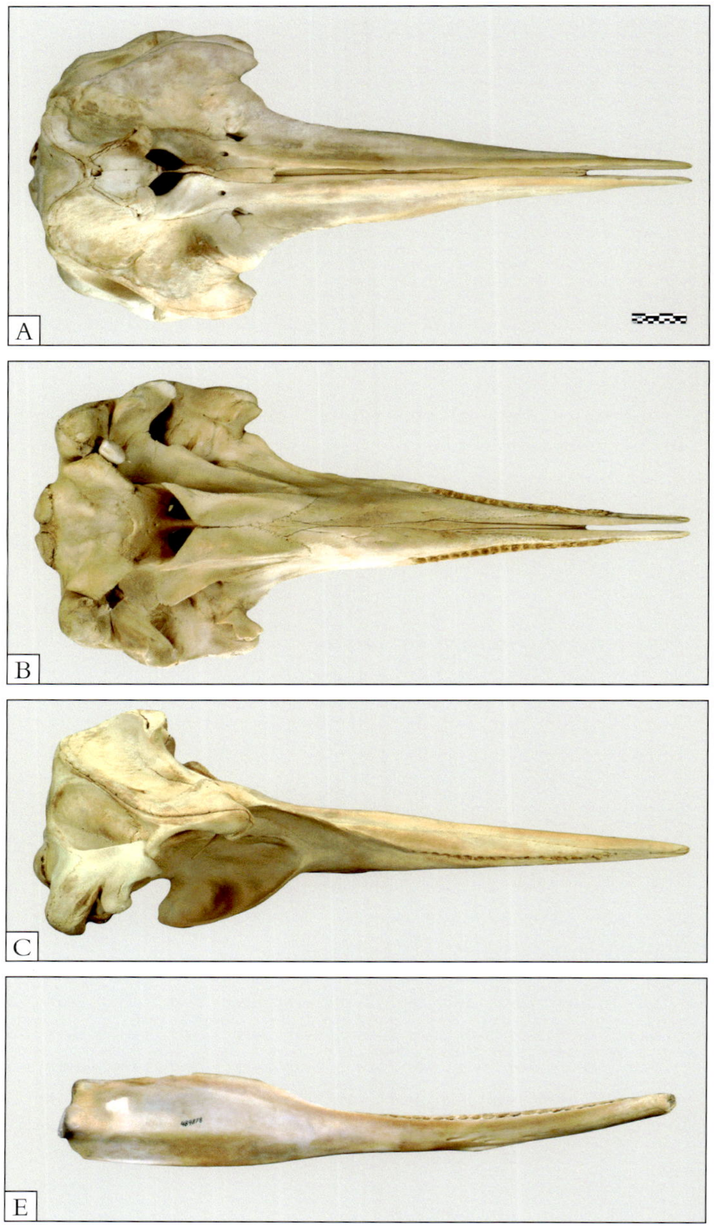

FIG. 61 Dorsal (A), ventral (B) and right lateral (C) views and apex of skull (D); right lateral (labial) (E) and dorsal (F) views of mandible (USNM 484878 – adult female); dorsal (G) and lateral (H) views of apical teeth (right), and lateral views of six posterior teeth (I) (ZM 40484).

References

1 Pitman *et al.* 2006
2 Mead and Payne 1975
3 Oliver 1937
4 Mead 1989a
5 Rice 1998
6 Reeves *et al.* 2002
7 Mead 2002
8 Cockcroft and Ross 1990c
9 Grandi *et al.* 2005
10 Fernández *et al.* 2005

Longman's Beaked Whale

Indopacetus pacificus

FIRST DISCOVERED IN 1882, there were still only six known specimens recorded of this species worldwide by 2004, qualifying it as one of the poorest known of the world's cetaceans. Small wonder then that there is still controversy over whether it deserves its own genus or should be classified as just another *Mesoplodon*. Genetic analysis has not yet been able to resolve the debate, and morphologically the beast seems to be different enough to be retained in its own genus for the time being.[1]

Appearance

Undoubtedly the biggest problem is distinguishing this species at sea from the southern bottlenose whale. Indeed, one of the reasons why it escaped attention for so long is probably because most sightings were incorrectly attributed to bottlenose whales.[2] Both have relatively short beaks and large melons (though that of Longman's is not as bulbous), are tan or grey-brown in colour, have prominent erect dorsal fins, and in both species some individuals have pale heads. The differences to look for are: 1) the posterior boundary of the pale melon extends back only as far as the blowhole in Longman's (but well behind it in the bottlenose); 2) there is a dark transverse band behind the pale melon in Longman's (but not in the bottlenose); 3) there may be a white spot within the dark transverse band in the region of the ear in Longman's (but not in the bottlenose); and 4) there is a whitish flank patch that extends up the side of Longman's (whereas the flanks of the bottlenose are dark).[1]

The longest specimen measured to date was a 6.48 m female stranded in Japan in 2002.[3] The largest male measured to date was 5.73 m long,[4] but the sample is too small to decide whether the species is dimorphic in size or not. The external appearance has not been well documented. The beak is long (from the tip of the snout to the eye being 15–17% of body length) and well differentiated from the melon, which is prominent but not as bulbous as in the southern bottlenose (although individuals with 'bulbous' melons have been recorded at sea). The characteristic V-shaped throat grooves of all ziphiids are present. The flippers are small (7–10% of body length), roughly oval in shape and situated in 'pockets' low down on the side of the whale.[1] The dorsal fin, described as tall and the largest (both absolute and relative) in all the ziphiids,[2] had a height that was 4–5% of body length in two juveniles. It was situated 62–65% of the body back from the snout.[1]

The colour pattern has only been described in detail for juveniles. From the blowhole backwards, the back is black becoming dark grey laterally and then white ventrally. A broad band connects the black dorsal pigmentation to the flipper, which is also dark grey on its outer surface, and another band runs from the top of the head to the eye. In the region of the external ear opening, there may be a small lighter patch contained within the dark pigment. Much of the upper jaw is black, and there is a streak of dark grey running back in the midline from the beak towards the blowhole. The rest of the face, lower jaw and throat are white. The underside of the tail flukes is pale grey with dark streaks radiating from the centre of the trailing edge towards the leading edge.[1] Features of the adult coloration are mentioned above, notably the pale melon anterior to the blowhole, the dark band behind it, and the whitish flank patch. Some individuals

Length:weight relationship

The partially eviscerated carcase of a 5.73 m male weighed 2.5 tonnes.[4]

Skeletal data

Vertebral formula (n = 2): 7 cervical, 10 thoracic, 9–12 lumbar and more than 7–15 caudal; total more than 36–41. A third specimen had 10 ribs, 7 of them double-headed, confirming the number of thoracic vertebrae, but the total vertebral count for the species remains uncertain owing to the missing caudal vertebrae of both specimens examined.[1]

The first five cervical vertebrae are fused (the fifth only partially), and the remaining two are free.[1]

Phalangeal formula: Unknown.

may also have a grey 'cape' extending over much of the anterior back, and carry scarring (presumably) from intra-specific fighting and cookie-cutter shark bites.[2]

There is a single pair of teeth, situated at the tip of the mandible. These presumably erupt in adult males only. In juveniles and adult females they are roughly conical in shape, slightly flattened transversely, and somewhat inclined forwards. Adult teeth may have a thick cementum sheath at the base. In adult females the teeth are about 3–4 cm in height,[1, 3] and about 1.3–1.7 cm in diameter.[1] The teeth of the adult male remain to be described.

Distribution and migrations

So far Longman's beaked whale is only known from the tropical and subtropical waters of the Indian Ocean and the Pacific Ocean. In the subregion it is known from two strandings in South Africa (one at Blythesdale Beach and the other at Sodwana Bay, both in KwaZulu-Natal), and a specimen from an unknown locality in Kenya.[1] A stranded specimen reported in the media from Plettenberg Bay, South Africa, in 2002 is considered unlikely to have been this species.[3] Several sightings attributed to Longman's beaked whale and two specimens (from Somalia and the Maldive Islands) have been recorded in the northern Indian Ocean.[2] A sighting of 10 animals at 30°05′ S 32°20′ E on 7 April 1973 may also have been of this species.[5]

Sightings have been made in every month except June, so the species may be resident in the tropics.[2]

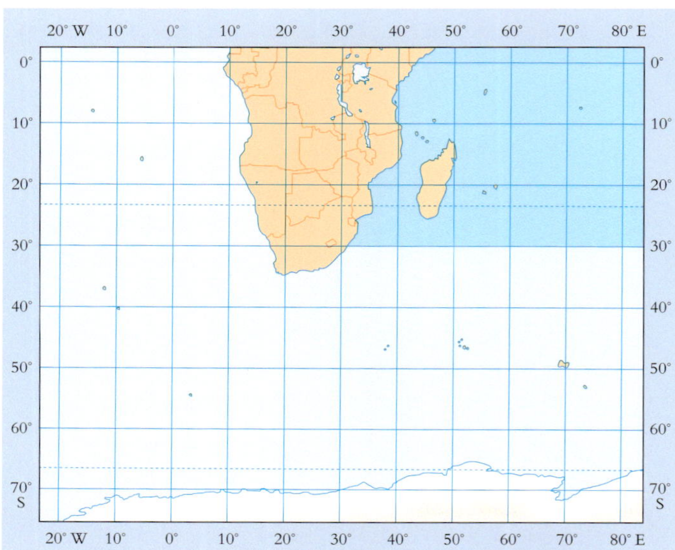

Postulated range

FIG. 62 Distribution map for Longman's beaked whale.

External commensals and parasites

Numerous cyamids (unidentified) were recorded from an adult female stranded in Japan.[3] Fresh wounds and scars attributable to a small shark such as the cookie-cutter *Isistius* were found on a juvenile and adult.[1, 3] Perhaps surprisingly, stalked barnacles have not been apparent on the lower jaws of any animals seen at sea, although these are frequently attached to the exposed tips of erupted teeth of other beaked whales.[2]

Food and feeding behaviour

Cephalopod beaks and eye lenses (unidentified) were recorded from the stomach of a Longman's beaked whale from Japan.[3] Otherwise there is no information on feeding habits.

Reproduction and growth

A 2.91 m male that stranded live in November was variously described as a juvenile or neonate; the umbilicus was completely healed but some foetal folds were still present, so it could have been several weeks old. A 3.63 m male stranded in August carried no such traces of recent birth.[1] Calves may therefore be just under 3 m at birth. A 5.96 m female (measured over the curve) was pregnant with a 1.04 m foetus in January,[1] and a 6.48 m female from Japan was physically mature, possible senile.[3]

Behaviour and vocalisations

The species seems fairly gregarious: sightings attributed to Longman's beaked whale consisted of groups of 1–100 with a mean of 18.5 animals (n = 41), and in five sightings made in the

FIG. 63 These two pictures illustrate the long beak, prominent melon and dorsal fin, and pale head of the 'tropical bottlenose whale', now believed to be Longman's beaked whale.

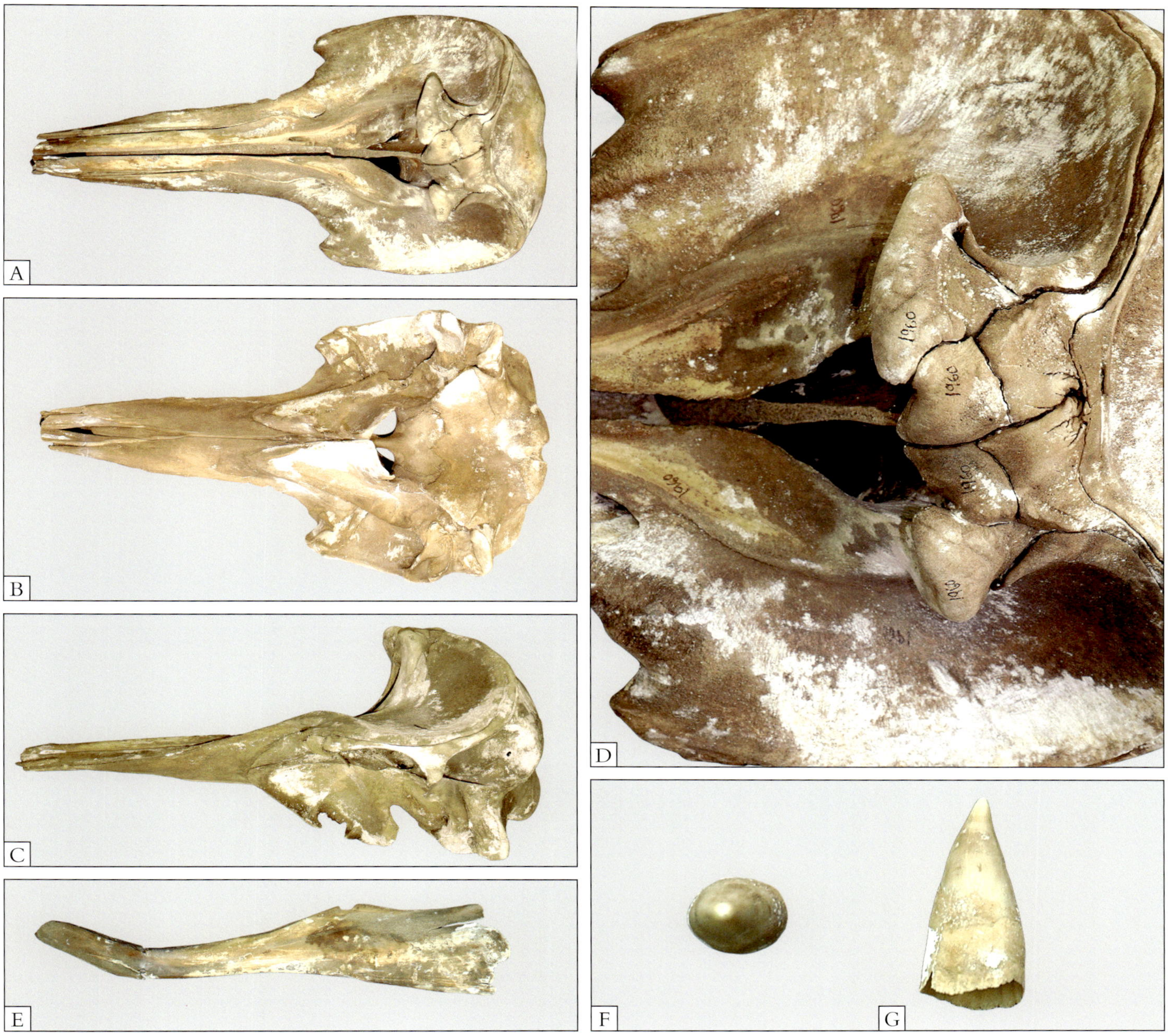

FIG. 64 Dorsal (A), ventral (B) and left lateral (C) views and vertex of skull (D); lateral (labial) view of left mandible (E); dorsal (F) and lateral (G) views of tooth (PEM 1960 – juvenile).

northern Indian Ocean the groups consisted of 2–40 with a mean of 17.4 animals. In the Pacific region they have been seen in association with short-finned pilot whales fairly often (and less regularly with other species).[2]

They are described as 'aggressive swimmers', often bringing the beak and melon out of the water as they surface. The blow is low, bushy and conspicuous (and may be canted forward). Breaching behaviour, and individual dives of 18 min. and 25 min. have been reported.[2]

No information on vocalisations has yet been published.

Conservation status

Longman's beaked whales have never formed the target of a directed hunt, and seem to have little direct involvement with fisheries. Although there are no published population estimates, the species is generally considered uncommon.[2] The only identifiable threat to it would seem to be from loud sounds associated with seismic exploration or military activity, which have been implicated elsewhere in mass strandings of other beaked whales.[6]

Longman's beaked whale is listed as Data Deficient by the IUCN and in the South African Red Data Book. It is included on Appendix II of CITES, but not listed by CMS.

References

1 Dalebout *et al.* 2003
2 Pitman *et al.* 1999
3 Yamada 2002
4 Acebes *et al.* 2005
5 MRI Whale Unit data
6 Fernández *et al.* 2005

Southern Bottlenose Whale

Hyperoodon planifrons

Adult ♂

Calf

UNLIKE THE NORTHERN and southern *Berardius* species, the northern and southern bottlenose whales – *Hyperoodon ampullatus* and *H. planifrons* respectively – are morphologically well-marked species.[1] Genetically, however, the situation is more complicated. The southern bottlenose whale has very high within-species variation, greater than found between some other species of beaked whale. This could indicate either a large population size or the presence of more than one 'evolutionary significant unit' within the species (is there more than one southern bottlenose whale species out there?).[2] Because of this complication, the genetic distinction between northern and southern bottlenose whales does not appear as clear-cut as it is between other species in the family.[3]

Considerably more is known about the northern bottlenose, not least because it once formed the basis of a substantial whaling industry, so sometimes in this account I have had to purloin information from the better-known species when it was lacking for the southern.

Appearance

The similarity between the external appearance of Longman's beaked whale and the southern bottlenose, and how to tell them apart, are discussed under the account of the former species. Cardinal points that distinguish southern bottlenose whales are: 1) there is a noticeable overhang of the melon over the beak in adults; 2) the pale coloration of the head extends back well past the blowhole; and 3) there is no dark transverse band behind the pale head.

The largest sexed southern bottlenose whales appear to be a 7.0 m male[4] and a 7.45 m female,[5] but the available sample of measured animals is small (13 males, 10 females). This may explain why there is no sign yet of the sexual dimorphism shown in northern bottlenose, where the largest recorded male is 1.1 m longer than the largest female.[6] The beak is prominent, forming 4.4–5% of body length (n = 3),[7,8] and in adults is overhung by the melon, which becomes especially bulbous in males (although probably not to the extent of the northern bottlenose, given that the maxillary crests of the skull do not develop to the same

Length:weight relationship

A 3.27 m male weighed 0.505 tonnes and a 6.43 m male 3.69 tonnes (both were weighed in pieces).[7] Combined with two others,[4] $W = 0.014L^{3.0734}$, where W = weight in tonnes and L = total length in metres. This implies that a 4 m whale would weigh 0.99 tonnes and a 6 m whale 3.4 tonnes (these weights should presumably be adjusted for weight loss during dissection).

Skeletal data

Vertebral formula (n = 5–7): 7 cervical, 8–9 thoracic, 10–12 lumbar and 17–21 caudal; total 44–47.[9,10,13–15]

All seven cervical vertebrae are fused.[9,10]

Phalangeal formula (including metacarpals, n = 2): $I_2, II_7, III_6, IV_5, V_3$.[9,13]

FIG. 65 Leaping clear of the water, this southern bottlenose whale (*left*) gives a good overall impression of its body shape, while even a snow storm cannot hide the short beak and pale head reaching beyond the blowhole typical of the species (*below*).

extent). A distinct dimple develops in the front of the melon of adult males. The flippers are small (10.3–11.5% of body length, n = 6), elliptical in shape and situated well down on the side. The dorsal fin is relatively impressive in size for a ziphiid, being about 4.6–6.7% of body length in height (n = 6),[7–10] falcate in shape and situated 66.5–68.6% (n = 3) of the body back from the snout.[7, 9]

The colour pattern of the southern bottlenose whale seems to vary considerably, possibly as a result of age. Juveniles are distinctly brownish-black above and white below, with the border between the two colours running back from the level of the eye as far as the caudal peduncle, where a tongue of the lighter ventral colour extends up onto the side. The pale ventral coloration also extends onto each side of the head, between the eye and the base of the beak, but is separated in the midline by a dark band running back from the beak.[7, 9] In subadults the colour merges into a greyish-brown all over, perhaps slightly lighter below but with no major distinction between back and belly, and with a uniformly coloured head. There may be small white spots and some scratch-like scars scattered over the body surface.[7] In adults the head lightens in colour as far back as the blowhole and well behind it, while the body becomes so scratched and scarred that it may appear white in very old animals.[11]

There is a single pair of teeth, situated at the tip of the mandible, which erupt in adult males only. In juveniles and adult females they are slender, conical and slightly curved in shape, 50–51 mm in height and 16–18 mm in diameter, with an unworn apical denticle projecting 6–8 mm above a thin investing coat of cement.[9, 10, 12] In adult males the teeth are slightly longer (57–59 mm) but much stouter, with a maximum diameter of 36–37 mm; the apical denticle may be worn but is still recognisable.[13] A series of 16–20 vestigial teeth on each side of the upper jaw and starting about 24 cm from the jaw tip has been described in one adult female. It is not clear how many, if any, were erupted. These teeth were 4–14 mm in height with diameters up to 2 mm.[10] In a juvenile male several minute rudimentary teeth were discovered embedded in the gum of the mandible.[9]

Distribution and migrations

Southern bottlenose whales are usually portrayed as occurring throughout the southern hemisphere south of about 30° S, although the type specimen was from the Dampier Archipelago off north-western Australia at *c.* 20° S,[6] and there was a stranding in December 1990 at Walvis Bay, Namibia (*c.* 23° S).[7] The species may well therefore range further north than 30° S on occasion.

Within the subregion, there have been specimen records from South Africa, Namibia and Heard Island.[16] These exclude two strandings in South Africa now attributed to *Indopacetus*.[17] In addition, there have been numerous sightings attributed to the species off the south-east coast of South Africa in summer and autumn,[18] and relatively high encounter rates for 'beaked whales' (the majority of which were believed to be southern bottlenose whales) in the Antarctic sector of the subregion between mid-December and mid-February. Other sightings in high latitudes have been made in October and March, and the species can occur at the ice edge and as far south as 74° S.[19] All the sightings off South Africa were in water more than 1 000 m deep.[18]

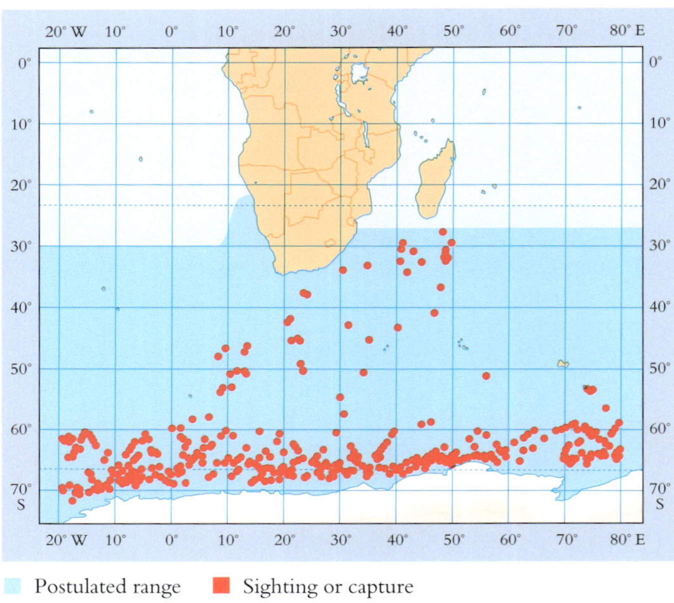

FIG. 66 Distribution map for southern bottlenose whale.

Evidence of seasonal movements is difficult to interpret. The apparent high abundance of southern bottlenose whales in Antarctic waters in summer has been mentioned above. During a cruise between 20° S and 40° S in summer in the south-west Indian Ocean, bottlenose-like whales were not seen between 20° S and 30° S, and were encountered four times more frequently between 35° S and 40° S than between 30° S and 35° S, confirming that the summer distribution of the species seems to be mainly in higher latitudes. In the Durban whaling grounds (*c.* 30° S) there was a marked seasonality of occurrence of whales identified as 'like-bottlenose' from the spotter aircraft, with the maximum sighting rate being at the start of the whaling season in February, falling rapidly to a minimum close to zero between June and August, and increasing again in October. This pattern could be interpreted as representing a movement north from higher latitudes in late summer, and a movement back south again in spring, and the relatively high incidence of Antarctic prey remains in two animals on the South African coastline in January (and the presence of cold-water skin diatoms on one) would tend to support the idea of a movement out of the Antarctic in late summer. However, the question remains: where do the animals go in winter? Previously sightings of bottlenose-like animals in tropical waters had been interpreted as possibly representing *Hyperoodon*,[20] but it now seems more likely that these animals were actually Longman's beaked whale,[17] so although there is evidence that southern bottlenose whales occur seasonally in some areas, the pattern of any migration remains unknown.

External commensals and parasites

Cyamids (*Platycyamus thompsoni*) were recorded from a young male stranded in Australia[9] and were recovered from a male harpooned at sea off South Africa.[7] Stalked barnacles (*Conchoderma* sp.) have been recorded from the teeth of adult male northern bottlenose, and also occur on their southern congener.[7] The validity of a record of the parasitic copepod *Pennella* from southern bottlenose is in doubt.[6] The cold-water diatom *Bennettella* (= *Cocconeis*) *ceticola* was identified on the skin of a young male from South Africa[21] and a lactating female in the Antarctic.[22]

Food and feeding behaviour

Only remains of squid were found in the stomachs of two animals (one male, one female) from South African waters, represented by beaks from 1 974 and 1 622 identified individuals. Such an accumulation of beaks is more typical of an animal that is a 'nibbler' rather than one that takes periodic 'meals'. In both stomachs combined, a total of 36 species of 14 families of squid were identified, all of them oceanic.

Using reconstructed weights, four Antarctic squids (*Kondakovia longimana*, *Galiteuthis glacialis*, *Alluroteuthis antarcticus*, *Mesonychoteuthis hamiltoni*) and four subantarctic squids (*Gonatus antarcticus*, *Moroteuthis knipovitchi*, *M. ingens*, *Histioteuthis eltaninae*) formed 72.4% of the squid eaten by the female and 43.7% by the male, presumably indicating a recent arrival from feeding grounds to the south. Of the other, more local squid, histioteuthids (*H. meleagroteuthis*) and *Taonius pavo* were among the most important eaten by both whales, with *Moroteuthis* sp., *M. robsoni*, *Taningia danae* and *Octopoteuthis* spp. also featuring in the male and *M. robsoni*, *T. danae* and *Onychoteuthis* spp. in the female. The sizes of the squids eaten were generally small, with individuals weighing more than 1 kg comprising only 0.7% by number in the male and 7.5% in the female.[20]

In a stomach from Heard Island (at 53°05′ S), 15 species of nine families of squid could be identified, of which nine species defined as Antarctic in origin (*Psychroteuthis glacialis*, *K. longimana*, *M. knipovitchi*, *G. antarcticus*, *A. antarcticus*, *M. hamiltoni*, *Brachioteuthis* sp., *Chiroteuthis* sp. and *Martialia hyadesi*) formed 87.2% by mass of those eaten – a very much higher proportion than found in South African stomachs, as might be imagined. Once again the sizes of the squid eaten were quite small, those of more than 1 kg comprising only 2.8% of prey. Interestingly, the remains of three Patagonian tooth fish (*Dissostichus eleginoides*) were also found.[16]

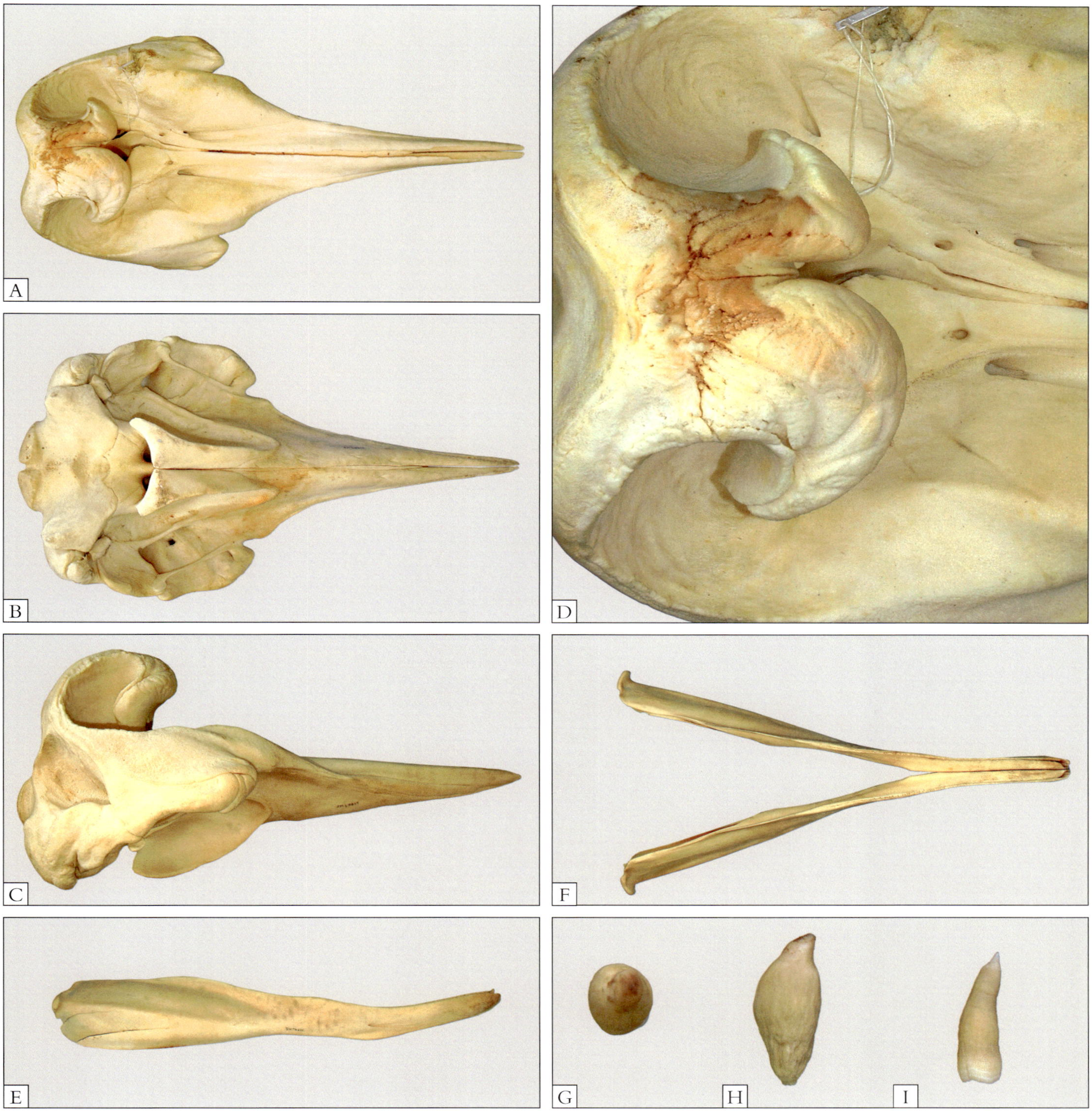

FIG. 67 Dorsal (A), ventral (B) and right lateral (C) views and vertex of skull (D); right lateral (labial) E and dorsal (F) views of mandible (ZM 40855 – adult female); dorsal (G) and lateral (H) views of teeth of adult male (PBB 07/05) and lateral view of tooth of adult female (I) (ZM 40855).

Reproduction and growth

Birth is believed to occur at about 3.5 m in the northern bottlenose whale,[6] but stranded southern bottlenose as small as 3.19 m[23] and 3.27 m[7] have been recorded (the record of a 2.91 m neonate was a misidentified Longman's beaked whale[17]). The 3.27 m individual (a male) had a healed umbilicus, so birth presumably takes place at a smaller size in southern bottlenose, possibly around 3.2 m or even less. Although a spring-early summer calving season has been proposed,[9, 24] this was strongly influenced by the 2.91 m misidentified neonate. In the northern bottlenose whale, gestation is estimated to last 12 months (although the data given would suggest a somewhat longer period),[25] lactation about a year and the calving interval about two years.[26] Caution has been expressed, however, about extrapolating such data to the southern bottlenose whale.[27] Female southern bottlenose as small as 5.7 m have been reported as lactating,[4] but there are no data on the size or age at sexual maturity in either sex.

Similarly, sizes and ages at physical maturity are unknown in this species. A 7.45 m female had 26 growth layers in the cementum of its teeth.[22]

Behaviour and vocalisations

Sightings attributed to the southern bottlenose whale off South Africa ranged from one to 10 with a mean of 3.9 animals (n = 47).[18] The mean school size of 'beaked whales' in Areas IIE and III of the Antarctic (equivalent to the high latitude sector of the subregion) was somewhat smaller, ranging from 1.7 to 3.6.[19]

Dive cycles recorded for 12 groups of beaked whales in the Antarctic (most of which were probably southern bottlenose whales) consisted of long submergences ranging from 11–46 (mean 25.3) min. in duration, interspersed with surfacing bouts in which 6–18 (mean 8.8) blows would be recorded, with an average duration at the surface of 3.7 sec.[19]

The northern bottlenose whale is known to be an excellent diver, one individual submerging about every 80 min. for a duration of 25–70.5 (mean 37) min., and reaching maximum depths of 493–1 453 (mean 1 060) m.[28]

The northern bottlenose whale produces clicks and whistles. The clicks range from below 500 Hz to over 26 kHz in frequency, and are produced in intermittent series of 3–50 clicks with repetition rates as high as 82 per sec. The frequencies of whistles range between 3 kHz and 16 kHz, and vary in duration from 115 to 850 ms. Some whistles stay at one frequency while others sweep through various frequencies.[29] Similar click trains have been recorded around southern bottlenose whales.[30]

Conservation status

Southern bottlenose whales have never formed the target of a directed hunt, although individual specimens have been taken out of curiosity in the commercial catch.[8, 10, 22] They also appear to have little direct involvement with fisheries. They are an abundant species: an estimate of 599 000 beaked whales south of the Antarctic Convergence in summer is considered to refer mostly to southern bottlenose whales.[19] The only identifiable threat to the species would seem to be from loud sounds associated with seismic exploration or military activity, which have been implicated elsewhere in mass strandings of other beaked whales.[31]

The southern bottlenose whale is listed as Lower Risk, Conservation Dependent by the IUCN and as Least Concern in the South African Red Data Book. It is also included on Appendix I of CITES, but not listed by CMS.

References

1 Rice 1998
2 Dalebout *et al.* 1998
3 Dalebout *et al.* 2004a
4 Zemsky and Budylenko 1970
5 Fraser 1964
6 Mead 1989b
7 MRI Whale Unit data
8 Tomilin and Latyshev 1967
9 Dixon *et al.* 1994
10 Fraser 1945
11 Pitman *et al.* 1999
12 Tietz 1966
13 Hale 1931
14 Gianuca and Castello 1976
15 Moreno 1895
16 Slip *et al.* 1995
17 Dalebout *et al.* 2003
18 Findlay *et al.* 1992
19 Kasamatsu and Joyce 1995
20 Sekiguchi *et al.* 1993
21 Nemoto *et al.* 1980
22 Clarke 2005
23 Baker 1983
24 Ross 1984
25 Benjaminsen 1972
26 Benjaminsen and Christensen 1979
27 Mead 1984
28 Hooker and Baird 1999
29 Winn *et al.* 1970
30 Leaper and Scheidat 1998
31 Fernández *et al.* 2005

Hector's Beaked Whale

Mesoplodon hectori

WHEN WRITING ABOUT this little-known whale, one has to be careful because much of the published information actually concerns a different species, *Mesoplodon perrini* (Perrin's beaked whale), which was first identified as Hector's beaked whale but later shown (genetically and then morphologically) to be a new species.[1] Fortunately, Hector's beaked whale is only known so far from the southern hemisphere, and Perrin's beaked whale from the North Pacific, so some selective culling of the literature can be made based on the geographic origin of the specimen.

Appearance

There appear to be only two published photographs of a living specimen,[2, 3] and descriptions of the colour pattern in two fresh animals.[3] Consequently it is difficult to provide much information about identification at sea.

The largest known specimens are a 4.34 m adult male and a 4.3 m female.[4] There is too little information to speculate about the relative size of the sexes. The beak appears to be slightly longer than in True's beaked whale (tip of snout to centre of eye being 12.9–13% of body length[3]), and the melon does not appear as prominent. The dorsal fin is small, with a distinct hook at the tip.[5] There are two V-shaped throat grooves as in other ziphiids, although a juvenile lacking throat grooves has been described.[6] The flippers have the normal ziphiid conformation, with a long 'forearm' and short 'hand', and in length are 10.7–11.4% of the body length.[3] There is no central notch in the tail flukes of adults.[5]

A subadult male had a grey body with a paler head, and the snout and lower jaws were white as far back as the gape and the throat grooves.[5] A physically immature female was light grey above and white below, with two possible ventral extensions of the darker colour of the back, one at about the level of the umbilicus and one at about the level of the genital aperture. The underside of the tail flukes was the same colour as the belly and there was a black eye patch.[3] A photograph of an animal of unknown sex or maturity (but from its estimated size, 3 m, and lack of visible teeth, likely to be subadult) suggests that the animal was dark above and pale below, with a possible extension of light pigmentation up the side between the flipper and the eye. The melon and beak appeared dark, with a conspicuous black eye patch, but the lower jaw and throat were the same pale colour as the belly.[2] A male with erupted teeth was dark grey, almost black, all over except for the beak, forehead and lower jaw, which were much lighter, and the belly, which was slightly lighter than the back. The body carried a number of white, linear, sometimes paired scars, presumably originating from the teeth of conspecifics, as well as oval scarring from the probable visitations of cookie-cutter sharks *Isistius*. Claims that the species has a sexually dimorphic colour pattern do not seem to have taken sufficient account of possible age-related changes in pigmentation.[3]

A single pair of mandibular teeth erupts in adult males only, situated about 1–2.5 cm from the tip of the lower jaw in all but the oldest males, where they may enlarge as far as the tip.[1] The teeth are roughly triangular in shape, with the leading edge almost vertical and the trailing edge forming an angle of about 45° with the vertical.[7, 8] The leading edge may have a series of 'steps' or flat areas in its outline. A small pointed denticle occurs at or close to the apex of the tooth, forming an angle of 85–90°.[1]

Length:weight relationship

There are no published weights for the species.

Skeletal data

Vertebral formula (n = 7): 7 cervical, 9–10 thoracic, 11–13 lumbar and 17–21 caudal; total 46–50 (but this sample contains both Hector's and Perrin's beaked whales).[5] The formula in a definite hectori was 7 cervical, 10 thoracic, 11 lumbar and 21 caudal = 49.[3]

The first two cervical vertebrae are fused and the remainder are free.[5]

No phalangeal formula is available.

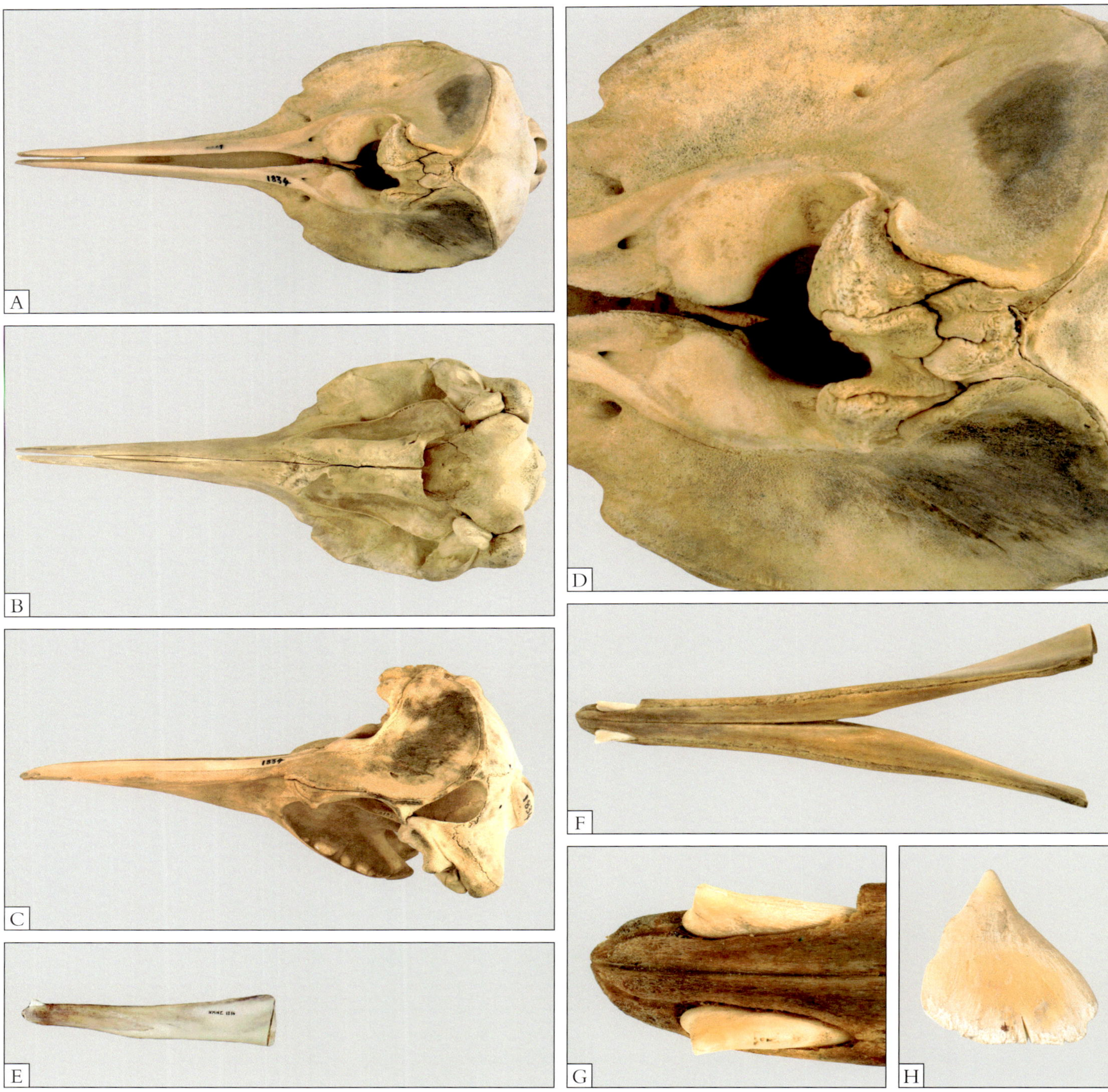

FIG. 68 Dorsal (A), ventral (B) and left lateral (C) views and vertex of skull (D); left lateral (labial) (E) and dorsal (F) views of mandible; dorsal (G) and lateral view (H) of tooth of adult male (NZNM 1834).

The teeth are quite flattened, with a height of 2.2–3.1 cm and a thickness of 0.5–0.8 cm in females and subadults,[3, 7] but larger and fatter in adult males (height about 6 cm and thickness about 1.75 cm).[1]

Distribution and migrations

Hector's beaked whale has only been described from the southern hemisphere, with stranding records between the latitudes of 55° S (Tierra del Fuego) and 32° S (Brazil),[9] suggesting that it may be a species inhabiting temperate rather than tropical regions.

In the subregion it is only known from a single stranding of two juveniles at Lottering River mouth, South Africa, in 1967.[10]

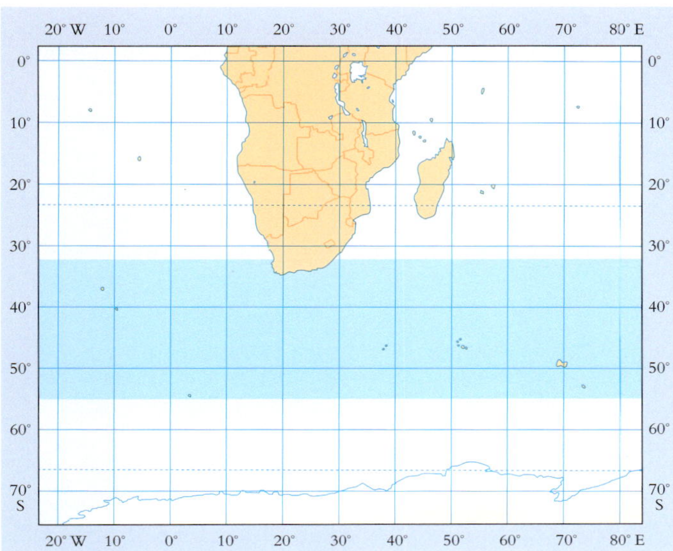

Postulated range

FIG. 69 Distribution map for Hector's beaked whale.

External commensals and parasites

No ectoparasites have yet been described, although one could expect stalked barnacles (*Conchoderma* sp.) to occur on the erupted teeth of adult males.

Food and feeding behaviour

No food items have yet been described for Hector's beaked whale (although one stomach was said to contain only the remains of invertebrates and some 'crystallines', probably from fish[3]).

Reproduction and growth

Very little is known. Calves 1.9 m and 2.02 m long have been reported.[11] From the presence of internal parasites it is assumed a 2.853 m female was already taking solid food.[6]

A 3.73 m male was classified as subadult[5] and a 4.3 m male as an adult[12] (both presumably on the basis of tooth eruption status), while a 3.94 m male had erupted teeth, a 3.84 m female had a juvenile colour pattern[3] and females 4.0 m and 4.15 m long were accompanied by calves.[11]

FIG. 70 This picture from Australia is one of the very few of a living Hector's beaked whale (identified genetically).

Behaviour and vocalisations

A single animal about 3 m long about 50 m from shore dived for periods of up to 4 min., swam around a boat and breached several times.[2]

No information is available on sound production.

Conservation status

Hector's beaked whales have never been reported taken in a fishery, either directly or indirectly. There are no estimates of population size or indications of abundance (although the rarity of strandings might be taken as indicating that the species is uncommon). The only identifiable threat to the species would seem to be from loud sounds associated with seismic exploration or mid-frequency sonar (see the account of Cuvier's beaked whale), but how much of those activities take place in the species' range is unknown.

Hector's beaked whale is listed as Data Deficient by the IUCN and in the South African Red Data Book. It is included on Appendix II of CITES, but not listed by CMS.

References

1 Dalebout *et al.* 2002
2 Gales *et al.* 2002
3 Cappozzo *et al.* 2005
4 Mead 1989c
5 Mead and Baker 1987
6 Baker *et al.* 2001
7 Fraser 1950a
8 Baker 1983
9 Zerbini and Secchi 2001
10 Ross 1970
11 Lichter 1986
12 Mead 1981

True's Beaked Whale

Mesoplodon mirus

Adult ♂

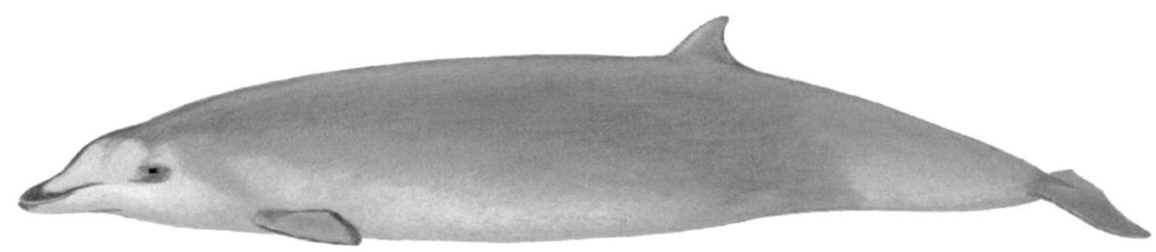

Calf

BECAUSE THE SPECIES had only been recorded from the North Atlantic since the year it was first described (1913), the news of a stranding of True's beaked whale at the tip of southern Africa in 1959[1] was greeted with some surprise, and initially designated as a 'curious exception',[2] until 10 years later, when a pregnant and lactating cow and its calf stranded about 300 km to the east of the 1959 stranding.[3] Subsequently more individuals have stranded in South Africa as well as Australia, and although a few morphological differences have been detected between northern and southern populations,[4] a proper genetic comparison still has to be made. For the time being, therefore, the two populations are regarded as belonging to the same species, although they may eventually be candidates for taxonomic recognition.

Appearance

Adult True's beaked whales in the southern hemisphere have a quite unique marking (still to be described for the northern population), which should enable groups at sea containing at least one adult to be identified. From just in front of the dorsal fin, a pale greyish-white band runs back along the midline before expanding to cover virtually the whole of the last quarter of the animal's body. A head shape resembling that of the bottlenose dolphin (with some animals having a well-rounded melon and distinct short rostrum) has been given as a field characteristic,[5] but this is not so apparent when the animal is encountered as a stranding.

The largest known male was 5.335 m long, but the end points of the measurement are not given. The next largest male was only 4.85 m (n = 16). The largest female was 5.1 m long (n = 12).[6,7] Females seem to be slightly larger than males.[8] The beak is relatively short, the distance from tip of snout to centre of eye being 9.9–13% of body length (mean 11.5%, n = 17).[6] Otherwise the body shape is characteristic of other beaked whales, having a pair of V-shaped throat grooves, small flippers low down on the side of the body, with associated flipper pockets, a small hooked dorsal fin about two-thirds back down the body, and no central notch in the tail flukes.

There is a marked change in colour pattern with age. Neonates are a dark grey above shading to pale grey on the sides and white on the belly. The white extends up onto the side of the head just behind and in front of the eye, with the eye itself in a small black patch. The melon is lighter in colour than the rest of the back, darkening almost to black at the tip of the snout. The white belly coloration also shifts dorsally behind the dorsal fin so that the dark dorsal coloration is confined to a narrow ridge on the caudal peduncle. The upperside of the flukes is dark grey, but the underside is white fading to dark brown along the trailing edge, with dark streaks radiating from where the fluke notch would be. A black patch may extend from the centre of the genital aperture to halfway to the umbilicus. As the animal ages, the belly darkens behind the throat region, becoming covered with dark flecking. A white anogenital patch develops, the tip of the beak whitens, and the rear section of the body from the dorsal fin backwards develops its characteristic white appearance. The underside of the tail flukes darkens owing to a whorl of dark streaks from the centre of the trailing edge. The timing of these changes relative to the onset of sexual maturity is not clear.[9]

A single pair of mandibular teeth erupts at the tip of the lower jaw in adult males only. In females and subadult males they are roughly conical and quite slender, measuring 2.4–3.1 cm in height with opposing diameters of 0.9–1.2 and 0.6–0.8 cm (being somewhat flattened laterally). The tip is slightly curved backwards. In adult males the teeth are enlarged and more robust, reaching 4.8–5.0 cm in height and 1.8–1.9 cm by 1.0–1.2 cm in diameter.[4] An alternative explanation for an apparently anomalous adult male that failed to develop any teeth[7] may be that the teeth were lost in old age.

Distribution and migrations

True's beaked whale seems to have an anti-tropical distribution. In the North Atlantic it is only known as far south as the Canary Islands (*c.* 28° N),[13] while in the southern hemisphere the most northerly record is from south-eastern Brazil (23°46′ S).[7] A stranding from northern KwaZulu-Natal (at 28°27′ S) has been attributed to an unusual oceanographic event.[14]

In the subregion, specimens of True's beaked whale have only been recorded from South Africa[4] and Tristan da Cunha.[6] There are no authenticated sightings.

Four sightings attributed to this species in the North Atlantic were made in water that was 1 100 to more than 4 100 m deep,[5, 15] indicating a deepwater habitat.

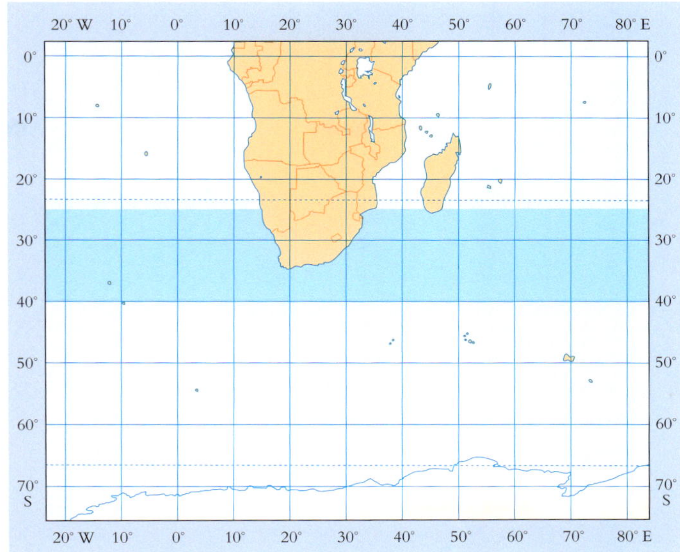

Postulated range

FIG. 71 Distribution map for True's beaked whale.

External commensals and parasites

Stalked barnacles (*Conchoderma* sp.) were found on the teeth of four of the six adult males examined, ranging from two to 18 in number. Three cyamids (*Platycyamus* n. sp.) were found on a subadult female, two on the dorsal fin and one on the tail fluke.[6] Three individuals carried stalked barnacles (*Xenobalanus* sp.) on the trailing edge of the tail fluke, ranging from 1 to 56 in number,[4, 6] and the parasitic copepod *Pennella* was recorded from an animal in Brazil.[7] White oval scars probably representing the attacks of cookie-cutter sharks *Isistius* sp. may also be found scattered over the flanks.[7]

Length:weight relationship

Combining four published weights[10] with five unpublished,[6] **$W = 13.258L^{2.7968}$**, where W = weight in kilograms and L = total length in metres. This implies that a 2.5 m whale would weigh 172 kg and a 4.5 m whale 890 kg. As at least seven of these animals were weighed in pieces, some correction should be made for weight loss during dissection. One animal cut up weighed 93.3% of its weight when entire, so this correction should be applied to the estimates derived from the equation above.

Skeletal data

Vertebral formula (n = 9): 7 cervical, 9–10 thoracic, 10–12 lumbar and 20–22 caudal; total 47–49.[6, 7]

The first three cervical vertebrae are fused and the remainder are free (although the fourth and fifth may be partly fused).[11, 12]

Phalangeal formula (n = 1), including metacarpals: I_2, II_6, III_6, IV_5, V_3.[6] This is somewhat more than given for two northern *M. mirus*, even allowing for the non-inclusion of metacarpals: I_{1-2}, II_{4-5}, III_{3-4}, IV_3, V_2.[11, 12] It is possible that some of the terminal bones were lost in preparation, the southern *M. mirus* count being determined radiographically.

Food and feeding behaviour

Four stomachs have been examined from South Africa. The remains of three unidentified fish and three squid (*Teuthowenia* sp.) were found in three animals,[16] and the remains of four squid (*Loligo reynaudi*) in the other.[4] This is the only information on feeding behaviour available for True's beaked whales.

Reproduction and growth

The largest recorded foetus was 2.185 m long and the smallest calf was 2.325 m. The length at birth is thus estimated at about 2.2 m.[4] A 2.4 m male calf had a partly healed umbilicus.[6] If the four known foetuses (three from South Africa and one from Tristan da Cunha) are plotted with the two calves from South Africa against date of occurrence, the resultant regression indicates a foetal/neonatal growth rate of 0.512 (±0.024 *SE*) cm/day, with conceptions occurring in summer. If the length at birth is correct, this implies a gestation period of about 430 days. A female simultaneously pregnant and lactating has been recorded.[4] Applying the foetal/neonatal growth rate above to the difference between the size of the foetus (10.5 cm) and the accompanying calf (3.4 m) of this animal, would indicate a conception interval closer to two years than one, with lactation therefore lasting at least 300 days.

The smallest male with erupted teeth was 4.47 m long and the largest with unerupted teeth was 4.62 m long (n = 11),[6] which (if eruption coincides with puberty) would indicate that sexual maturity occurs at about 4.5 m. The smallest mature female was 4.55 m long and the largest immature 3.67 m (n = 9),[6] which is not inconsistent with the size of males at sexual maturity.

Age determination in this species has not yet been successful.[4]

Behaviour and vocalisations

Four sightings attributed to this species in the North Atlantic involved groups of 1–3 individuals.[5, 15]

During a typical surfacing, the rostrum is lifted clear of the water first. The blow is described as short and columnar but

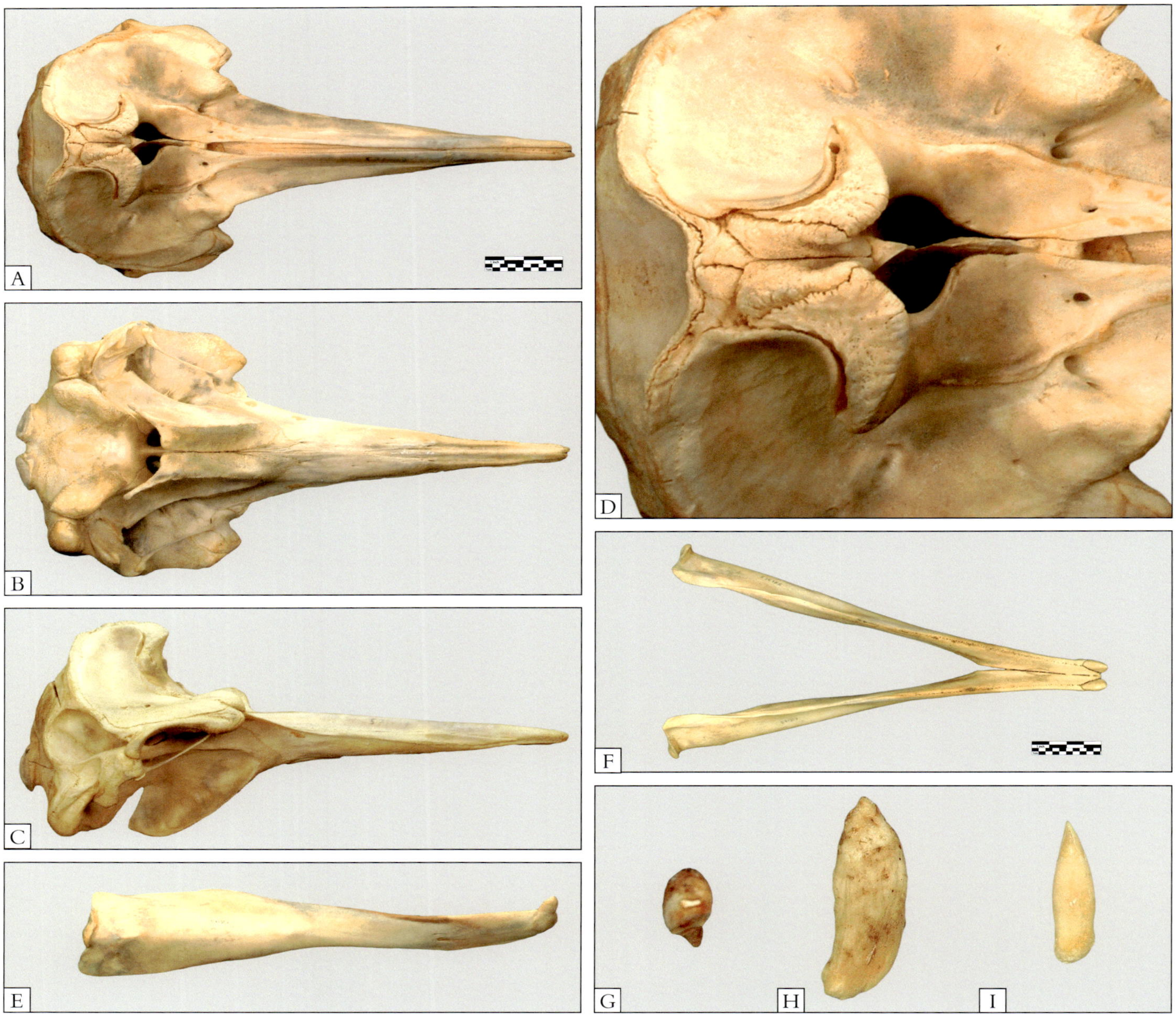

FIG. 72 Dorsal (A), ventral (B) and right lateral (C) views and vertex of skull (D); right lateral (labial) (E) and dorsal (F) views of mandible (USNM 571357 – adult male); dorsal (G) and lateral (H) views of teeth of adult male (ZM 39837) and lateral view of tooth of adult female (I) (ZM 38219).

relatively indistinct, being no taller than the length of the head.[15] In three of the sightings, animals breached repeatedly, up to 24 times at intervals of 20–60 sec., emerging from the water to well past their dorsal fin at angles of 60–80°, before falling back on their side or belly.[5] This behaviour, however, may have been the reason why the animals were detected, and so may not be typical.

No information is available on sound production.

Conservation status

True's beaked whales have never formed the target of a directed fishery, although a few have been recorded as fishery by-catches in the North Atlantic. There are no estimates of population size or indications of abundance, although the species is considered to have a more limited distribution (and therefore may have a smaller population size) than any of the other *Mesoplodon* species in the North Atlantic.[5] The only identifiable threat to the species would seem to be from loud sounds associated with seismic exploration or mid-frequency sonar (see the account of Cuvier's beaked whale). True's beaked whale is listed as Data Deficient by the IUCN and in the South African Red Data Book. It is included on Appendix II of CITES, but is not listed by CMS.

References

1 McCann and Talbot 1963
2 Moore 1966
3 Ross 1969
4 Ross 1984
5 Weir *et al.* 2004
6 MRI Whale Unit data
7 Souza *et al.* 2005
8 MacLeod 2006
9 Ross *et al.* 1988
10 Mead 1989c
11 Raven 1937
12 Ulmer 1941
13 MacLeod 2000
14 Ross *et al.* 1985
15 Tove 1995
16 Sekiguchi 1994

Gervais' Beaked Whale

Mesoplodon europaeus

GERVAIS' BEAKED WHALE is included here on the basis of three specimens from Ascension Island in the tropical Atlantic Ocean (*c.* 8° S),[1] and a stranding in northern Namibia (*c.* 17°20′ S) that has been provisionally identified as this species,[2] but strandings in Brazil (at 2°54′ S and 23°58′ S)[3, 4] suggest that it may be more widespread in the South Atlantic.

Appearance

As there have been very few at-sea sightings of Gervais' beaked whale, and the colour pattern is not well documented, it is difficult to give many clues to identification. Compared to True's beaked whale, it is said to have a more dorsally flattened head, with only a slightly rounded melon sloping almost diagonally in profile to the tip of the rostrum.[5]

The length of the largest known female is often given as 5.2 m, but this seems to refer to a True's beaked whale. The next largest female is 4.85 m long.[1, 6, 7] The largest male is given alternatively as 4.56 m[6] or 4.8 m,[8] but the latter seems an error. This is one of the few beaked whales for which sexual dimorphism in size seems well established.[7] As mentioned above, the melon is not very prominent in Gervais' beaked whale, giving the profile of the head a somewhat flattened appearance and resulting in a lack of definition between the melon and the beak. The jaw line is more or less straight, not strongly arched. The flippers are relatively small (7–8% of body length, n = 2) compared to those of True's beaked whale (7–11%, n = 5),[9] but larger samples need to be compared. In other respects the body shape is characteristic of beaked whales, having a pair of V-shaped throat grooves, flippers low down on the side of the body, with associated flipper pockets, a small hooked dorsal fin about two-thirds back down the body, and no central notch in the tail flukes.[8]

Coloration is said to be dark grey-black above and on the sides, becoming lighter on the belly with a pink hue in some specimens. The flippers and flukes are dark grey above and below. The belly is white in juveniles, but much darker in adults, where an irregular white blaze may develop in the region of the mammary glands and genital slit (at least in adult females), the throat region may be somewhat lighter than the rest of the underside, and a dark patch may surround the eye.[1, 8] Linear and oval scarring may develop in adults, particularly males.[8]

A single pair of mandibular teeth erupts in adult males only: they are situated opposite and near the back of the symphysis joining the two mandibles, which means they erupt slightly in front of the middle of the gape. The erupted tooth in the adult male is roughly triangular in shape, with the tip inclined forwards. In the entire tooth the root section is more or less straight, but oriented in such a way that the leading edge of the tooth is much straighter than the trailing edge. The teeth are greatly flattened laterally, and the roots tend to curve inwards at the base. The maximum dimensions of adult male teeth are a height of about 8.0–9.6 cm, and opposing diameters of 5.1–5.4 cm and 1.2–1.5 cm.[10] Female and subadult teeth are much smaller, roughly triangular in shape and flattened, with a maximum height of 4.4 cm and a diameter of 3.4 cm from front to back.[11]

Length:weight relationship

Using five published weights (and excluding one 'outlier'),[1] **$W = 13.349L^{2.6684}$**, where W = weight in kilograms and L = total length in metres. This implies that a 2.5 m whale would weigh 154 kg and a 4.5 m whale 739 kg. It is not known how many of these animals were weighed in pieces, so some correction should probably be made for weight loss during dissection.

Skeletal data

Vertebral formula (n = 3): 7 cervical, 9–10 thoracic, 11–12 lumbar and 19–20 caudal; total 47.[12, 13]

The first three or four cervical vertebrae are fused and the remainder are usually free, but the fifth and sixth are sometimes joined.[10, 12, 13]

Phalangeal formula (n = 1), including metacarpals: I_2, II_6, III_6, IV_5, V_4.[12]

FIG. 73 A Gervais' beaked whale photographed in the North Atlantic.

Distribution and migrations

The distribution of Gervais' beaked whale is confined to the Atlantic Ocean. In the North Atlantic the species (along with Blainville's beaked whale) seems to occur in more southerly regions than other beaked whales, and its distribution is probably trans-equatorial.[14] It may frequent all the warm and temperate-warm waters of the North Atlantic.[11]

In the subregion the species is only known from three strandings (two males and a pregnant female) on Ascension Island in 1980,[1] and an adult male that stranded at the Kunene River mouth in Namibia in 1997.[2]

There is no information on seasonal movements.

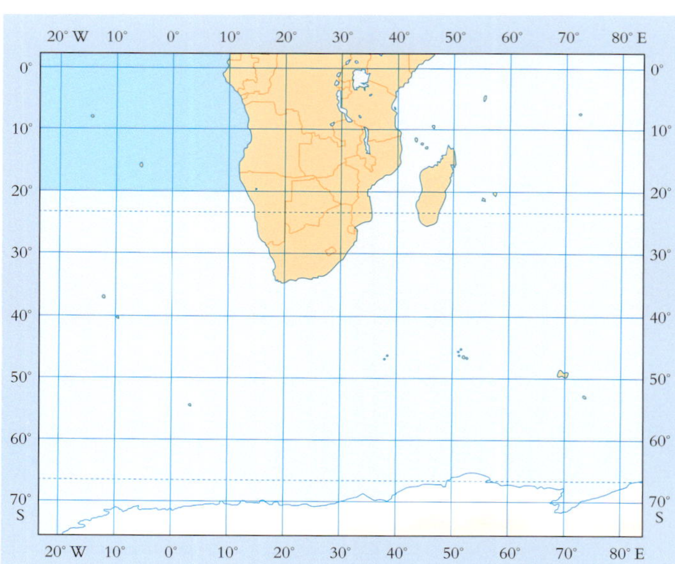

Postulated range

FIG. 74 Distribution map for Gervais' beaked whale.

External commensals and parasites

Stalked barnacles (*Conchoderma auritum*) occur on the erupted teeth of adult males,[1, 13, 15] and numerous cyamids (*Isocyamus delphinii*) have been found in a healing wound.[1] Oval wounds or white scars caused by the cookie-cutter shark *Isistius* may also be present.[8]

Food and feeding behaviour

No data are available for the subregion. In the North Atlantic, one specimen contained 10 prey species, including the remains of squid (*Octopoteuthis* sp., *Mastigoteuthis* sp., *Taonius pavo* and possibly *Leachia* sp.), fish (viper fish *Chauliodus sloani*, and snake mackerel *Nesiarchus nasutus*) and crustaceans (the mysid *Gnathophausia*). The mass composition of the remains was 42% cephalopod, 12% fish and 43% crustacean (although the importance of the crustaceans may have been exaggerated by their relative indigestibility). The average size of prey items was about 15 cm.[13] Only squid beaks were recorded from another three specimens,[1] while only *Gnathophausia* were reported from a fourth.[15]

Most of the identified prey items are primarily mesopelagic in distribution, being found at depths ranging from at least 200 m (*Octopoteuthis* and *Mastigoteuthis*) down to a maximum of 4 000 m (*Gnathophausia*).[13] This suggests that Gervais' beaked whale normally forages at considerable depth.

Reproduction and growth

With the longest reported foetus being 2.18 m and the smallest calf 1.96 m, the length at birth has been estimated at 2.1 m. The mean length at sexual maturity in females has been estimated at 4.5 m, although there are three recorded females 4.3 m long that were either pregnant or accompanied by a calf, and this would seem a more appropriate length.[6] Details of the reproductive cycle and its timing are unknown.

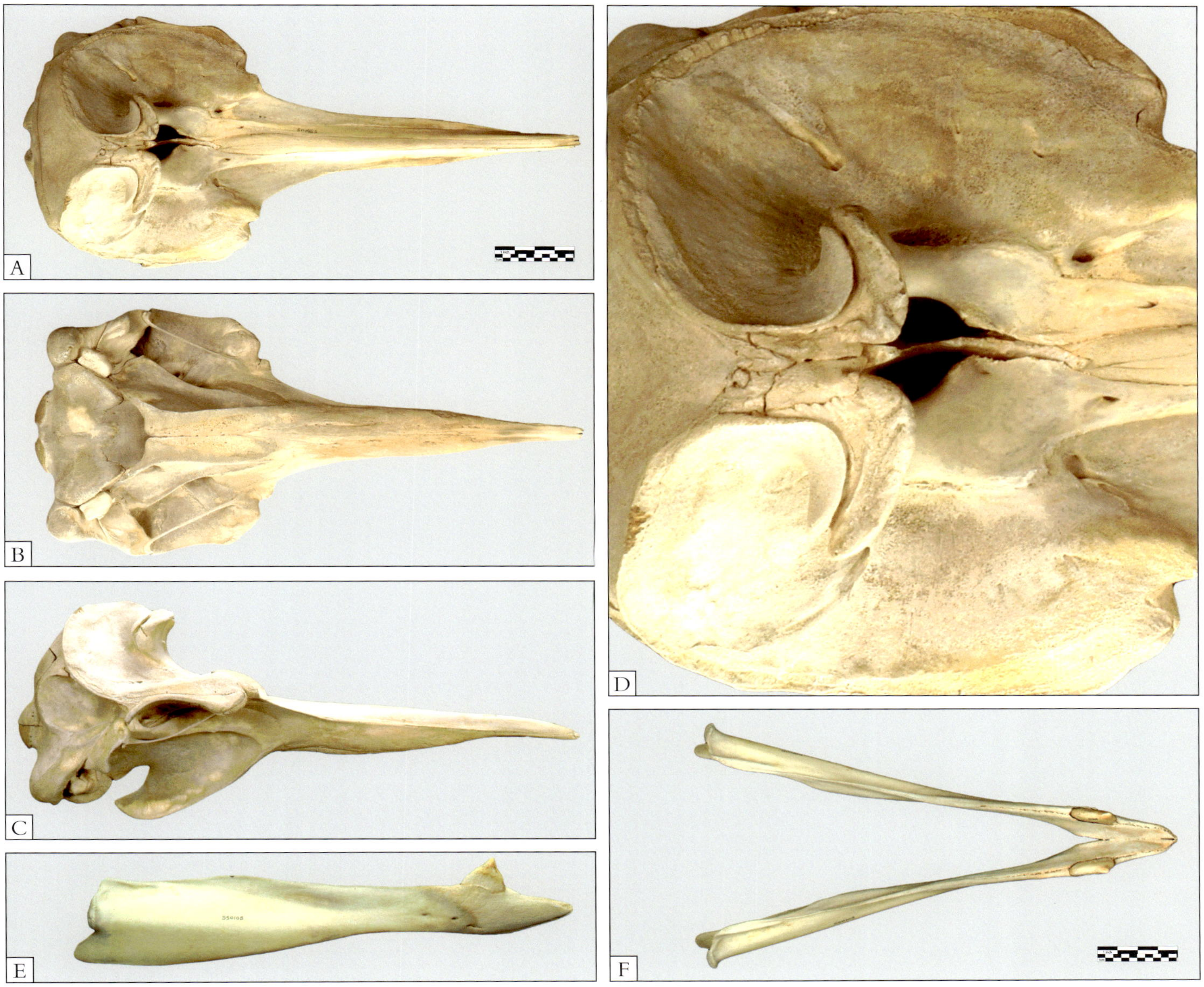

FIG. 75 Dorsal (A), ventral (B), right lateral (C) and vertex (D) views of skull; right lateral (labial) (E) and dorsal (F) views of mandible (USNM 550105 – adult male).

An individual with at least 48 growth layers in the cementum of its teeth has been reported, but the rate of growth layer formation is unknown.[16]

Behaviour and vocalisations

Gervais' beaked whale is said to thrust its beak prominently above the water when surfacing.[8] Otherwise the general lack of sightings in its presumed distributional range has led to speculation that the species may be particularly cryptic in behaviour.

A stranded animal was said to produce frequent vocalisations at a high amplitude: slow and fast clicks and a narrow-band sound were recorded.[8]

Conservation status

Gervais' beaked whales have never formed the target of a directed fishery, although it may feature as a by-catch in the North Atlantic. There are no estimates of population size or indications of abundance. The only identifiable threat to the species would seem to be from loud sounds associated with seismic exploration or mid-frequency sonar, and it has been involved in mass strandings associated with military activities (see the account of Cuvier's beaked whale).

Gervais' beaked whale is listed as Data Deficient by the IUCN, but not included in the South African Red Data Book. It is included on Appendix II of CITES, but not listed by CMS.

References

1 Mead 1989c
2 Griffin and Coetzee 2005
3 Santos *et al.* 2003
4 Martins *et al.* 2004
5 Weir *et al.* 2004
6 Mead 1984
7 MacLeod 2006
8 Norman and Mead 2001
9 Moore and Wood 1957
10 Moore 1960
11 Robineau and Vely 1993
12 Raven 1937
13 Debrot and Barros 1992
14 MacLeod 2000
15 Rosario-Delestre *et al.* 1999
16 International Whaling Commission 1980

Gray's Beaked Whale

Mesoplodon grayi

Adult ♂

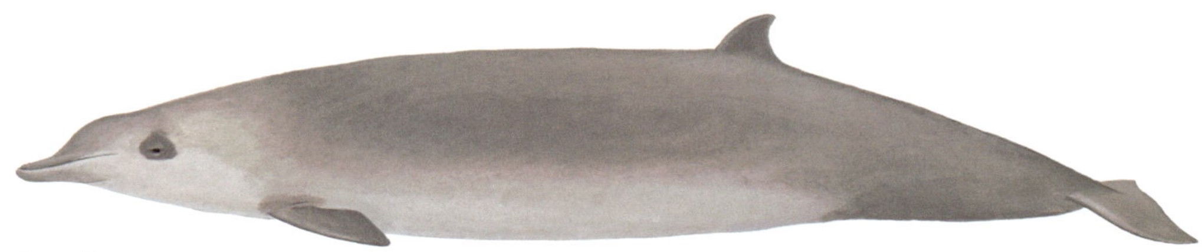

Juvenile

Appearance

Adult Gray's beaked whales can be distinguished at sea by their long white beak and poorly defined melon, giving the head an extremely streamlined appearance. When surfacing, the beak is thrust well clear of the water.

Because mandibular teeth sometimes erupt in adult female Gray's beaked whales[1] (which does not happen in any other *Mesoplodon* species), and because some specimens were sexed simply on the basis of whether or not they had erupted teeth, it is not easy to establish the relative sizes of the sexes in this species. For example, the largest reported specimen of either sex, 5.64 m, was not sexed in the original publication[2] but has been assumed to be a male because it was said that there were two teeth projecting 'very slightly' from the side of the mouth. However, inspection of photographs of the animal suggests it may well have been a female.[1] The next largest recorded individual was a female 5.33 m long,[3] while the largest male seems to have been 4.84 m long (n = 10).[1,4] It seems likely, therefore, that females grow bigger than males.

The beak is long and slender, the distance from tip of snout to eye ranging from 12.4–15.6% of body length (n = 8), and this impression is accentuated by the low melon, which merges almost imperceptibly into the beak. The animal's head almost seems 'upside-down' in profile. The jaw line is straight, not arched, and there are the usual V-shaped throat grooves. The flippers are relatively large for a beaked whale (length 10.2–12.2% of body length, n = 5), and are situated low down on the side of the body, with associated flipper pockets. The dorsal fin is small (height 4.3–5.6% of body length, n = 5), hooked, and situated about two-thirds of the way down the body.[4]

As in some other beaked whales, the colour pattern of Gray's beaked whales changes with age. Calves are a dark grey above, fading to white below, with pale sides to the melon, a dark eye patch, and a dark beak and tip to the lower jaw. The outer surface of the flipper is light grey with dark grey margins, while the underside of the flukes is off-white margined with dark grey. In adults of both sexes the belly darkens to resemble the colour of the back, although white blazes may develop in the anogenital region. The most prominent feature of adults, however, is the striking white front to the head, extending from the beak as far back as the blowhole dorsally, and as the beginning of the throat grooves and angle of the gape ventrally. White scarring may also appear in adults, including small irregular scars of unknown origin, thin curvilinear scars (especially in males), which probably represent tooth rakes by conspecifics, and oval scars, particularly in the anogenital area, which are probably the result of the attentions of the cookie-cutter shark *Isistius*. Adoption of the adult colour pattern may coincide with the attainment of sexual maturity.[1]

Length:weight relationship

Three individuals from the subregion have been weighed, all in pieces. A 4.74 m male weighed 1 075 kg,[7] a 4.84 m male 1 156.8 kg and a 4.94 m female 997.3 kg.[4]

Skeletal data

Vertebral formula (n = 7): 7 cervical, 9–11 thoracic, 10–12 lumbar and 19–20 caudal; total 47–49.[5,8]

The first two cervical vertebrae are fused; the remainder are free.[8]

Phalangeal formula (n = 2), including metacarpals: I_1, II_{5-6}, III_6, IV_5, V_3.[4]

A single pair of mandibular teeth, situated at about the centre of the gape, erupts in adults. Unlike most other beaked whales, it seems that these teeth may erupt in adults of both sexes, although they may be more prominent in males.[1] In adults these teeth are greatly flattened laterally but onion-shaped in outline, with a symmetrically-placed central point. Most of the tooth is buried in the gum. Their height ranges from 7.0–8.4 cm, depending to some extent on wear.[5] In juveniles and subadults the teeth are more triangular in shape and greatly flattened, reaching heights of 3.4–4.3 cm and 'lengths' (front to back) of 3.4–4.9 cm. Gray's beaked whales are also unusual among members of their genus in commonly having a series of smaller erupted teeth in their upper jaw: all seven fresh specimens examined in the subregion had such teeth, ranging from four to 19 on each side (others were found buried in the gum in some of these individuals). These teeth are dolphin-like in shape, the largest reaching about 1.2 cm in height and 0.3 cm in diameter. Some individuals may also possess such teeth, but smaller, in the lower jaw behind the main mandibular tooth.[6] The stage at which these vestigial teeth erupt is not known, but seems to precede that of the mandibular teeth (and therefore probably takes place before puberty).[4]

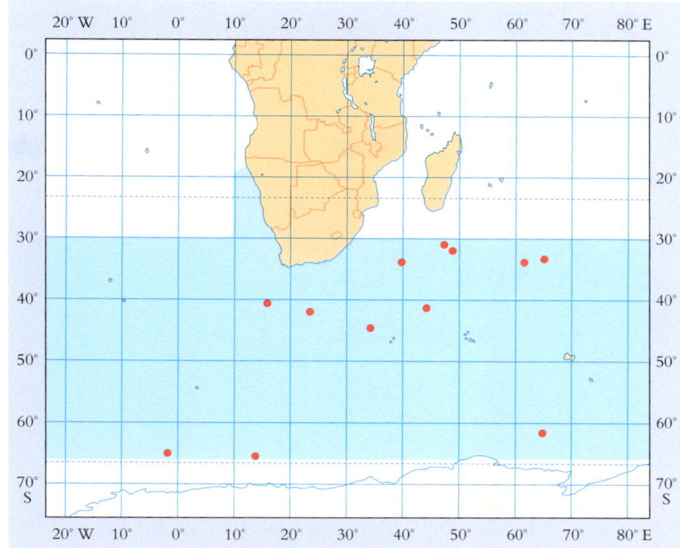

FIG. 76 Distribution map for Gray's beaked whale.

Distribution and migrations

Apart from one apparent 'stray' in the North Atlantic,[9] all Gray's beaked whale records have come from the southern hemisphere. It is believed to have a circumpolar distribution over a wide latitudinal range, having been contacted at sea between the latitudes of 31° S and 65° S.[10, 11]

In the subregion the species has been recorded from Namibia and South Africa,[12] and sightings have been made over the same latitudinal range as given above. Gray's beaked whales are said to be one of the most frequently stranded beaked whales in Namibia,[13] and a stranding there at 21°35′ S seems to represent the most northerly record in the subregion,[12] although it has been found at 13°49′ S in Peru.[14] This suggests that the species' distribution may extend further north under the influence of cold current systems such as the Benguela.

All the sightings in the New Zealand region (apart from the brief visit of a cow-calf pair to a harbour) were in water deeper than 1 000 m,[11] and this also seems to be the case for the sightings made in the subregion. There is no information on seasonal movements, although in both the South African and New

FIG. 77 The long white beak of the adult Gray's beaked whale (*left*) is stained yellowish with skin diatoms, while that of its calf (*right*) is much shorter and dark in colour.

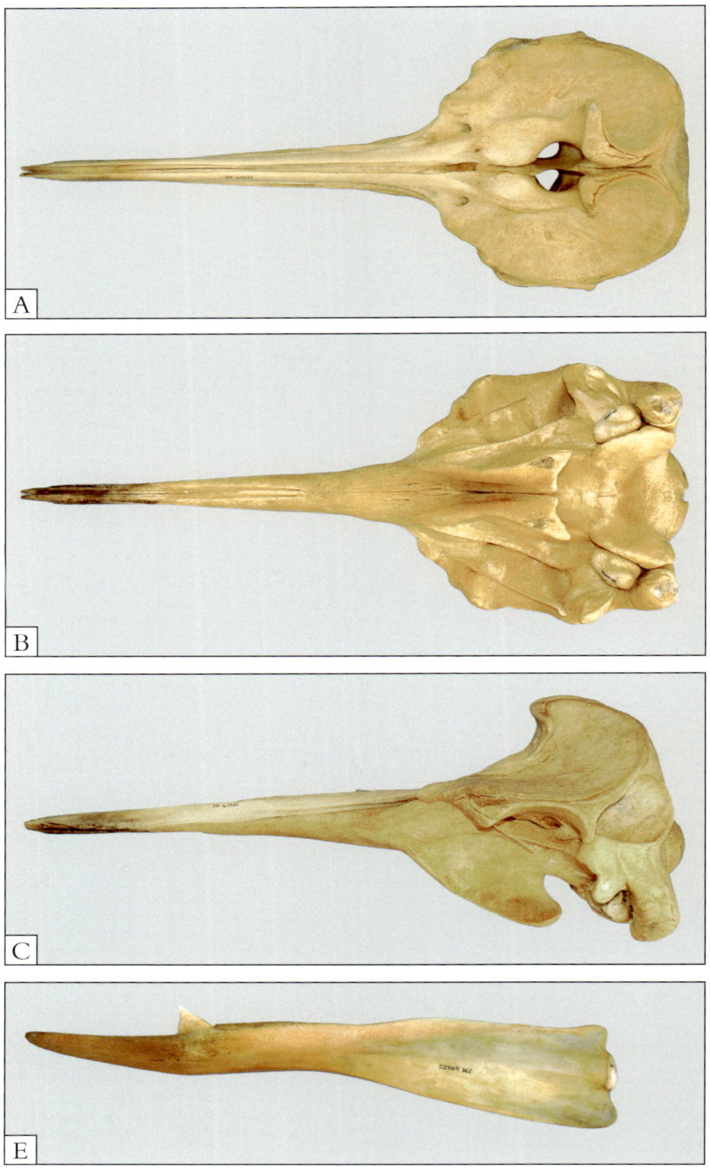

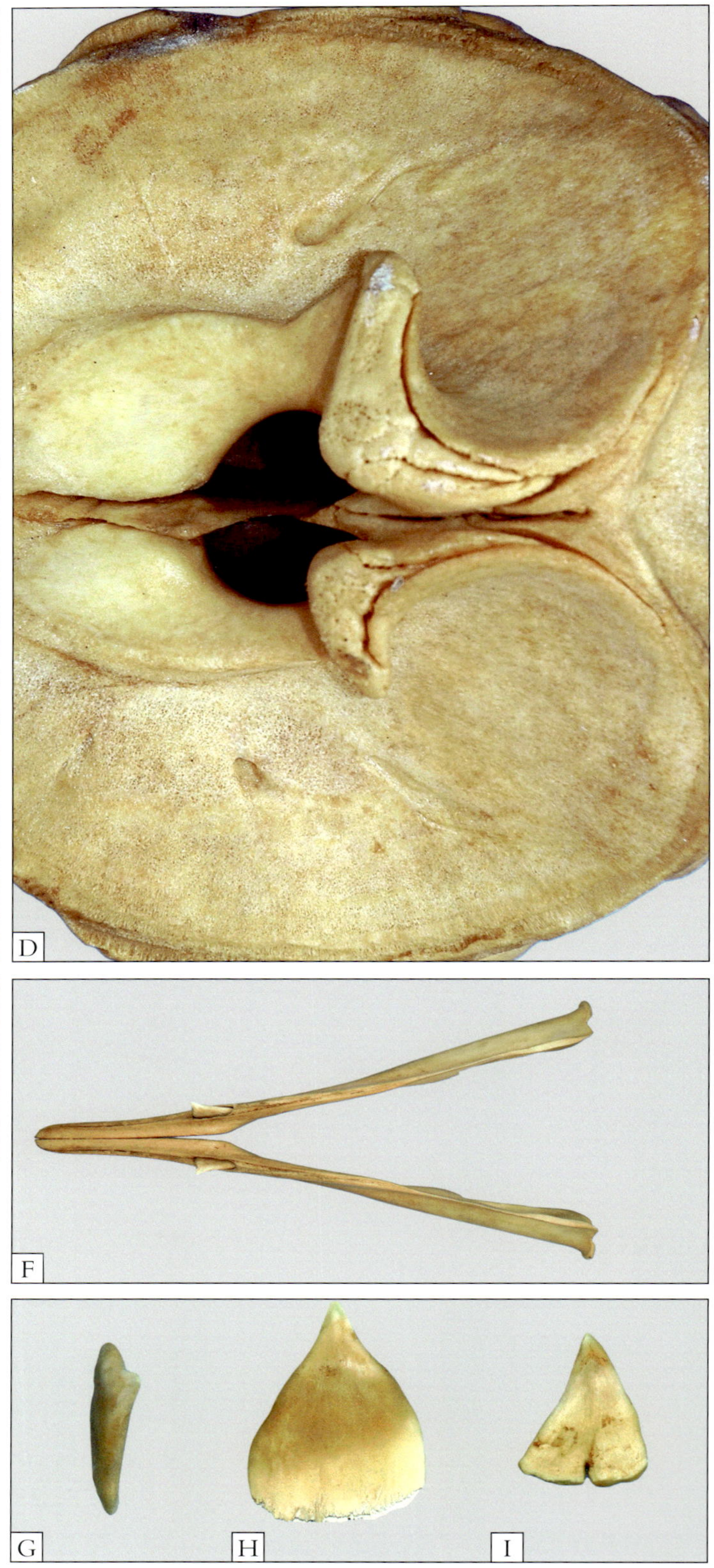

FIG. 78 Dorsal (A), ventral (B) and left lateral (C) views and vertex of skull (D); left lateral (labial) (E) and dorsal (F) views of mandible (ZM 40622 – adult male); dorsal (G) and lateral (H) views of teeth of adult male (ZM 40622) and lateral view of tooth of adult female (I) (ZM 36846).

Zealand regions most strandings seem to occur in summer (December–February or April respectively), from which it has been suggested that there might be some inshore movement at that time.[5, 11]

External commensals and parasites

Stalked barnacles (*Conchoderma* sp.) occurred on the erupted teeth of two of the three adult males examined, and a single *Xenobalanus* on the trailing edge of the tail flukes of an adult female. Oval wounds or white scars caused by the cookie-cutter shark *Isistius* may also be present.[4]

Food and feeding behaviour

The stomachs of only two Gray's beaked whales in the subregion have been examined. Otoliths from one hake (*Merluccius* sp.) and from at least three other unidentified fish were recovered from one, while in the oesophagus of a second the remains of two mesopelagic fish (one *Photichthys argenteus* and one *Lampadena* sp.) were found.[15] Elsewhere unidentified fish remains have also been recorded,[16] suggesting that this species may be largely piscivorous.

Reproduction and growth

A foetus 2.13 m long has been reported[1] and the smallest calf was 2.42 m,[3] which presumably covers the likely size range of the animal at birth. A live calf estimated at 2.5 m had no foetal crease marks and carried scars that were probably caused by the cookie-cutter shark *Isistius*, indicating that it was not a recent birth.[11] There is no information on the reproductive cycle or seasonality of breeding. From testis sizes, a male 3.55 m long was sexually immature and one at 4.31 m was mature,[4] while a 3.989 m male had unerupted teeth.[6] Sexual maturity in the male may therefore occur at 4.0–4.3 m. A female 4.72 m long had only one corpus luteum in its ovaries.[1]

There is no information on age and growth.

Behaviour and vocalisations

Sightings attributed to Gray's beaked whale have ranged in size from one to eight individuals (n = 31), with most groups (90.3%) being five or less. The overall proportion of calves in schools where it was possible to make an assessment was 19.5% (n = 20).[11] A mass stranding of about 28 animals apparently occurred on the Chatham Islands south of New Zealand in 1875, although only three individuals were positively identified.[17]

The surfacing behaviour of a cow-calf pair that frequented a harbour for five days consisted of 3–5 blows in succession followed by a 10–15 min. submergence. The head and beak were often brought clear of the surface when breathing, and the blow was described as low and diffuse. At other times the pair spent some time logging at or just below the surface. When submerging, the back was arched high but the flukes were never lifted clear of the water. On the last day their behaviour was less stereotyped, and flipper slapping, lobtailing, spy-hopping (calf) and frequent breaching by both animals were seen.[11]

Sounds produced by this species have not been documented.

Conservation status

Gray's beaked whales have never formed the target of a directed fishery, and their involvement as a by-catch in any southern fishery is unknown. There are no estimates of population size or indications of abundance.

The only identifiable threat to the species would seem to be from loud sounds associated with seismic exploration or mid-frequency sonar (see the account of Cuvier's beaked whale), but how frequently these activities occur in its range is unclear.

Gray's beaked whale is listed as Data Deficient by the IUCN and in the South African Red Data Book. It is included on Appendix II of CITES, but not listed by CMS.

References

1 Ross *et al.* 1988
2 Courtenay-Latimer 1963
3 Mead 1984
4 MRI Whale Unit data
5 Ross 1984
6 Robson 1975
7 Mead 1989c
8 Oliver 1922
9 Boschma 1951
10 Gambell *et al.* 1975
11 Dalebout *et al.* 2004b
12 Findlay *et al.* 1992
13 Griffin 1998
14 Reyes 1990
15 Sekiguchi 1994
16 Goodall and Galeazzi 1985
17 Von Haast 1876

Andrews' Beaked Whale

Mesoplodon bowdoini

Adult ♂

PREVIOUSLY KNOWN ONLY from the Australasian region, this species is included here on the basis of two juvenile specimens that were found stranded on Tristan da Cunha in the South Atlantic in October 1988.[1]

Appearance

As there are no recorded sightings attributed to this species, it is difficult to provide field characters for identification. Adults may possibly be recognised by the white front of the head and the grey blaze behind the blowhole (and distinguished from adult Layard's beaked whales by the latter's longer beak and prominent black melon).

Only 16 Andrews' beaked whales have been accurately measured.[1,2] Females grow to 4.87 m (estimated) or 4.36 m (measured) in length, and males to 4.40 m (measured), so it appears to be one of the smaller beaked whales. Adult Andrews' beaked whales are described as having a very low melon and a short thick beak, the distinction between beak and melon being poor. The jaw line is strongly arched upwards towards the back of the gape, and in males the teeth erupt at the apex of this arch, so that their tips may project above the level of the beak. There is the usual ziphiid pair of V-shaped throat grooves. In contrast to the short head, the body is described as quite robust. The dorsal fin is small, triangular and blunt-tipped in adults, but with a long curved leading edge and a short steep trailing edge in juveniles. The flippers have rounded tips.[1]

The coloration of juveniles and adults differs. Juveniles resemble the young of other beaked whales in being dark greyish-brown above and white below, including a white lower jaw and white extensions up onto the side of the head in front of and behind the eye. Adult males, on the other hand, are dark bluish-black all over apart from the front half of the snout and lower jaws, which are white or pale grey, and a white patch on either side of the melon at the base of the beak. A grey blaze also extends back from the blowhole for about half the length of the back. Females are reported to have less white on the beak, a slate grey back, and pale greyish-white sides and belly (but whether this refers to adults or subadults is unknown). Adults often carry scarring, usually single linear white scars from tooth rakes, as well as white oval scars.[1]

A single pair of large, flattened mandibular teeth, situated towards the rear of the gape, erupts in adult males only. The exposed part of the tooth tends to be convex on the rear edge and slightly concave on the front edge, so that the apical denticle seems closer to the front of the tooth than the back, and it points forwards, upwards and outwards. When extracted from the jaw, the maximum dimensions of a male tooth are 13.8 cm high, 7.7 cm in width from front to back, and 1.7 cm from side to side. An imaginary straight line drawn from the tip of the denticle to the base of the tooth at the front will barely touch the tooth at any point in between (another way of saying that it is quite asymmetrical in profile). The female teeth and those from subadult males are quite different in shape, still flattened but roughly triangular, with a longer leading edge than trailing edge, so that the apical denticle tends to point backwards (rather than forwards as in males). They reach about 4.7–5.5 cm in height, and increase in width with age, to a maximum of 6.4 cm (front to back) and 0.9 cm (side to side).[1,2]

Distribution and migrations

All records of Andrews' beaked whale are confined to the southern hemisphere, and all known records before 1988 were from the Australasian region (a specimen from the Kerguelen Islands, originally described as this species, was subsequently re-identified as Layard's beaked whale[4] and then as the fossil

> **Length:weight relationship**
>
> No weights are available for this species.
>
> **Skeletal data**
>
> Vertebral formula (n = 7): 7 cervical, 10–11 thoracic, 9–11 lumbar and 19–20 caudal; total 46–48.[1,2]
>
> The first three cervical vertebrae are fused and the remainder are free.[3]
>
> Phalangeal formula (n = 1), including metacarpals: I_1, II_5, III_4, IV_4, V_3.[1]

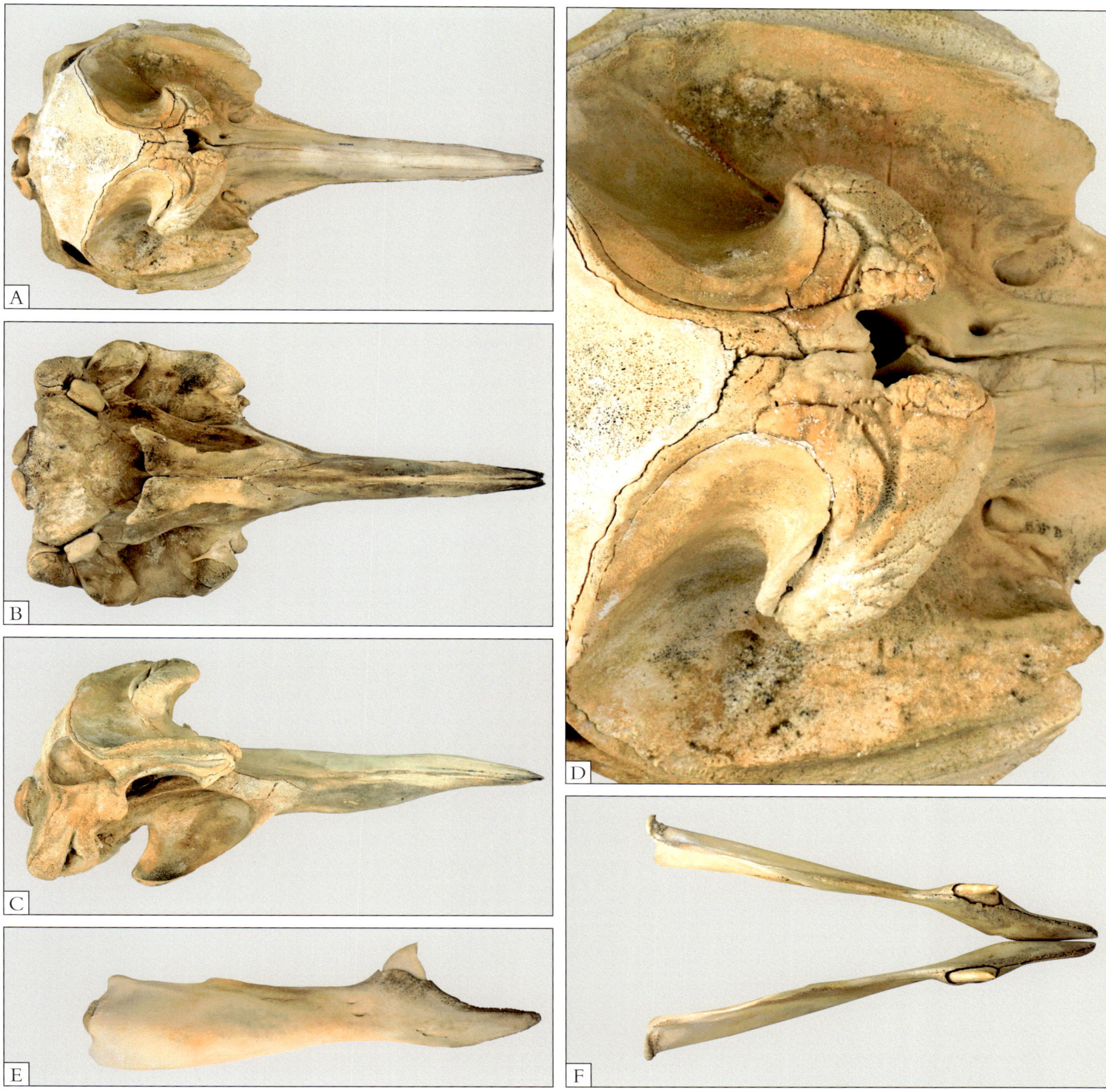

FIG. 79 Dorsal (A), ventral (B) and right lateral (C) views and vertex (D) of skull; right lateral (labial) (E) and dorsal (F) views of mandible; dorsal (G) and lateral (H) views of tooth of adult male (NZNM 2133).

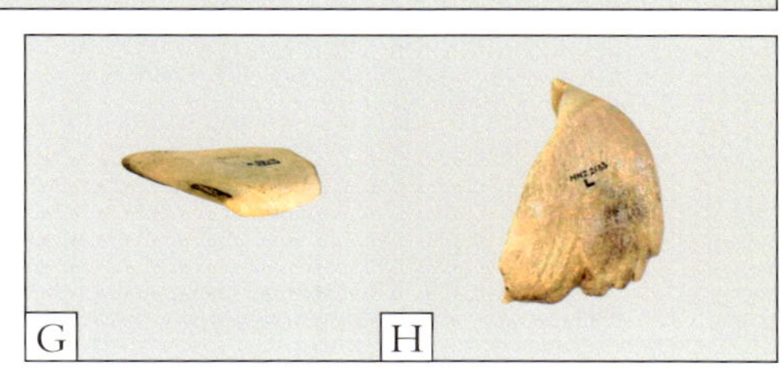

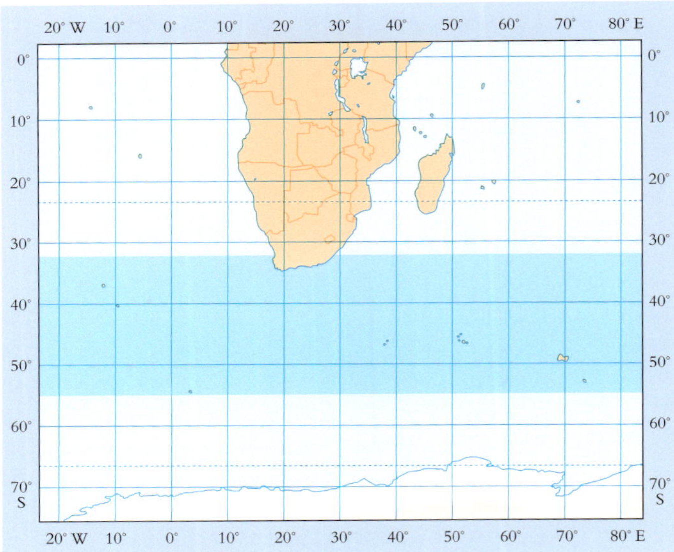

Postulated range

FIG. 80 Distribution map for Andrews' beaked whale.

Mesoplodon longirostris[5]). In 1988 the remains of two juveniles were found on Tristan da Cunha in the South Atlantic, and two skulls were located on the Falkland Islands. More recently, specimens have been recorded from Tierra del Fuego and Uruguay,[2] so the species is now considered to have a circumpolar distribution between the latitudes of 32° S and 54°30′ S.[1] The Tristan da Cunha specimens are the only records from the subregion.

There is insufficient information from which to investigate seasonality of movement.

External commensals and parasites

Stalked barnacles (probably *Conchoderma* sp.) are figured on the erupted teeth of an adult male. Oval wounds or white scars caused by the cookie-cutter shark *Isistius* are present, especially around the posterior ventral region of adults.[1]

Food and feeding behaviour

There appears to be no information available on the feeding habits of this species.

Reproduction and growth

From a perinatal 2.3 m calf with foetal creases still evident, the length at birth is estimated at about 2.2 m, and the season of birth as either summer or autumn.[1] Sizes at first reproduction, and details of the reproductive cycle, are unknown.

There is no information on age and growth.

Behaviour and vocalisations

Given the lack of recorded observations of Andrews' beaked whale, there is no information available on school size or behaviour, and no recordings of vocalisations attributed to this species have been made.

Conservation status

Andrews' beaked whales have never formed the target of a directed fishery, and their involvement as a by-catch in any southern fishery is unknown. There are no estimates of population size but the fact that only 34 specimens of this species were recorded between its discovery in 1904 and 1999 (all but four of which were in the Australasian region)[1] suggests that it may be only locally abundant. The only identifiable threat to the species would seem to be from loud sounds associated with seismic exploration or mid-frequency sonar (see the account of Cuvier's beaked whale), but how frequently these activities occur in its range is unclear.

Andrews' beaked whale is listed as Data Deficient by the IUCN, but not included in the South African Red Data Book. It is included on Appendix II of CITES, but not listed by CMS.

References

1 Baker 2001
2 Laporta *et al.* 2005
3 Andrews 1908
4 Robineau 1989
5 Mead 1989c

Layard's Beaked Whale

Mesoplodon layardii

Adult ♂

Ventral view

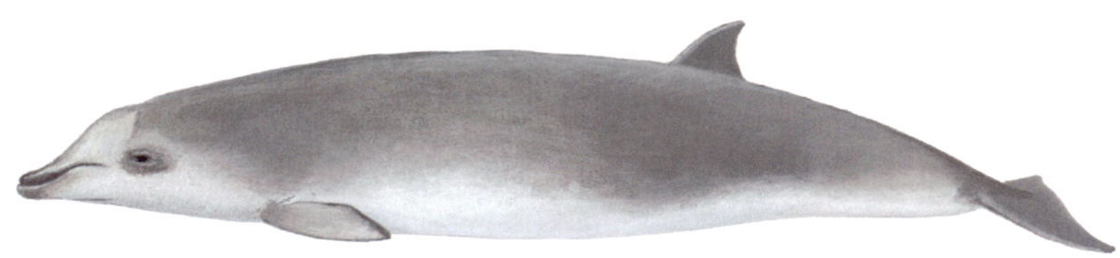

Calf

ALTHOUGH THIS SPECIES is often known (for obvious reasons) as the strap-toothed whale, I prefer to acknowledge its South African origins by retaining the common name of Layard's beaked whale (see box on page 139).

> **Length:weight relationship**
>
> Using unpublished data for 12 Layard's beaked whales, where three were weighed whole and two both whole and in parts, and applying a correction factor from these two (7.4%) to the remaining seven that were only weighed in pieces,[6] **$W = 24.573L^{2.5176}$**, where W = weight in kilograms and L = total length in metres. This implies that a whole 3 m whale would weigh 390 kg and a 5.5 m whale 1 796 kg.
>
> **Skeletal data**
>
> Vertebral formula (n = 11): 7 cervical, 8–10 thoracic, 10–12 lumbar and 17–20 caudal; total 45–47.[6]
>
> The first three cervical vertebrae are fused and the remainder are free.[2]
>
> Phalangeal formula (n = 2), including metacarpals: I_1, II_{6-7}, III_6, IV_{5-6}, V_{3-4}.[6]

Appearance

An adult Layard's beaked whale should be readily distinguishable at sea by its long white beak, contrasting prominent black melon, and dorsal grey blaze extending back from behind the blowhole. Although it might seem that the teeth should also be obvious in adult males of this species, they are often coated with greenish-brown diatoms and so are relatively inconspicuous.

Despite the fact that its external appearance is unknown at the time of writing, and that it is only known so far from the South Pacific, the spade-toothed whale *Mesoplodon traversii* might be confused with Layard's beaked whale because of its similar tooth shape. Nevertheless, the teeth of Layard's beaked whale taper towards the tip, while those of the spade-toothed whale are broader and of even width throughout. The apical denticle of the

FIG. 81 The contrasting white beak, black melon and grey cape on the back make an adult Layard's beaked whale almost unmistakable.

spade-toothed whale's tooth is also larger[1] and lies in the same plane as the principal axis of the tooth, whereas in Layard's beaked whale it is at a definite angle to the tooth axis.[2]

Adult female Layard's beaked whales as large as 6.12 m[3] or 6.15 m,[4] and males as large as 5.84 m[3] have been recorded, but some of these pre-date the adoption of standardised cetacean measurement, and the end points used are not always clear. The largest of 19 females accurately measured in South Africa since 1965 was 5.87 m long, and the largest of 15 males was 5.65 m long. It seems that females grow slightly larger than males.[5] The beak is long and tapered, the distance from snout to eye ranging from 11.8% to 16.3% (mean 14.6%, n = 31) of body length, and the melon is fairly pronounced. The jaw line is straight in females and juveniles and only slightly arched at the base of the teeth in adult males, and there are the usual ziphiid V-shaped throat grooves. The flippers are relatively large for a beaked whale (length 9.5–12.0% of body length, average 11%, n = 24), and are situated low down on the side of the body, with associated flipper pockets. The dorsal fin is small (height 3.2–6.0% of body length, average 4.9%, n = 28), hooked, and situated about two-thirds of the way down the body. Interestingly, the tail flukes seem to develop an upwards curl at the tip in adults of both sexes, but especially males. This may have something to do with the conspicuous colouring at the fluke tips (see below).[6]

As in some other beaked whales, the colour pattern of Layard's beaked whale changes with age. Calves are dark brownish-grey above (including the upper surface of the flukes) shading to brownish-grey on the sides and white on the belly. The sides of the head and melon, as far back as just behind the blowhole, are a light yellowish-brown, except for a dark eye patch. The upper jaw and chin are dark brownish, and this colour extends back in the mid-line as far as the blowhole. The outer surface of the flipper is a pale grey with a dark margin to the trailing edge. The undersurface of the flukes is pale grey at the base deepening to dark brownish-black in a semi-circle around the centre of the trailing edges. In adults, the beak becomes white as far back as the base of the melon. The melon itself darkens and a bright grey blaze develops on the back from the blowhole as far back as the dorsal fin. The belly also darkens, except for a distinctly white throat patch connected behind the eye to the grey blaze on the back, and a separate elliptical white blaze in the anogenital region. Both surfaces of the flukes darken, except for light grey patches at the extreme tips. Adoption of the adult colour pattern may coincide with the attainment of sexual maturity.[4] Linear scarring, usually paired, occurs in adult males, particularly on the lower belly and sides.[6]

A single pair of mandibular teeth erupts towards the back of the gape in adult males only. When fully grown they are long and strap-like, curving back and over the upper jaw and almost meeting in the midline at the front of the melon. The tip of each tooth carries a small denticle, which (owing to the curvature of the teeth) comes to project almost vertically upwards. It is these denticles that are presumably responsible for the paired linear scarring seen in adult males. In adult males the extracted tooth reaches a length of 31–34 cm, a width (where it emerges from the jaw) of 5–6 cm, and a thickness of 1–1.5 cm. In adult females and juveniles of both sexes the teeth are much smaller, reaching a maximum length of 10–13 cm, a width of 3–4 cm, and a thickness of about 1 cm.[3] They are triangular in shape in juveniles, but with age the front edge elongates and the back edge becomes convex, and as a result the tooth becomes strongly asymmetrical about its tip.[7] As in adult males, the denticle is situated at the tip and directed outwards at right angles to the main axis of the tooth.

Distribution and migrations

Layard's beaked whale is only known from the southern hemisphere. It is believed to have a circumpolar distribution in cold–temperate waters.[8]

In the subregion the species has been recorded from Namibia, South Africa,[9] Marion Island[6] and Heard Island.[10] A specimen from Kerguelen[11] has been re-identified as a fossil species, *Mesoplodon longirostris*.[8]

Sightings in the subregion attributed to Layard's beaked whale have been made between 30° S and 58° S (n = 8), while the most northerly record in the subregion appears to be a stranding at 22°53′ S in Namibia[9] (it is said to be, along with Gray's beaked whale, one of the most frequently stranded beaked whales in Namibia[12]).

The most northerly record in the southern hemisphere is from 12°47.67′ S (north-eastern Brazil),[13] and the furthest south at 63.3° S.[14]

All the sightings in the subregion have been in water over 2 000 m deep,[9] suggesting that Layard's beaked whale is generally an open ocean animal.

There is no information on seasonal movements, although all strandings in southern Africa have taken place between January and June, with a peak in April.[3, 6, 9] This could indicate a seasonal inshore movement[9] or a northward movement from higher latitudes[3] (see Food and feeding behaviour).

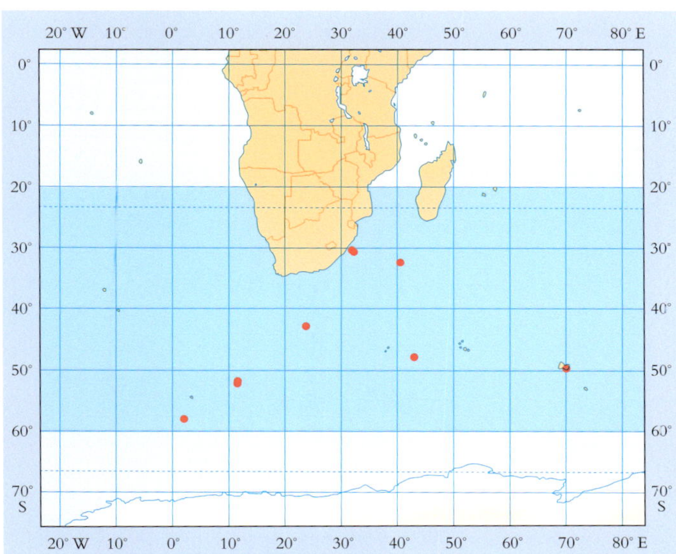

Postulated range ■ Sighting or capture

FIG. 82 Distribution map for Layard's beaked whale.

External commensals and parasites

Stalked barnacles (*Conchoderma* sp.) occurred on the erupted teeth of nine of 10 adult males examined, ranging from two to 69 in number. The teeth were also recorded as coated with greenish diatoms on five occasions:[6] a sample from one animal proved to include the pennate diatoms *Stauroneis olympica*, *Navicula* sp., *Gomphonema aestuarii* and *Synedra* sp.[15] (the specimens identified as *G. aestuarii* have subsequently been assigned to a new species, *Tripterion philoderma*[16]).

Food and feeding behaviour

In an examination of the stomachs of 13 Layard's beaked whales from the subregion, 24 species of oceanic squids accounted for 94.8% of prey items by number, the balance being made up of unidentified fish and crustacean remains. Of the squid, *Histioteuthis* sp. (22.6%), *Taonius pavo* (21.1%), *H. macrohista* (9.84%), *Gonatus antarcticus* (9.08%) and *Mastigoteuthis* sp. (7.62%) formed 70% of the mass of prey consumed. Some of the species eaten (*G. antarcticus*, *Moroteuthis knipovitchi*) were likely to be subantarctic in origin, indicating that the whales concerned must have moved into South African waters from higher latitudes.[17]

In general, the sizes of squid eaten were small, most weighing less than 200 g – similar to those eaten by much smaller cetaceans such as spotted dolphins and dwarf sperm whales. A comparison of the size of prey eaten by mature males and females/immature males indicated that although females/immature males ate somewhat longer individual squid, the masses of the squid eaten were similar for the two groups. So although the presence of fully erupted teeth restricted the gape of adult males more than that of females or immatures, this seemed to have no effect on the size of prey eaten.[17] Presumably a wide gape is of no particular advantage to an animal using suction-feeding (see the account of Arnoux's beaked whale).

There is no information on seasonality of feeding.

Reproduction and growth

There is little published information for this species. The largest foetus measured was 0.76 m long,[18] and the smallest calf (a male) 2.9 m[6] (a calf reported as 2.8 m[3] was actually 2.95 m long). The umbilicus of the 2.9 m animal (as well as that of a 2.95 m male) was healed, but a female 3.18 m long had a partly healed umbilicus and still carried traces of foetal folds.[6] A length at birth of 2.2–2.4 m has been estimated,[3] but judging by the calf sizes above, this may be an underestimate. If the lengths of the foetus and five calves 2.9–3.42 m long (and all stranding between 28 March and 14 April) are plotted against dates of death, a foetal/neonate growth rate of 0.84 (±0.10) cm/day is obtained, which would mean (given a birth length in the vicinity of 2.4 m) that pregnancy might last somewhat less than a year, and conceptions would occur in autumn.

Ten females 5.26 m or more in length were sexually mature, and five females 5.18 m or less in length were sexually immature (although the 5.18 m animal was described as pubertal). Sexual maturity in the female therefore seems to occur around 5.2 m. The smallest of nine males with erupted teeth was 5.1 m, and the largest of six males with unerupted teeth was 5.14 m long. A male 5.12 m long had erupted teeth but these were not fully grown. Tooth eruption in the male therefore seems to take place around 5.1 m. Males with fully erupted teeth had combined testes weights of 305 g or more (up to 392.5 g, n = 5), whereas males with unerupted or partly erupted teeth had combined testes weights of 7.7–262 g (n = 5). Tooth eruption therefore seems to take place around or shortly after puberty. Of the 10 sexually mature females, eight were lactating, but none was pregnant. If pregnancy lasts about a year, this would suggest that the reproductive cycle is likely to occupy at least two years.[6]

There is no information on age and growth.

Behaviour and vocalisations

Sightings attributed to Layard's beaked whale in the subregion involved groups of 2–6 in number, with an average of 4.1 (n = 8). A mass stranding of four whales near Cape Town in 1966

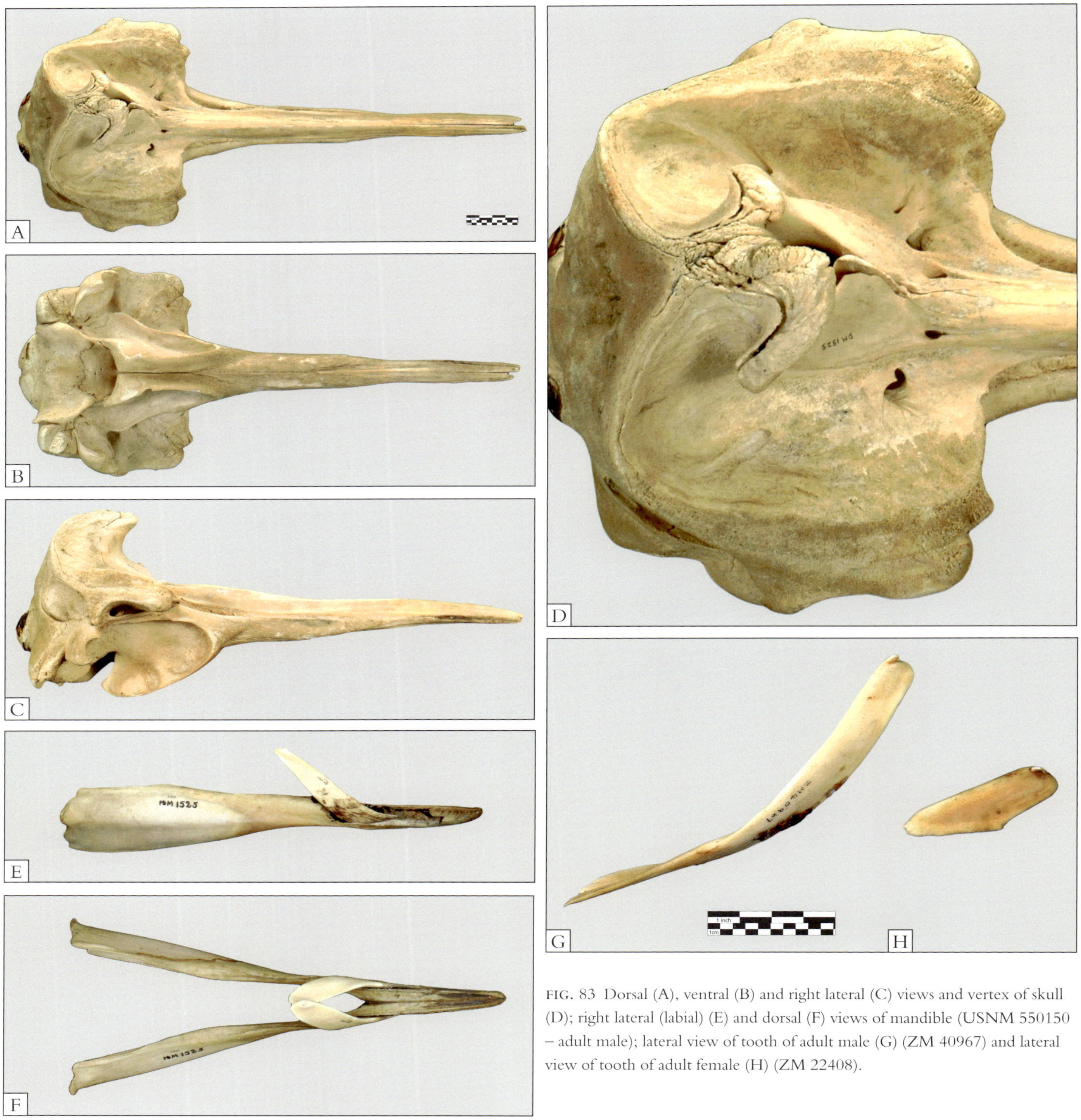

FIG. 83 Dorsal (A), ventral (B) and right lateral (C) views and vertex of skull (D); right lateral (labial) (E) and dorsal (F) views of mandible (USNM 550150 – adult male); lateral view of tooth of adult male (G) (ZM 40967) and lateral view of tooth of adult female (H) (ZM 22408).

involved two immature animals of each sex, while four whales found dead on Marion Island in March 1989 consisted of a presumed cow-calf pair, another presumed adult and a presumed juvenile/calf of the latter. Only the cow-calf pair was specifically identified.[6] The species therefore seems to be found in small social groupings with a variety of compositions.

Although knowledge of the vertical distribution and migrations of oceanic cephalopods is far from comprehensive, the presence of some species (e.g. *Vampyroteuthis infernalis*) in the diet of Layard's beaked whale suggests that it can dive to considerable depths, perhaps to at least 300 m.[17]

Sounds produced by this species have not been documented.

Conservation status

Layard's beaked whales have never formed the target of a directed fishery, and their involvement as a by-catch in any southern fishery is unknown. There are no estimates of population size or indications of abundance, although they are the most frequently stranded of the beaked whales in the subregion. The only identifiable threat to the species would seem to be from loud sounds associated with seismic exploration or mid-frequency sonar (see the account of Cuvier's beaked whale), but how frequently these activities occur in its range is unclear. Layard's beaked whale is listed as Data Deficient by the IUCN and in the South African Red Data Book, and is included on Appendix II of CITES, but is not listed by CMS.

References

1 Van Helden *et al.* 2002
2 Turner 1880
3 Ross 1984
4 Ross *et al.* 1988
5 MacLeod 2006
6 MRI Whale Unit data
7 McCann and Talbot 1963
8 Mead 1989c
9 Findlay *et al.* 1992
10 Guiler *et al.* 1987
11 Robineau 1989
12 Griffin 1998
13 Maia-Nogueira and Nunes 2005
14 MacLeod *et al.* 2006
15 Nemoto *et al.* 1980
16 Holmes *et al.* 1993
17 Sekiguchi *et al.* 1996
18 Gaskin 1968

The father of African cetology?

Edgar Leopold Layard was Curator of the South African Museum in Cape Town from 1855 to 1872. By all accounts his appointment came at just the right time, when the museum (founded in 1825) was at a low ebb. Layard was an extremely energetic worker and, despite having to divide his attentions between working at the Colonial Office and the Museum, at the age of 31 he set about building up the latter's collections and ensuring that they were displayed to the public. His particular interests were shells and birds (in 1867 he produced the pioneering *Birds of South Africa*), but he believed the museum should contain 'something for everybody' and amassed an amazing variety of mineral, fossil, ethnographic and other material through donations, purchases and exchange. Having no formal scientific training himself, he frequently sent interesting specimens to specialists for identification. Thus it was that when he came across a strange whale skull in 1865, he asked Roland Trimen, the Museum's entomologist, to prepare a sketch of it, which was sent with measurements to Dr Gray at the Natural History Museum in London. Gray described this as a new genus and species, recognising the collector by naming it *Ziphius layardii* (later transferred to the genus *Mesoplodon*).[1] Layard subsequently sent the specimen to the museum in London, along with examples of six other local cetacean species, some of which Gray, in his indefatigable way, also described as new species or genera (e.g. *Steno capensis*, now believed to be *Stenella attenuata*, and *Petrorhynchus capensis*, now *Ziphius cavirostris*).[2] Although these specimens were originally sent on loan, their apparent importance for comparative purposes with the British museum's collection was such that in 1869 the Trustees of the South African Museum were induced to donate them. This meant effectively that the South African Museum's collecting of cetaceans had to start from scratch again, and more than a century was to elapse before further examples of some of these species were obtained.[3]

FIG. 84 Edgar Layard.

References

1 Summers 1975
2 Gray 1865
3 Barnard, K.H., n.d.

Blainville's Beaked Whale

Mesoplodon densirostris

Adult ♂

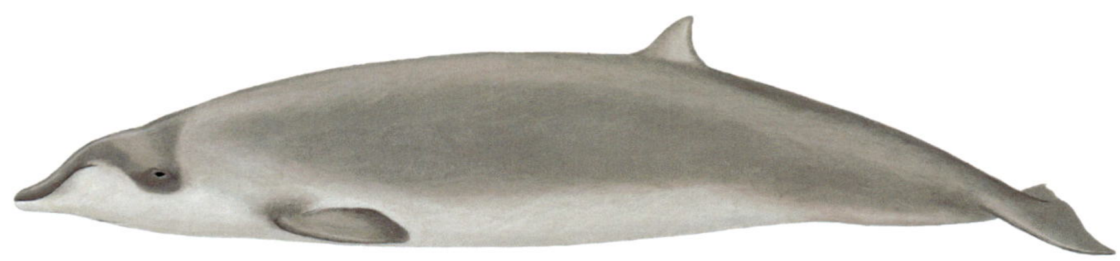

Calf

THIS IS SOMETIMES known as the dense-beaked whale, for reasons that are obvious if you ever get to see the skull of an adult male (see box on page 143).

Appearance

The relatively flat forehead and heavily arched lower jaw of an adult Blainville's beaked whale should make it readily distinguishable at sea from other beaked whales in the subregion.[1] Adults may also be particularly afflicted with white oval scars, probably from bites from cookie-cutter sharks *Isistius*.

Adult females as large as 4.725 m and males as large as 4.71 m have been recorded,[2] indicating that there does not seem to be much difference in size between the sexes. The beak is long (snout to eye being 12–16% of body length, average 14.3%, n = 18) and the melon is not particularly prominent. The jaw line is characteristically arched up towards the angle of the gape (a 'smile'), and in adult males this is where the teeth erupt, their tips being elevated higher than the intervening rostrum or melon. There are the usual ziphiid V-shaped throat grooves. The flippers are relatively large for a beaked whale (length 10.0–13.2% of body length, average 11.3%, n = 15), and are situated low down on the side of the body, with associated flipper pockets. The dorsal fin is small (height 3.9–5.7% of body length, average 4.8%, n = 16), hooked, and situated about two-thirds of the way down the body.[3]

The contrast between the colour pattern of calves and adults seen in many other beaked whales does not seem to be so marked in Blainville's beaked whale. Calves are dark bluish-grey above and off-white below, with the intergradation between the two occurring slightly below the centre-line of the body except behind the anus, where it descends to the midline, and on the side of the head, where it ascends onto the side of the melon. The outer surface of the flippers and the upper surface of the flukes are dark grey, and the undersurface of both are off-white. A broad stripe links the flipper with the dark coloration above. A dark eye patch on each side is connected with a dark band over

Length:weight relationship

Combining unpublished data for eight whales[3] with published data for another eight whales,[2,6] and applying a correction factor of 5.2% to allow for fluid loss for those specimens weighed in pieces (obtained from two individuals weighed whole and in parts), **$W = 15.261L^{2.7667}$**, where W = weight in kilograms and L = total length in metres. This implies that, weighed whole, a 2.5 m whale would weigh 193 kg and a 4.5 m whale 979 kg.

Skeletal data

Vertebral formula (n = 9): 7 cervical, 9–11 thoracic, 10–12 lumbar and 18–21 caudal; total 45–48.[3]

The first three cervical vertebrae are fused and the remainder are free.[7]

The phalangeal formula probably differs little from that of other beaked whales.

FIG. 85 An adult male Blainville's beaked whale sufaces nearer the camera, exposing its prominent teeth and the scarring on its upper back, while a second adult is relatively unscarred apart from pale oval marks probably indicating past visitations by the cookie-cutter shark.

the head that extends forward in the midline over the melon to the tip of the upper jaw. In adults the entire upper half of the head and body is dark apart from the beak, which may be white as far back as the angle of the gape. The throat and belly are white from the tip of the lower jaw to just behind the anus, and the border between the upper and lower body pigmentation is fairly defined but irregular. Much of the belly's pigmentation is heavily affected by scarring as a result of bites by the cookie-cutter shark *Isistius*, which seems particularly severe on this species of beaked whale (probably because of its warm water habitat). Hundreds of bites must occur in the lower belly area, but are also scattered over the back and throat, leaving either white or grey scars. The underside of the flukes is a uniform dark grey, with a pattern of dark streaks radiating out from the centre of the trailing edge. In adult males, the upper surface of the head and front half of the back become covered with linear white scars running more or less parallel with the body's axis.[4]

A single pair of mandibular teeth erupts in adult males only, towards the back of the gape where the tip appears at the apex of the arched upper jaw. The exposed portion of tooth is relatively small, and if not worn it has a point directed slightly backwards. The extracted tooth of an adult male is large, broad and somewhat flattened. The maximum dimensions are: height 13.0–15.4 cm, width front to back 6.9–9.6 cm, and thickness side to side 3.2–4.9 cm. It can weigh as much as 489 g.[2, 5] In profile, the tooth has a characteristic shape, with a convex leading edge and a somewhat S-shaped trailing edge. The adult female tooth is much smaller and very different in shape, being more wedge-shaped with a convex leading edge and a longer, straighter trailing edge. The maximum dimensions are: height 5.9–6.8 cm, width front to back 3.1–5.5 cm and thickness side to side 1.0–1.4 cm.[2, 5]

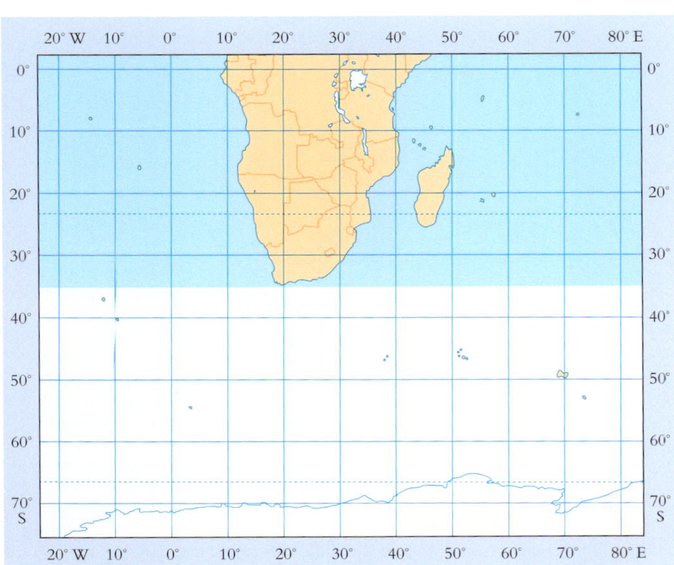

Postulated range

FIG. 86 Distribution map for Blainville's beaked whale.

Distribution and migrations

Blainville's beaked whale has the widest distribution of any beaked whale, occurring worldwide in tropical and warm–temperate waters.[6] In the subregion the species has been recorded from Namibia,[3] South Africa,[8] the Seychelles and Mauritius.[9]

In addition, a specimen was recorded from 5° S 65° E.[10] There are no sightings attributed to this species in the subregion.

Elsewhere it seems to be frequently sighted around oceanic islands, although this may partly reflect the distribution of effort. Off the Bahamas, the species showed a distinct preference for water depths between about 250 m and 600 m in association with a submarine canyon, although it was found over the range of 136–1 319 m.[11] It has been found in similar depths off the Canary Islands (100–850 m)[12] and around Hawaii (700–1 000 m).[6] This may indicate a preferred habitat related to continental and island slopes. Off the Bahamas and Hawaii their distribution contrasted with that of Cuvier's beaked whale, which was generally found in significantly greater water depths.[13, 14]

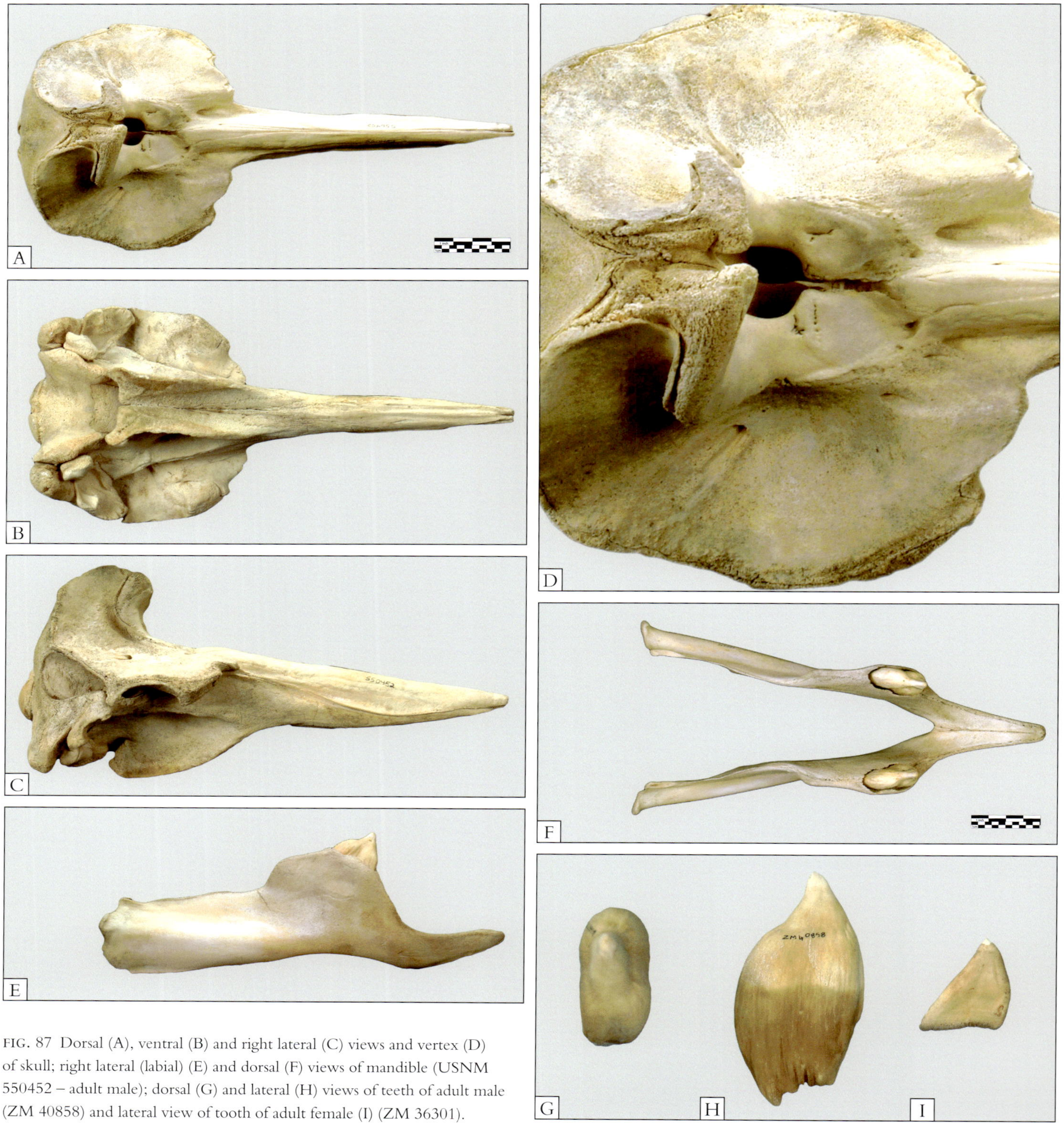

FIG. 87 Dorsal (A), ventral (B) and right lateral (C) views and vertex (D) of skull; right lateral (labial) (E) and dorsal (F) views of mandible (USNM 550452 – adult male); dorsal (G) and lateral (H) views of teeth of adult male (ZM 40858) and lateral view of tooth of adult female (I) (ZM 36301).

There is no information on seasonal movements. In the Western Cape, South Africa, strandings have taken place throughout the year, although there is a slight tendency for greater numbers in autumn (eight) and winter (eight) than in spring (two) and summer (three).[3] Off the Canary Islands, Blainville's beaked whales were encountered in nine months of the year with no particular sign of seasonality.[12]

External commensals and parasites

Stalked barnacles (*Conchoderma* sp.) occurred on the erupted teeth of three of five adult males examined in the subregion, ranging from one to 40 in number. A stalked barnacle (*Xenobalanus* sp.) was also found on the dorsal fin of one adult male, and four (plus 19 scars) on the trailing edge of the flukes of another. Cyamids (*Isocyamus delphinii*) were found in considerable numbers on a subadult male, 494 being collected. These were mostly on the rear half of the body, especially on the flanks, where they were concentrated in uneven surface areas associated with healing scar tissue.[3] An unidentified balanoid barnacle has been reported deeply embedded in the flank of a specimen from South Africa.[6]

Food and feeding behaviour

Very little has been published about the prey of Blainville's beaked whale. Food items have been identified from three stomachs in the subregion. One individual contained single beaks of the cephalopods *Todarodes sagittatus* and *Octopoteuthis* sp.;[15] another had 21 otoliths from the lantern fish *Lampanyctus* (19), bigscale fish *Scopelogadus* (1) and bandfish, possibly *Cepola* (1);[2] and the third contained 13 otoliths from hake *Merluccius* sp. (6), buttersnoek *Lepidopus caudatus* (2) and unidentified fish (5), plus three mantis prawns *Pterygosquilla armata*.[16] Elsewhere three individuals contained the remains of squid but not fish.[6]

The presence of midwater and deepwater prey items such as the lantern fish and buttersnoek suggests that some feeding takes place at depth, possibly below 400 m.[16]

There is no information on seasonality of feeding.

Reproduction and growth

There is little published information for this species. The largest foetus measured was 1.9 m long,[2] and the smallest calf (a male) also 1.9 m. A second calf 2.25 m long had a healed umbilicus. Birth presumably occurs at a length of about 1.9 m. Two additional foetal records are of a 37 cm male (15 March 1984) and a 68 cm male (31 July 1984). If the lengths of the three foetuses and two calves are plotted against dates of death, a foetal/neonate growth rate of 0.48 (±0.04 SE) cm/day is obtained, which would mean (given a birth length of 1.9 m) that pregnancy might last about 13 months, with birth taking place in autumn.[3] The calf may suckle at least into its second year of life, and may still be accompanying its mother in its third year.[17]

The largest male with unerupted teeth was 4.15 m (n = 6),[2, 3] and the smallest male with erupted teeth was 3.76 m (n = 9).[2–4, 7] The latter animal seems exceptionally small, since the next largest individual with erupted teeth was 4.13 m long. Ignoring the apparently precocious individual, this suggests that tooth eruption occurs around 4.14 m. The combined testes weights of four animals with unerupted teeth varied from 9.1 g to 42 g, compared to 102.8–225.0 g for four with erupted teeth. Tooth eruption may therefore coincide with sexual maturity. The

Old hard nose

FIG. 88 The scar that extends from the tip of the snout, between its raised teeth, and to the top of the head of this Blainville's beaked whale bears testimony to the intensity of inter-male disputes.

Running down the centre of the front of the skull of beaked whales (the part of the skull that forms the rostrum or 'beak') is a deep groove that tends to fill up with bone as the mesethmoid cartilage ossifies. In no species does this occur to the extent that it does in Blainville's beaked whale, where in adult males the entire groove fills with bone throughout its length, and the rostrum becomes deeper, wider and shorter than in females.[1] The bone produced is the stiffest and hardest and has the highest mineral content of any bone so far tested in the animal kingdom – it is indeed 'extreme bone'.[2]

Three functions have been suggested for this peculiar structure: for transmission of sound, as ballast to aid in deep diving, and for the protection of the rostrum during physical contests between adult males. The most likely of these appears to be the last.[3] Adult male Blainville's beaked whales typically carry linear scarring on the head and upper back that is not seen in adult females or juveniles of either sex. These scars start to appear about the time of tooth eruption, tend to be deeper on the head than on the back, and run more or less parallel to the long axis of the body. These clues all suggest that the scars are caused by the tusks of opposing males as they swim past each other in opposite directions and back to back, with the principal impact occurring as they butt heads. Some of the most spectacular scars occur on the melon right between the raised teeth, bearing witness to the strength of the initial strike. The physical structure of the rostrum, in which the 'grain' in the bone runs longitudinally, is well designed to resist impacts coming from above.

References

1 Besharse 1971
2 Zioupos *et al.* 1997
3 MacLeod 2002

largest immature female was 4.005 m long (n = 5),[3] and the smallest mature female 3.84 m (n = 11),[2–4] suggesting that maturity occurs at just under 4 m.

None of the three pregnant females examined was also lactating,[2, 3] suggesting that the reproductive cycle is likely to occupy at least two years.

Attempts at determining the age of Blainville's beaked whale using growth layers in the dentine and cementum of the teeth were not particularly successful, resulting in minimum ages only, with physically mature individuals having only 9–13 layers.[2]

Behaviour and vocalisations

Blainville's beaked whales have been encountered in groups of 1–9 individuals.[12, 14] Groups usually comprised a number of females, calves and/or juvenile animals. If an adult or adolescent male was present it was always the only one, and single animals are usually adult/adolescent males.[12, 17]

The species is believed to be among the most proficient of air-breathing divers. Telemetered individuals made foraging dives to an average depth of 835 m for an average of 46.5 min. The maximum depth attained was 1 251 m and the maximum duration was 57.4 min. Rates of descent (averaging 1.6 m/sec.) were always greater than rates of ascent (averaging 0.7 m/sec.).[18] An adult male in a social grouping telemetered for 2.09 hours made one deep dive to 890 m for at least 23.3 min., close to the sea floor.[14] These data confirm suggestions from prey remains in the stomach that the species frequently forages at depth.

Blainville's beaked whales produce high-frequency clicks of 25 kHz to at least 48 kHz.[19] Such clicking only starts when the whale is below 200–570 m, and is of two types: regular clicking (long click trains with inter-click intervals of 200–500 ms) and buzzes (short, 2–5 sec. bursts of clicks with inter-click intervals of 5–20 ms). The former are of higher intensity than the latter, and are believed to represent prey detection and inspection, while the buzzes represent attempts to locate the prey accurately immediately before capture, when it is approximately one body length away. On average 23 buzzes occur per foraging dive, but represent only about two-thirds of the total clicks produced, suggesting a considerable degree of prey selection.[20]

Conservation status

Blainville's beaked whales have never formed the target of a directed take, although they have been recorded as a by-catch in the Indian Ocean tuna fishery.[10]

There are no estimates of population size or indications of abundance, although as they are the most widespread of the beaked whales they cannot be considered rare. They have featured in mass strandings associated with seismic exploration or mid-frequency sonar (see the account of Cuvier's beaked whale), and this would seem to be the only identifiable threat to the species.

Blainville's beaked whale is listed as Data Deficient by the IUCN and in the South African Red Data Book. It is included on Appendix II of CITES, but not listed by CMS.

References

1 Borsa and Robineau 2005
2 Ross 1984
3 MRI Whale Unit data
4 Ross *et al.* 1988
5 Besharse 1971
6 Mead 1989c
7 Raven 1942
8 Findlay *et al.* 1992
9 Michel and Van Bree 1976
10 McCann 1963
11 MacLeod and Zuur 2005
12 Ritter and Brederlau 1999
13 MacLeod *et al.* 2004
14 Baird *et al.* 2004
15 Goodall and Galeazzi 1985
16 Sekiguchi 1994
17 MacLeod and D'Amico 2006
18 Tyack *et al.* 2006
19 Johnson *et al.* 2004
20 Madsen *et al.* 2005

Ginkgo-toothed Beaked Whale

Mesoplodon ginkgodens

THE SPECIES' COMMON name is based on the shape of the tooth of the first specimen described, which closely resembled the leaf of the ginkgo tree *Ginkgo biloba*.[1] Although the species has not yet been positively recorded from the subregion, there are enough possible records in the vicinity for the species to be included as a potential member of the fauna.

Appearance

As there are no authenticated sightings of ginkgo-toothed beaked whales, it is difficult to provide field characters for identification. Seemingly by default, beaked whales characterised as having 'a nondescript colour pattern, short beak, no visible teeth, and no prominent scarring' have been tentatively assigned to this species.[2]

Female ginkgo-toothed beaked whales have measured up to 4.9 m in length and males up to 5.1 m, but the sample is very small (13 in total).[3] The overall body conformation is similar to that of other beaked whales. Although the beak is fairly prominent, in profile the forehead slopes smoothly without a distinct 'melon'. The line of the mouth curves up at about its mid-length, and then continues more or less straight back towards the eye, forming a raised area that carries the erupted tooth in adult males. There is a single pair of throat grooves convergent at the front.[4] The flippers are small (10.6–11.3% of body length) with an almost straight leading edge and a weakly V-shaped trailing edge. The dorsal fin is small (4.4–4.8% of body length in height), strongly or moderately hooked, and placed about two-thirds of the way back from the tip of the snout. The flukes have no central notch and in width comprise 23.7–24.9% of body length.[5]

By all accounts the colour pattern is pretty bland. The body has been described as dark all over, including the underside of flippers and flukes. An adult male had indications of a faint light patch on the front of the rostrum and at the tip of the lower jaw, and the belly (particularly around the anogenital region) was peppered with white oval scars. A female was described as medium grey above and lighter below, without signs of the white scarring.[6] Neither sex seems to exhibit the linear scarring typical of many other beaked or bottlenose whales, possibly because the teeth never protrude very far from the gum.[5] There is no information on whether the colour pattern differs between juvenile and adult.

A single pair of large flattened teeth is situated in the lower jaw towards the back of the gape (and behind the symphysis of the mandible). In adult males the tooth is shaped like the leaf of a gingko tree, but it seems that only the central 'stalk' of the

Length:weight relationship

No weights are available for this species.

Skeletal data

Vertebral formula (n = 2): 7 cervical, 10 thoracic, 11 lumbar and 21 caudal; total 49.[5]

The first three cervical vertebrae are fused and the remainder are free.[5]

Phalangeal formula (n = 1), including metacarpals: I_1, II_7, III_6, IV_5, V_4.[5]

The following skull characteristics distinguish the species from other beaked whales:

1. When viewed from the front, with the skull upright and the beak horizontal, at least one of the premaxillary foramina lies on or below a line joining the centres of the maxillary foramina.
2. The length of vomer visible on the palate is greater than 100 mm.
3. In side view, the lacrymal is relatively thick and does not curl up around the front end of the frontal bone.
4. The maximum length of the right nasal bone on the vertex is more than twice the minimum distance between premaxillary foramina.
5. The maximum width of the premaxillae at the middle of the rostrum is between 40 mm and 60 mm.
6. The maximum width of the temporal fossa roughly perpendicular to its long axis is less than 60 mm.[7]

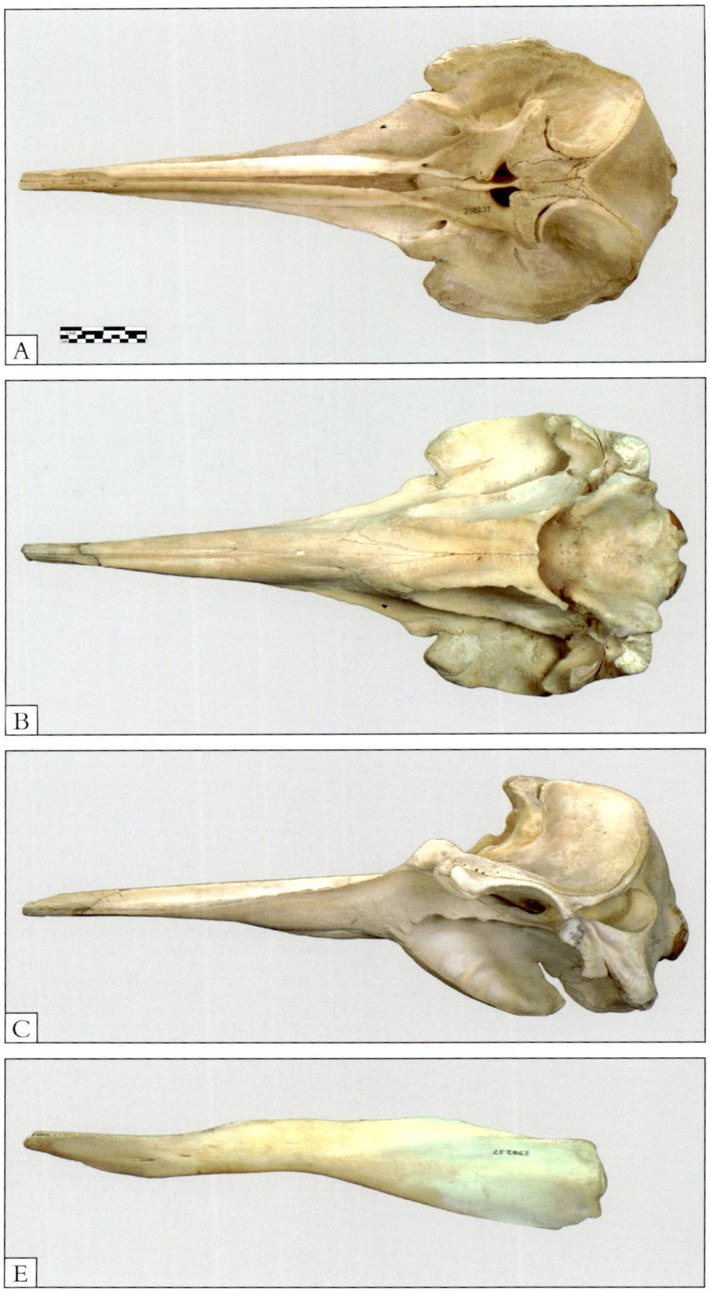

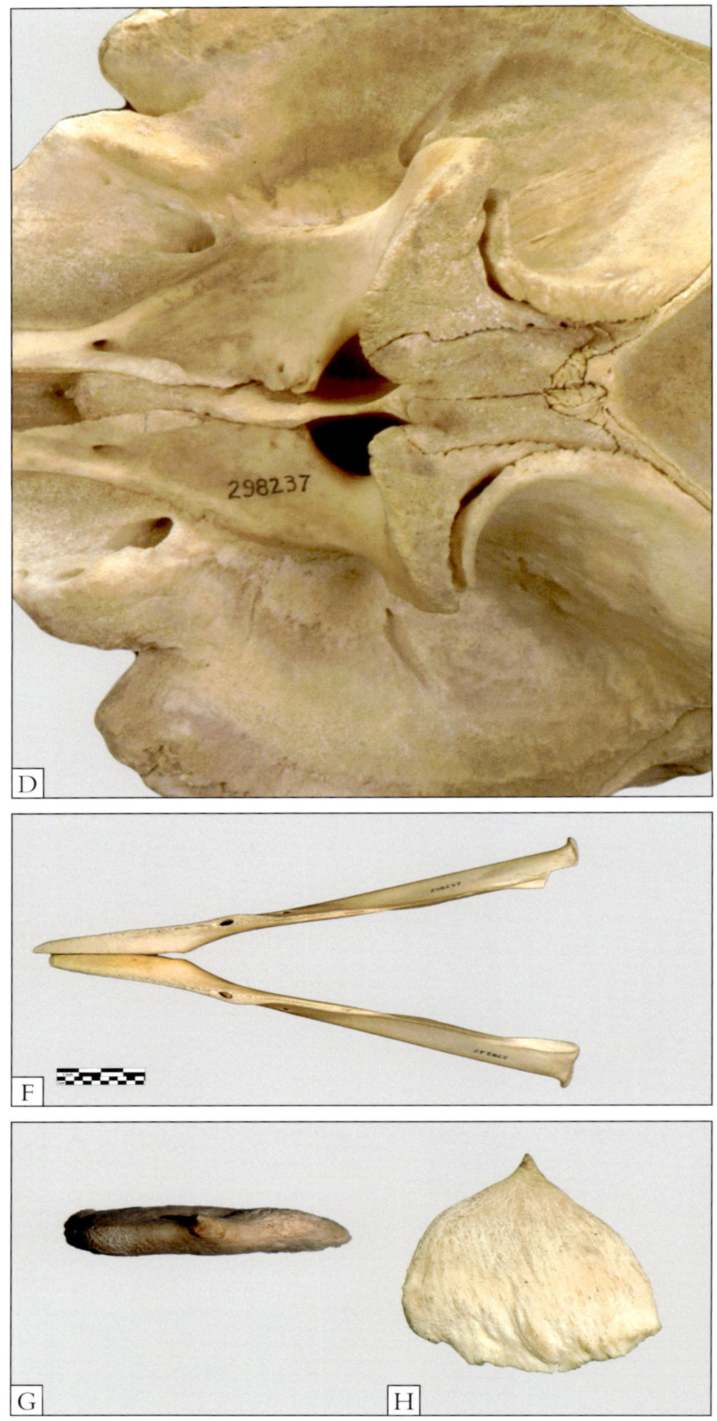

FIG. 89 Dorsal (A), ventral (B) and left lateral (C) views and vertex (D) of skull; left lateral (labial) (E) and dorsal (F) views of mandible (USNM 298237); dorsal (G) and lateral (H) views of tooth (MNM 329/362).

tooth erupts. The height of the tooth (79.0–94.5 mm) is almost the same as its anterior-posterior dimension (77–106 mm), and its thickness is 16–17 mm.[5] It is much more symmetrical than that of Andrews' beaked whale at all ages, so that a line drawn straight down from the tip of the 'stalk' will bisect the tooth into two approximately equal halves. The tooth presumably does not erupt in adult females.

Distribution and migrations

From the few records available, ginkgo-toothed beaked whales are assumed to be confined to warm-temperate and tropical waters of the Indo-Pacific.[6] There are no authenticated records yet from the subregion, the nearest being a tooth of uncertain provenance from the National Museum in the Maldives.[8] Sightings tentatively attributed to the species have been reported from east of Madagascar[2] and the Comoro Islands.[9]

Nothing is known about seasonality of movement.

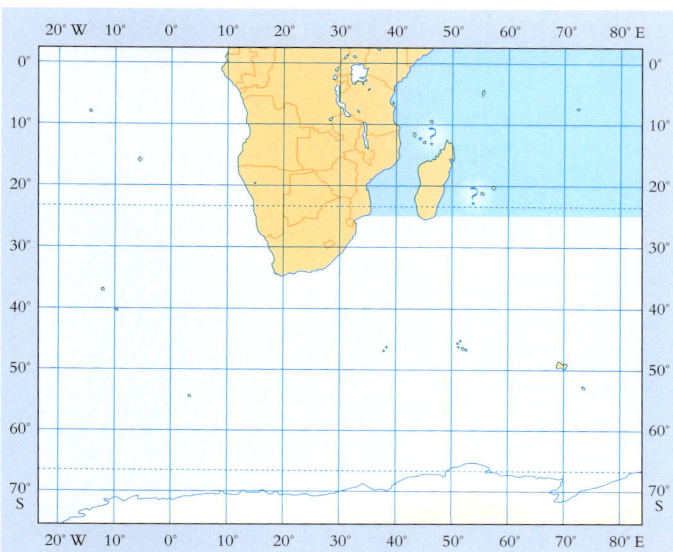

Postulated range ? Questionable origin

FIG. 90 Distribution map for gingko-toothed beaked whale.

External commensals and parasites

There are no records from the species. Stalked barnacles (*Conchoderma* sp.), which sometimes occur on the teeth of adult males of other beaked whales, might find it hard to settle on the small amount of tooth exposed in this species. White scars possibly caused by healed bites from the cookie-cutter shark *Isistius* are found on the belly of adults, especially on the more posterior region.[1]

Food and feeding behaviour

There appears to be no published information on the feeding habits of this species.

Reproduction and growth

The testes of a 4.77 m male were described as pubertal.[5] Ginkgo-toothed beaked whale milk consists of 14.6% fat, 17.6% protein, 0.1% carbohydrate and 1.2% ash; the rest is water.[10] No other information on reproduction, age or growth exists.

Behaviour and vocalisations

Sightings of unidentified beaked whales in the western tropical Indian Ocean, some of which were tentatively identified as ginkgo-toothed beaked whales, ranged in size from 1 to 5 with a mean of 2 (n = 19).[2] No recordings of vocalisations attributed to this species have been made.

Conservation status

Ginkgo-toothed beaked whales have never formed the target of a directed fishery, and their involvement as a by-catch in any southern fishery is unknown. There are no estimates of population size. The only identifiable threat to the species would seem to be from loud sounds associated with seismic exploration or mid-frequency sonar (see the account of Cuvier's beaked whale), but how frequently these activities occur in its range is unclear. The ginkgo-toothed beaked whale is listed as Data Deficient by the IUCN. It is unlisted in the South African Red Data Book, is on Appendix II of CITES, but is not listed by CMS.

References

1 Nishiwaki and Kamiya 1958
2 Ballance and Pitman 1998
3 MacLeod 2006
4 Leatherwood *et al.* 1982
5 Nishiwaki *et al.* 1972
6 Mead 1989c
7 Moore and Gilmore 1965
8 Anderson *et al.* 1999
9 Kiszka *et al.* 2006b
10 Sasaki *et al.* 2004

Family **DELPHINIDAE**

Heaviside's Dolphin

Cephalorhynchus heavisidii

Dorsal view

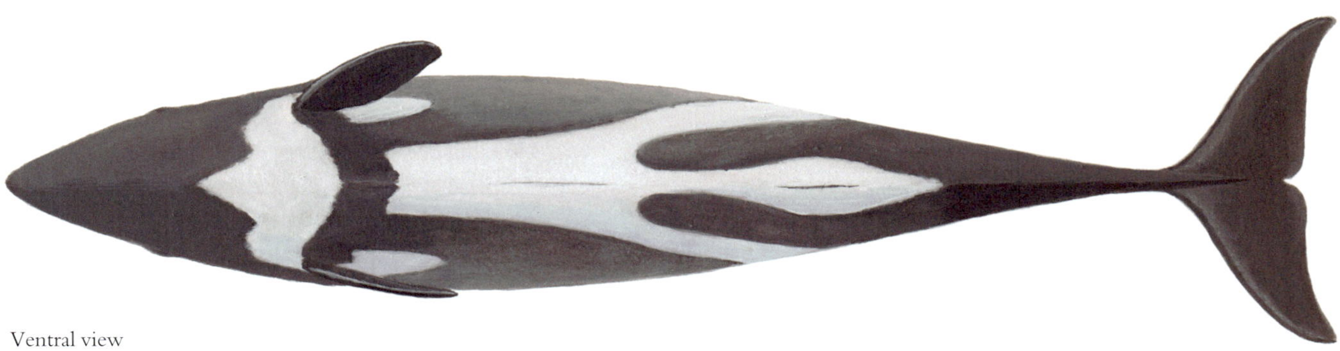

Ventral view

THE SPECIES WAS originally supposed to be named in honour of the captain of an East Indiaman who brought a collection of skulls, including an example of Heaviside's dolphin, from the 'Cape of Good Hope' to England in 1827. This collection was purchased by the Royal College of Surgeons in 1828. However, the captain's name was actually Haviside, not Heaviside. The mistake apparently arose because there was a surgeon, Captain Heaviside, who owned a collection of anatomical specimens that was sold at about the same time.[1] Despite this mistake and a subsequent suggestion that the species' common name should be changed to the Benguela dolphin,[2] most texts (and local scientists) continue to use the original name.

Length:weight relationship

Mass data from 33 Heaviside's dolphins weighed whole[3,5] have been used to produce the following relationship: $W = 15.799L^{2.8224}$, where W = weight in kilograms and L = total length in metres. This implies that a 1 m dolphin would weigh 15.8 kg and a 1.5 m dolphin 49.6 kg.

Skeletal data

Vertebral formula (n = 27): 7 cervical, 12–14 thoracic, 14–20 lumbar and 27–32 caudal; total 65–68.[5,6]

The atlas and axis are fused and the remaining cervical vertebrae are free.[6]

Phalangeal formula, including metacarpals (n = 15): I_{2-3}, II_{7-8}, III_{5-6}, IV_{3-4}, V_{1-2}.[5,6]

The cranial characteristics that separate Heaviside's dolphin from Commerson's dolphin include:

1 Total tooth count (both jaws combined) 102–109, compared to 114 or more.
2 Vomer visible (50–70 mm) on palate, compared to only a trace (maximum 40 mm).
3 Right premaxilla 13–16 mm wide at level of postorbital process, compared to 10 mm or less.
4 Pterygoid bones separated by finger of palatine bone capped by vomer, compared to no projection of palatines.[6]

Appearance

Heaviside's dolphin can be readily distinguished from most other small delphinids occurring in the subregion by its low, almost triangular dorsal fin. Locally only the spectacled porpoise (*Phocoena dioptrica*) has a similarly-shaped fin, but its back is completely dark, whereas Heaviside's dolphin has a conspicuous grey cape marking on each side of the thorax.

This is a small animal, reaching a maximum body length of only 1.75 m, with little obvious difference in size between the sexes. It is rather porpoise-like in shape, with a conical head and no pronounced beak, although there is enough demarcation between the rostrum and melon to produce something anteriorly that could be interpreted as a small 'beak'. The animal is robust, its maximum girth being about 62.5% of the body length and occurring just in front of the dorsal fin. The flippers are small (12.7–17.8% of body length) and characteristically blunt in shape, with leading and trailing edges more or less parallel for most of their length but with the leading edge curving strongly backwards at the extremity. In adults (particularly males) it is this curved part of the leading edge of the flipper that carries irregular serrations. These look for all the world like imperfections at a mould joint and project up to 5 mm from the flipper edge. The dorsal fin has an overall triangular shape, but the leading edge tends to be slightly convex and longer than the trailing edge, which may be slightly concave. In height the fin is about 9% (7.7–10.3%) of body length. The tail is crescent-shaped in outline when viewed from above, and its span averages 26.8% (24.7–29.5%) of body length.[3]

The coloration of Heaviside's dolphin is unmistakable: it is a dark blue-black above with a grey cape over the head and thoracic region, a light grey flank blaze and four well-defined white areas on the belly. The grey cape is separated in the middle of the back by a narrow strip of dark pigment that runs back from the blowhole and then widens about one-third of the way down the body, completely displacing the grey cape at about the level of the back of the dorsal fin. The four white areas on the belly consist of a roughly diamond-shaped mark across the chest that wraps around the front of the flippers, small ovoid patches in the 'armpit' region of each flipper, and a large trident-shaped marking that begins just behind the chest patch and continues in the midline as far as the anus, with a separate tongue extending up each side onto the flank. There is considerable variation in the outlines of these marks, particularly the chest patch and the beginning of the trident marking, and a distinct difference between the sexes in the way the trident marking terminates at the anus. In males it tends to taper to a point, whereas in females the marking widens to include the mammary slits, and then terminates more abruptly. Occasional dramatic coloration variants are seen, including at least three individuals in the Cape Town area[3] (and one in the Lüderitz area of Namibia[4]) that were almost completely white.

Tooth eruption commences shortly after birth, and is still incomplete when the dolphin is just over 1 m long. In adults there are 21–28 pairs of erupted teeth in each jaw (n = 33): these are small and peg-like with a maximum diameter of about 2–3 mm.[5]

Distribution and migrations

Heaviside's dolphin is endemic to the subregion, being distributed along the west coast of South Africa, Namibia and southern Angola.[7] The southern (and eastern) limit to distribution appears to be Cape Point, although there are sightings from False Bay to the east of Cape Point, and an odd vagrant has been reported from Plettenberg Bay.[8] The most northerly limit to distribution is not well known but sightings have been made up to 16°30′ S.[9] It is a coastal species, occurring from the surf zone as far as 45 nautical miles offshore (and in water as deep as 180 m), but it is most abundant in water less than 100 m deep.[6]

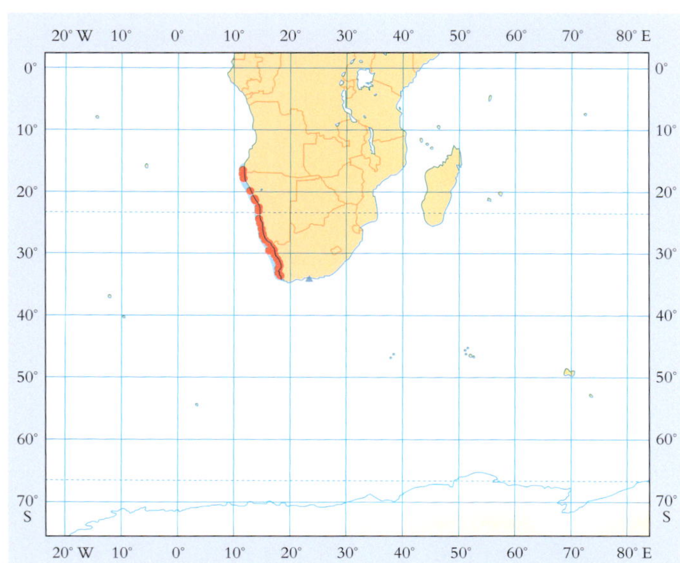

Postulated range Sighting or capture Extralimital record

FIG. 91 Distribution map for Heaviside's dolphin.

FIG. 92 Although the similarity between the white belly markings of Heaviside's dolphin and the killer whale was once thought to be an example of protective mimicry, our knowledge of the dolphin's full colour pattern now makes this seem less likely.

Observations from the shore and the results of satellite tagging in the Western Cape have shown a distinct diel pattern of movement, with animals being closest inshore in the early morning and furthest offshore around midnight. As Heaviside's dolphin feeds mainly on bottom-living species such as hake that migrate towards the surface at night, it is believed that the offshore movement may be timed to coincide with this increased availability of fish.[10, 11] Animals may then move back inshore to rest and socialise.

Locations received via satellite telemetry from five dolphins over 44–55 days indicated that they used home ranges of between 875 and 2 000 km^2, and that towards the end of this period the ranges seemed to be stabilising in size. Although these were all females, and males might be expected to range wider, the information does suggest that at least over this time scale the species shows limited dispersal.[11] A similar conclusion was reached from the distribution of sightings of one or more distinctively-coloured animals over three years[12] and the recovery of a tagged individual after 17 months.[6]

External commensals and parasites

Cyamids (*Isocyamus delphinii*) occurred on 21.4% of 14 Heaviside's dolphins examined shortly after death. In each case a single cyamid was present, all located within recent tooth rakes, so it is perhaps not surprising that all three hosts were male. Another of the dolphins carried three stalked barnacles (*Xenobalanus* sp.), one each on the dorsal fin, tail flukes and flipper. Patches of discoloured skin reminiscent of diatom films have been seen but not sampled.[6]

There are no records of wounds, either open or healed, on Heaviside's dolphins that could be attributable to attacks by small sharks such as *Isistius*; this is presumably a consequence of their nearshore habitat.

Food and feeding behaviour

Fish comprised the majority (73.7% of the reconstituted mass) of the identified prey items in the stomachs of 23 Heaviside's dolphins. Shallow water hake (*Merluccius capensis*) and *Merluccius* sp. constituted 39.1%, and kingklip (*Genypterus capensis*) 15.5%, together making up more than half of the mass of prey consumed. Other minor fish species eaten included bearded goby (*Sufflogobius bibarbatus*, 10.8%), horse mackerel (*Trachurus capensis*, 4.4%) and gurnard (*Chelidonichthys capensis*, 3.4%). Of the cephalopods, which made up the remaining 26.3% of the reconstituted mass, *Octopus* sp. predominated (17.9%), with *Loligo reynaudi* (8.4%) the only other species of importance.[5, 6] Benthic or demersal species seem to play an important role in the diet of Heaviside's dolphin, and several of the principal prey items are also of commercial importance, implying that there is a potential for indirect interactions with fisheries. The hake eaten were mostly juveniles, ranging in total length from 7 cm to 28 cm, with an average of 19.5 cm.[13] Shallow-water hake in this size range migrate vertically from the bottom into subsurface layers at night to forage,[14] where presumably they are more vulnerable to dolphin predation (see Distribution and migrations above).

Four dolphins had also eaten small amounts of crustaceans (up to a maximum of 30 g). One contained at least two Cape mantis shrimps (*Pterygosquilla armata*), a second 30 hyperiid

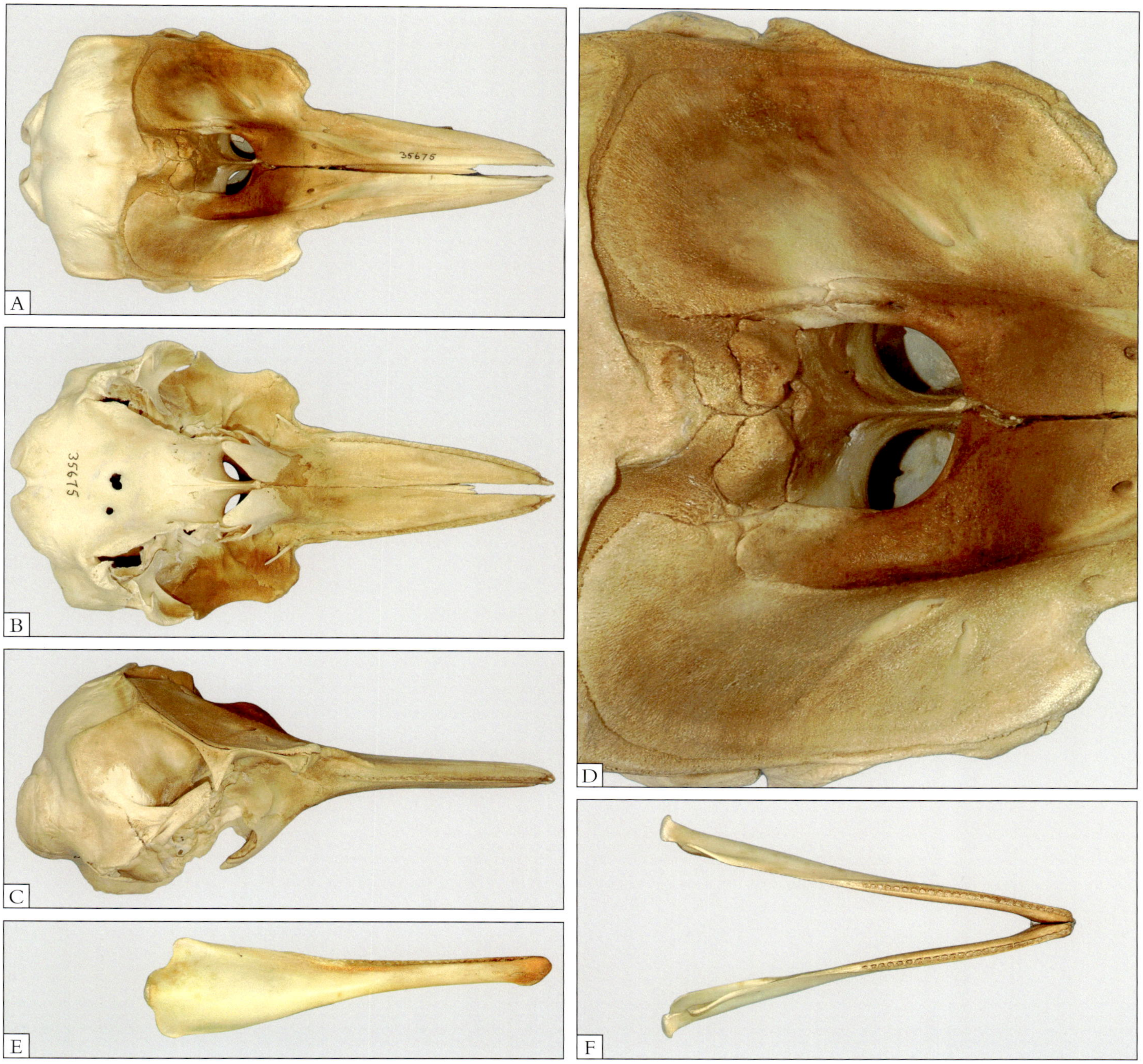

FIG. 93 Dorsal (A), ventral (B) and right lateral (C) views and vertex of skull (D); right lateral (labial) (E) and dorsal (F) views of mandible (ZM 35675).

amphipods and 14 glass shrimps (*Pasiphaea* sp.), a third 17 unidentified euphausiids, and a fourth an unidentified amphipod.[5] Some of these may have been ingested 'accidentally' while feeding on hake, but crustaceans have been recorded as an important item in the diet of other *Cephalorhynchus* spp.[15]

Reproduction and growth

Five calves with an unhealed umbilicus ranged in size from 78.7 cm to 89.5 cm. The smallest of these was considered likely to be an aborted foetus; the next largest was 84.7 cm long. Averaging the remaining four gives a size at birth of about 87 cm. One record was from August, two from December and one from January, and a fifth calf of a similar size (86 cm) but with a damaged umbilicus also occurred in December. Two slightly larger calves (90 cm and 91.8 cm) with partly and completely healed umbilici occurred in January and April respectively, and live neonates (with foetal folds still visible) have been seen in November and December. The balance of evidence suggests that births occur mainly in summer (December/January), though the season may be much more protracted. For example, the only two pregnant animals examined had very similar-sized foetuses (22.2 cm and 28.8 cm) but were examined six months apart (in October and April respectively).[5,6]

The length of gestation and the duration of lactation are unknown. Dolphins less than 110 cm long failed to contain any solid food (or nematodes in the stomach to indicate past prey ingestion), one 118.3 cm individual contained three squid beaks (but no nematodes), and one 138 cm individual contained cephalopod and crustacean remains and nematodes. Weaning may therefore start at a length of 110–118 cm.[5]

All 12 females less than 159 cm were sexually immature, one animal 159 cm long was immature and another was mature, and all but one of eight animals 161 cm or more in length were mature. The exception was a 165 cm female weighing only 46.7 kg, compared to all the other mature females, who weighed 62 kg or more.[5,6] If dolphins with combined testes weights of less than 100 g, 100–160 g, and more than 160 g are immature, pubertal and mature respectively,[6] the largest of six immature males was 155 cm long and the smallest of eight mature males was 155.5 cm. As concluded previously, puberty in both sexes seems to occur at a body length of 1.56–1.59 m.[6]

Very little is known about the frequency of reproduction. Two out of seven mature females examined (28.6%) were pregnant, while the overall proportion of calves seen at sea has been estimated as 9.9%.[6]

Behaviour and vocalisations

Group size in Heaviside's dolphin is generally small, from one to 10 with an average of 3.2 animals, although occasionally larger amalgamations occur.[6]

Aerial activity and social interactions are more prevalent in the larger groups. In particular, percussive leaps (when the dolphin re-enters the water on its side or back) and near-vertical leaps (when the dolphin re-enters the water head-first and almost vertical) are characteristic of such groups. The near-vertical leaps can involve 2–5 individuals leaping sequentially, often towards each other. Other conspicuous behaviours include spy-hopping and tail-slapping. Heaviside's dolphins also display an unusual respiratory pattern, in which several explosive exhalations or 'coughs' are produced in rapid succession as the animal swims slowly at the surface.[6] Whether this is related to the presence of lungworm (with which the adults are almost universally infested) is not known.

While inshore in the early morning, these dolphins seem to favour areas with high surf activity, where they may spend time riding the waves inshore and leaping out of the back as they break. Although feeding behaviour inshore is rarely observed, Heaviside's dolphins have been seen harassing cormorants (probably *Phalacrocorax carbo lucidus*, the white-breasted cormorant) fishing in shallow water, either in play or in an attempt to purloin fish the birds have caught.[5]

Two satellite-tagged dolphins monitored for 50–112 days provided information on diving behaviour, but as the information was binned by 6-hour periods it is not easy to interpret. Both dolphins made most (more than 90%) of their dives to less than 30 m, and for less than 2 min. in duration. The maximum depths reached were around 100 m and the maximum durations were about 6–10 min. Diel patterns were not consistent between the two dolphins.[16]

Vocalisations have only been recorded in animals held temporarily captive, and then with equipment that could not record sounds above 10 kHz. As other *Cephalorhynchus* species produce narrow-band, high-frequency pulses around 120–125 kHz, it is likely that the sounds recorded were not fully representative of the dolphins' repertoire. Sequences of low-frequency clicks and some 'cry' sounds (consisting of a high repetition of pulsed sounds) were detected. No pure tones (whistles) were recorded, and *Cephalorhynchus* species in general are believed to be non-whistling dolphins.[17]

Conservation status

A mark-recapture estimate using re-sightings of individually identified dolphins indicates a population of some 6 300 animals south of about 32° S. As the proportion of marked animals in this population is low, this estimate includes a large extrapolation factor (*c.* x 7).[10] Estimates of genetic diversity also suggest a large effective population size,[18] and published figures of 800–1 000 for the total population seem unrealistically low.[19]

Nevertheless, several factors make Heaviside's dolphin, like its congeners, susceptible to mortality through entanglement in fishing gear. Its nearshore habitat means that it is likely to encounter nets set for mullet, sharks and other fish on the west coast, where the generally turbid water severely restricts visibility. What is worse, being a non-whistling dolphin, its sonar may resemble that of a porpoise in being quite limited in range, and apparently unable able to detect a gill net at a distance of more than 3–6 m.[20] Entanglements and subsequent mortalities do occur, but owing to the informal nature of most of the set-net fisheries and the legal ban on the capturing of dolphins, it is difficult to obtain reliable estimates of the level of such take. In the dolphins' favour may be the regional abundance of Cape fur seals *Arctocephalus pusillus pusillus*, whose purloining of fish from set nets makes long soak times inadvisable.

Heaviside's dolphin is listed as Data Deficient by the IUCN and in the South African Red Data Book. It is included on Appendix II of CITES and CMS.

References

1 Fraser 1966
2 Watson 1981
3 Best 1988b
4 Fertl *et al.* 2004
5 MRI Whale Unit data
6 Best and Abernethy 1994
7 Findlay *et al.* 1992
8 Cockcroft, V.G., pers. comm.
9 Roux, J-P., pers. comm.
10 Elwen PhD thesis
11 Elwen *et al.* 2006
12 Rice and Saayman 1984
13 Sekiguchi *et al.* 1992b
14 Pillar and Barange 1993
15 Bastida *et al.* 1988
16 David *et al.* 1998
17 Watkins *et al.* 1977
18 Jansen van Vuuren *et al.* 2002
19 Carwardine 2002
20 Kastelein *et al.* 2000a

Commerson's Dolphin

Cephalorhynchus commersonii

Ventral view

TWO DISJUNCT POPULATIONS of this species exist, one on the eastern seaboard of southern South America and adjacent waters, and the other around the Kerguelen Islands in the southern Indian Ocean. They differ in size, colour pattern, vocalisations and some skull characteristics, and are genetically isolated. The suggestion has been made that they are different enough to be separate subspecies,[1] but no formal names have yet been proposed. Because the one from the Kerguelen Islands occurs in the subregion, the description given here will refer principally to this population.

Appearance

Commerson's dolphin is unlikely to be confused with any other small cetacean in the subregion because of its characteristically rounded and backward-pointing dorsal fin (resembling a Mickey Mouse ear), and its striking black-and-white colour pattern.

In the Kerguelen Islands population, Commerson's dolphins grow as large as 166–174 cm in length, or about 30 cm longer than those from the South American population.[2] Off South America adult females are larger than males by about 5 cm,[3] and it is probable that similar sexual dimorphism applies to the Kerguelen population, where the largest male measured was 166.5 cm and the largest female was 174 cm. The general body shape is similar to that of Heaviside's dolphin, with a conical head and no well-defined beak, although a rostral groove separates the melon from the rostrum on each side of the head. It is a robust species, with the girth in front of the dorsal fin being 60.4–67.2% of the body length. The flippers are small (14.8–16.9% of body length)[2] and are described as elliptical and smoothly rounded at the tips. As in Heaviside's dolphin, serrations at the tips of the leading edge of the flippers have been described in dolphins from the South American population, where these develop in mature animals of both sexes and especially males. Unlike its congener, however, these serrations develop mainly on the left flipper in Commerson's dolphin, and only rarely on the right.[4] Animals from the Kerguelen population have not been described with such serrations, and none can be seen in a photograph of the left flipper of a large male.[2] The dorsal fin is rounded or lobate at the tip. Its shape can vary from being acutely angled on the leading edge and slightly undercut on the trailing edge to more upright on the leading edge and deeply S-shaped on the trailing edge: there is no sign of sexual dimorphism. Fin height seems to be related to body length,[5] and is relatively less than that of Heaviside's dolphin, being 6–7% of body length.[2] The tail flukes are strongly concave in their posterior margin (more so apparently than in Heaviside's dolphin),[5] and their width ranges from 22.0% to 24.6% of body length (or somewhat less than in Heaviside's dolphin).[2]

The colour pattern of Commerson's dolphin is striking. On the upperside the head is black as far back as about the level of the eyes but this colour continues down as a broad stripe connecting the flippers, which are also black. There is a flask-shaped white throat patch, while a broad black band across the chest joining the flippers extends back in the midline as a sharp point. The dorsal fin is black and surrounded by a black shield-like marking in front that extends back as a broad spinal field in the midline to connect with the black posterior quarter of the body. The genital region carries a black blaze, which (as in Heaviside's dolphin) varies in outline between the sexes, tapering to a point posteriorly in males but ending in a bracket-shape in females. In the South American population the rest of the body is pure white,[5] but in the Kerguelen population there is a grey cape over the anterior part of the back and onto the flanks, and only the abdomen and lower chest are pure white.[2] Individual variation occurs in the shape of the border between black and white/grey areas, such as behind the blowhole, or where the spinal field descends to cover the last quarter of the body, but major anomalies have not been seen. Calves in the South American population are not born with the adult coloration but with a largely muted colour pattern of grey on darker grey and black, the grey areas corresponding essentially to those that are white in the adult. Some individuals develop the adult colour pattern as early as five months of age, while others (particularly females) maintain a grey tinge to the lighter areas of the body until one or two years of age.[5,6]

Length:weight relationship

Mass data from 19 Commerson's dolphins from Kerguelen[2,7] have been combined with those from 12 individuals from South American waters,[3] as the size range for the Kerguelen animals was limited. All animals are believed to have been weighed whole, to produce the following relationship: $W = 17.439L^{2.835}$, where W = weight in kilograms and L = total length in metres. The overall fit is good ($r^2 = 0.968$), indicating that combining the data was a valid procedure. This equation implies that a 1 m dolphin would weigh 17.4 kg and a 1.5 m dolphin 55 kg.

Skeletal data

Vertebral formula, Kerguelen population (n = 11): 7 cervical, 12–14 thoracic, 14–16 lumbar and 29–32 caudal; total 64–66.[8]

The atlas and axis are fused; the remaining cervical vertebrae are free, although the fourth and fifth are occasionally fused.[5]

Phalangeal formula, including metacarpals, for the Kerguelen population (n = 11): I_2, II_{7-8}, III_5, IV_{2-3}, V_1.[8]

For the cranial characteristics that separate Commerson's dolphin from Heaviside's dolphin, see the latter's species account.

South American animals have 24–33 pairs of erupted teeth in each jaw (n = 81), less than 3 mm in diameter,[5] while animals from Kerguelen seem to have 1–2 pairs more in each jaw.[3]

Distribution and migrations

Commerson's dolphin occurs in two distinct populations in the southern hemisphere. One is centred on the east coast of South America between about 40° S and Cape Horn, with sightings further south in the South Shetland Islands, and to the west at the Falkland Islands. The second is situated 8 500 km to the east, at Kerguelen Islands in the southern Indian Ocean.[6] Until recently, the Kerguelen population provided the only records of the species in the subregion. In November 2004, however, a sighting attributed to this species was made at 35°54.26′ S 20°32.69′ E, or about 130 km south-east of the southern tip of Africa.[9] At the time the dolphin was in a water depth of 130 m and a sea surface temperature of 19.9 °C, close to the edge of the Agulhas Bank, and about 4 200 km west of Kerguelen and 6 300 km east of South America. Although closer to the Kerguelen population, it is more likely that this individual travelled with, rather than against, the prevailing current system (the West Wind Drift), and so originated from South America. This would be consistent with the genetic evidence, which suggests the Kerguelen population originated from a small founder group from the South Atlantic.[10]

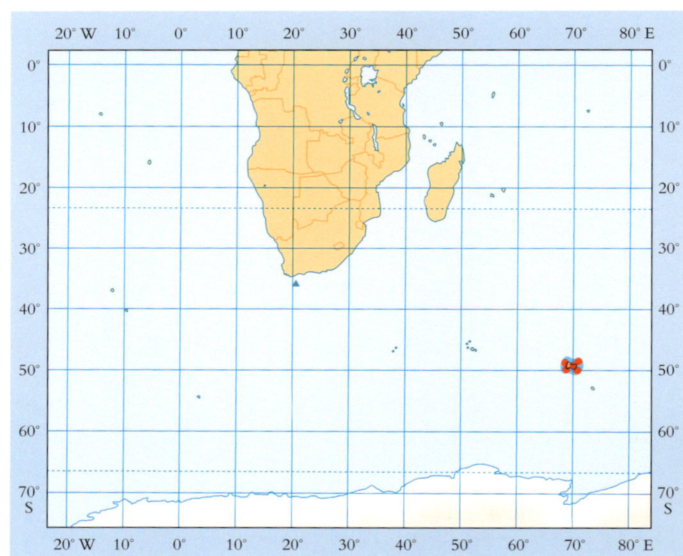

Postulated range ■ Sighting or capture ▲ Extralimital record

FIG. 94 Distribution map for Commerson's dolphin.

All members of the genus *Cephalorhynchus* are normally coastal animals, and Commerson's dolphin is no exception. In South America most sightings have been made near the mouths of bays and estuaries, or over the continental shelf (in water less than 200 m deep), and have occurred as far as 30 km up Patagonian rivers that are tidal. Kelp beds are also frequented.[6] The few pelagic records in deeper water (in the Drake Passage and near the South Shetlands) have been regarded with some reservation.[5,11] At Kerguelen, most sightings have been made within the sheltered waters of the Gulf of Morbihan and also on the outer coast, usually in water less than 100 m deep, but

FIG. 95 These Commerson's dolphins were photographed at the Falkland Islands: those at Kerguelen tend to have more grey within the white areas.

sightings away from the islands but still over the plateau (and so in water less than 200 m deep) have been made.[12, 13]

Although there are anecdotal accounts of a seasonality of occurrence of Commerson's dolphins in South America, these are not supported by the incidence of sightings.[5] At Kerguelen there was a noticeable decrease in the number of dolphins seen per trip in the Gulf of Morbihan between June and December, suggesting an offshore movement in winter.[12] This was confirmed by the seasonal occurrence of 16 tagged individuals that left the Gulf in winter and were re-sighted there again the following summer. While within the Gulf of Morbihan the movements of these tagged animals were quite restricted, the maximum distance between re-sightings being 24 km, with the dolphins mainly occupying an area 20 x 3 km, or about 5% of the available space in the gulf.[7]

External commensals and parasites

No external parasites have been recorded from Commerson's dolphin. Small circular depressions in the skin 2–4 cm in diameter have been reported from individuals in the Kerguelen population, and tentatively attributed to past attachments of barnacles such as *Coronula*.[2]

There are also no records of wounds, either open or healed, that could be attributable to attacks by small sharks such as *Isistius*. This is presumably a consequence of their nearshore habitat.

Food and feeding behaviour

Commerson's dolphins at Kerguelen are largely fish-eaters, with the dominant species being the semi-pelagic mackerel ice-fish (*Champsocephalus gunnari*), of which juveniles 15–20 cm long are taken. Other fish eaten include the more benthic Antarctic cod (*Notothenia*). Although of relatively minor importance, crustaceans, especially hyperiid amphipods, the flat spider crab (*Halicarcinus planatus*), and euphausiids were frequently found in stomachs. The predation on *Halicarcinus* is revealing in that it implies that the dolphins must forage on the sea floor at times. This is supported by the presence of ascidians (sea squirts) and the tubes of annelid worms in one stomach. Pieces of algae were also frequently found, but were considered to have been ingested accidentally.[14]

Based on their frequency of occurrence in the stomachs of 45 dolphins taken accidentally in inshore net fisheries, mysid shrimps (*Arthromysis magellanica*, 22.5%) and fish – silverside (*Austroatherina*), sardine (*Sprattus fuegensis*) and tailed hake (*Macruronus magellanicus*) – at 29.4% were the most important food items for the South American population. Patagonian squid (*Loligo gahi*) and other cephalopods accounted for 14.1%. Algae (10%), free-living benthic invertebrates such as the isopod *Cirolana* and a polychaete worm (almost 10%), and sessile benthic invertebrates such as the colonial tunicate *Polyzoa* (6.7%), were also considered as prey items. Some of the benthic species, such as *Cirolana*, are usually buried in soft sediments, suggesting that the dolphins may stir up the bottom to locate them. Alternatively, the surge in shallow water may cause some of these organisms to become suspended above the bottom.[15] A separate study of the stomachs of nine dolphins taken in offshore trawls suggested that hake (*Merluccius hubbsi*) was the preferred species, with short-fin squid (*Illex argentinus*) about one-third as important.[16]

Cooperative feeding behaviour, where up to 15 dolphins herd fish against the shore (including stranding themselves in some instances), has been recorded in both populations.

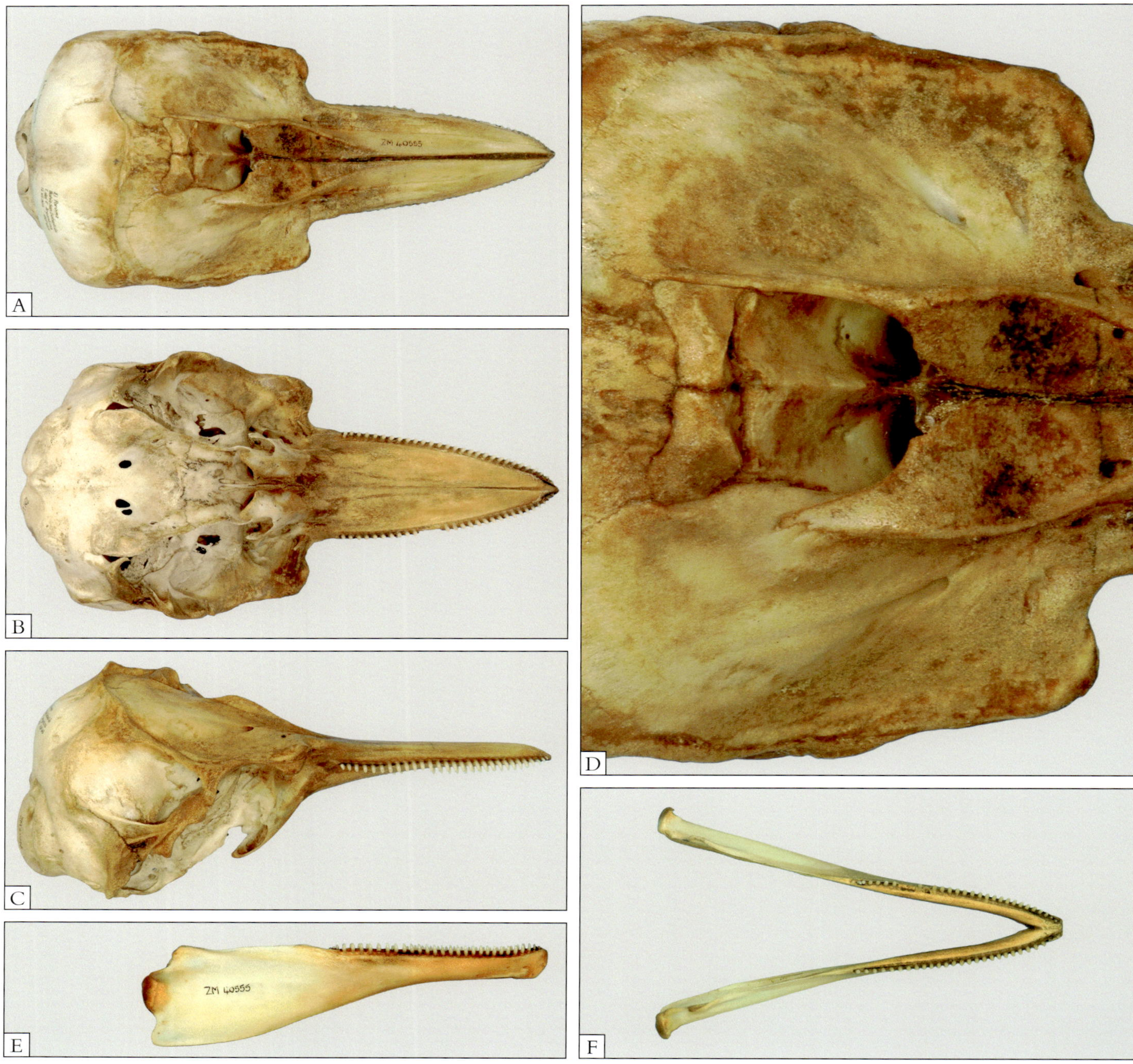

FIG. 96 Dorsal (A), ventral (B) and right lateral (C) views and vertex of skull (D); right lateral (labial) (E) and dorsal (F) views of mandible (ZM 40555).

Encirclement of fish shoals and reciprocal feeding by individuals from groups of 2–6 dolphins has also been observed away from the shore.[5]

In captivity, adult Commerson's dolphins consume between 9% and 12% of their body weight per day, depending on a variety of factors. Food intake is reduced only slightly during pregnancy, but increases greatly during lactation.[17]

Reproduction and growth

The first three calves (from the South American population) born in captivity were said to be 55–65 cm long at birth,[6] but this compares with a 64.3 cm 'nearly full-term' foetus and a 73.4 cm near-term foetus in the wild[5] and a 91 cm captive calf that died shortly after birth.[18] The earlier sizes were possibly estimates of unrestrained calves. Through monitoring captive animals, it has been established that gestation lasts for 345 ±20 days (n = 8).[19] Births in the South American population seem to occur in summer, between November and February.[5] A captive calf nursed about every 40 min. during the first four days of its life, each nursing bout consisting of an average of eight attachments, each lasting 5.5 sec. Full dependence on milk lasted 73–236 days (mean 153 days) in six captive calves, although they continued to suckle for some months after milk dependence ceased.[18] In less than a year one calf had grown to 80% of its mother's size,[6] and after two years calves are approaching adult size.[17] In the wild,

South American dolphins in their first year are on average 99.1 cm and in their second year 111.7 cm in length. Sexual maturity in males is reached at 5–6 years of age and at a body length of 127–131 cm. In females it is less well established and occurs between four and eight years.[3] Frequency of reproduction is unknown in the wild, but intervals between births in captive animals ranged from 16 to 37 months, with an average of 25.4 months (n = 5).[19]

Less information is available for the Kerguelen population. Four out of five mature females taken in January had recently ovulated, suggesting that breeding takes place in summer. Sexual maturity in females occurred at the age of five years and at a body length of about 165 cm. Information for males was less conclusive. The annual ovulation rate was estimated as 0.5–1.2.[20] Plotting the corpora counts against age, including the two immature animals, produces an ovulation rate of 0.83, or one ovulation every 14 months.

Behaviour and vocalisations

Commerson's dolphins in both the South American and Kerguelen populations are normally found in small schools of fewer than 10 animals, with a mode at two or three animals. Occasionally larger groupings of 100 or more have been encountered in both populations, but whether these are temporary accumulations of smaller groupings or more permanent associations is unknown.[4, 12, 13] Re-sightings of tagged individuals in the Kerguelen population suggest that the composition of schools is variable and is not based on stable relationships between individuals.[7] In a captive situation, males appear dominant over females and contest with one another for dominance and access to females. At any one time a single male may be dominant and is often challenged by lower-ranked males; less aggressive males may seek refuge in an adjacent pool.[21]

During courtship behaviour in captivity, Commerson's dolphins frequently use their flippers for tactile stimulation, vibrating the front edge across the body, genital slit or flippers of other animals. This behaviour is almost always performed by adult males, and almost invariably using the left flipper, supporting the idea that the flipper serrations serve to enhance tactile stimulation during social contact.[22]

In the wild, Commerson's dolphins are fond of riding waves of all sorts, including bow waves of boats. While doing so they may spin on their longitudinal axis. Swimming upside down is also a behaviour commonly seen in the wild and in captivity.

Other play activities observed include repeatedly nudging gulls (*Larus dominicanus*) that are resting on the surface, causing them to lift off briefly, and pushing objects (even inflated inner tubes) around. The species is capable of varied and repeated aerial acrobatics.[5]

Like other members of the genus, the acoustic repertoire of Commerson's dolphin contains no pure tones (whistles), only clicks. These are mainly short, narrow band, and ultrasonic in nature, with the dominant frequency about 125 kHz, and with high repetition rates (up to 500–600/sec.). The highest click rates generate separate, lower-frequency 'cry-like' sounds audible to the human ear (the dominant component at about 1 kHz). These seem to be produced particularly when the animal is under stress or excited. Separate, low-frequency (to about 6 kHz), broadband clicks are also sometimes produced. Some differences in the nature of the sounds produced by dolphins from Kerguelen and South America have been proposed as additional evidence of intraspecific divergence between the two populations.[23–25]

Conservation status

Estimates of the size of the Commerson's dolphin population exist for the eastern or north-eastern parts of the Straits of Magellan, ranging from 718 (summer 1989/90)[26] and 1 206 (June 1996)[27] to 3 211 (January/February 1984).[28] These were all aerial surveys using line transect methodology, and the differences between their results may reflect genuine variations in dolphin abundance owing to seasonal movement or inter-annual shifts in distribution, or may simply be an expression of sampling variability (the 95% confidence interval for the 1996 estimate, for instance, which is the most precise, ranged from 711 to 2 049 dolphins). None of these estimates corrected for dolphins that may have been over-flown or not detected, so are presumably underestimates. There are no published estimates for other parts of the species' range, including Kerguelen.

Commerson's dolphins are believed to be vulnerable to entanglement in set nets, in part because of the characteristics of their sonar.[29] At least 5–30 dolphins die each year in nets set close inshore in eastern Tierra del Fuego alone, and it is known that the species is taken in similar fisheries further north in Argentina, in the eastern Strait of Magellan and in southern Chile.[6] A preliminary estimate of the annual take in an area in southern Argentina is 179 (CI 141–212).[30] Some dolphins may be deliberately taken for crab bait, and although there are no reliable estimates of the size of this catch, it has been proposed that as many as 2 350 dolphins of all species are taken annually just to supply the crab fishery.[5] The annual take of Commerson's dolphin in midwater and bottom trawls off Argentina has been estimated as 25–171 per year.[16] At least 35 animals have been captured for display in oceanaria worldwide since 1978. No incidental mortality is known for the Kerguelen population.[6]

Commerson's dolphins are described as Data Deficient by the IUCN and are not listed in the South African Red Data Book. The species is included on Appendix II of CITES and the South American population is listed on Appendix II of CMS.

References

1 Robineau 1986
2 Robineau 1984
3 Lockyer *et al.* 1988
4 Goodall *et al.* 1988b
5 Goodall *et al.* 1988a
6 Goodall 1994
7 De Buffrénil *et al.* 1989
8 Robineau and de Buffrénil 1985
9 De Bruyn *et al.* 2006
10 Pichler *et al.* 2001
11 Brownell 1974
12 Robineau 1985
13 Robineau 1989
14 Robineau and Duhamel 1984
15 Bastida *et al.* 1988
16 Crespo *et al.* 1997
17 Kastelein *et al.* 1993b
18 Kastelein *et al.* 1993a
19 Asper *et al.* 1992
20 Collet and Robineau 1988
21 Cornell *et al.* 1988
22 Johnson and Moewe 1999
23 Dziedzic and De Buffrénil 1989
24 Kamminga and Wiersma 1982
25 Dawson 2002
26 Lescrauwaet *et al.* 2000
27 Venegas 1996
28 Leatherwood *et al.* 1988
29 Evans *et al.* 1988
30 Iñíguez *et al.* 2003

Rough-toothed Dolphin

Steno bredanensis

THIS SPECIES GETS its common name from the fine vertical ridging on the sides of its teeth, making them rough to the touch, but these are really only obvious in juveniles as they tend to get worn away with age and may be almost completely absent in old adults.[1] For many years it was not clear whether rough-toothed dolphins really occurred in southern African waters, as early records were of museum specimens of dubious origin. A skull sent from the South African Museum to the Natural History Museum in London in 1865, for instance, had been presented by a Simon's Town resident, and the catalogue annotation said that it was 'probably obtained from some vessel putting into Simon's Bay'.[2] A second skull purchased from a London dealer by the Australian Museum early in the twentieth century was simply labelled 'South Africa'. Eventually the species could be added to the fauna list when an adult female stranded in KwaZulu-Natal in 1984[3] and an adult male at Hermanus in 1993.[4] Because there are relatively few records within the subregion, much of this account is based on rough-toothed dolphins from other areas.

Appearance

Rough-toothed dolphins are relatively easy to recognise at sea when they are close enough, as they are the only long-beaked dolphins with no visible distinction between beak and melon – the one just merges smoothly into the other (hence the somewhat irreverent name 'slopehead'). They also have a distinctive dark cape down the back with a constriction (or 'neck') in front of the dorsal fin.

The largest male measured was 2.65 m and the largest female 2.55 m[1] (although an unsexed individual 2.75 m long has been recorded[5]). Asymptotic lengths (at which growth ceases) are said to be 2.31 m for both sexes in the Japanese population (although the information on which this statement is based is not clear), and adult females may have proportionately longer beaks.[1] Apart from their characteristic head shape, they are distinctly stocky ('barrel-shaped') in the front half of the body and more slender towards the rear. The eyes have also been described as unusually protruding ('bug-eyed'). The dorsal fin is tall (9–10% of the body length in height)[5] and only slightly recurved at the back, while the flippers are relatively large (17–19% of body length)[6] and broad, although strongly hooked at the tip. Some adult males may possess a pronounced keel or hump behind the anus.

Rough-toothed dolphins tend to be somewhat variable in coloration. Consistently, a grey cape on the side runs from just above the eye up towards the back above the flipper, and then

Length:weight relationship

Mass data from 12 rough-toothed dolphins weighed whole are available.[4,8,9] Because the data only cover a limited size range (2.21–2.5 m), the following relationship is of limited value for extrapolation to animals outside this range: $W = 21.728L^{1.9641}$, where W = weight in kilograms and L = total length in metres. This implies that a 2.3 m dolphin would weigh 111.6 kg.

Skeletal data

Vertebral formula (n = 28–9): 7 cervical, 13 thoracic, 13–18 lumbar and 28–33 caudal; total 65–67.

The atlas and axis are fused; the remaining cervical vertebrae are free.[1]

Phalangeal formula, including metacarpals (n = 2): I_{2-4}, II_{7-8}, III_{5-6}, IV_{2-3}, V_{2-3}.[4,10]

Apart from the roughened teeth, the characteristics of the skull that differentiate *Steno* from *Sousa* (the only species it is likely to be confused with in the region) include: 1) the lower tooth count; and 2) a cylindrical ridge on the ventral surface of the skull, straight or slightly curved anteriorly, running diagonally forwards from the base of the pterygoid towards the antorbital process. In *Sousa* this ridge is not cylindrical or as prominent and is strongly curved posteriorly.[11]

sweeps downwards at the middle of the body: between the capes on each side the back is dark. The distinctiveness of this cape, however, may vary between individuals. In some the throat and lower jaw are distinctly paler, having grey streaks or even being completely white, and in some the tip of the upper jaw may also be white. There is usually an irregularly-shaped white blaze in the genital region that varies considerably in size and shape, and may extend partially up the side. Where the belly is pale this may be flecked with dark grey spots. Large white spots, some of which could represent the attentions of cookie-cutter sharks *Isistius* sp., can also occur over the belly and sides. In general, juveniles are more muted in coloration and have less spotting than adults.[1,7,8]

There are 19–26 pairs of teeth in the upper jaw and 19–28 in the lower jaw (n = 81–82). They are stout (about 7–8 mm in maximum diameter) and conical, with vertical rugosities on the crown, especially in juveniles.[1]

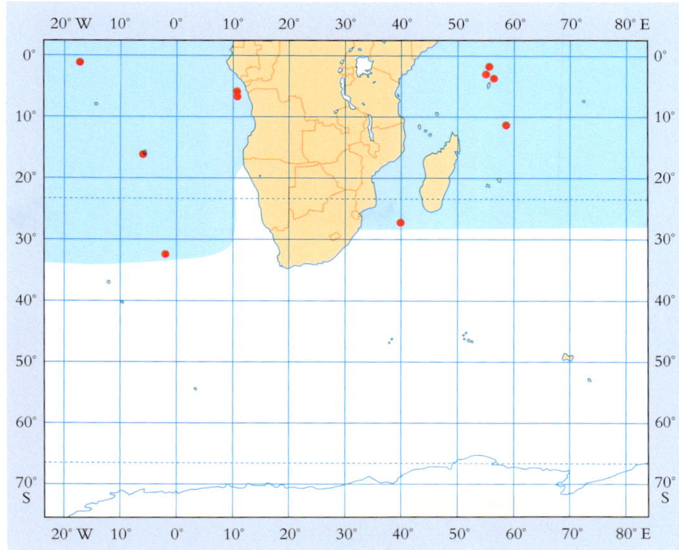

FIG. 97 Distribution map for rough-toothed dolphin.

Distribution and migrations

Rough-toothed dolphins occur worldwide in tropical and warm-temperate waters, including the Mediterranean. In the subregion they have been recorded from St Helena,[12] Angola,[13] Namibia, South Africa,[14] Tanzania[15] and various offshore localities. All the records to date from South Africa and Namibia have been strandings. As there have been no authenticated sightings of rough-toothed dolphins in either South African or Namibian waters, the strandings have been attributed to individuals at the extreme southern limits of their range.[14] A reported stranding at Tristan da Cunha[16] seems unlikely given the prevailing oceanography, and probably refers to a specimen taken at sea in 1874 some 1 000 km ENE of the island group, which (at 32°29′ S) is still the most southerly at-sea record in the subregion.[17]

Rough-toothed dolphins are generally considered to be deepwater animals, most sightings occurring in depths of 500 m or more.[7, 18, 19] In the eastern tropical Pacific, rough-toothed dolphins are the only cetaceans known to be associated with floating debris, 17% of sightings being made in association with flotsam such as logs, dead whales, wooden pallets, etc., possibly because these attract certain species of large fish.[20]

FIG. 98 From their long tapering beak and lack of a distinct 'forehead' it is easy to see why rough-toothed dolphins are sometimes referred to as slopeheads.

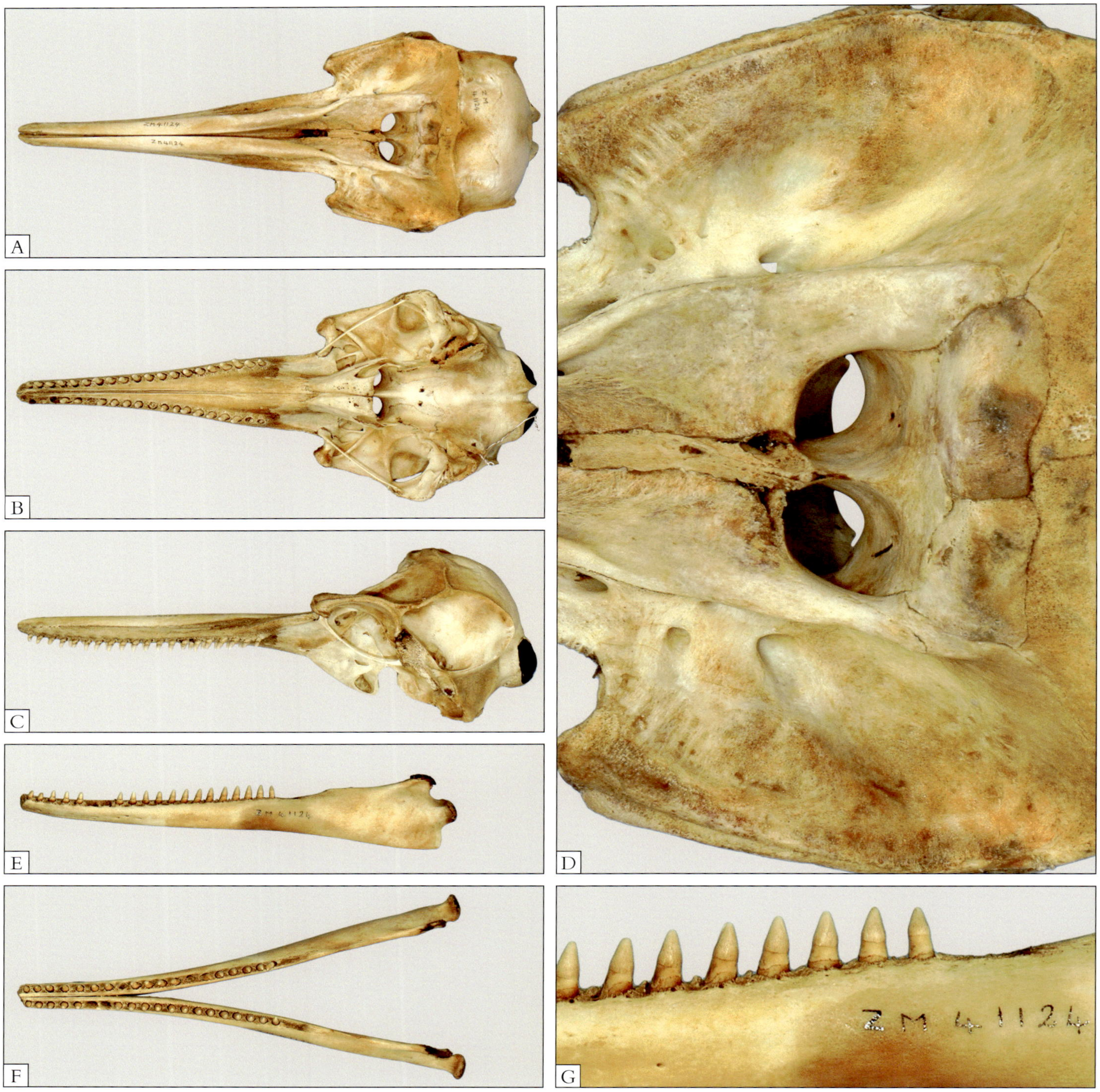

FIG. 99 Dorsal (A), ventral (B) and left lateral (C) views and vertex of skull (D); left lateral (labial) (E) and dorsal (F) views of mandible; and tooth detail (G) (ZM 41124).

There is no information suggesting that the species undertakes seasonal migrations. Sightings in the Canary Islands and Windward Islands, for instance, are made year-round.[7, 19]

External commensals and parasites

Cyamids (*Isocyamus delphinii*) have been recorded from rough-toothed dolphins elsewhere.[21] Given the suggestion that *Isocyamus* is metabolically dependent on scar tissue,[22] *Steno* must be a popular host as it frequently carries wounds attributed to the bites of the cookie-cutter shark *Isistius* sp. Stalked barnacles (*Xenobalanus globicipitis*) have been photographed on the tip of the dorsal fin of a live animal.[23]

Food and feeding behaviour

There is no information from the subregion. Elsewhere, stranded specimens have been recorded as feeding on fish – silverside (*Pranesus insularum*), saury (*Cololabis adocetus*) and houndfish (*Tylosurus crocodilus*)[1] – and cephalopods, including the cranchiid

Dolphins with attitude

When the first rough-toothed dolphins were brought into captivity, it soon became clear that they were dolphins with a difference. Apart from their somewhat reptilian looks, their behaviour was not as docile as that of the bottlenose dolphin. Described as 'insisting upon a degree of dignity and equality with humans', newly caught rough-toothed dolphins tended to bite people without compunction. If handled carefully, however, they became the tamest of all trained dolphins, but continued to pick and choose their human friends with care. They proved to have extraordinary manipulative abilities, opening gates with their mouths and pulling hypodermic needles from other dolphins being given antibiotics. On tank-cleaning day, recently captured individuals would pick up the nets with their teeth and duck under them.

They were full of curiosity and utterly careless about hurting themselves, wedging themselves through closing gates if they felt they had a right to be in a particular tank! Above all, they were remarkably quick learners, with a tremendously long attention span and desire to solve puzzles, sometimes even carrying on working for a solution after becoming satiated with fish rewards.[1,2] Two rough-toothed dolphins were the subject of experiments that demonstrated higher-order learning, or the ability to combine facts to learn a principle. If a behaviour pattern was novel it was reinforced by the trainers, following which the dolphins spontaneously produced up to 16 behaviours, including four that the trainers had never observed previously in four years of working with 50 individuals of seven cetacean species. Eventually the aerial behaviours, although novel, became too complex to be described.[3] These captive studies supported the impression gained from field observations that these were dolphins that exhibited highly flexible and opportunistic behaviour with an unusual degree of coordination and cooperation between individuals.

References

1 Pryor 1975
2 Norris 1976
3 Pryor *et al.* 1969

squid *Teuthowenia*, blanket octopus (*Tremoctopus violaceus*) and tropical arrow squid (*Loligo plei*).[14, 24, 25] The alga *Sargassum filipendula* has also been recorded in stomachs of stranded animals, possibly ingested accidentally.[24] At-sea observations have been made of rough-toothed dolphins feeding on epipelagic fish such as flying fish,[19] on snipe fish (*Macrorhamphosus scolopax*),[26] sharksucker (*Echeneis naucrates*)[27] and even on dorado (*Coryphaena hippurus*) up to 1 m long, where the dolphins fed by peeling off the flesh. This behaviour could lead to such large fish being under-represented in stomach content analyses based on trace remains such as otoliths, while smaller prey items in the dorado's stomach (if consumed) could be transferred to the dolphin.[20]

Several reports refer to the opportunistic and likely cooperative nature of predation by rough-toothed dolphins. They have been seen feeding near trawlers,[23] trapping fish against the shore or possibly using the side of the vessel to block the escape of prey,[28] catching sharksuckers while interacting with humpback whales,[27] and sharing large items of prey they have caught.[20, 29]

Newly captured rough-toothed dolphins gutted and de-headed all fish, irrespective of size, by smashing them against the water's surface,[30] and similar behaviour has been seen in wild dolphins.[28]

Captive animals are reported to consume 59.5 kcal per kg of body weight per day.[1]

Reproduction and growth

The length at birth is unknown, but (as foetuses up to 86.8 cm have been recorded[5]) is probably about 1 m.[1] However, a female rough-toothed dolphin in captivity gave birth to a calf estimated to be 60 cm long, which was a hybrid conceived with a bottlenose dolphin.[31] There are no reliable estimates of the length of gestation, and it is not known how seasonal conceptions are. The hybrid calf started to play with fish at two months of age, by which time it had increased in length by about 15 cm, and by 10 months it was carrying fish around and may have swallowed some.[31] A male 1.67 m long may still have been nursing,[8] but there is no estimate of how long lactation lasts. All males 2.23 m or more in length (n = 11) and 14 years or more in age were sexually mature.[8, 9] Females attain sexual maturity at 2.1–2.2 m and at about 10 years of age.[1] The frequency of reproduction is unknown, but ovarian examination suggests that the species may be seasonally polyoestrous.[8] A captive female that miscarried shortly after arrival in captivity gave birth again (to a hybrid calf) about two years later.[31] Calves were present in 16% of sightings in the Canary Islands[7] and in 29% of sightings in the Windward Islands.[19]

Hybridisation between a female rough-toothed dolphin and a male bottlenose dolphin has occurred in captivity, producing a female calf that survived about four years.[32]

Behaviour and vocalisations

Rough-toothed dolphins usually occur in groups of one to 35–50 animals, with averages of 10.8 (n = 38) and 16.8 animals (n = 137) in the Windward Islands and Canary Islands respectively. Occasionally larger groupings (of up to 160 or even 300 animals) may occur,[1] and some strandings have involved as many as 118 animals.[5] They are sometimes seen in mixed groups with other species, such as melon-headed whales, Fraser's dolphins, bottlenose dolphins, Atlantic spotted dolphins, pantropical spotted dolphins and spinner dolphins.[7, 19, 33]

An outstanding feature of rough-toothed dolphin schools is the formation of tight subgroups of 2–7 animals that tend to surface synchronously.[7, 23] It has been proposed that such behaviour might be an energetic adaptation for travelling in a pelagic dolphin that facilitates 'eavesdropping', in which non-vocalising members of a subgroup can gain information on targets ahead by listening to echoes from signals produced by the vocalising member.[34] Satellite telemetry of three rough-toothed dolphins rehabilitated from a mass stranding indicated that after release they stayed together for the 20–23 days that they were monitored.[35]

In open-ocean diving tests a rough-toothed dolphin reached 30.5 m, but the tests were discontinued when the dolphin absconded, apparently in response to circling sharks. Nevertheless the impression was gained (from the animal's physical conformation and behaviour) that this was primarily a diving species that could probably go deeper.[36] Subsequently a rough-toothed dolphin was reported rubbing against a hydrophone at a depth of 70 m.[37] However, two individuals rehabilitated from a mass stranding and monitored for up to 23 days after release rarely dived deeper than 100 m or for longer than 3 min.[35]

Rough-toothed dolphins produce pure-tone whistles with maximum energy between 2 kHz and 14 kHz, and broad-spectrum clicks between 0.1 kHz and 200 kHz. The clicks are very brief (50–250 μsec in duration) with the highest frequencies directed forward in a very narrow beam. Very loud, low-frequency-rich clicks may be produced when the animal is in an orientation mode.[38, 39]

The deep body temperature of a captive rough-toothed dolphin ranged from 36.5 °C to 37.6 °C, depending on activity, with a mean of 36.9 °C.[40]

Conservation status

There are no estimates of the number of rough-toothed dolphins in the subregion, but from the few records available the impression is gained that the species is not very abundant.

It has not been identified as being involved in any fisheries interaction, though rough-toothed dolphins have been implicated as a by-catch in fisheries off north-eastern Brazil.[41]

Rough-toothed dolphins are listed as Data Deficient by the IUCN and in the South African Red Data Book. The species is included on Appendix II of CITES, but is not listed by CMS.

References

1 Miyazaki and Perrin 1994
2 Layard 1862
3 Ross *et al.* 1985
4 MRI Whale Unit data
5 Cadenat 1949
6 Jefferson 2002
7 Ritter 2002
8 Miyazaki 1980
9 Cadenat 1959
10 True 1889
11 Van Waerebeek *et al.* 1999
12 MacLeod and Bennett 2007
13 Weir, C., pers. comm.
14 Findlay *et al.* 1992
15 Chande *et al.* 1994
16 Watson 1981
17 Peters 1876
18 Davis *et al.* 1998
19 Gannier and West 2005
20 Pitman and Stinchcomb 2002
21 Martin and Heyning 1999
22 Sedlak-Weinstein 1992
23 Addink and Smeenk 2001
24 Layne 1965
25 Dos Santos and Haimovici 2001
26 Steiner 1995
27 Wedekin *et al.* 2004
28 Lodi and Hetzel 1999
29 Perkins and Miller 1983
30 Defran and Pryor 1980
31 Dohl *et al.* 1974
32 Sylvestre and Tasaka 1985
33 Perrin and Walker 1975
34 Götz *et al.* 2006
35 Wells and Gannon 2005
36 Norris *et al.* 1965
37 Watkins *et al.* 1987
38 Popper 1980
39 Norris and Evans 1967
40 Whittow *et al.* 1978
41 Monteiro Neto *et al.* 2000

Atlantic Humpback Dolphin

Sousa teuszii

ALTHOUGH THERE IS some confusion about the number of forms of humpback dolphin that exist and whether they should be assigned specific or subspecific status,[1,2] most studies seem to show that the dolphins occurring in the Atlantic Ocean off West Africa are both geographically isolated and morphologically different from those in the Indo-Pacific region, and this distinction is recognised by assigning them to the species *S. teuszii*.[3]

Appearance

Humpback dolphins are relatively easily distinguished from all other small cetaceans occurring in the subregion by the characteristic fleshy hump on the back that bears the dorsal fin. Interestingly, one account of the behaviour of Atlantic humpback dolphins states that they surface without lifting their head above water,[4] which (if true) would be a behavioural distinction from the Indo-Pacific population in the subregion, in which the head and beak are usually thrust clear of the water on surfacing.

Very few individuals have been measured: the largest seems to be a 2.61 m animal of unknown sex.[5] It is likely that there is sexual dimorphism as in *S. plumbea*,[1] in which case adult males will be some 10 cm longer than females. The beak is long (5.9–6.9% of body length) and well differentiated from the melon. The flippers are large (14–15% of body length), broad and pointed, with a straight trailing edge. The dorsal fin is most conspicuous, being situated in non-calves on a broad fleshy hump that occupies 26–32% of the animal's back. The fin itself appears to have a much longer leading than trailing edge, and is only slightly recurved. In height it forms 8.5–9.4% of the body length. The tail flukes are broad (23.6–25.7% of body length) with an almost straight trailing edge.[6,7]

Atlantic humpback dolphins are dark grey on the back, head and flippers, becoming progressively slate-coloured on their sides and fading to a light grey below. The throat is also light grey and sharply demarcated from the adjacent areas of the head.[6,8] There may be dark flecks on the tail stock but nowhere else.[5] Young animals are said to be grey in colour.[4]

There are 27–32 pairs of teeth in the upper jaw (n = 14) and 28–31 in the lower jaw (n = 7). They are stout (5.8–7.4 mm in maximum diameter)[3] and conical, but may become heavily chipped and worn in older animals.

Distribution and migrations

Atlantic humpback dolphins are only known from the coast of West Africa between 23°50′ N in the Western Sahara and 15°47′ S in Angola, but there are conspicuous gaps in the

Length:weight relationship

Mass information is available for five Atlantic humpback dolphins.[6,7] Despite its paucity, the information happens to cover a reasonable size range (1.865–2.48 m), so the following relationship may not be too far from reality: $W = 11.83L^{2.8761}$, where W = weight in kilograms and L = total length in metres. This implies that a 2 m dolphin would weigh 87 kg and a 2.4 m animal 147 kg.

Skeletal data

Vertebral formula (n = 2): only the total is known (52–53 vertebrae).

The atlas and axis are fused; the remaining cervical vertebrae are free.[6]

Phalangeal formula, including metacarpals (n = 1): I_1, II_6, III_{5-6}, IV_4, V_3.[8]

The characteristics of the skull that differentiate *S. teuszii* from *S. plumbea* include : 1) the lower tooth count (upper row 27–32 cf. 33–39; lower row 28–31 cf. 32–37); and 2) a relatively wider skull, the condylobasal length/zygomatic width being less than 2.4 compared to 2.4 or more in 90% of *S. plumbea* skulls.[3]

FIG. 100 A cow-calf pair of the rarely-photographed Atlantic humpback dolphin on the coast of Gabon.

distributional record, some of which may be real. Eight 'stocks' have been recognised, only two of which occur in the subregion: off Gabon and Angola, where sightings have been recorded 'about 150 km north of Luanda'[5] (or *c.* 7°30′ S), and between Namibe (15°03′ S) and Tombua (15°47′ S).[5,9]

Within their range, Atlantic humpback dolphins occur in coastal and estuarine habitat where the water is shallow (probably less than 20 m deep).[1] Favoured habitats seem to be regions with sand banks or mangrove swamps exposed at low tide and a channel system carrying highly turbid, saline (even hypersaline) water that allows access into the interior at high tide.[1,2] Despite anecdotal reports of Atlantic humpback dolphins being found up rivers, there are no definite records of their occurrence in genuine freshwater bodies, although individuals may move upstream on a rising tide when there is a strong seawater intrusion.[5]

Movements into and out of channels with the rising and falling tide have been described, as well as (if the tide is right) a regular diurnal patrolling movement between banks. Major seasonal movements or migrations are not seen, though from apparent differences in the seasonality of occurrence, a movement between Senegal and Mauritania has been proposed[4] but considered unfounded.[5]

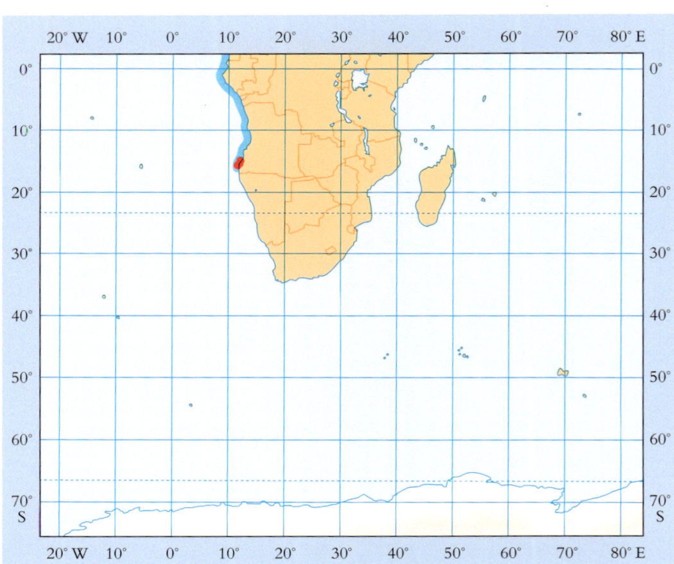

FIG. 101 Distribution map for Atlantic humpback dolphin.

External commensals and parasites

None have been described, and individuals in the Saloum Delta of Senegal were virtually devoid of notches, scars and other marks.[5]

A vegetarian dolphin?

Ex Africa semper aliquid novi (from Africa, always something new) – so said Pliny the Elder in AD 77. So it was perhaps no surprise when a strange dolphin with tubular extensions of the nostrils, 'like the erect ears of a hippopotamus', and a stomach full of leaves, mangrove fruits and grass, was caught in the naval harbour at Cameroons some time before 1892, and described as a new species.[1] The discoverer, Wilhelm Kükenthal, named it after the director of the botanical gardens in Cameroun, Mr E. Tëusz. The skull was sent to the Natural History Museum in London, where it lay as the type (and only known) specimen of the species for more than 50 years, while the legend of a vegan dolphin living in Darkest Africa persisted. It was only in the 1950s, when more specimens of what turned out to be the same species surfaced, that it was revealed that there were no 'hippo ears' and that the dolphin fed on fish, not vegetation. So how did the story arise? Kükenthal never saw the dolphin, and only received the skull (and a verbal description of the animal) from Tëusz. One possibility is that an Atlantic humpback dolphin and a manatee *Trichechus senegalensis* were captured at the same time in the same net (as still happened in the 1950s), and that the stomach contents of the manatee (a known vegetarian) were incorrectly attributed to the dolphin.[2] However, only sharks were recorded from the same net. Another possibility is that the vegetation was ingested accidentally, as has been seen occasionally in other dolphins and porpoises.[3] Where the tubular extensions of the nostrils originated is less clear, but as the carcase was partially eaten by sharks it has been suggested that these were the result of damage to the head.[3]

References

1 Kükenthal 1892
2 Cadenat and Paraiso 1957
3 Van Bree and Duguy 1965

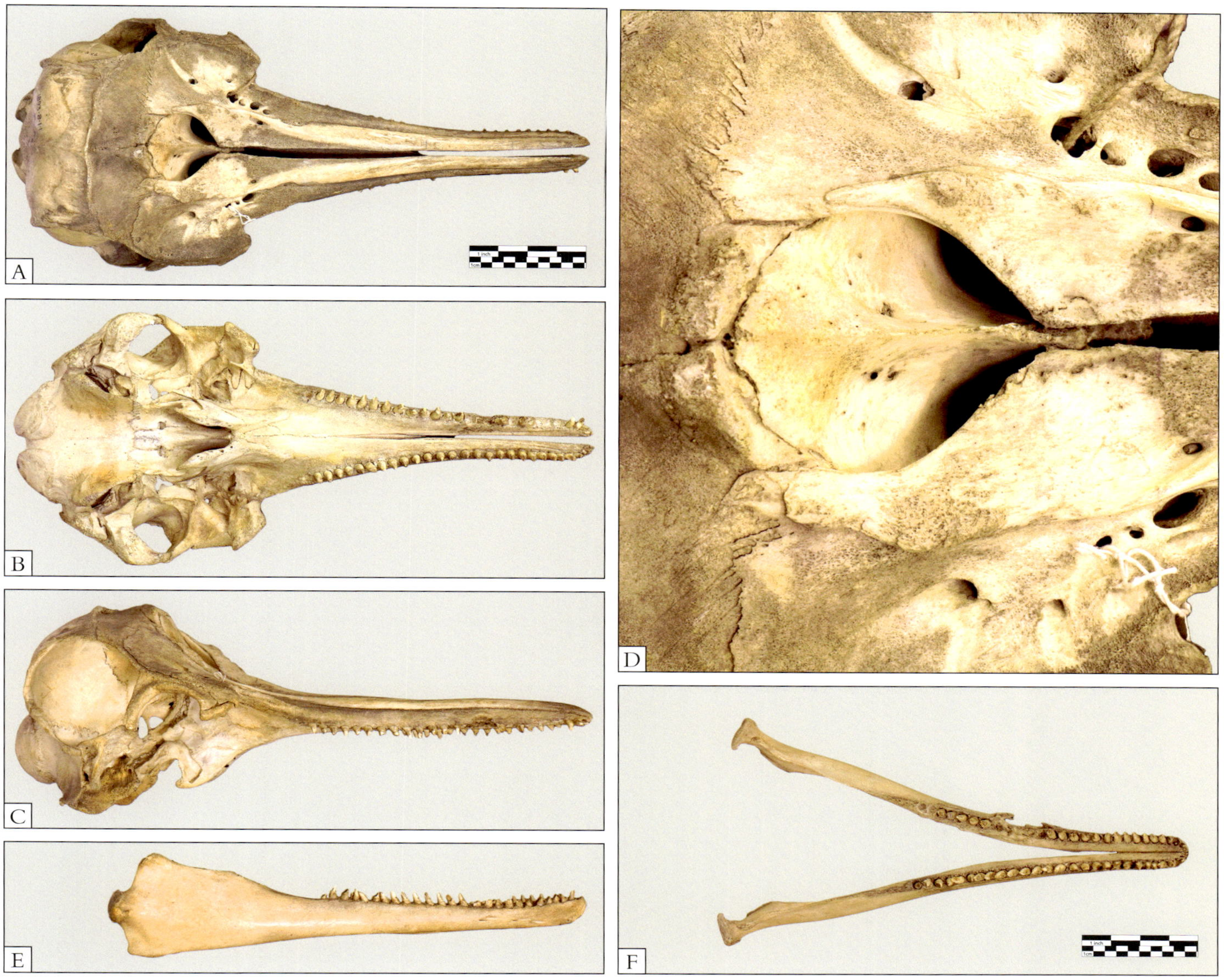

FIG. 102 Dorsal (A), ventral (B) and right lateral (C) views and vertex of skull (D); right lateral (labial) (E) and dorsal (F) views of mandible (NHM ZD 1893.8.1.1).

Food and feeding behaviour

There is no information from the subregion. Of eight stomachs examined from Senegal, one was empty and the others contained only fish remains. The fish identified were Sompat grunt (*Pristipoma jubelini*), bonga shad (*Ethmalosa dorsalis*) and mullet (*Mugil* sp.).[6–8]

In Mauritania, humpback dolphins have been described participating in a cooperative fishing venture with humans, whereby they contain mullet against the shore so that fishermen can set nets from the beach.[10] The main participants appear to be bottlenose dolphins, and the inclusion of humpback dolphins may be merely fortuitous.[1]

Reproduction and growth

Very little is known. The season of birth in Senegal is said to be March–April, possibly a little later in Mauritania.[4] The size at birth and the size and age at sexual maturity in either sex, or the frequency of reproduction, are unknown.

Behaviour and vocalisations

Atlantic humpback dolphins have been recorded in groups of 1–22 animals, but rarely alone and most often in groups of 4–8 individuals.[4, 11] Four groups in southern Angola consisted of two animals in one group and four each in the other three.[9] The size of groups in Senegal seemed to be larger on average than those in Mauritania, with groups of 10 or more animals occurring more often.

These larger groups tended to be composed of subgroups of 4–6 dolphins, possibly representing a familial structure, and might represent temporary aggregations of smaller groups during feeding. Young have been recorded within groups of three, seven and 25 animals. Mixed groups with bottlenose dolphins are sometimes seen.[4]

These dolphins appear to be shy of boats, and groups tend to break up into smaller units of 2–3 individuals when approached, carrying on with what they have been doing but keeping a distance of 15–20 m from the boat.[4]

Vocalisations have not been described, but presumably they are not very different from those of *S. plumbea*.

Conservation status

Although no survey-based estimates exist, Atlantic humpback dolphins do not seem to be abundant. Populations in Baie d'Arguin, Mauritania, and the Saloum Delta, Senegal, were said not to exceed a hundred individuals in each area,[4] while later estimates suggest that the former might be 'at least high hundreds' and the latter 'low hundreds, and maybe less'.[5] There are no estimates of abundance for populations within the subregion.

Because the species is highly coastal, and tends to live in areas of high human population growth and protein deficit, the potential for both direct and indirect mortalities exists. Although there are no authentic figures for such takes, incidental observations confirm that they do occur.[5]

Given the low apparent abundance of the species, and the lack of effective conservation measures over most of its range, there is some concern about its future.

Atlantic humpback dolphins are listed as Data Deficient by the IUCN, but are not included in the South African Red Data Book. The species is included on Appendix I of CITES and on Appendix II of CMS.

References

1 Ross *et al.* 1994
2 Ross 2002
3 Jefferson and Van Waerebeek 2004
4 Maigret 1980
5 Van Waerebeek *et al.* 2004
6 Cadenat and Paraiso 1957
7 Cadenat 1959
8 Cadenat 1956
9 Weir, C., pers. comm.
10 Busnel 1973
11 Duguy 1976

Indo-Pacific Humpback Dolphin

Sousa chinensis

Adult

Calf

CONFUSION CURRENTLY REIGNS over how many species of humpback dolphins may occur in the Indo-Pacific region. On external morphology there seem to be two distinct forms, one dark brownish-grey with a definite hump on the back beneath the dorsal fin (the '*plumbea* form'), and the other very light or even pinkish-white in colour (sometimes with freckling) and lacking a hump (the '*chinensis* form'). The former is distributed mainly from the Bay of Bengal westwards, and the latter from the Gulf of Thailand eastwards, but there may be some overlap in the Bay of Bengal and Andaman Sea. Despite the apparent clear separation into two forms externally, this is not so neatly reflected in skull morphometrics,[1] nor in genetic comparisons to date.[2] Some authors have considered the forms west and east of the Bay of Bengal as being two separate species, to which the names *plumbea* and *chinensis* have been applied,[3, 4] while others consider this premature and would rather stick with one, *S. chinensis*, at least for the time being.[1]

It seems that only the *plumbea* morphotype occurs within the subregion, but I have followed the most recent local researchers in using *chinensis* as the species name until further systematic research has been carried out. The account below is heavily based on the South African population, being the most extensively researched, but it is possible that differences (in pigmentation, morphometrics and behaviour, for instance) may exist in other populations in the subregion.

Appearance

Humpback dolphins are relatively easily distinguished from all other small cetaceans occurring in the subregion by the characteristic fleshy hump on the back that bears the dorsal fin. The Indo-Pacific species has a characteristic surfacing routine, in which the long rostrum rises steeply out of the water before the rest of the head. When the blowhole does break surface, most of the body remains submerged except for a small part of the upper back and the anterior portion of the hump.[5]

Males up to 2.79 m long and females up to 2.49 m are known from the subregion, and this seems to reflect a real dimorphism of the sexes. The beak is long (6.3–10.1% of body length) and well differentiated from the rest of the head.[4] Deformed beaks (misaligned mandibles, for instance) appear to be commoner than in most other dolphins in the subregion.[6, 7] The body is robust (particularly so in adults[7]), but becomes increasingly compressed towards the rear, and there is a pronounced keel to the upper and lower surfaces of the caudal peduncle in animals of all ages. The dorsal fin is situated on a fleshy ridge 4–8 cm thick, the length of which increases from about 25% of body length in neonates to 30–37% in adults. The fin itself is small and hooked with a long leading edge, and in total height varies from 6% to 9% of body length. The flippers are broad with a rounded tip and comprise 13.1–16.8% of the body in length. The flukes are 'full' (or broad in a front-back direction) with a deep notch, and in width constitute 22.9–28.1% of the body length.[4,8]

Calves are born a pale grey on the sides shading to white below, with a contrasting area of dark grey on the back and the upper surface of the flukes. An indistinct stripe of grey extends from the flipper towards the eye, and the upper and lower keels of the caudal peduncle are off-white. This pattern seems to darken with age, so that adults are a dark lead-grey above shading to lighter grey on the flanks and off-white on the belly. Some adults are greyish on the belly and may be lightly flecked with a few dark spots. In some of the largest adults, parts of the dorsal fine and hump become white, as do the tips of the snout and flukes in some individuals. Adults may also carry prominent scars on the flanks and dorsal surface, particularly behind the dorsal fin.[7,8] Many of these may be caused by unsuccessful attacks by large sharks.[9]

There are 33–38 pairs of teeth in the upper jaw and 30–37 pairs in the lower jaw.[4] They are stout, about 7 mm in diameter at the gum, and may be expanded at the base in an antero-posterior direction so that they become wedge-shaped.[8]

Distribution and migrations

As their name suggests, Indo-Pacific humpback dolphins are found throughout the Indian Ocean and the south-west Pacific, from South Africa in the west to eastern Australia, Indonesia, Korea and China in the east. In the subregion they have been recorded from South Africa, Mozambique, Tanzania (including Zanzibar), Kenya, the Comoro Islands, and Madagascar.[11] The most westerly published specimen is a skull from Muizenberg, False Bay, collected in 1896,[12] but in 2004/5 there were reports (and some photographs) of a single animal in the Langebaan lagoon on the west coast of South Africa (33°09′ S 18°03′ E): presumably this was a vagrant *chinensis* rather than *teuszii*, but no material was available to confirm this[6] (examples elsewhere of such vagrancy include individual humpback dolphins that traversed the Suez Canal in order to move from the Red Sea into the Mediterranean[13]). The most western locality where humpback dolphins occur with any regularity appears to be either side (but particularly to the south) of the Danger Point peninsula (34°38′ S 19°17.5′ E).[6] East of this there seems to be a more or less continuous distribution around the coast as far as southern Mozambique, and possibly further to the north, but sighting effort thereafter is patchy.[11] Around Madagascar, humpback dolphins appear to occur mainly on the west coast, and have been recorded from the Anakoa area in the south and Nosy Mitsio in the north.[14]

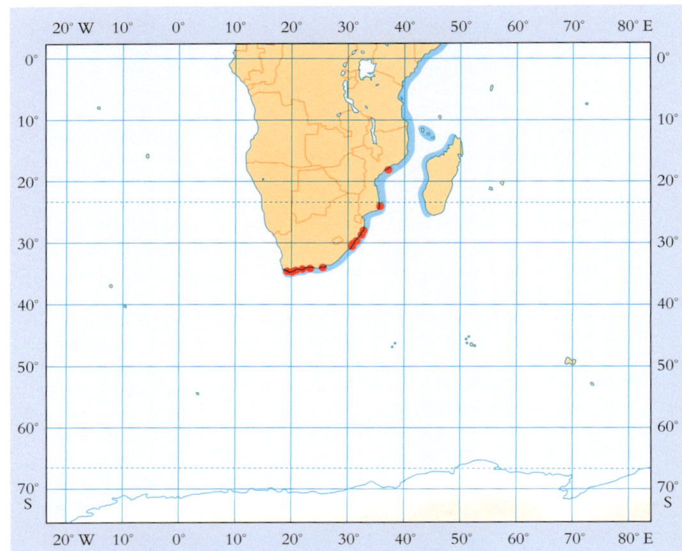

Postulated range ■ Sighting or capture

FIG. 103 Distribution map for Indo-Pacific humpback dolphin.

Throughout the subregion the Indo-Pacific humpback dolphin is predominantly a shallow-water species, generally being restricted to water depths of less than 50 m,[12] with an apparent preference in Algoa Bay for water less than 15 m deep[15] and off Richards Bay for water less than 20 m deep.[16] Although this means that in most areas the dolphins are found relatively close to the shore (within 250–400 m[7,15]), on occasion they can be found several kilometres offshore if the water is still shallow.[15–17] Whereas in the Plettenberg Bay and Algoa Bay regions the dolphins appeared to concentrate in the vicinity of rocky reefs or sandy gullies,[7,15] off KwaZulu-Natal they seemed to prefer large estuarine systems, and in Maputo Bay, Mozambique, they utilised coral reefs, seagrass beds and sheltered shallow areas.[18] An apparent preference for turbid water in some areas is not seen in others, and may be an incidental association arising from the distribution of prey and the dolphins' preference for shallow water.[15]

Although concerted seasonal movements or migrations are not characteristic of this species, seasonality in humpback dolphin occurrence was recorded in Algoa Bay, with group sighting rate and group size being significantly higher in summer (January–April) and late winter (September) than at other times of the year.[19] This was accompanied by an apparent influx of previously unidentified individuals into the bay.[20] A summer influx of dolphins into eastern Maputo Bay, Mozambique, has also been suggested.[21] In Plettenberg Bay, however, no significant trend in the sighting rate between seasons was detected, although the dolphins tended to stay longer in the study area, and group sizes tended to be larger, in winter than in spring.[7]

Length:weight relationship

This has been calculated for 42 *S. chinensis* from southern Africa as $\mathbf{W = 0.00000457L^{3.183}}$, where W = weight in kilograms and L = total length in centimetres.[4] This implies that a 1.5 m dolphin would weigh 39 kg and a 2.5 m animal 196 kg.

Skeletal data

Vertebral formula (n = 22): 7 cervical, 11–12 thoracic, 9–12 lumbar and 20–24 caudal; total 49–52.[4]

The atlas and axis are fused; the remaining cervical vertebrae are free.[10]

Phalangeal formula, including metacarpals (n = 8): I_1, II_{7-8}, III_{6-7}, IV_4, V_3.[4]

The characteristics of the skull that differentiate *S. chinensis* from *S. teuszii* include the higher tooth count and the relatively narrower skull (see the account of *S. teuszii* for details).

FIG. 104 Patrolling the backline on the south coast of South Africa, a group of Indo-Pacific humpback dolphins (*top*) illustrates variations in beak length, dorsal fin shape and scarring that are probably age related; while some large individuals (*bottom*) carry irregular white markings on the dorsal fin.

External commensals and parasites

The only external parasite recorded is the cyamid *Syncyamus aequus*, said to have been identified from two Indo-Pacific humpback dolphins from KwaZulu-Natal.[4]

Food and feeding behaviour

Fish dominated the diet of Indo-Pacific humpback dolphins examined in South African waters. Of 503 identified prey remains from the stomachs of 17 dolphins taken in shark nets off KwaZulu-Natal, the major species numerically were orange-mouth glassnose (*Thryssa vitrirostris*, 46.4%), cutlass fish (*Trichiurus lepturus*, 9.2%), piggy (*Pomadasys olivaceum*, 8.6%), snapper kob (*Otolithes ruber*, 7.2%) and blacktail (*Diplodus sargus*, 3.6%). The diversity of the diet is shown by the fact that there were another 28 species in the remaining 24% of the prey items (although some of these may well have been 'secondary' prey, i.e. eaten by the fish first).[4]

The stomachs of two stranded individuals from the Eastern Cape, South Africa, contained the remains of 20 fish, including 12 blue hottentot (*Pachymetopon aneum*), four unidentified sea bream (Sparidae), one stumpnose (*Rhabdosargus* sp.), one spotted grunter (*Pomadasys commersonii*), and one mackerel (*Scomberomorus japonicus*). In addition there were five cephalopods, comprising four squid (*Loligo* possibly *reynoldi*) and one octopus.[22]

In Mozambique, Indo-Pacific humpback dolphins have been recorded feeding on bone fish (*Albula vulpes*).[23]

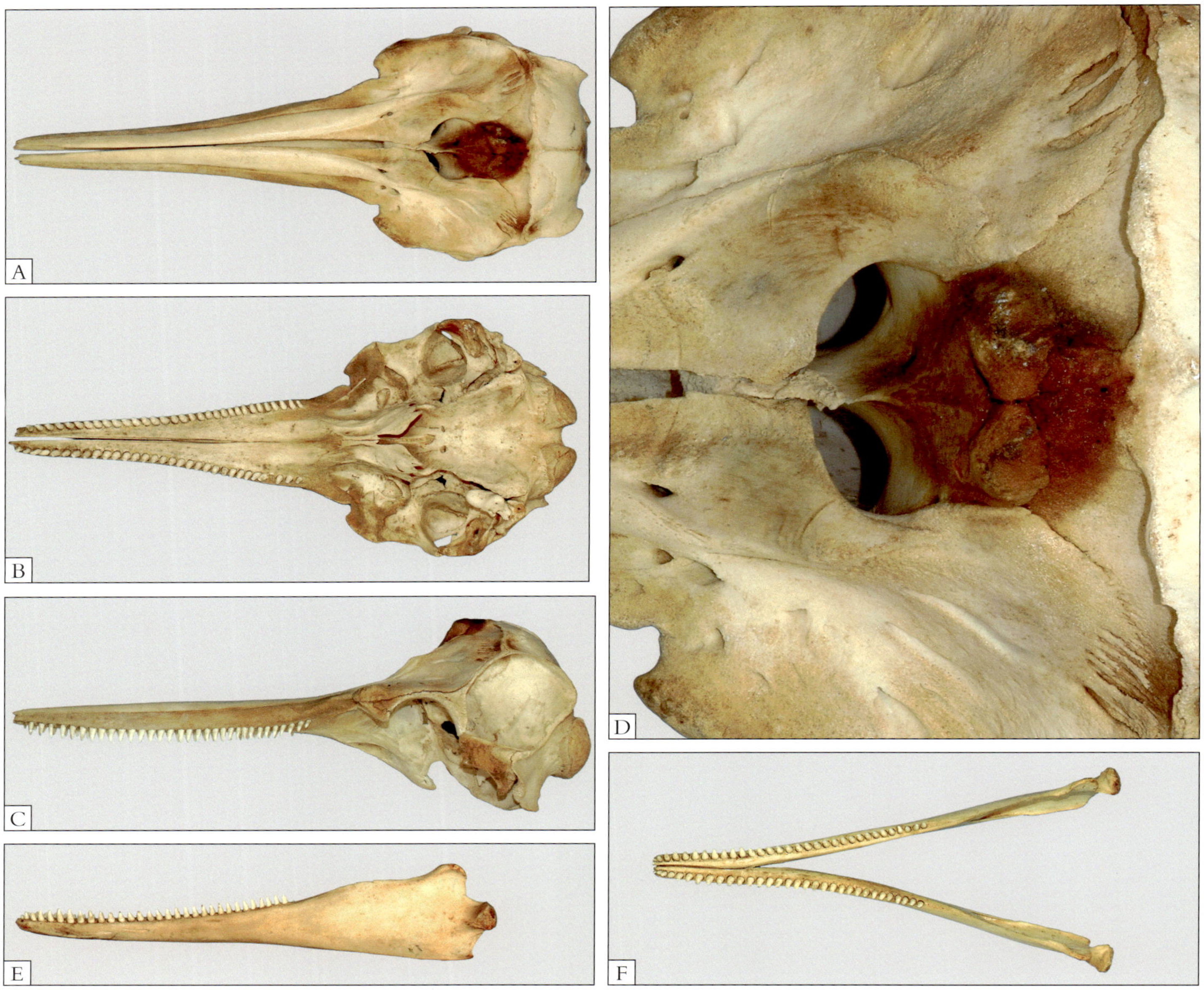

FIG. 105 Dorsal (A), ventral (B) and left lateral (C) views and vertex of skull (D); left lateral (labial) (E) and dorsal (F) views of mandible (ZM 35100).

Two feeding strategies have been described for humpback dolphins in Algoa Bay. Generally the feeding groups were widely dispersed, with frequent and asynchronous diving, and a lack of directional movement. However, smaller groups (six or fewer individuals) remained in closer proximity to each other (about 1–20 m apart) and moved up and down several hundred metres of coastline. Both here and in Plettenberg Bay, animals appeared to capture fish on an individual basis, with limited if any cooperation between school members.[5, 24]

Fish are seized behind the neck with a sideways movement of the head and then manipulated so as to swallow them head first. One juvenile was seen catching seven fish about 15 cm long in 13 min.[7] A specialised beach-hunting technique, in which dolphins trapped fish against a sandbank and then charged at them, causing a wave of water to wash the fish onto the sandbank where the dolphins could beach themselves to secure their prey, has been described in the Bazaruto Archipelago, Mozambique. Such feeding was done both individually and cooperatively.[23]

Reproduction and growth

From lack of air in the lungs or other features suggesting a recent birth, three neonates 0.97 m, 1.04 m[6] and 1.08 m in length were classified as stillborn or as having died shortly after birth, while from its dentinal growth a fourth, 1.01 m long, was believed to have survived for some time after birth. This would suggest that calves are about 1 m long when born.[8] Births can occur throughout the year, but in Algoa Bay most (73.3%, n = 11) took place between December and April.[25] A similar pattern was shown for Plettenberg Bay, with most births (65.3%) estimated to occur between December and February,[7] but in Maputo Bay, Mozambique, births are said to occur year-round with no defined season, although four out of the six births recorded were in January and February.[21] A female brought into captivity in Port Elizabeth with a male on 18 January 1963 was pregnant with a foetus 7.5 cm long when it died about two months later; the male (with which it frequently mated) had died on 19 February.[6]

This also suggests a summer conception. Ovarian activity in captive females was observed in every month of the year, with an irregular pattern of cycling, one to four consecutive cycles (each with a mean length of 28–36 days) being interspersed with periods of anoestrus of one to eight months. The length of pregnancy in a captive animal was determined as 11–12 months. Birth intervals have been recorded as 21 months (after a stillbirth),[26] approximating three years (two births) and at least 3.5–4 years (one birth).[25] The frequency of ovulation in mature females has been estimated as 0.3 per growth layer in the teeth, which sets an upper limit to calf production of three in 10 years. The length of lactation is unknown but probably occupies a large part of the interval between successful births: four calves, each with about two growth layers in the teeth, were still suckling,[27] and the stomach of a stranded 1.63 m female over two years of age contained both solid food and milk.[22] Observations suggest that associations between mother and calf may persist for up to four years.[25]

Sexual maturity in females is reached at an age of about 10 growth layers in the teeth,[27, 28] and in males about 2–3 years later.[27] Equivalent sizes are not known, but a recently mature female (from the Hong Kong population) was 2.35 m long.[28] Age determination studies using dentinal growth layer groups suggest that longevity in both sexes may exceed 46 years.[27]

Behaviour and vocalisations

Solitary individuals were the most frequently observed group size for humpback dolphins in both Plettenberg Bay (16.1%, n = 211) and Algoa Bay (15.4%, n = 104). The remaining groups ranged in size from two to 25 (Plettenberg Bay) or three to 24 (Algoa Bay), but groups of 10 or more were uncommon (23.2% and 16.3% of sightings respectively).[7, 19] Group sizes recorded in Maputo Bay, Mozambique, and around Anakao, Madagascar, were distinctly larger: although the maximum group sizes (25 in each case) were similar to those in South Africa, there was a complete absence of solitary animals, and the groups averaged 14.9 and 13 animals respectively.[14, 21]

In Maputo Bay the larger groups were believed to represent the coalescence of small groups in the deeper waters of the bay during low tide. Animals found as singles or pairs in South African waters were mostly adults.[7, 25] Larger groups contained 2–16 adults (mean 4.4), or 63% of group members. Groups consisting solely of juveniles were rarely seen, and most groups larger than two animals contained at least one juvenile. The number of calves in a group varied from none to five, with 1–2 being most frequently seen. The mean size of groups containing calves (10) was significantly larger than that of groups without calves (4.7).[25] Overall, calves constituted 10% of the population in Plettenberg Bay,[7] while in Algoa Bay the proportion of calves in groups of three or more was 14%. Apart from mother-offspring pairs, associations between individuals in Algoa Bay were generally weak, with the majority of identified individuals associating loosely with a large number of other individuals. Similarly, site fidelity (as determined by the number of re-sightings) was generally low, indicating high motility along the coast, although females immediately pre- and post-parturition seemed more resident.[25] Temporary alliances between nursing females with a possible mutual calf-care function have been suggested.[5]

A strong diurnal pattern of behaviour was seen in Algoa Bay, with 'feeding' peaking in the morning and declining thereafter, and 'travelling' and 'opportunistic feeding' increasing. In winter there was a secondary increase in feeding in the late afternoon.[24] Such a marked pattern was not seen in Plettenberg Bay, although feeding bouts were longer in the early part of the day than thereafter.[7]

The tidal cycle seemed to have no influence on humpback dolphin behaviour in Algoa Bay,[24] but in Plettenberg Bay feeding activity increased and travelling decreased on a rising tide and vice versa on a falling tide.[7]

Behaviour patterns were also influenced by habitat, with foraging/feeding being more intensive in the vicinity of shallow reefs and relatively sheltered areas, and travelling being more prevalent along open stretches of coastline, sandy shores, and areas with high human activity.[15] The average speed of travel has been estimated at 4.8 km/hr (n = 58).[7]

During courtship and mating behaviour in Algoa Bay, pairs of humpback dolphins would become temporarily isolated from other groups. Initially they would swim rapidly side by side in a spiral-like movement, occasionally interrupted by an aerial display, then swim rapidly to another location where the behaviour would be repeated several times. Apparent copulation would follow this sequence, in which the two animals would come into close apposition belly-to-belly for 25–30 sec. and swim slowly together, rolling beneath the surface. Such copulation episodes would be repeated several times in succession.[5] Captive studies have shown that while it is usually the male that is inverted beneath the female during copulation, the roles can be reversed.[7] When several adults were involved in courtship at the same time, the initial phase was swum at high speed in a circle about 5–8 m in diameter.[5]

Despite the overlap in their distribution, mixed groups of humpback and bottlenose dolphins were only occasionally seen. Single humpback dolphins appeared to be fully integrated in bottlenose schools,[7] but larger numbers always formed a distinct peripheral subgroup. Such groups tended to follow the bottlenose dolphin school, and were segregated from them when feeding at the same time.[5] Aggression towards single humpback dolphins was displayed by bottlenose dolphins in Plettenberg Bay, but not in Algoa Bay.[5, 7]

Indo-Pacific humpback dolphins in Algoa Bay seemed to be susceptible to disturbance caused by inshore boat traffic, particularly power boats. The species never rode the bow or stern wave of such boats, and indeed seemed actively to avoid them when they were in motion. If harassed, they tended to make a long dive, change direction and swim at right angles away from the boat.[5] Similar behaviour was recorded for humpback dolphins in the Indus Delta.[29]

The *plumbea* and *chinensis* forms of humpback dolphin apparently produce very similar sounds.[29, 30] These have been classified into five types; broadband clicks (produced in trains at about 23/sec.); barks (highly variable burst pulses with strong harmonics, between 0.2 and 7.4 sec. in duration and with a frequency range between 0.6 kHz and more than 22 kHz); quacks (similar to a bark but much shorter in duration and lower in frequency, 0.6–3.7 kHz); grunts (low-frequency narrow-band sounds between 0.9 kHz and 1.4 kHz and only 0.09 sec. in duration); and whistles (narrow-band, frequency-modulated sounds that could be divided into 17 types). Broadband clicks and whistles were the most frequently recorded sounds while foraging; grunts

Two South African pioneers

This is an appropriate place to recall the contributions to cetology made by an unlikely pair of South Africans in the 1960s and 1970s. Graham Saayman had a PhD in psychology from the University of London and a background in primate behaviour, while Colin Tayler was an electrical technician with an interest in dolphin acoustics and behaviour (sparked by his exposure to two wild bottlenose dolphins that interacted with bathers at his home town, Fish Hoek, in 1953). Coming together at the Port Elizabeth Museum in the 1960s, where Colin was initially the dolphin trainer and then Curator of the Oceanarium, and Graham was a professional officer on the Museum's staff, the two formed a team that was responsible for a series of pioneering papers on dolphin ecology and behaviour. Their field studies of bottlenose and humpback dolphins were among the first to demonstrate the feasibility of studying cetacean behaviour in the wild through careful observation of the undisturbed animal.[1] At the same time, they appreciated the value of captive studies that allowed them to interpret their field observations.[2] Graham's earlier primate experiences led to the first extensive comparison of dolphin-primate behaviour in the wild.[3]

FIG. 106 Colin Tayler and friend.

FIG. 107 Graham Saayman.

Colin's interest in dolphin phonation was less well published, but included one of the first descriptions of signature whistles, and he was an early protagonist of the view that these sounds were generated in the nasal sacs.[4] In 1971 the collaboration was broken when Colin left the Oceanarium for his other passion, motor cycle racing. He wrote a book on the sport of observed trials and invented a power valve to be used in the exhaust of 2-stroke motor cycles. Sadly, as I wrote this species account in March 2006, I learned of his death in Port Elizabeth at the age of 75. Graham, too, left the Museum in 1973 to return to an academic career in psychology at UCT and in 1989 he emigrated to Canada to work as a clinical psychologist and family therapist at a psychiatric hospital for children and adolescents.

References

1 Saayman and Tayler 1979
2 Saayman *et al.* 1973
3 Tayler and Saayman 1972
4 Tietz and Tayler 1964

and whistles were most often recorded during socialising; and broadband clicks most regularly recorded during travelling.[30] The clicks produced by the *chinensis* form have been found to have a strong ultrasonic component, up to at least 200 kHz,[31] and the same is probably true for the *plumbea* form.

Boat traffic passing within 1.5 km of a group of humpback dolphins was correlated with an increase in whistling activity after the boat had passed, possibly to re-establish vocal contact within the group.[32]

Conservation status

Indo-Pacific humpback dolphins seem to occur at low densities throughout their range, but there are relatively few actual population estimates and all are based on mark-recapture methodology using photo-identified individuals. As the populations to which these estimates refer are all open, it is difficult to assess to what proportion of the species' range they might refer. In Algoa Bay, South Africa, the adult (marked) population was estimated at 270 individuals, from which a 'total' population estimate (including unidentified adults and all other classes in the population) of 466 was derived.[20] Modelled population growth rates for this population vary from −3% to +2% a year, suggesting some stability.[18] There are estimated to be 160 (95% CI 134, 229) humpback dolphins off KwaZulu-Natal,[16] and no matches of individuals are known between this population and that in Algoa Bay.[20] Some estimates place the entire South African population at 'not more than 1 000 animals'.[18] In Maputo Bay, Mozambique, a population size of 105 dolphins (95% CI 31, 151) was estimated using the same methodology as for Algoa Bay.[21] Mark-recapture estimates indicate a population of 58–65 animals on the south coast of Zanzibar for the years 1999–2002.[33] Between 1980 and 1998, 129 humpback dolphins were caught and died in anti-shark nets off KwaZulu-Natal, or at a rate of 6.8 a year on average. More than half (54.1%) of these were taken at Richards Bay.[16] Although this take is small in absolute terms, it may be of local significance, especially off Richards Bay. In Madagascar, humpback dolphins are hunted directly (22 are known to have been taken near Anakao between 1985 and 1999) and are caught incidentally in nets set for sharks (three are known to have been taken near Mahajanga in 1997). Another 30 were reported 'stranded' near Anakao during the same 15-year period:[14] as the species rarely strands naturally in South Africa, some of these may represent human intervention. Five humpback dolphins were reported as a by-catch in Zanzibar between 1995 and 1999,[34] and 11 between 2000 and 2003.[35]

Apart from the South African coastline, the Indo-Pacific humpback dolphin inhabits areas of generally high human population growth and protein deficit throughout much of its range in the Indian Ocean. Being found so close to shore, it is naturally vulnerable to entanglement in nets set as part of artisanal fisheries in the region, and is liable to be targeted deliberately for human consumption. It is also exposed to any pollutants that enter the coastal environment through river run-off:[18] off KwaZulu-Natal the species has been recorded with the highest organochlorine concentrations of any marine mammal in South African waters.[36] Given all these factors, and the generally low population sizes (where known), there is considerable concern over the future of this species in the subregion.[18] Indo-Pacific humpback dolphins are listed as Vulnerable in the South African Red Data Book, but as Data Deficient by the IUCN, and are listed on Appendix I of CITES and Appendix II of CMS.

References

1 Jefferson and Van Waerebeek 2004
2 Rosenbaum *et al.* 2002b
3 Rice 1998
4 Ross *et al.* 1994
5 Karczmarski *et al.* 1997
6 MRI Whale Unit data
7 Saayman and Tayler 1979
8 Ross 1984
9 Cockcroft 1991
10 Flower 1869
11 Jefferson and Karczmarski 2001
12 Findlay *et al.* 1992
13 Kerem *et al.* 2001
14 Razafindrakoto *et al.* 2004
15 Karczmarski *et al.* 2000
16 Atkins *et al.* 2004
17 Corkeron 1990
18 Karczmarski 2000
19 Karczmarski *et al.* 1999a
20 Karczmarski *et al.* 1999b
21 Guissamulo and Cockcroft 2004
22 Barros and Cockcroft 1991
23 Peddemors and Thompson 1994
24 Karczmarski and Cockcroft 1999
25 Karczmarski 1999
26 Brook *et al.* 2004
27 Cockcroft 2002
28 Jefferson 2000
29 Zbinden *et al.* 1977
30 Van Parijs and Corkeron 2001b
31 Goold and Jefferson 2004
32 Van Parijs and Corkeron 2001a
33 Stensland *et al.* 2006
34 Amir *et al.* 2002
35 Amir *et al.* 2005b
36 Cockcroft 1999

Common Bottlenose Dolphin

Tursiops truncatus

THE TAXONOMY OF bottlenose dolphins worldwide is a subject of some debate. In several regions outside the Indian Ocean there seem to be separate coastal and offshore populations that to some extent differ from each other morphologically and ecologically,[1–3] but whether these 'morphotypes' are analogous is not known, so they have usually all been lumped together as the common bottlenose dolphin, *Tursiops truncatus*.[4] Meanwhile, coastal bottlenose dolphins in the Indian Ocean (and the south-west Pacific) are distinctive enough morphologically from the offshore and other coastal populations to have been assigned to a separate species, *T. aduncus*, by some researchers,[5] while others (sometimes the same people!) have disagreed[6] or sat on the fence.[7]

When genetics is brought into the picture, the situation does not become much clearer. An analysis of nuclear markers suggests that, at least in the western North Atlantic, the coastal and offshore morphotypes are divergent enough to be considered separate species.[8] A wider geographical analysis, however, using both mitochondrial and microsatellite DNA, showed that there was significant differentiation among all the proposed regional populations of bottlenose dolphins examined, implying that there was little genetic interchange between them of individuals of either sex (and thus a high potential for speciation). Phylogenetic reconstructions supported the separation of the Indian Ocean coastal dolphins from South Africa and those from China previously assigned morphologically and genetically to *aduncus* as two lineages separate from the rest of the *truncatus* types.[9] On this analysis, there would be three bottlenose dolphin species, *truncatus*, *aduncus*, and a third as yet unnamed species from China. Just to complicate matters further, this analysis also lent some support to an earlier cytochrome *b* study, which proposed that the *aduncus* dolphins were more closely related to species in the genera *Stenella* and *Delphinus* than to *T. truncatus*.[10]

Both morphotypes of bottlenose dolphin occur in the subregion, although their distributions do not completely correspond to the coastal vs. offshore pattern described above. While the *aduncus* form occurs inshore on the southern African coast east of Cape Town, the *truncatus* form occurs not only offshore on all coasts but also inshore on the coast of Namibia (and possibly further north). For this reason I have preferred not to use the terms 'coastal' and 'offshore' when referring to the morphotypes in these accounts.

This account concerns the larger morphotype, here referred to as the common bottlenose dolphin, *T. truncatus*. The smaller morphotype will be referred to as the Indo-Pacific bottlenose dolphin, *T. aduncus*.

A cautionary note: the following information is derived mainly from studies in South African waters, and because of the high regional variability of bottlenose dolphins noted above, the descriptions may not be equally applicable to all populations in the subregion. For similar reasons, I have been loath to ascribe information collected for other populations, especially behavioural, to those in the subregion.

Appearance

Bottlenose dolphins are distinguishable from other delphinids in the subregion from their short, stout beak, thick teeth and large dorsal fin. Common bottlenose dolphins can be distinguished from their Indo-Pacific counterparts externally by their shorter beak and (in the south-east Atlantic at least) a brighter (more contrasted) colour pattern, especially a light grey blaze that runs back from the side of the head and has a branch curving up towards the dorsal fin that is muted or absent in Indo-Pacific bottlenose dolphins. The tip of the beak in older animals can also become distinctly white (possibly from scarring), and the caudal peduncle very grey. Common bottlenose dolphins also do not show the extensive dark speckling on the belly and sides that develops with age in Indo-Pacific bottlenose dolphins.[7]

Animals of the *truncatus* type are supposed to reach lengths of 3.3 m,[11] but individuals found inshore on the Namibian coastline have measured up to 3.65 m.[12] The largest female measured (n = 10) was 3.16 m long, whereas there are three males over that size (n = 12),[5, 11–13] reflecting sexual dimorphism.[7] The beak is short, measuring 3.1–4.7% (mean 3.7%, n = 19) of body length, and is well differentiated from the melon, like the neck of a bottle. The body is robust, with a large, centrally placed dorsal fin. Its height is 8.2–12.3% (mean 9.7%, n = 18) of body length and its shape is recurved, but not strongly so. The flippers are moderate in size, their length ranging from 13.8% to 17.9% (mean 15.1%, n = 17) of body length, with a strongly convex leading edge and a concave trailing edge. The width of the flukes ranges from 21–26.2% (mean 23.3%, n = 16) of body length, with a well-defined central notch.[5, 11, 12]

Length:weight relationship

Weights of 16 common bottlenose dolphins are available from the subregion, six of which were weighed whole and 15 in pieces.[5, 11, 12] Using an adjustment factor of 1.03 (obtained from the five animals that were weighed both whole and in parts) to account for weight loss during dismemberment, the length: weight relationship for whole animals has been calculated as $W = 11.32L^{2.9869}$, where W = weight in kilograms and L = total length in metres. This implies that a 2 m dolphin would weigh 90 kg and a 3 m animal 301 kg.

Skeletal data

Vertebral formula (n = 13): 7 cervical, 12–14 thoracic, 16–19 lumbar and 25–29 caudal; total 64–66.[5, 11, 12] The first vertebra with an arterial canal through the transverse process varies between the 43rd and the 46th from the head.[5]

The atlas and axis are fused, and sometimes the third cervical is also joined to them, but the remaining cervical vertebrae are free.[14]

Phalangeal formula, including metacarpals (n = 4): I_{1-2}, II_{8}, III_{6-7}, IV_{3-4}, V_{1-2}.[5, 12]

The characteristics of the skull that differentiate *T. truncatus* from *T. aduncus* in the subregion include the following:

1. A larger skull – up to 57.8 cm in condylobasal length (*aduncus* up to 50.7 cm).
2. A wider rostrum – at the base and at mid-length averages 26.1% and 16.1% of skull length, compared to 23.8% and 13.8% in *aduncus*.
3. A wider cranium – the preorbital width averages 46.3% of skull length compared to 43% in *aduncus*.
4. The posterior portion of the vomer is wide and flaring (compared to narrow with parallel or sub-parallel margins in *aduncus*).
5. The premaxillae are strongly convex in the middle third of the rostrum, when viewed from the front (as opposed to not strongly convex in *aduncus*).
6. The cancellous bone on the side of the premaxillae in the middle third of the rostrum is broad (more than 3 mm wide), but is absent or very narrow in *aduncus*.[11]

FIG. 108 Variation in coloration of common bottlenose dolphins is substantial: a dark individual from off the Wild Coast of South Africa (*top*) contrasts with the brightly-coloured animals photographed from the air in St Helena Bay on the west coast (*bottom*).

The coloration of common bottlenose dolphins varies in intensity (if not in patterning) throughout the subregion. Those found offshore on the west coast of South Africa seem to have particularly 'bright' coloration, whereas those close inshore on the Namibian coast (and offshore in the Indian Ocean) are more muted in comparison. The beak is a dark grey to black (although it may whiten at the tip in larger animals in some populations), and a similarly coloured band runs from the base of the beak back towards the blowhole, before expanding behind the blowhole to form a dark 'cape' covering most of the upper back as far as halfway between the dorsal fin and the tail. Below this cape there is a zone of lighter grey extending from the side of the head back along the side of the body, sweeping up to cover most of the rear half of the caudal peduncle behind the level of the anus. A separate tongue of this light grey pigmentation runs up towards the centre of the dorsal fin base. Below the lateral zone of grey, from the throat back to the anus, the body is white, and there is no sign of the extensive dark freckling seen in the Indo-Pacific bottlenose dolphin. There is a small dark eye patch and a dark stripe running forward from it towards the base of the beak. The flippers are dark grey to black above and lighter below, as are the flukes. A thin dark band extends from the flipper towards the angle of the gape, and there may also be thin dark lines extending on each side from the rear of the flipper base towards the midline of the chest, where they end a few centimetres apart at about the level of the eye. Calves and juveniles are somewhat paler in coloration than adults.[12]

There are 20–27 (mean 23.5, n = 24) pairs of teeth in the upper jaw and 20–25 (mean 22.1, n = 24) pairs in the lower jaw.[5, 11, 12] They are stout, 6–8 (mean 7.1, n = 8) mm in maximum diameter at the gum, and conical in shape, but may become worn down, chipped or even lost in older animals.[12]

Distribution and migrations

Bottlenose dolphins in general occur worldwide in temperate and tropical waters of all oceans, including the Black Sea, with populations in coastal waters, around islands and atolls, and over shallow banks, as well as offshore in deep water.[4] Within the subregion the same situation applies, with bottlenose dolphins occurring coastally in Angola,[15] Namibia, South Africa,[16] Ascension Island,[17] St Helena,[18] Mozambique,[19] Tanzania (including Zanzibar[20]), Kenya,[21] the Seychelles,[22] Madagascar,[19] Mauritius, Réunion,[23] as well as offshore in the Indian Ocean and the South Atlantic.[16] It is not always possible to determine to which form/species some of these records apply, but it has been assumed that all those occurring coastally in the Indian Ocean should be referred to the *aduncus* form (and so are not included here).

Within the south-west Atlantic a coastal population of the common bottlenose dolphin occurs off Namibia. Its range is uncertain: it has been sighted frequently between Walvis Bay and Cape Cross,[16] but given the remote nature of much of the coast it may range wider. There are two sightings of 15 and 20 bottlenose dolphins inshore at Baia dos Tigres (16°37.58′ S) and Rio Chambala (16°10.21′ S) in southern Angola,[15] but it may also occur further north (off the Congo and even Gabon). To the south, despite a substantial amount of cetacean survey effort close inshore for other species, there have been no sightings of these dolphins off the west coast of South Africa, although incidental sightings (and the occasional stranding) do occur. It is assumed that these represent animals from an offshore population, rather than longshore strays from a coastal population, and stomach contents generally support that assumption (see below).

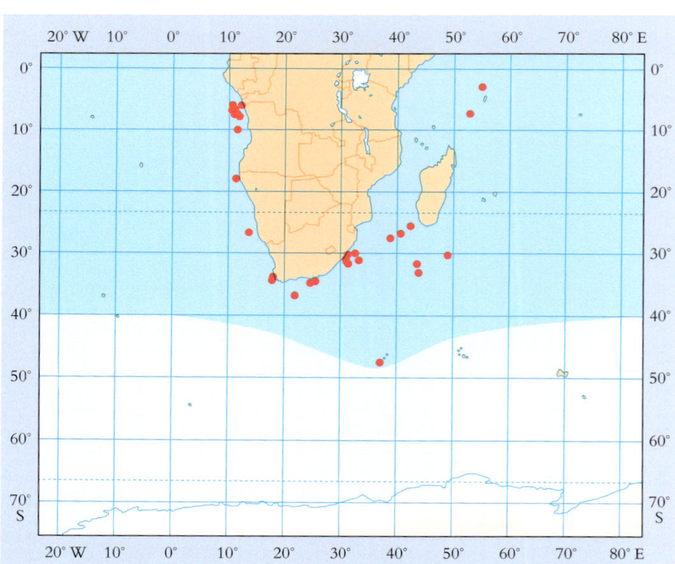

Postulated range Sighting or capture

FIG. 109 Distribution map for common bottlenose dolphin.

Offshore sightings of bottlenose dolphins in the subregion are relatively infrequent, probably reflecting the overall lack of effort, but have occurred from 3° S to 47° S, and in the South Atlantic and the Indian Ocean.[12]

The Namibian coastal population of common bottlenose dolphins is normally encountered in water less than 10 m deep.[24] In most areas of the South African coast, the sighting rate of bottlenose dolphins showed two peaks: one in water less than 100 m deep and the other in water 500–1 000 m deep. These presumably refer to inshore populations (including *aduncus*) and offshore dolphins of the *truncatus* type.[16] This pelagic habit does not prevent the offshore form of the common bottlenose dolphin from frequenting areas of shallower water occurring over banks or reefs. Walter Shoal is an isolated seamount 700 km south of Madagascar, about 40 km in diameter and reaching within 20 m of the surface in places. Sightings of bottlenose dolphins have been made there each time it has been visited (once in January 1974 and twice in December 1996), and a specimen collected there in 1974 confirmed this was the *truncatus* form.[12, 25] Offshore animals may also visit coastal waters occasionally: in October 2003, a mixed school of 100 common bottlenose dolphins and 60 false killer whales was photographed in St Helena Bay within 2 km of the coastline and in about 10 m of water. Within 24 hours all the animals had left (or been escorted from) the bay.[12]

There seemed to be no marked pattern of seasonal abundance for the inshore population of common bottlenose dolphins in Namibia. Although individuals could be seen in the Walvis Bay area over several days, they might not be re-sighted there for several months, but between 70% and 81% of known individuals were seen in each of the four seasons.[24] Whether there is any seasonality of movement for the offshore population is not known, though off eastern Australia *truncatus*-type dolphins appeared to be present in both summer and winter.[7]

Two rehabilitated stranded individuals from the offshore population in the North Atlantic ranged widely, one covering 2 050 km in 43 days and the other 4 200 km in 47 days. The latter animal moved over 12° of latitude and 22° of longitude, and in the last half of the period spent nearly all of its time in water more than 5 000 m deep.[26]

External commensals and parasites

In 14 common bottlenose dolphins examined in the subregion, stalked barnacles (*Conchoderma* sp.) were recorded attached to the right lower jaw of one animal, where its teeth had become damaged, and two *Xenobalanus* sp. were found attached to the right flipper and five to the notch region of the tail flukes in another. Barnacles resembling *Xenobalanus* were also reported

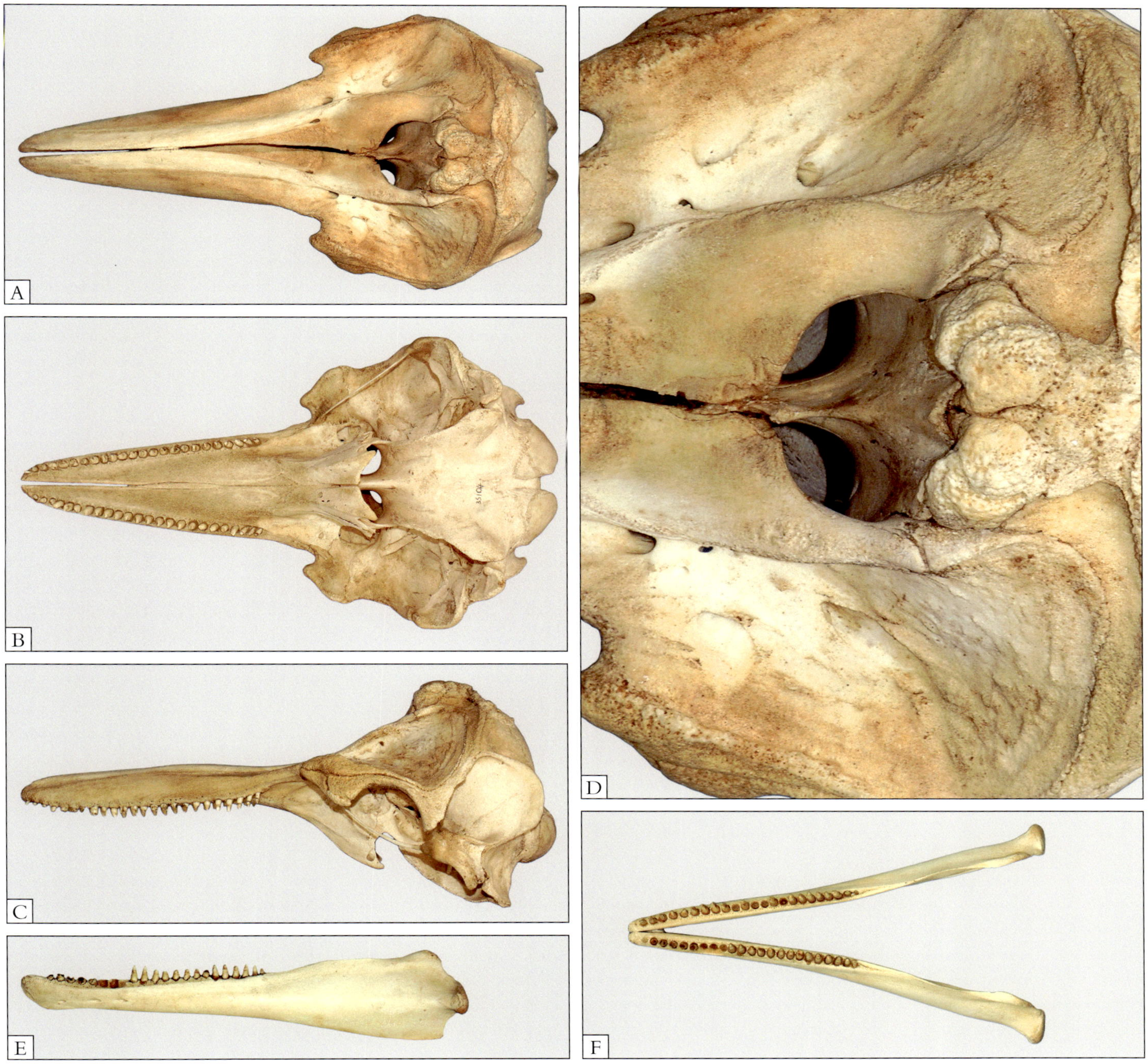

FIG. 110 Dorsal (A), ventral (B) and left lateral (C) views and vertex of skull (D); left lateral (labial) (E) and dorsal (F) views of mandible (ZM 35104).

from a stranded female that was refloated. No cyamids have been recorded from the species in the subregion.[12]

Food and feeding behaviour

The stomachs of three common bottlenose dolphins that stranded near Walvis Bay, Namibia (and so were presumably from the inshore population occurring there), all contained only fish remains: two had been feeding predominantly on southern mullet (*Liza richardsonii*) and one on horse mackerel (*Trachurus capensis*). Of the 95 identified prey items in total, 57.8% were southern mullet, 35.8% horse mackerel, 2.1% flathead mullet (*Mugil cephalus*), 2.1% Cape silverside (*Atherina breviceps*) and 2.1% may have been galjoen (Coracinidae). Not unexpectedly, these indicate a diet of nearshore fish.[12]

A male taken south of Madagascar (on Walters Shoal) in January 1974 contained trace remains from 374 lightfish (*Photichthys argenteus*), 92 lanternfish (including 22 *Ceratoscopelus warmingii*, 20 *Diaphus* sp., eight *Diaphus hudsoni*, two *Lampanyctus hectoris* and two possible *Myctophum aurolaternatum*) and 13 unidentified fish, plus the remains of two histioteuthid squids and one *Lycoteuthis diadema*. This indicates a diet based on mesopelagic fish and squid.[12]

The stomach contents of five of six common bottlenose dolphins stranded on the South African coast tended to resemble that of the Walters Shoal animal, containing in total 277 identified prey items, of which lanternfish (including *Ceratoscopelus*) comprised 28.1%, hake (*Merluccius capensis*) 6.9%, buttersnoek (*Lepidopus caudatus*) 6.1%, hatchet fish (*Maurolicus muelleri*) 3.6%, argentine (*Nansenia* sp.) 1.1%, and cephalopods 53.4% (90% of which were *Lycoteuthis diadema*).[5, 11, 12] The sixth animal contained the remains of 105 white seacatfish (*Galeichthys feliceps*).[12] Unlike most of the other prey species listed above, which tend to be mesopelagic or even bathypelagic in nature, *Galeichthys* is a coastal species occurring in estuaries or muddy/sandy areas down to 60 m. Its dorsal and pectoral fins also carry sharp spines covered with a venomous mucus that can inflict painful wounds,[27] and it would not seem like a normal prey species. The dolphin concerned stranded twice over two days in company with a smaller animal, was seen alive on the second and fourth days but found dead on the beach five days after originally stranding. Presumably the seacatfish were consumed some time during this period, when the animal may have been in extremis.

Little is known about the feeding behaviour of the *truncatus*-type bottlenose dolphins in the subregion. Observed feeding bouts in the inshore population off Namibia were always very brief, possibly because most observations took place mainly within the dolphins' resting areas.[24]

The average daily food intake by captive *truncatus* from the northern hemisphere declined as a percentage of body mass from about 5% in a 100 kg animal to 1.75% in a 250 kg animal, although there was much variation in females, depending on their reproductive status. Average annual food consumption in adults was around 1 900 kg.[28]

Reproduction and growth

A common bottlenose dolphin neonate stranded at Walvis Bay measured 1.35 m in length,[12] while a male calf from the same population born in captivity was estimated at 1.15 m.[29] Pregnancy in captive *truncatus* lasts 370 ± 7 days.[30] Births in the Walvis Bay population are not sharply seasonal, but may take place mainly in late summer. The calf born in captivity (but conceived in the wild) arrived in January,[12] newborns have been sighted off Walvis Bay in February, May and June,[31] and the neonate stranding occurred in March.[12] The calf born in captivity continued to suckle for at least 20 months, with suckling bouts falling rapidly over the first two months from 139 to 42 a day, but thereafter declining more slowly, so that by 20 months it was still indulging in 11 bouts a day. The mean duration of a suckling bout remained remarkably constant throughout lactation, at between 3 and 6 sec., and took place equally from each of the mammary glands.[29] A milk sample taken from the mother after two years of lactation comprised 72.8% water, 13.2% fat and 12.1% protein.[32] Other studies earlier in lactation have produced average fat values ranging from 20.7% to 29.4%, and protein values of 9.8% to 12.2%.[33] The captive calf began taking solid food in a concerted way around 11 months of age and for the first two months thereafter its intake of solid food increased greatly, followed by a period of fluctuation. By 30 months it had already reached a length of 2.75 m.[29]

Of eight females examined in the subregion, four with a length of 2.795 m or less were sexually immature, while four that were 2.81 m or longer were mature. This suggests that sexual maturity in females is attained at a length of about 2.8 m. Combined testes weights (or histological data) are available for 11 males from the subregion: seven with a length of 2.86 m or less all had testes weighing 35.4 g or less or were histologically immature, while four that were 2.93 m or more in length had testes weighing 459 g or more. This implies that maturity is reached at about 2.9 m in males.[5, 11, 12, 20] A female that conceived in the wild gave birth in captivity to a calf, which may have been her first, at an age determined as 6.5 years from tooth layers.[13] Ovulation in bottlenose dolphins appears to be spontaneous,[34] and calving intervals in captive animals ranged from 18 to 67 months with an average of 32 months.[30] Annual ovulation rates in wild *truncatus* off Japan averaged 0.44–0.46/year and pregnancy rates 33.3–40.8%, and neither changed with increasing age. According to growth layers in the teeth, both sexes could live up to 45 years, although older males may have a lower survival rate than females.[35] A female named Nellie, born in captivity at Marineland of Florida in February 1953, has so far lived for 54 years.[36]

Behaviour and vocalisations

The population of common bottlenose dolphins inshore on the Namibian coast occurs in groups of 1–39, although rarely in groups of more than 20, the average size being seven. Groups tend to be larger in winter and spring than in summer and autumn.[24] Offshore bottlenose dolphins in the subregion have been encountered in schools of 1–300 animals (mean 35, n = 37), although most (78%) were less than 30. A characteristic of these groups is that several (24%) were in association with other species: two with false killer whales, four with pilot whales, one with Risso's dolphins, one with melon-headed whales and Fraser's dolphins, and one with unidentified dolphins and toothed whales.[12] Such associations could provide the advantage of safety in numbers, while providing more options in feeding strategies.[37]

The dolphins in the Walvis Bay population spent much of their time slowly and repeatedly patrolling certain short stretches of the coastline. These movements followed a clear diurnal pattern, with the number of 180° switches in direction increasing from 1–3 an hour in the early morning to 7–10 an hour in the late afternoon.[24] A diurnal pattern in surfacing behaviour has been described for coastal common bottlenose dolphins elsewhere, in which the fewest, longest dives (and thus the highest proportion of time underwater) occurred in the afternoon/evening, and the shortest, most frequent dives (and thus the least time underwater) occurred in the early morning.[38]

The home range of a population of inshore common bottlenose dolphins in Florida was about 85 km², in which female-calf pairs and subadult males had ranges of 40 km² and other adult females, subadult females and adult males had ranges of 15–20 km². Some parts of the range were used more than others, depending on the season.[39]

'Displacements' of individual coastal bottlenose dolphins from their usual range by up to 600 km have been recorded, possibly in response to environmental change, and while some animals have returned to their original range others remained in their new area.[40]

In captivity, common bottlenose dolphins formed dominance hierarchies, with adult males dominant over all other tank mates. This could include aggressive behaviour such as biting, fluke-slapping and ramming. The incidence of tooth rakes present on the skin of larger animals in the wild demonstrates

Hybridisation?

FIG. 111 Spot the odd man out – three *truncatus/aduncus* 'hybrid' males and one pure *truncatus* male at U'Shaka Marine World, Durban, 2006.

Hybridisation in captive cetaceans is not uncommon. There are at least 25 reported instances, all involving bottlenose dolphins as either the dam or the sire: 13 with Risso's dolphins, six with false killer whales, two with short-finned pilot whales, two with short-beaked common dolphins, and one each with long-beaked common dolphins and rough-toothed dolphins. Only one of these offspring is reported to have been fertile: a calf with bottlenose dolphin/false killer whale parentage produced two calves of its own, both sired by bottlenose dolphins. Hybridisation has also been reported in the wild, most frequently (at least 11 cases) between blue and fin whales, but also three instances between bottlenose and Risso's dolphins, and single cases between a long-beaked common and a dusky dolphin, and between a beluga and a narwhal. Only one of these was reported to be fertile: a blue/fin whale cross was pregnant with a foetus for which the father was identified genetically as a blue whale. Hybrid sterility may be more characteristic of male offspring, but this is generally more difficult to demonstrate: two male blue/fin hybrids examined proved to be sexually immature, even though one of them was estimated to be 24 years old.[1,2]

The distinction between the *truncatus* and *aduncus* morphotypes is one of the principal issues in the debate over the taxonomy of bottlenose dolphins. Some morphological and genetic evidence seems to indicate that these may well be separate species (and the fact that their distributions at least partially overlap in some areas would seem to support this), while other genetic analysis suggests they shouldn't even be in the same genus!

U'Shaka Marine World in Durban, South Africa, has kept bottlenose dolphins of both morphotypes together since 1976. In particular, a male (Gambit) from the inshore common bottlenose dolphin population at Walvis Bay has sired five calves with his favourite female Frodo, an Indo-Pacific bottlenose dolphin captured off Durban, and two with a female called Sandy, also from Walvis Bay. One of the five 'hybrid' offspring, Freya, has herself given birth to a calf, sired by Kelpie, an inshore common bottlenose dolphin born in captivity.[3] This shows that the first generation 'hybrids' between the two morphotypes are fertile, at least if they are female.

References

1 Bérubé 2002
2 Árnason *et al.* 1991
3 U'Shaka Marine World, pers. comm.

that this behaviour is not confined to captivity. When a female is receptive, males spend prolonged periods of time in courtship, rubbing, mouthing, nuzzling, jaw-clapping, and adopting a particular S-shaped body posture. Pregnant females tend to isolate themselves or, towards the end of pregnancy, associate with a particular adult female, and often continue this relationship after the birth of the calf. Epimeletic or care-giving behaviour is frequently seen in both wild and captive common bottlenose dolphins, and involves the supporting of sick individuals and the carrying of their own dead young for days or weeks.[39]

Evidence of infanticide (and a possible association with the killing of porpoises) has been gathered for at least one population of common bottlenose dolphins.[41]

In many places throughout the world, individual wild bottlenose dolphins have interacted socially with human beings on a voluntary basis, such associations lasting from a few months to several years. These dolphins can be of either sex and of various ages and may be victims of circumstances or opportunists, rather than social misfits. The behaviours exhibited are similar to those seen in captivity, and include acts of aggression and sexuality towards humans.[42] In one instance a male bottlenose dolphin injured 29 bathers and killed a 30-year-old man after repeated instances of harassment.[43] In 1953 two female bottlenose dolphins (morphotype unknown) interacted with bathers at Fish Hoek, South Africa, allowing frequent physical contact including sitting astride them and being towed through the water while hanging on to their dorsal fin. This episode was unusual in that more than one dolphin was involved.[44]

There are no published studies of the acoustic repertoire of common bottlenose dolphins in the subregion. Elsewhere the repertoire has been divided into tonal (whistles and squeals) and pulse-based calls (coughs, quacks, buzzes, creaks, etc.). The tonal calls are narrow-band, frequency-modulated signals that are typically either less than 300 ms in duration with most energy below 5 kHz (squeals) or longer than 500 ms with most energy between 7 kHz and 26 kHz (whistles). Pulse-based calls can be divided into 'single-burst' (short impulsive sounds with most energy below 5 kHz, e.g. coughs and quacks), 'click-burst' (regular broadband pulses with rapid onset/offset times, which differ mainly in the length of interval between bursts, e.g. click trains, buzzes and creaks) and 'repeat burst' (sequences of single pulses typically below 2 kHz, e.g. 'hiccups' and 'laughter'). Only click trains and buzzes were exclusively or principally ultrasonic in nature:[45] these have been shown to be used in echolocation, whereby the animal can detect objects ahead of it, including prey, from returning echoes. A common bottlenose dolphin can detect a 7.62 cm diameter fluid-filled sphere at a distance of 113 m and possibly further, depending on background noise.[46]

Some of the whistles have been described as being individually distinctive and stereotypic in nature ('signature whistles'), developing when the calf is between about two months and a year old and persisting for several years afterwards (unless the individual was kept in prolonged isolation).[47] Although the signature whistle hypothesis has been criticised,[48] playback experiments with restrained animals have shown that mothers responded more strongly to the whistles of their independent offspring than to those of similar-aged non-offspring, while independent offspring responded more strongly to the whistles of their mother than to those of a familiar, similar-aged female.[49] Furthermore, signature whistles and probable signature whistles formed 52% of all whistles produced by free-ranging groups of bottlenose dolphins, increased in frequency of occurrence as group size increased, and were more frequent during socialising than feeding, but less frequent during travelling than when socialising.[50] Regional differences in whistle contours have been described.[51]

Conservation status

After 20 months of photo-identification surveys in 1993/4, 83 dolphins were individually identified among the inshore population of common bottlenose dolphins at Walvis Bay, and few new individuals were discovered over the final six months.[24] This suggests that virtually all the identifiable dolphins had been accounted for: what is not known is the exact proportion of identifiable animals in the population, although 'the great majority' were apparently well marked.[31] Consequently, there may be about 100–150 dolphins in this population. Between 1975 and 1983, 36 common bottlenose dolphins were live-captured at Walvis Bay: 26 were released alive, one died at capture and nine were taken into captivity. This represents a total removal of 10 animals out of the population over eight years. Direct exploitation of these animals has now ceased, although they have recently become the subjects of a regular dolphin-watching industry.[12]

There are no estimates of population size for common bottlenose dolphins elsewhere in the subregion. Levels of incidental mortality are also unknown. An animal that stranded near Gansbaai, South Africa, in 2003 came ashore with its abdomen (but not thorax) cut open, suggesting that it had been part of a fisheries by-catch in the vicinity.[12] Common bottlenose dolphins have been recorded taken in gill-net fisheries at Zanzibar,[20] and while it has been estimated that 43% of a minimum number of 93 dolphins taken there in a year may have been bottlenose dolphins,[52] most of these may have been the *aduncus* morphotype.[20]

Levels of organochlorine residues in common bottlenose dolphins from the subregion have only been monitored in five individuals from the west coast,[53] and the highest Total DDT and PCB levels detected were lower than in 65% and 85% respectively of Indo-Pacific bottlenose dolphins from KwaZulu-Natal.[54]

Bottlenose dolphins in general are listed as Data Deficient by the IUCN and are included on Appendix II of CITES. The *truncatus* form is listed as Data Deficient in the South African Red Data Book. No southern hemisphere population is listed by CMS.

References

1 Mead and Potter 1995
2 Walker 1981
3 Van Waerebeek *et al.* 1990
4 Rice 1998
5 Ross 1977
6 Ross and Cockcroft 1990
7 Hale *et al.* 2000
8 Kingston and Rosel 2004
9 Natoli *et al.* 2004
10 LeDuc *et al.* 1999
11 Ross 1984
12 MRI Whale Unit data
13 Peddemors 1989
14 Rommel 1990
15 Weir, C., pers. comm.
16 Findlay *et al.* 1992
17 Pelembe, T., pers. comm.
18 Perrin 1985
19 Peddemors *et al.* In press
20 Amir *et al.* 2005b
21 Borobia 1997
22 Keller *et al.* 1982
23 Corbett 1994
24 Praetsch 1995
25 Gambell *et al.* 1975
26 Wells *et al.* 1999
27 Taylor 1986
28 Kastelein *et al.* 2003b
29 Peddemors 1992
30 Asper *et al.* 1992
31 Praetsch 1994
32 Peddemors *et al.* 1989
33 Oftedal 1997
34 Ozharovskaya 1990
35 Kasuya *et al.* 1997b
36 W. Hurley, pers. comm.
37 Scott and Chivers 1990
38 Mate *et al.* 1995
39 Shane *et al.* 1986
40 Wells *et al.* 1990
41 Patterson *et al.* 1998
42 Lockyer 1990
43 Santos 1997
44 Tayler and Saayman 1972
45 Boisseau 2005
46 Au 2002
47 Caldwell *et al.* 1990
48 McCowan and Reiss 2001
49 Sayigh *et al.* 1998
50 Cook *et al.* 2004
51 Bazúa-Durán 2004
52 Amir *et al.* 2002
53 De Kock *et al.* 1994
54 Cockcroft *et al.* 1989

Indo-Pacific Bottlenose Dolphin

Tursiops aduncus

THE UNCERTAIN TAXONOMY of bottlenose dolphins is discussed in some detail in the species account of the common bottlenose dolphin. Suffice it to say that this account deals with the smaller morphotype of bottlenose dolphin that is found inshore throughout the Indian Ocean and into the south-west Pacific, and which is here recognised as a separate species, *T. aduncus*.

Appearance

Bottlenose dolphins are distinguishable from other delphinids in the subregion by their short, stout beak, thick teeth and large dorsal fin. Indo-Pacific bottlenose dolphins can be distinguished from common bottlenose dolphins by their longer beak and more uniform dorsal coloration. In particular, a light grey blaze that runs back from the side of the head (which has a branch curving up towards the dorsal fin in common bottlenose dolphins) is muted or even absent in Indo-Pacific bottlenose dolphins. The belly and sides of larger individuals of *T. aduncus* may also become freckled with dark spots, something that does not occur in common bottlenose dolphins.[1]

Indo-Pacific bottlenose dolphins reach physical maturity at average lengths of 2.43 m in males and 2.38 m in females, although individual males attain 2.57 m and females 2.49 m. The species is therefore barely sexually dimorphic.[2] The beak is prominent, measuring 4.4–6.0% (mean 5.2%, n = 30) of body length (and therefore proportionately longer than in *truncatus*),[3] and is well differentiated from the melon in the traditional 'bottle' shape. The body is not as robust as in the common bottlenose dolphin, but there is a similar-sized dorsal fin, in height 7.5–12.8% (mean 10%, n = 34) of body length, situated about mid-body length. The flippers are proportionately somewhat larger than in the common bottlenose dolphin, in length 15.5–23.4% (mean 17.4%, n = 38) of body length, but are similar in shape, with a strongly curved leading edge and a concave trailing edge. The width of the flukes is similar to that in common bottlenose dolphins, being 17.9–29.4% (mean 23.2%, n = 34) of body length, with a well-defined central notch.[3, 4]

The coloration of Indo-Pacific bottlenose dolphins is far less striking than that of common bottlenose dolphins, although there is a basic similarity to the overall pattern. The contrast between the dark dorsal cape and the grey area beneath it is much less, and the separate tongue of pigment that runs up towards the dorsal fin is either completely lacking or present only as a remnant. Although the lateral grey area extends up again behind the dorsal fin onto the dorsal peduncle, because of its muted nature it is not very distinctive. The belly is white from throat to anus in the calf and juveniles, but small dark freckles develop with age and can cover much of the abdomen

Length:weight relationship

The length:weight relationship for 41 animals from the subregion weighed whole[3–5] has been calculated as **$W = 12.365L^{2.9495}$**, where W = weight in kilograms and L = total length in metres. This implies that a 1.5 m dolphin would weigh 41 kg and a 2.5 m animal 184 kg.

Skeletal data

Vertebral formula (n = 28): 7 cervical, 12–14 thoracic, 13–18 lumbar and 22–27 caudal; total 57–63[3, 5] (compared to 64–66 in common bottlenose dolphins). The first vertebra with an arterial canal through the transverse process varies between the 38th and the 41st from the head[3, 4] (cf. 43rd and 46th in common bottlenose dolphins).

The atlas and axis are fused, and sometimes the third cervical is also joined to them, but the remaining cervical vertebrae are free.[8]

Phalangeal formula, including metacarpals (n = 1): I_2, II_8, III_6, IV_4, V_3.[4]

The characteristics of the skull that differentiate *T. aduncus* from *T. truncatus* in the subregion are listed in the latter's species account.

and sides in adults. There is a small dark eye patch and a streak running forward to the apex of the melon. The flippers and flukes are dark grey above and below, and there is an indistinct stripe from flipper to eye. There is no trace of the dark lines on the throat extending forwards from the base of the flippers seen in the common bottlenose dolphin, but there may be a light grey 'anchor mark' running across the throat at roughly eye level.[1, 5] Calves can be born a very light grey or nearly black, and the underside can be pinkish (this colour disappearing within a few months).[6]

There are 24–29 (mean 25.6, n = 35) pairs of teeth in the upper jaw and 22–29 (mean 25.8, n = 35) pairs in the lower jaw.[3–5] They are stout, 6–8 mm (mean 6.6 mm, n = 4) in maximum diameter at the gum, and conical in shape.[5] Tooth eruption begins in the upper jaw at about three months of age, and most teeth in both jaws are visible at about 5.5 months.[7]

Distribution and migrations

Indo-Pacific bottlenose dolphins probably occur throughout the coastal and shallow offshore waters of the temperate and tropical regions of the Indian Ocean and the south-west Pacific, from South Africa in the west to south-east Asian waters from the East China Sea southwards, New Caledonia and the east coast of Australia, in the east.[1]

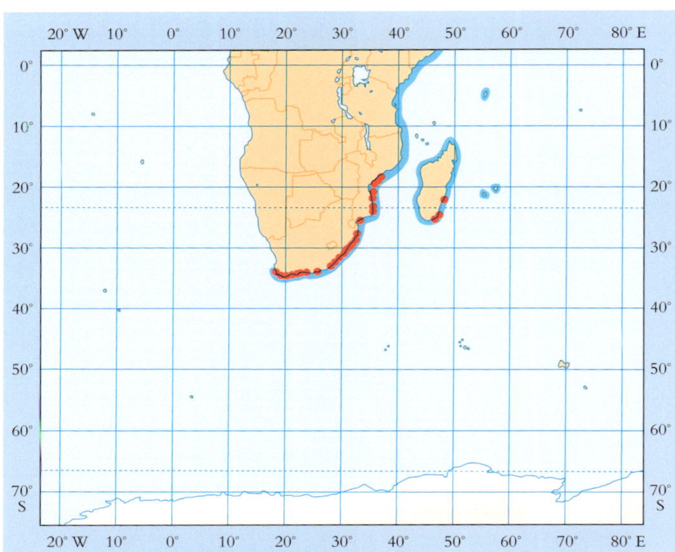

Postulated range   Sighting or capture

FIG. 112 Distribution map for Indo-Pacific bottlenose dolphin.

Within the subregion, bottlenose dolphins that might be referred to this form have been recorded from South Africa,[9] Mozambique,[10] Tanzania (including Zanzibar[11]), Kenya,[12] the Seychelles,[13] Madagascar,[10] Mauritius, and Réunion.[14] It is not always possible to determine to which form/species some of these records apply, but it has been assumed that all those occurring coastally in the Indian Ocean should be referred to the *aduncus* form. On the southern African coast, Indo-Pacific bottlenose dolphins have been sighted as far west as False Bay in the Western Cape, South Africa.[15] Their distribution is essentially continuous from Cape Agulhas eastwards as far as southern Mozambique (and possibly further north).

In the Eastern Cape and KwaZulu-Natal, South Africa, Indo-Pacific bottlenose dolphins occurred almost exclusively in water less than 30 m deep, at least during daylight surveys.[16] Along the KwaZulu-Natal coast, the dolphins seem to have 'preferred areas', about 30 km apart, where they are more frequently encountered, but whether these correspond to discrete home ranges is not known.[17, 18] Differences in organochlorine levels between dolphins in three adjacent areas, each 70–80 km in extent, along the coast of KwaZulu-Natal suggested some form of population structure, but it is not clear whether possible differences in the age compositions of these samples were fully taken into account.[19] The practice of bottlenose dolphins stealing fish from hooks off Margate, KwaZulu-Natal, apparently persisted for 25 years without the behaviour spreading to adjacent areas, also suggesting that home ranges may be quite limited in extent.[20] On the other hand, return movements of known individuals have occurred between Algoa Bay and Plettenberg Bay on the south coast of South Africa, or a round trip of about 420 km.[3] Furthermore, there may be a seasonal movement of bottlenose dolphins into the waters of southern KwaZulu-Natal in association with the 'sardine run'.[17] These individuals may originate from as far west as Plettenberg Bay on the south coast of South Africa (or a coastwise distance of about 800 km).[21]

Average rates of travel have been recorded as 2.79 km/hr on the coast of KwaZulu-Natal and 3.85 km/hr in the Eastern Cape, South Africa.[20] Progression rates (when the dolphins were in transit) were much higher, averaging 9.87 km/hr.[22]

External commensals and parasites

In nine Indo-Pacific bottlenose dolphins examined in the Western Cape, one individual had stalked barnacles (*Xenobalanus* sp.) attached to the left flipper (1) and flukes (3).[5] These barnacles have been frequently noted on the dorsal fin and flukes of bottlenose dolphins in Algoa Bay, and 18 *Xenobalanus* have been recorded attached to the trailing edge of the flukes in one individual.[3] They appear to be more abundant on younger, smaller individuals, and may show a seasonal cycle of infestation.[23] Three cyamids (unidentified) were collected from the lip of the blowhole of two dolphins from the south-east coast of South Africa.[3]

Food and feeding behaviour

Although more than 72 prey species were found among 5 610 items from the stomachs of 165 Indo-Pacific bottlenose dolphins caught in anti-shark nets in KwaZulu-Natal, six species predominated: these were the fish piggy (*Pomadasys olivaceum*, 41.5% by number), red tjor-tjor (*Pagellus bellotti*, 8.8%), African maasbanker (*Trachurus delagoa*, 5.7%) and mackerel (*Scomber japonicus*, 3.3%), the common cuttlefish (*Sepia officinalis*, 7.4%) and the squid (*Loligo* spp., 4.4%). This suggests that a wide variety of prey is consumed, including reef and sandy-bottom benthic prey and their associated predators (pelagic shoaling fish and cephalopods), and deeper-water fish. Calves tended to eat smaller size-classes of prey (and concentrated more on *P. olivaceum*) than other dolphins, while adults ate larger individuals than adolescents. The only seasonal changes apparent were a relative increase in the importance of *Loligo* spp. and *T. delagoae* (and a relative decrease in *S. officinalis*) between October and December.[24] Stranded dolphins from the Eastern Cape, South Africa, seemed

FIG. 113 Indo-Pacific bottlenose dolphins, seen here at Plettenberg Bay on South Africa's south coast (*above*), are a feature of the nearshore environment of the western Indian Ocean. A captive female at least 25 years old (*left*) shows the heavy ventral spotting typical of older animals in the southern African population.

to have a greater proportion of cephalopods in their diet, but this could be because cephalopod beaks tend to remain longer than fish otoliths in the stomachs of stranded animals.[3] In the Western Cape, trace remains from five stranded individuals included (in total) five shallow-water hake (*Merluccius capensis*), five baardman (*Umbrina canariensis*), one geelbek (*Atractoscion aequidens*), one anchovy (*Engraulis japonicus*), one seabream (Sparidae), three unidentified fish, and four common squid (*Loligo vulgaris*).[5] Stomach content analysis for 26 dolphins taken in gill nets off Zanzibar also revealed a high diversity of prey items (50 species of fish and three of cephalopods), but eight species comprised about 65% of the diet by weight. In order of importance these were slender conger (*Uroconger lepturus*), Kaup's arrowtooth eel (*Synaphobranchus kaupii*), plain cardinalfish (*Apogon apogonides*), bigfin reef squid (*Sepioteuthis lessoniana*), yellowtail emperor (*Lethrinus crocineus*), broad club cuttlefish (*Sepia latimanus*), Indian squid (*Loligo duvauceli*), and blacktail snapper (*Lutjanus fulvus*). Fish constituted 84% of the prey by mass and cephalopods 16%, and the species eaten indicate that the dolphins foraged nearshore and over reef and soft-bottom substrata.[25]

Several coordinated feeding strategies have been described for Indo-Pacific bottlenose dolphins. Herding of prey was carried out by all groups within the school and commonly involved high speed chasing followed by encirclement or entrapment and feeding from the periphery by individuals.[15, 26] Such coordinated behaviour allowed the dolphins to feed rapidly to satiety, after which they were sometimes seen to play with uneaten fish.[27] The use of beach hunting, as practised in Shark Bay, Western Australia,[28] has been described for Indo-Pacific humpback dolphins in the subregion (see the species account), but not yet for bottlenose dolphins.

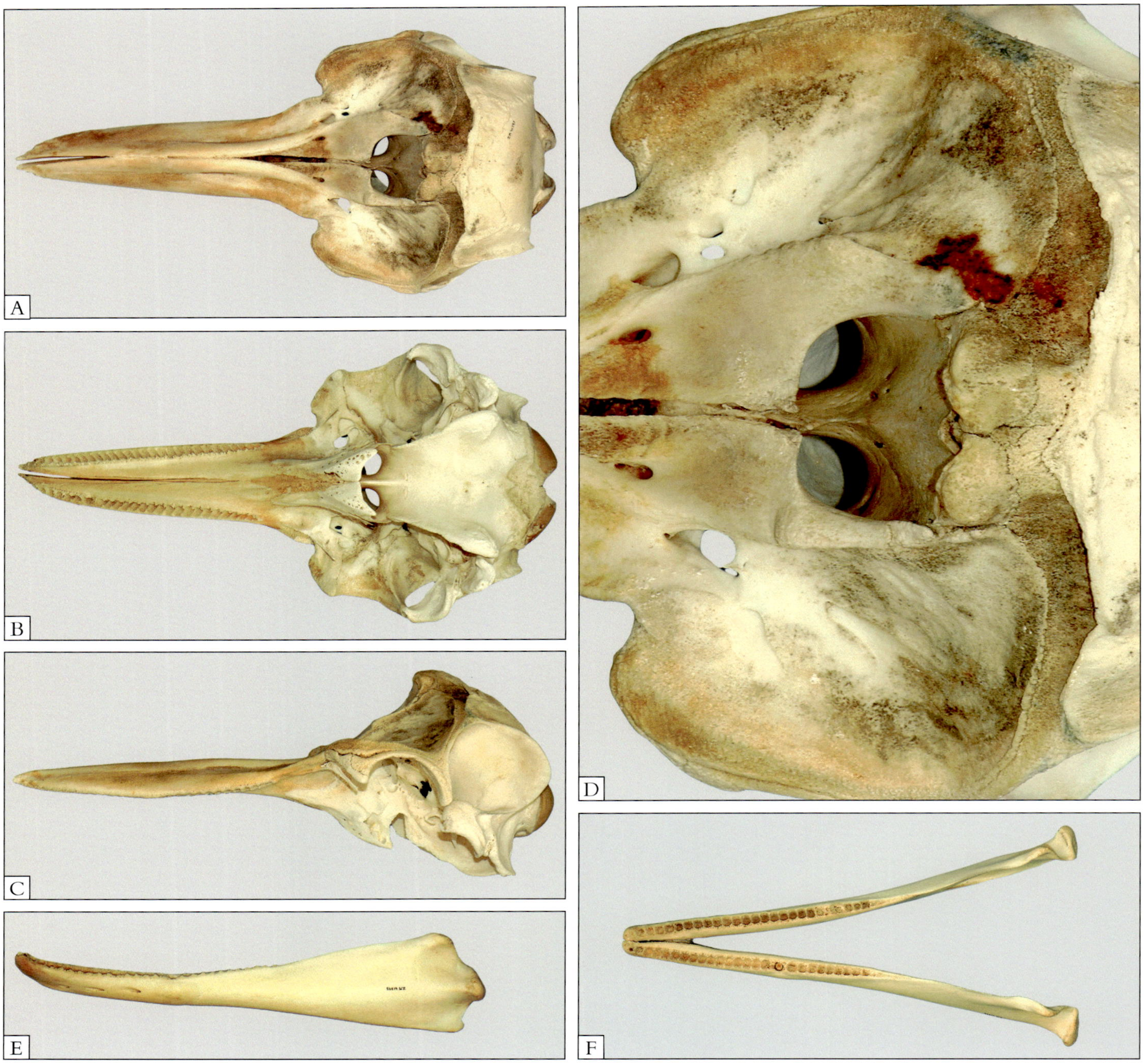

FIG. 114 Dorsal (A), ventral (B) and left lateral (C) views and vertex of skull (D); left lateral (labial) (E) and dorsal (F) views of mandible (ZM 41395).

Daily feeding rates in captive Indo-Pacific bottlenose dolphins ranged from 6.7% of body weight (in a 1.93 m, 83 kg animal) to 3.87% of body weight (in a 2.35 m, 156 kg animal).[3] Ambient water temperature had a marked effect on food intake, with higher rates at lower temperatures.[29]

Reproduction and growth

Indo-Pacific bottlenose dolphins are born at an average length of 1.03 m (range 86–115 cm, n = 15),[2] after a gestation period of 357–384 (mean 371, n = 9) days.[30] On the KwaZulu-Natal coast, most (76%, n = 25) births occur in summer and autumn (November–April),[2] whereas in Western Australia the timing is a little earlier (73% between October and December).[31] Suckling is recorded as persisting for at least 23 months in captivity,[7] and up to three years of age judging by stomach contents,[2] but in the Western Australian population may last up to nine years, though 66.7% of individuals (n = 42) were weaned before they were four years old.[31] Off Mikura Island, Japan, weaning occurred at 3–6 years (mean 3.5 years, n = 22) in both sexes.[32] In a captive calf, suckling occurred equally from either nipple, with an average bout duration of 4.7 sec. The number of suckling bouts per hour was high (about 30) immediately after birth but decreased with age, particularly over the first six months, by which time the calf was suckling about six times an hour. The first observed intake of solid food occurred 321 days after birth, and regular feeding 10 days later.[7] Stomach content analysis indicated that prey intake in dolphins off KwaZulu-Natal commenced at about six months of age. Females may ovulate for the first time between 9.5 and 11 years of age,[2] and produce their first viable calf at 12–15 years,[29, 31] while puberty in males may occur at nine years but mainly between 10 and 12 years[2, 29] (sexual 'maturity' being attained at about 14.5 years[2]). Sizes at sexual maturity range between 2.13 m and 2.3 m in females,[3] and at 2.4 m^2 (or a little less[29]) in males.

The ovulation rate in females off KwaZulu-Natal averaged 0.29 per year, with no sign of a decline with age.[2] This assumes that all ovulations are recorded in the ovaries as scars (corpora albicantia), an assumption that has been challenged.[33] It compares with a birth rate (number of neonates per year/number of adult females) of 0.27 for dolphins off Mikura Island, Japan.[32] Such a rate would imply a minimum inter-birth interval of three to four years. The interval between surviving calves ranged from three to 6.2 years with an average of 4.55 years (n = 33) off Western Australia,[31] and from three to five years (mean 3.5 years, n = 19) off Mikura Island, Japan.[32] Females may continue to suckle their calves until halfway through their next pregnancy. If the calf is lost within three months of birth, however, the mother will become pregnant again within a year. Calf mortality is high, 29% dying within the first year, 18% in the second and 3% in the third.[31]

Age readings indicate that both sexes can live to 42–43 years.[2] A female born in captivity in Port Elizabeth lived for 37 years, while another female that lived for 32 years conceived four years after capture and so was presumably at least 38 years old at the time of her death.[5]

Behaviour and vocalisations

Estimates of group size made on aerial surveys of the Transkei and KwaZulu-Natal coast in South Africa varied from one to 287, with little difference between the estimates made in April/May and September, and an overall mean of 40 dolphins per group (n = 203). Most groups (79.8%) were 70 or fewer in size, although groups of 100 or more animals were not uncommon (10.8%).[16–18, 34] Groups in the Eastern Cape, South Africa, seemed to be much larger, in one study ranging from three to 1 000, with a mean size of 140, and 63% of the groups (n = 93) contained more than 50 individuals.[27] In a second study, groups of up to 1 000 were recorded, but 52.7% (n = 129) contained 50 or fewer individuals, and it was suggested that the 'functional' group size lay between 20 and 50 animals and that the larger groups were amalgamations of these units.[3]

In both Algoa Bay and Plettenberg Bay, South Africa, the dolphins are seen mainly in the early morning and (to a lesser extent) in the evening.[26, 35] In the early morning 'feeding' was the predominant activity, gradually declining in occurrence to a minimum around 14h00, and then increasing again in the evening. Physical interactions ('rubbing') followed a similar pattern but with the peaks about an hour later, while 'mating' was closely correlated with rubbing but failed to show an evening peak.[26] This pattern of daylight activities does not vary seasonally, except that mating behaviour is more prevalent in summer.[35]

In captivity an older female assumed dominance over a younger, and maintained this after the introduction of two males except when the younger female was being courted by a male. The larger of the two bulls (captured from the same school) was dominant over all animals in the tank, and the smaller took over this role after the death of the larger. Courtship involved high-speed swimming, leaping with flukes upturned when airborne, and sharp, exaggerated turns and dives. This was followed by the 'rubbing' behaviour mentioned above, including massaging with flippers and flukes, and copulation occurred exclusively with the favoured cow for 2–40 days. The bull might resume displaying to a different cow 1–8 days later. This pattern of copulatory behaviour continued throughout the year but increased in frequency in spring and early summer.[15]

Patterns of individual association among Indo-Pacific bottlenose dolphins in Shark Bay, Western Australia, showed that 'parties' of dolphins that were encountered contained on average 4.8 animals but were unstable in size and associations. The only long-term consistent associations were between animals of the same sex, and the most consistent were between males and between mothers and young.[36] Inter-male associations took the form of alliances, where 2–14 males formed associations to sequester and control females in breeding condition. Some alliances were stable, involving pairs or trios of males that associated for up to 17 years; others were larger, more fluid alliances in which partners frequently switched. Males in the more stable alliances tended to be more closely related than would occur by chance, whereas those in the larger alliances were not. Although males from alliances were responsible for most conceptions, juvenile males without partners were also successful.[37] Such associations have not been seen in South African waters,[38] although some observations are suggestive.[26] Mothers and their young associated consistently for at least four years,[36] and were more likely to be found away from other animals than any other grouping. When they were in a group they were most likely to be found with other females or (*c.* 6% of the time) in a mixed group with at least one other adult female and an adult male present, but never with an adult male alone.[6]

The incidence of tooth rakes in the Western Australian population suggests that males are more aggressive than females, that cycling females receive more aggression than non-cycling females, and that calves receive less aggression than either juveniles or adults. Adult female aggression is almost non-existent, and juvenile and adult males are responsible for most of the attacks on other animals in the population.[39]

Although it has not been as intensively studied, the acoustic repertoire of Indo-Pacific bottlenose dolphins contains the same basic elements as that of common bottlenose dolphins, namely pulsed calls and whistles. A comparison of the whistle contours of bottlenose dolphins from several populations revealed differences,[40] but as some of the samples may have been from *aduncus*-type and some from *truncatus*-type dolphins, it is not clear whether the differences were geographical or systematic. Differences in the whistle contours among three Japanese populations of Indo-Pacific bottlenose dolphins have been demonstrated,[41] but no such study has yet been undertaken in the subregion.

Conservation status

Aerial surveys for Indo-Pacific bottlenose dolphins along the KwaZulu-Natal coast were carried out in 1984, 1985 and 1989, producing counts for the coast from Durban to 80 km to the north of 367, 433 and an estimate of 520 dolphins (95% CI 156, 970) respectively. Only the last estimate was corrected for the proportion of groups missed (69%) as a result of over-flying. A survey of the coast from Durban 100 km southwards produced uncorrected counts of 219–249 dolphins in 1985 and 98–132 in 1990.[17, 18, 34] Since 1980 (when reliable statistics start), between 14 and 62 (mean 37) Indo-Pacific bottlenose dolphins have died annually in nets set along the KwaZulu-Natal coast to protect bathers from sharks[42, 43] (see box on page 187). The impact of these catches on the population depends to some extent on the contribution of the 'migratory' substock, as catches are seasonally higher in midwinter at the time that this substock arrives in KwaZulu-Natal waters. However, a straight comparison of these catches with the population figures available suggests that if there is a problem it may be more serious south than north of Durban.[18] Indo-Pacific bottlenose dolphins are also taken incidentally in artisanal gill-net fisheries in Zanzibar, 43 being recorded between 1995 and 1999, and another 68 between 2000 and 2003.[11, 44] These compare with population estimates for the south coast of Zanzibar of 136–179 between 1999 and 2002, and the by-catch is very likely unsustainable.[45] The species may also feature as direct and indirect takes in other inshore fisheries in the sub-region,[46] but the levels of such takes are largely undocumented.

Concentrations of organochlorines (such as PCBs, DDT and dieldrin) in the blubber of Indo-Pacific bottlenose dolphins off KwaZulu-Natal were closely correlated with age until the attainment of sexual maturity, when levels dropped markedly in females (presumably as they offloaded material to their young during pregnancy and lactation), while males continued to accumulate residues throughout life. Age classes principally at risk were therefore neonates (especially first-borns, that might receive as much as 80% of their mother's residue load) and mature males. While no deleterious effects were apparent, the PCB levels were twice as great as (and the dieldrin levels similar to) those found in other small cetaceans where they were correlated with decreased levels of testosterone in males.[19]

Bottlenose dolphins in general are listed as Data Deficient by the IUCN and are included on Appendix II of CITES. In the South African Red Data Book the *aduncus* form in general is listed as Vulnerable, and the migratory subpopulation as Endangered. The populations in the Timor Sea and the Arafura Sea are included on Appendix II of CMS.

References

1 Hale *et al.* 2000
2 Cockcroft and Ross 1990a
3 Ross 1984
4 Ross 1977
5 MRI Whale Unit data
6 Mann and Smuts 1999
7 Cockcroft and Ross 1990c
8 Rommel 1990
9 Findlay *et al.* 1992
10 Peddemors *et al.* In press
11 Amir *et al.* 2005b
12 Borobia 1997
13 Keller *et al.* 1982
14 Corbett 1994
15 Tayler and Saayman 1972
16 Ross *et al.* 1987
17 Cockcroft *et al.* 1991
18 Cockcroft *et al.* 1992
19 Cockcroft *et al.* 1989
20 Cockcroft *et al.* 1990b
21 Peddemors 2004
22 Saayman *et al.* 1972
23 Orams and Schuetze 1998
24 Cockcroft and Ross 1990b
25 Amir *et al.* 2005a
26 Saayman *et al.* 1973
27 Saayman and Tayler 1973
28 Sargeant *et al.* 2005
29 Cheal and Gales 1992
30 Brook and Kinoshita 2005
31 Mann *et al.* 2000
32 Kogi *et al.* 2004
33 Brook *et al.* 2002
34 Ross *et al.* 1989
35 Hoffmann and Best 2005
36 Smolker *et al.* 1992
37 Krützen *et al.* 2004
38 Karczmarski, L., pers. comm.
39 Scott *et al.* 2005
40 Wang *et al.* 1995b
41 Morisaka *et al.* 2005
42 Peddemors *et al.* 1998
43 Progress Report(s), South Africa, International Whaling Commission, 1998–2004
44 Amir *et al.* 2002
45 Stensland *et al.* 2006
46 Razafindrakoto *et al.* 2004

Catches in shark nets off KwaZulu-Natal

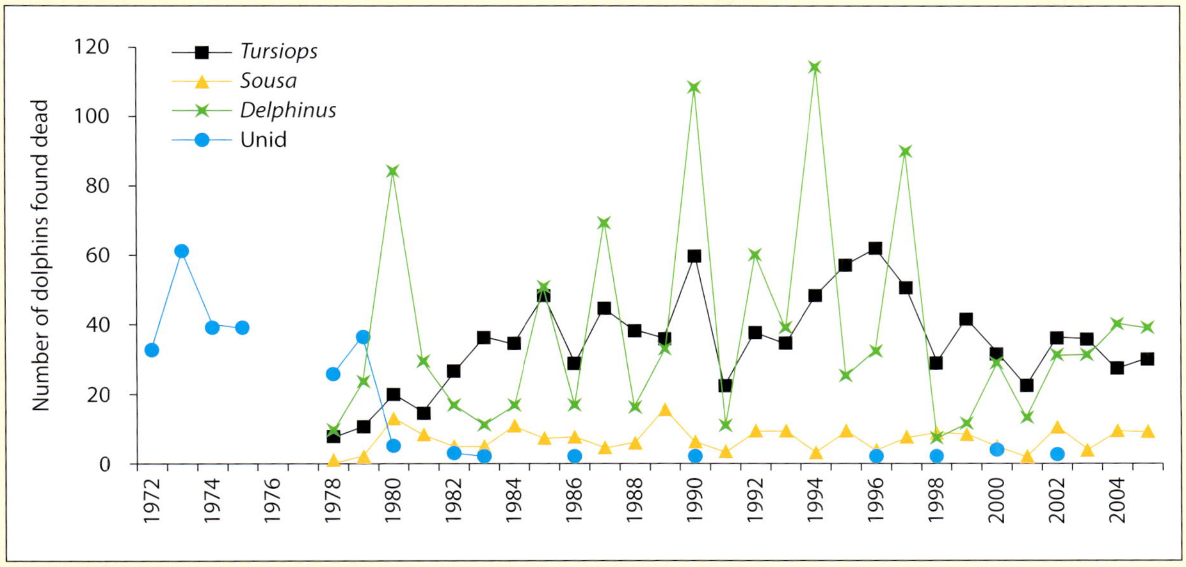

FIG. 115 Annual catches of dolphins in shark nets off KwaZulu-Natal.

The first nets set to protect bathers from sharks along the coastline of KwaZulu-Natal were established at Durban in 1952, after 21 shark encounters (seven of which were fatal) between 1943 and 1951.[1] Following another spate of attacks in 1957/58, when five people died in 107 days, other municipalities attempted to erect shark barriers, with mixed success. In 1962 extra nets were laid to the north and south of Durban, and in 1964 the Natal Anti-Shark Measures Board (now known as the Natal Sharks Board) was established. Today these nets in total cover 27.5 km, along a coastline of some 320 km, and all are maintained by the Natal Sharks Board. Most nets are 212 m long and 6.1 m deep, with a 25.5 cm mesh and anchored at each end. They are laid in two rows about 20 m apart parallel to the shore, about 500 m offshore, and up to 1998 were staggered so that there was about a 20 m overlap at each end. Since 1999 the stagger has been dispensed with as part of a 30% net reduction programme. The nets are inspected each weekday, weather permitting, and any catch removed. Apart from sharks, the nets unfortunately catch a number of other large vertebrates, including dolphins. Mitigating measures include releasing all live animals, and lifting the nets during the winter sardine run, when the presence of shoals inshore attracts a raft of predators to the vicinity of the nets.[2] Even so, on average 76 (range 36–175) dolphins become entangled and die each year, of which 46% are common dolphins, 42% are bottlenose dolphins and 8% are humpback dolphins.[3–5]

References

1 www.shark.co.za
2 Davis *et al.* 1989
3 Peddemors *et al.* 1998
4 IWC 2001b
5 Dudley, S., pers. comm.

Pantropical Spotted Dolphin

Stenella attenuata

Adult

Juvenile

SPOTTED DOLPHINS WORLDWIDE have been divided into two species, an Atlantic endemic species, *Stenella frontalis*, and a worldwide pantropical species, *S. attenuata*. The two species differ in coloration, and have non-overlapping vertebral counts.[1] Both are believed to occur in the subregion. Because relatively little research has been carried out on spotted dolphins locally, this account has had to incorporate information from other regions where necessary.

Appearance

Pantropical spotted dolphins can be distinguished from other dolphins in the subregion by their long, narrow beak and conspicuous dark cape extending well above the eye and down the sides of the thorax, with no spinal blaze curving up towards the dorsal fin. They have a bi-coloured peduncle, and a dark stripe from the flipper to the angle of the gape.

The largest individuals measured in the subregion are a 2.25 m male and a 2.15 m female (n = 14).[2, 3] Elsewhere body size varies geographically, with the largest males ranging from 2.20 m to 2.46 m and the largest females from 2.06 m to 2.27 m,[1] indicating the sexual dimorphism of the species.[4] The animal is slender in build, especially when compared with a bottlenose dolphin, for instance, or even an Atlantic spotted dolphin. The beak is long (4.6–6.9% of body length, mean 5.4%, n = 13) and slender, and well differentiated from the melon. As in other species of *Stenella*, the appendages are somewhat diminutive: the dorsal fin height is 6.7–9% (mean 7.9%, n = 12), the flipper length 11.1–13.2% (mean 12.2%, n = 12) and the fluke width 18.2–24.7% (mean 20.8%, n = 12) of body length.[3] The flippers are narrowly pointed with strongly curved leading edges, while the dorsal fin is recurved but somewhat 'slender' in appearance. A bulge may develop on the caudal peduncle behind the anus in adult males.

Length:weight relationship

The length:weight relationship for 11 animals from the subregion weighed whole[2, 3] has been calculated as **$W = 10.742L^{2.7412}$**, where W = weight in kilograms and L = total length in metres. This implies that a 1.5 m dolphin would weigh 33 kg and a 2.2 m animal 93 kg.

Skeletal data

Vertebral formula (n = 11): 7 cervical, 13–17 thoracic, 17–24 lumbar and 32–39 caudal; total 77–81[2, 3] (compared to 72 or fewer in Atlantic spotted dolphins[1]).

The atlas and axis are fused, and the third cervical may partly fuse with them in older animals, but the remaining cervical vertebrae are usually separate and free.[5]

Phalangeal formula, including metacarpals (n = 3): I_{2-3}, II_{8-9}, III_{5-6}, IV_{2-3}, V_{1-2}.[3]

Because the skulls of this species and the Atlantic spotted dolphin overlap in every character, distinguishing between them requires using a combination of characters. Generally, skulls of pantropical spotted dolphins have a proportionately narrower rostrum tip, a longer prenarial triangle and more teeth.[1]

The colour pattern of pantropical spotted dolphins, particularly the extent of spotting, can vary with age and between populations. At birth the young is completely unspotted. It has a well-defined cape passing high above the eye and down the side to its maximum expression ventrally at about the level of the front of the dorsal fin, before it sweeps up again to end in the midline about one-third of the way back from the dorsal fin. There is a dark stripe from the flipper to the angle of the gape, and from around the eye towards the apex of the melon. The belly is white but the upper half of the caudal peduncle behind the termination of the cape is two-tone, first pale grey and then towards the base of the flukes dark grey ('two-tone' stage).

With age the belly first darkens to light grey, then dark grey spots appear in the throat and abdominal regions ('speckled' stage). These are followed by white spots on the back. The dark flecks on the underneath of the animal gradually increase in size and number until they almost run together, while the white spots also increase in size and number ('mottled' stage). Finally the ventral spots become completely convergent so that the belly appears to be a uniform medium to dark grey, the tip of the rostrum and the lower jaw become white, and the edges of the gape may be similarly coloured ('fused' stage).[1,5,6] A 'chin-strap' effect may arise in older individuals, extending as pale flecking from the corner of the mouth forward and down onto the throat region.[7] Some of these changes in spotting are apparently correlated with reproductive maturation: 90% of fused and 50% of mottled females were sexually mature, compared to only 4% of speckled animals,[8] and similar trends were shown in males.[9]

There are 35–43 (mean 38.2, n = 11) pairs of teeth in the upper jaw and 34–40 (mean 38, n = 11) pairs in the lower jaw.[2,3] They are slender, 2.8–3.0 mm (mean 2.9 mm, n = 2) in maximum diameter at the gum, and conical in shape.[3]

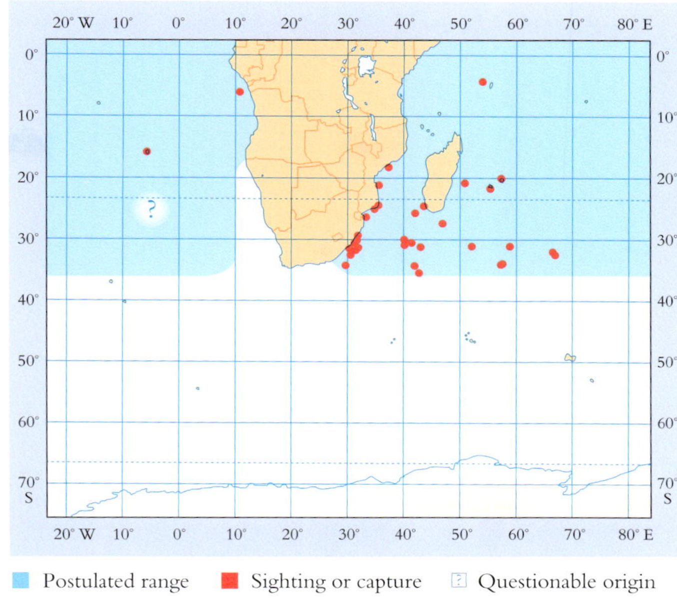

FIG. 116 Distribution map for pantropical spotted dolphin.

Distribution and migrations

Pantropical spotted dolphins occur throughout the tropical, subtropical and some warm-temperate regions of the world, between about 40° N and 40° S.[10] They can occur in both coastal waters and the open ocean, and in the eastern tropical Pacific the coastal and offshore forms are to some extent morphologically distinct.[1] Within the subregion, pantropical spotted dolphins have been recorded from Ascension Island,[11] St Helena,[12] Angola,[13] South Africa,[2] Mozambique,[14] Tanzania (including Zanzibar[15]), the Seychelles,[16] the Comoro Islands,[17] Madagascar,[14] Mauritius,[18] and Réunion.[3]

FIG. 117 This group of pantropical spotted dolphins was photographed off Port St Johns on the east coast of South Africa, about the furthest south that the species ranges in the Agulhas Current system.

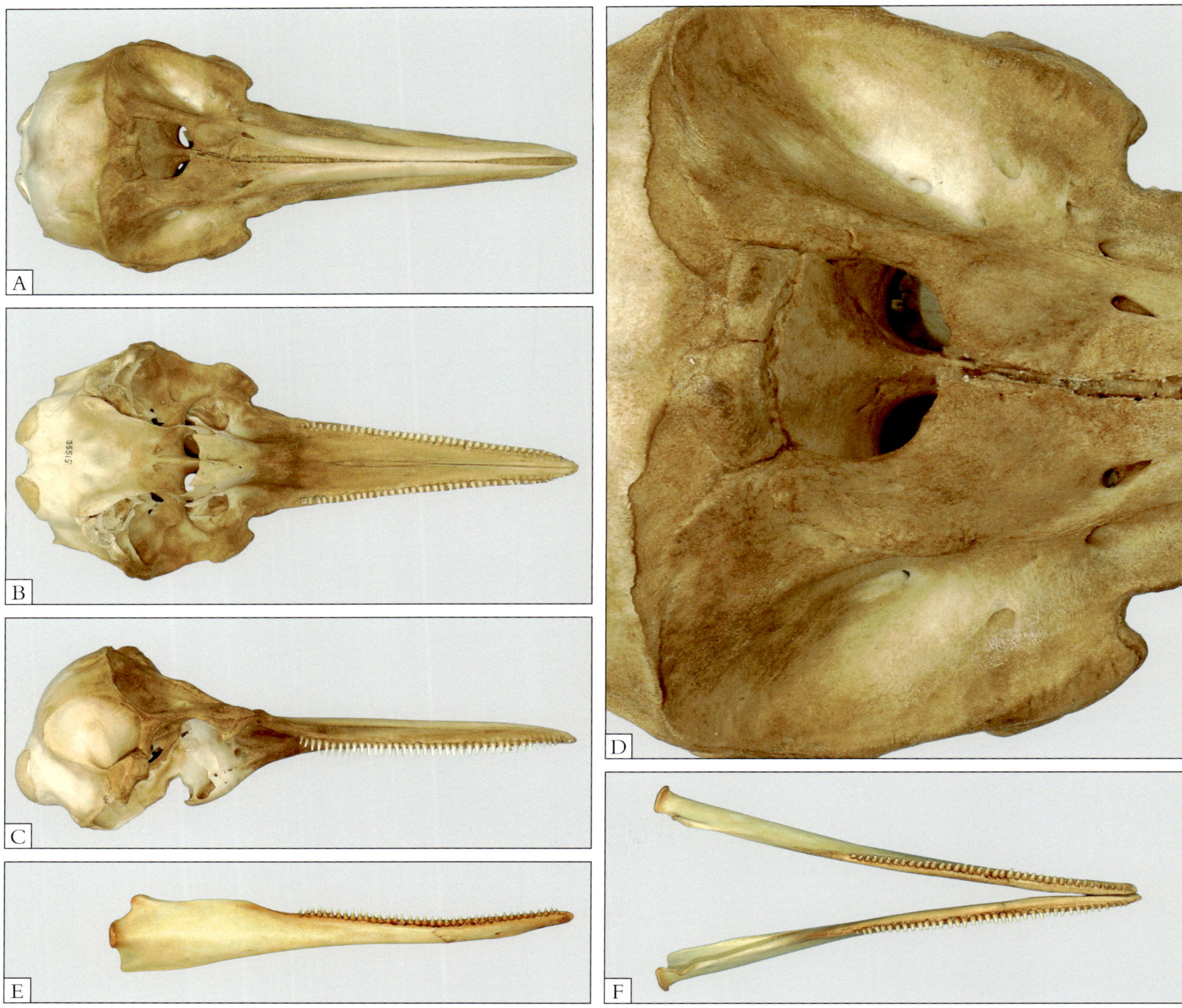

FIG. 118 Dorsal (A), ventral (B) and right lateral (C) views and vertex of skull (D); right lateral (labial) (E) and dorsal (F) views of mandible (ZM 35515).

In the Indian Ocean, pantropical spotted dolphins have been sighted in inshore waters only as far west as 29° E,[19] but strandings have occurred sporadically in the Western Cape, including up the west coast as far as St Helena Bay.[3] It is most likely that these are strays from the Indian Ocean population, but the possibility cannot be ruled out that they originate from the south-east Atlantic, where they occur to at least 16° S (St Helena).[12] Although sightings have been made in water 45 m deep off the KwaZulu-Natal coast of South Africa,[20] most sightings in the subregion have been in water more than 200 m (and up to 3 000 m or more) in depth.[19] The information is too sparse to consider the possibility of seasonal movements, although elsewhere seasonal shifts in distribution in relation to environmental change have been documented.[21]

Speed of travel for a telemetered dolphin ranged from 0 to 12 knots (average = 6.5 knots).[22] The maximum speed attained by a trained dolphin over a short distance (25 m) was 21.4 knots.[23]

External commensals and parasites

In 12 pantropical spotted dolphins examined in the subregion, three had stalked barnacles (*Xenobalanus* sp.) attached to the trailing edge of the flukes, comprising 1, 2 and 17 individuals.[2, 3] One individual had a cyamid (*Syncyamus* sp.) on the tip of the right flipper.[3] Elsewhere the stalked barnacle (*Conchoderma auritum*) has been recorded attached to the teeth.[24]

Food and feeding behaviour

The stomach contents of six individuals from the subregion have been examined and 407 prey items identified, of which 68.6% were fish and 31.4% cephalopods. Lantern fish (*Symbolophorus* 25.1%, *Myctophum* 16.5%, *Ceratoscopelus* 6.9% and *Diaphus* sp. 6.1%) were numerically the most abundant prey items, followed by the oceanic squids *Chiroteuthis* (12.5%) and *Sthenoteuthis*

(6.6%).[2, 3] Similar results were obtained from much larger samples of the species elsewhere: in the eastern tropical Pacific and off the coast of Taiwan, fish (especially myctophids) predominated numerically and enoploteuthid or ommastrephid squid (such as *Sthenoteuthis*) featured among the cephalopods eaten.[25, 26] The predominant prey species by number changed seasonally off Taiwan, with the contribution of enoploteuthid squid, in particular, declining in spring/summer compared to autumn/winter.[25] In the eastern tropical Pacific, the composition of the diets of pregnant and lactating individuals differed, although not in a consistent manner.[26, 27] Squid also tended to occur more often and as larger individuals in the diet of calves as weaning progressed.[28]

Pantropical spotted dolphins are believed to feed mainly on organisms associated with the deep scattering layer that rises towards the surface at night. In an analysis of stomach fullness for dolphins in the eastern tropical Pacific, indices were highest in the early morning.[26] In telemetered dolphins off Hawaii, the frequency of deep dives increased just after sunset, to depths of more than 150 m, apparently as exploratory dives made down to the deep scattering layer, while shallower dives were made later in the night. Overall, dives were deeper (a mean of 57 m and a maximum of 213 m) at night than during the day (mean 12.8 m, maximum 122 m), and swimming speed was higher after dark.[29]

Reproduction and growth

Two neonates from the subregion were a female estimated at 83 cm and a male 88.5 cm long.[2] In the eastern tropical Pacific the length at birth was estimated as 83.2–85.4 cm, depending on the population concerned.[30] The calving season in the same region was protracted, but with a low point in winter and high points in spring and autumn. Gestation is estimated to occupy 11.3–11.5 months.[4] Calves may begin to take solid food around six months of age (and a length of 1.15 m) but can continue suckling for nearly two years and up to a length of 1.60 m. On average, weaning takes place at eight months and a length of 1.22 m.[28] Milk contains on average 25.3% fat, 8.3% protein, and 1.1% lactose (n = 8).[31] In the eastern tropical Pacific, sexual maturity is attained at lengths of about 1.86 m in males and an age of 13.2–16.3 years,[9] and in females at a length of 1.81 m[4] and an age of 10.7–12.2 years,[8] although there may be differences in the age of sexual maturity in females from different populations[32] and in response to exploitation.[33] Off Japan, sexual maturity was achieved at ages of 8.6–10.1 years in females.[34] Ovulation can be spontaneous,[35] and occurs more frequently in younger than older females. The average calving interval was estimated as 26–36 months in the eastern tropical Pacific,[4, 8] and 36–43 months off Japan,[34] and may increase with the age of the female in both localities (although reproductively senescent individuals formed only 0.4% of females examined).[8]

Age readings indicate that males off Japan live to 42.5 years and females to 44.5–45.5 years,[34] and females in the eastern tropical Pacific to 39 years.[32] Annual mortality rates range from 5–6% in subadults to 13–16% in adults over 20–24 years of age.[34]

Behaviour and vocalisations

Schools of pantropical spotted dolphins seen in the subregion ranged in size from four to 1 000 (mean 88.4, n = 39), but most (87.2%) contained less than 150 animals. Mixed schools were seen on two occasions, one with striped dolphins and one with unidentified dolphins. Pantropical spotted dolphins were in the minority in either case (five of 205 and 23 of 230 respectively).[3] In the eastern tropical Pacific, schools tended to be of two sorts: breeding schools of under 300 animals, and smaller schools (mostly male and usually juveniles) that were found in association

What price tuna?

In the last half of the twentieth century, the eastern tropical Pacific, a wedge-shaped piece of ocean extending out from the west coast of central America between the latitudes of 30° N and 15° S, and stretching as far west as the Hawaiian Islands, was the scene of the largest recorded kill of cetaceans by humans. Between 1959 and 1997, an estimated 6 515 115 dolphins were removed from this system, as a by-catch in a purse-seine fishery for tuna. It was the misfortune of the dolphins (principally pantropical spotted and spinners) to be found in association with shoals of yellowfin tuna, so that by chasing, herding and setting their net around a school of dolphins, the fishermen could secure the fish swimming beneath while (hopefully) releasing the dolphins alive. Although attempts to release dolphins were made, these were manifestly not always successful, and some dolphins would inevitably become trapped in the net and suffocate. The number of animals taken in this way became one of the major issues leading to the passage of the US Marine Mammal Protection Act in 1972, which *inter alia* called for the dolphin by-catch to be reduced to 'insignificant levels approaching zero'. It took another 20 years (and innumerable legal actions) before this goal would finally be reached. Success was achieved through a combination of technical innovation (the incorporation of a small mesh panel to reduce chances of entanglement), better fishing practices (use of a back-down procedure by the tuna vessel to assist in separating dolphins and tuna) and, to a lesser extent, changes in fishing strategy (so-called dolphin-safe methods of setting on things other than dolphin schools).[1]

Despite this success in reducing mortality, two major populations of dolphins, the north-eastern stock of the offshore pantropical spotted dolphin and the eastern spinner dolphin, had been so heavily impacted by the catches that by 1998 they were still only at 15% and 34% respectively of their sizes when the fishery began.[2] Surveys carried out from 1979 to 2000 have also failed to detect any significant recovery in these two populations. The suggested reasons for this are a higher level of by-catch than reported, a decline in habitat quality, possible interspecific effects, or some effect of the fishery beyond direct mortality (such as the stress of chasing). It is a frightening statistic that individual pantropical spotted dolphins from this stock may encounter a purse-seiner 2–50 times each year.[3]

References

1 Gosliner 1999
2 Gerrodette 2002
3 Gerrodette and Forcada 2005

with very large schools of spinner dolphins. The breeding schools were composed of female-calf pairs, female-young subgroups, young adult subgroups, adult male subgroups and sometimes juvenile subgroups (although such animals were usually conspicuously absent). These subgroups seemed to be stable in composition in the short term, and in the case of the adult males might represent long-term, exclusive associations. Inter-school variation (but intra-school consistency) in pigmentation patterns, combined with the behavioural evidence of affiliation, has suggested that the breeding schools might have long-term stability in composition.[36] When in mixed schools, interspecific aggression, copulation and mutual coexistence can occur between pantropical and spinner dolphins,[37] and a possible hybrid spinner: pantropical spotted dolphin calf was seen several times with a spinner dolphin female over a two-year period.[38] At St Helena in the 1980s, pantropical spotted dolphins were found each morning within 0.5 km of the shore, and then passed the harbour going north and offshore. The school usually consisted of several hundred animals, dispersed while travelling over 0.5 km in subgroups of 5–10 animals. Aerial activity, including high leaps, forward flips, pitch-poling and tail-lobbing, was frequent. Once offshore, the school apparently broke up into several subunits and spent the rest of the day several kilometres offshore.[12] By 2003, however, this behavioural pattern had changed, with animals tending to remain inshore for most of the day rather than moving north. This change was attributed to a possible cessation of hunting by local fishermen.[39]

Pantropical spotted dolphins produce pulsed sounds (clicks) up to 150 kHz in frequency and with a duration of 0.075–0.2 ms, as well as pure-tone whistles.[40] Thirteen whistle types have been identified in the eastern tropical Pacific, which differed from those of the Atlantic spotted dolphin in having many more upsweeps and downsweeps.[41]

Conservation status

There are no estimates of population size for pantropical spotted dolphins in the subregion. The major known source of human-induced mortality was the harpoon fishery at St Helena, in which some 75–200 dolphins, nearly all *S. attenuata*, were estimated to have been taken annually. This fishery probably began towards the end of the nineteenth century but was officially terminated in 1979. A few individuals may have been taken annually thereafter. Its impact on the local population of pantropical spotted dolphins is unknown, although no long-term trend in catches or availability was apparent to the fishermen concerned.[12] Elsewhere, a few individuals are known to have been taken in the shark nets off KwaZulu-Natal,[3,7] and a low proportion (4%) of the dolphins taken in gill nets off Zanzibar were of this species.[15] The rarity of this species in shallower water suggests that it may have a reduced vulnerability to entanglement in many artesanal and other gill-net fisheries in the subregion.

Pantropical spotted dolphins are listed as Lower Risk: Conservation Dependent by the IUCN (largely because of their status in the eastern tropical Pacific – see box on previous page) and are included on Appendix II of CITES. In the South African Red Data Book they are listed as Data Deficient. The eastern tropical Pacific and south-east Asian populations are included on Appendix II of CMS.

References

1 Perrin *et al.* 1987
2 Ross 1984
3 MRI Whale Unit data
4 Perrin *et al.* 1976
5 Perrin 1975
6 Perrin 1969a
7 Best 1969a
8 Myrick *et al.* 1986
9 Hohn *et al.* 1985
10 Perrin 2001a
11 Pelembe, T., pers. comm.
12 Perrin 1985
13 Weir, C., pers. comm.
14 Peddemors *et al.* In press
15 Amir *et al.* 2005b
16 Ballance *et al.* 1996
17 Kiszka *et al.* 2006a
18 Corbett 1994
19 Findlay *et al.* 1992
20 Ross *et al.* 1987
21 Fiedler and Reilly 1994
22 Leatherwood and Ljungblad 1979
23 Lang and Pryor 1966
24 Perrin 1969b
25 Wang *et al.* 2003
26 Robertson and Chivers 1997
27 Bernard and Hohn 1989
28 Archer and Robertson 2004
29 Baird *et al.* 2001
30 Hohn and Hammond 1985
31 Pilson and Waller 1970
32 Chivers and Myrick 1993
33 Barlow 1985
34 Kasuya 1985
35 Benirschke *et al.* 1980
36 Pryor and Shallenberger 1991
37 Psarakos *et al.* 2003
38 Silva-Jr. *et al.* 2005b
39 MacLeod and Bennett 2007
40 Popper 1980
41 Moore 1990

Atlantic Spotted Dolphin

Stenella frontalis

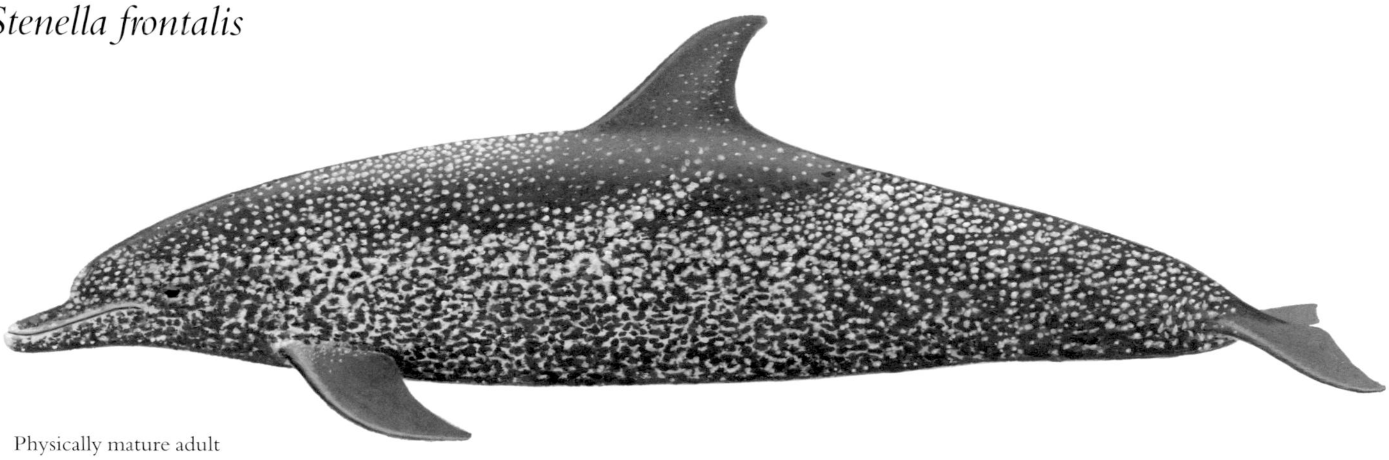

Physically mature adult

Sexually mature adult

Calf

THIS, THE SECOND of two spotted dolphin species recognised worldwide (the other being the pantropical spotted dolphin, *S. attenuata*), was until recently known in the subregion only from records from St Helena (that were considered likely to have originated elsewhere) and Gabon.[1] It now appears that Atlantic spotted dolphins are commonly seen off Angola (as well as Gabon),[2] so the occurrence of the species in the subregion is firmly established. However, very little local research has been carried out on spotted dolphins, so this account has had to use information from other regions where necessary.

Appearance

Atlantic spotted dolphins differ from pantropical spotted dolphins in being heavier-bodied animals, with a definite spinal blaze (a streak of light colour extending back from the cape marking on the head towards the dorsal fin) but no sign of the two-tone caudal peduncle. The white spots or dark flecks on adults are also much larger than in the pantropical spotted dolphin.

There are no published sizes for animals from the subregion. Body size tends to vary geographically, with adults ranging in length from 1.66 m to 2.29 m with no major size difference between sexes. The animal is similar in shape to the pantropical spotted dolphin but more robust in build, adult Atlantic spotted dolphins weighing 10–30 kg more than pantropical spotted dolphins of the same length. The flippers are also larger, ranging in adults from 26–37 cm in length, compared to 23–30 cm in the pantropical species. The dorsal fin is taller (height in adults 16–25 cm, compared to 13–20 cm) and the flukes are wider (35–66 cm, compared to 36–49 cm). As pantropical spotted dolphins are somewhat longer, expressed as proportions of body length these differences would presumably be greater.[1] A prominent post-anal swelling (or 'hump') ventrally in males does not seem to have been described in this species.

As with the pantropical spotted dolphin, young are born unspotted, and similar in basic coloration to the neonatal pantropical spotted dolphin. However, unlike that species, the margin between the dark grey dorsal and light grey lateral

FIG. 119 Atlantic spotted dolphins encountered off Angola. Note the difference in colour pattern between a cow and her calf in the top photograph, and the clear spinal blaze in the lower photograph.

pigmentation that runs back from the side of the head is interrupted by a separate streak diverging towards the dorsal fin (the 'spinal blaze'). In addition, the light belly pigment ends behind the anus, so that the whole caudal peduncle is uniformly grey, rather than dark above and light below, as in the pantropical spotted dolphin. At about the mid-length of the animal there may be irregular excursions of the white belly colour ('blotches') up into the grey colour of the side ('two-tone' coloration). The intensity of spotting varies geographically. It usually starts to appear before puberty. Dark grey or black spots first appear in the throat and belly regions, followed by light spots on the back and light and dark spots on the sides ('speckled' phase). The spots on the belly enlarge and coalesce, and the light spots on the back spread ventrally, coming to overlie some of the dark spotting ('mottled' phase). Finally the light spots on the back enlarge and coalesce in some areas, notably above the eyes, near the spinal blaze and on the upper half of the caudal peduncle, and the tip of the snout may whiten ('fused' phase).[1] In the Bahamas, the two-tone phase lasted from birth to 3–4 years, the speckled phase from four to nine years, the mottled phase from 10 to 16 years, and the fused phase from 16 years onwards. This pattern seemed to be the same in both sexes.[3]

There are 32–42 (mean 36.1, n = 109) pairs of teeth in the upper jaw and 31–40 (mean 34.5, n = 101–103) pairs in the lower jaw.[1] They are slender (the largest being 3.2–5.3 mm in diameter) and conical in shape.

Distribution and migrations

Atlantic spotted dolphins, as their name suggests, are confined in distribution to the tropical and warm-temperate Atlantic Ocean, from about 45° N to 35° S on its western side and between the coasts of Mauritania and Angola on its eastern side. They can occur in both coastal waters and the open ocean, and there are differences in size, pigmentation and some morphological features between the coastal and offshore forms.[5]

Until recently, the presence of the species in the subregion was only represented by a specimen in the Zoological Museum, University of Amsterdam, from off Point Noire, Gabon (about 5° S), and two museum skulls collected in the nineteenth century with the locality given as the island of St Helena. All records of spotted dolphins at St Helena in the twentieth century, however, are of the pantropical species, and it was speculated that these old skulls may have been collected by ships on their way to or from the island.[1] Nevertheless, recent sighting records indicate that the Atlantic spotted dolphin is the most frequently sighted delphinid off Angola, and also confirm the species' presence off Gabon.[2] There are as yet no records of the species off Namibia or South Africa.

Off Angola the species is found in a wide range of water depths, from 200 m to over 2 200 m.[2] In the south-west Atlantic Ocean, Atlantic spotted dolphins appear to prefer a nearshore habitat, being restricted to water less than 1 000 m deep.[6] In the Gulf of Mexico, Atlantic spotted dolphins occurred in water depths of 102–589 m (mean 197 m), or essentially on the shelf edge, and over a less sloping sea floor than other species.[7] A rehabilitated adult male in the same area that was satellite-tagged before release spent the next 24 days between the 20 m and 60 m isobaths, or over the shelf itself.[8] In the Bahamas, these dolphins occur over shallow sandbanks 6–16 m deep,[8] as well as out to water almost 2 200 m deep.[9]

Length:weight relationship

The length:weight relationship for 16 males weighed whole has been calculated as **$W = 0.0000069214L^{3.1087}$**, where W = weight in kilograms and L = total length in centimetres.[4] This implies that a 140 cm dolphin would weigh 32.5 kg and a 200 cm animal 98.5 kg.

Skeletal data

Vertebral formula (n = 30–52): 7 cervical, 13–15 thoracic, 15–20 lumbar and 28–35 caudal; total 67–72 (compared to 74 or more in pantropical spotted dolphins[1]).

As with pantropical spotted dolphins, the atlas and axis are probably fused, and the third cervical may partly fuse with them in older animals, but published statements to this effect seem to be missing.

Phalangeal formula, including metacarpals (n = 29–31): I_{1-3}, II_{7-10}, III_{6-8}, IV_{2-4}, V_{0-2}.[1]

Because the skulls of this species and the pantropical spotted dolphin overlap in every characteristic, distinguishing between them requires using a combination of characteristics. Although there is considerable geographical variation, generally the skulls of Atlantic spotted dolphins have a proportionately broader rostrum tip, a shorter prenarial triangle, and larger and less numerous teeth. To classify individual skulls as belonging to one or the other species requires a statistical (maximum likelihood) approach.[5]

Sightings of Atlantic spotted dolphins off Angola have occurred throughout the year, and it is believed to be a 'resident' species.[2] No seasonality of occurrence was detected in the Gulf of Mexico,[10] although inshore-offshore movements have been suggested.[11]

The swimming speed of a satellite-tagged male averaged 2.88 km/hr over 24 days following release: this is likely to be a minimum estimate as it assumed the animal swam in straight lines between successive locations.[8]

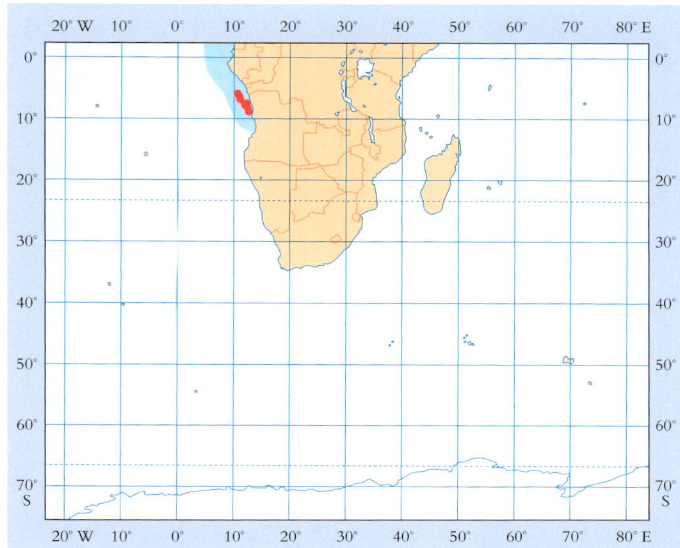

Postulated range   Sighting or capture

FIG. 120 Distribution map for Atlantic spotted dolphin.

External commensals and parasites

There are no records from the subregion. Elsewhere, stalked barnacles (*Conchoderma auritum*) have been found attached to the teeth.[12] The barnacle *Xenobalanus globiciptis*, an unidentified cyamid,[1] and the remora or whale-sucker *Remilegia australis*[13] have been found on the body surface.

Food and feeding behaviour

There is no information from the subregion, and relatively little documented from elsewhere. In 19 animals stranded in the USA, nine contained only squid, six had both fish and squid, and four contained only fish. Identified fish included the kobs *Cynoscion* and *Stenotomuschrysops* and the clupeid *Anchoa*, plus representatives of the Congridae (conger eels), Gadidae (cods), Trichiuridae (frostfishes), and Triglidae (gurnards).[1] The stomachs of animals harpooned at St Vincent in the West Indies apparently contained midwater fish and squid, while dolphins have been seen chasing and taking halfbeaks (*Hemiramphus*) at the surface.[4] In the Bahamas, feeding takes place on a variety of benthic fish, including flounders (Bothidae), lizard fish (Synodontidae), wrasse (Labridae), blennies (Tripterygiidae), klipfish (Clinidae), and conger eels (Congridae).[14]

Atlantic spotted dolphins have been observed feeding in a coordinated way, involving a large school of at least 100 individuals split into small subunits, each herding small clupeid fish into a dense ball. This would then be forced to the surface by a multi-tier formation of dolphins. Dolphins passing beneath the

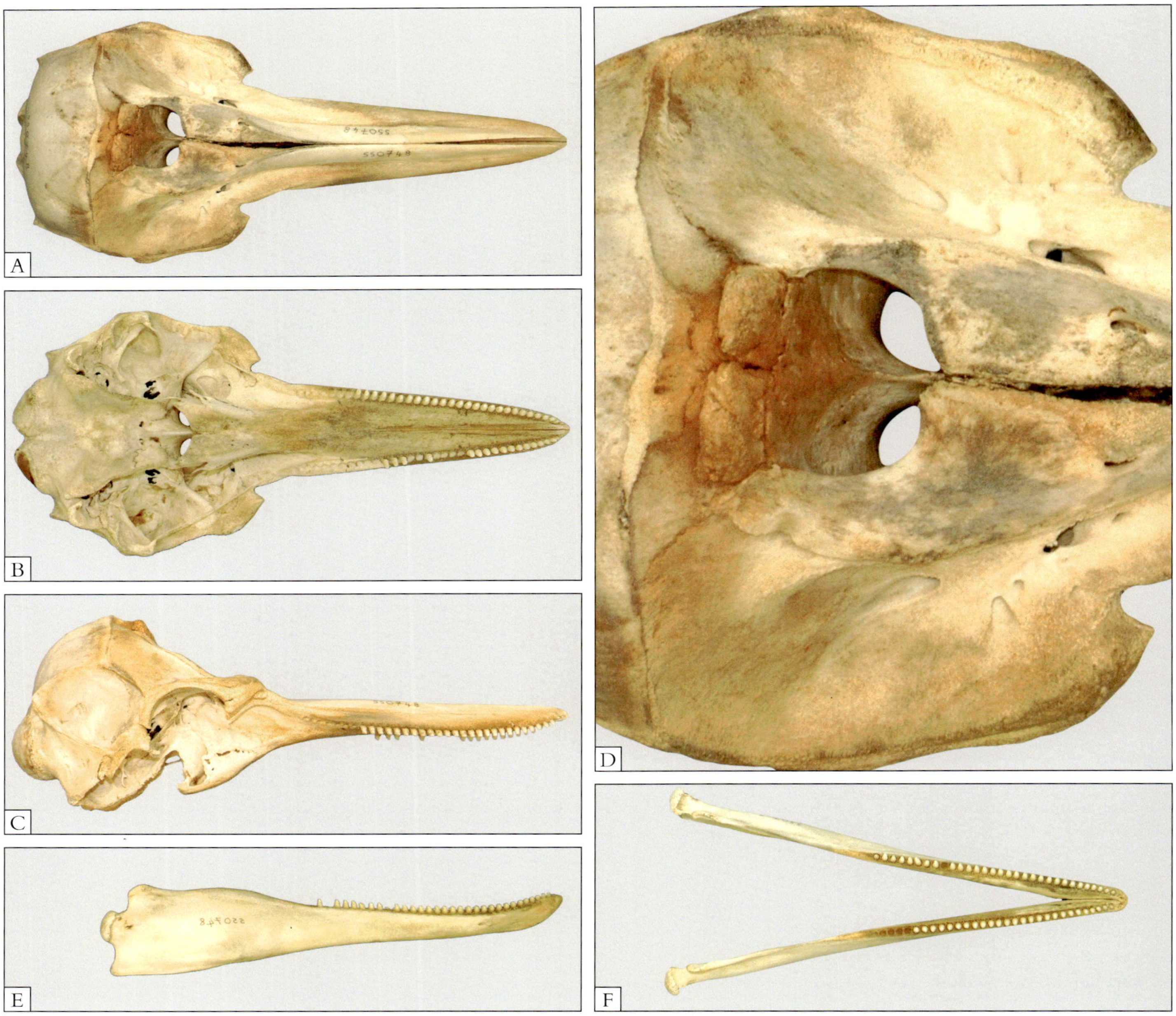

FIG. 121 Dorsal (A), ventral (B) and right lateral (C) views and vertex (D) of skull; left lateral (labial – reversed) (E) and dorsal (F) views of mandible (USNM 550748).

school released bursts of bubbles that served to displace individual fish near the surface from the rest of the shoal, and to maintain the bunching of subsurface fish through a collapsing effect. Individual dolphins passed through the contained shoal, presumably feeding. One adult was seen catching fish and tossing them one at a time towards the outskirts of the group, where they were apparently consumed by other, smaller individuals.[15] In some areas Atlantic spotted dolphins have developed the habit of following shrimp trawlers to feed on the catch in the net or discards from it.[16]

A rehabilitated dolphin foraging in water 20–60 m deep, showed no diel pattern to its diving behaviour. On average, 58.1% of dives were to 4–10 m, while the deepest dives were to 40–60 m. Throughout the period that the satellite tag functioned, the dolphin consistently made dives deep enough to reach the sea floor.[8]

Reproduction and growth

There is no local information. Size at birth is not well established but lies between 88 cm and 120 cm.[4] In the Bahamas, calves are born mainly in early spring and late autumn, after an assumed gestation of 12 months. Lactation lasts on average for over three years, but up to five years in some individuals.[3] However, successful foraging by the calf has been seen as early as its second year.[17] Milk may contain 18% fat, 9% total protein, and 0.6% sugar (n = 1).[18] First births occurred at ages of 10–12 years.[3] Small sample sizes (and possible geographical variation) prevent accurate estimation of sizes at sexual maturation.[4] Calving intervals in the Bahamas ranged from one to five years, with an overall average of 2.96 years (n = 23), although successful cycles were longer, averaging 3.56 years (n = 16). First-year mortality averaged 24%.[3]

Behaviour and vocalisations

Off Angola, schools of Atlantic spotted dolphins ranged in size from one to 500 with a mean of 106 animals (n = 37).[2] This is substantially higher than in the Gulf of Mexico, for instance, where school size varied from two to 137 animals with a mean of 27 (n = 35), or in the Bahamas, where group size averaged between nine and 10 animals.[19] Groups encountered in the Bahamas were composed on average of 20.9% calves, 20.4% juveniles, 26.7% young adults and 31.9% adults (n = 382). Sex ratios were roughly equal in calves and adults but skewed towards males in juveniles and young adults.[3]

Associations between mother and calf remain high until the birth of the next offspring. Juvenile females display strong associations with other females of the same age class, but these ties weaken markedly at sexual maturity. At first parturition, a female may form new associations with older females in the same birth year, and older females tend to form associations with females in a similar reproductive state.[20]

Behavioural activities of Atlantic spotted dolphins in the Bahamas were categorised as social (31.4%), play (18.4%), inquisitive (9.7%), travel (29.2%) and foraging (11.4%). Social activities were commonest among groups of mixed sex and age, and rarest among mother-calf pairs, while play was most frequent among adult females with juveniles and least frequent among mixed groups. The percentage of time spent foraging was lowest in mother-calf pairs and highest in groups of mixed sex and age. Sand-rubbing (involving all or part of the body) was the most frequently observed contact behaviour (28%), followed by rubbing of one dolphin's body against another (18%), simple body contact without rubbing (13%), petting, namely moving a flipper along the body of another dolphin (12%), and petting/rubbing (9%). Each of these behaviours varied according to the dolphin's age, sex, activity and group type. In general, contact behaviour occurred most frequently during social activity, and mixed and adult female/juvenile groups exhibited more contact than other groups. Contact could also be modified by posture, which is used to indicate intent. In the case of the spotted dolphin, the various degrees of spotting in combination with posture may provide information on the meaning of particular signals.[21]

Mixed schools with bottlenose dolphins have been recorded from the Gulf of Mexico,[22] and commonly occur in the Bahamas, where bottlenose dolphins were present in up to 15.8% of encounters with Atlantic spotted dolphins. Over 30% of such encounters were of an aggressive/sexual nature, principally involving aggressive male-male interactions, including chasing, genital nuzzling, mounting and actual intromission, in which usually the bottlenose dolphins were the aggressor. In one instance (involving offshore bottlenose dolphins) a spotted dolphin was pushed clear of the water during a high speed pursuit. Pursuit of female Atlantic spotted dolphins and attempted copulation by male bottlenose dolphins was not uncommon, but usually deterred by coalitions of male spotted dolphins. Attempts by male spotted dolphins to mate with female bottlenose dolphins were rare. Nevertheless, a calf born to a female Atlantic spotted dolphin appeared morphologically to be a hybrid with a bottlenose dolphin. In other instances, interpecific interactions were cooperative, including male coalition, travelling associations of pregnant females, alloparental care and foraging behaviour.[19, 23]

The vocalisations of Atlantic spotted dolphins have been categorised into seven types: signature whistle (frequency modulated, 4–18 kHz), excitement vocalisation (burst-pulsed call with overlapping signature whistle), genital buzz (high repetition rate clicks, 1.2–2.5 kHz), squawk (broadband burst-pulsed vocalisation, 0.2–12 kHz), scream (overlapping whistles, 5.8–9.4 kHz), bark (burst-pulsed vocalisation, 0.2–2 kHz) and synchronised squawk (burst-pulsed vocalisation, 0.1–15 kHz). Signature whistles are used primarily in mother-calf reunions, during courtship and alloparental care; excitement vocalisations during distress or excitement; genital buzzes during courtship and when imposing discipline; and all the other calls during agonistic and aggressive interactions (or, in the case of squawks, sexual play).[14]

Conservation status

There are no estimates of population size for Atlantic spotted dolphins in the subregion, and no indications of any substantial human-induced mortality.

Atlantic spotted dolphins are listed as Data Deficient by the IUCN and are included on Appendix II of CITES. They are not listed in the South African Red Data Book or by CMS.

References

1 Perrin *et al.* 1987
2 Weir, C., pers. comm.
3 Herzing 1997
4 Perrin *et al.* 1994a
5 Perrin 2001b
6 Moreno *et al.* 2005
7 Davis *et al.* 1998
8 Davis *et al.* 1996
9 MacLeod *et al.* 2004
10 Mullin *et al.* 2004
11 Caldwell and Caldwell 1966
12 Van Bree 1971
13 Mahnken and Gilmore 1960
14 Herzing 1996
15 Fertl and Würsig 1995
16 Fertl and Leatherwood 1997
17 Miles and Herzing 2003
18 Eichelberger *et al.* 1940
19 Herzing *et al.* 2003
20 Herzing and Brunnick 1997
21 Dudzinski 1998
22 Mullin *et al.* 1994b
23 Herzing and Johnson 1997

Spinner Dolphin

Stenella longirostris

THE COMMON NAME of this dolphin is derived from its habit of frequently spinning around its long axis when leaping from the water (see box on page 202) – a behaviour also exhibited by the clymene dolphin (otherwise known as the short-snouted spinner dolphin). Geographical variation in appearance is said to be greater in spinner dolphins than in any other cetacean,[1] and four subspecies have been described: the Hawaiian or Gray's spinner dolphin (*S. l. longirostris*), the Eastern Pacific spinner dolphin (*S. l. orientalis*), the Central American spinner dolphin (*S. l. centro-americana*),[2] and the dwarf spinner dolphin (*S. l. roseiventris*).[3] Only Gray's spinner dolphin is known to occur in the subregion,[4] but there are unconfirmed reports that the dwarf form could occur in the south-west Indian Ocean.[5] Because so little is known about spinner dolphins in the subregion, most of this account has been borrowed from studies elsewhere, particularly in the eastern tropical Pacific.

Appearance

The spinner dolphin is most likely to be confused at sea with the clymene dolphin, but apart from their different beak lengths the best field characters for distinguishing between the two species are: 1) The lower margins of the dark back and grey side fields of coloration are more or less parallel in spinners, but in the clymene dolphin the dark coloration of the back dips down towards the white belly at about its mid-length. 2) The beak of the spinner dolphin is edged with black but does not carry the distinctive black 'moustache' marking in the mid-line seen in clymene dolphins.[6]

Lengths of adults (from the North Atlantic) averaged 1.89 m (maximum 2.04 m) in females and 1.93 m (maximum 2.08 m) in males.[7] Adult dwarf spinners may be 30–40 cm shorter.[3] The body shape is similar to that of other stenellids, but somewhat less robust than in Atlantic spotted and other similar-sized dolphins. The relative slimness of the spinner dolphin actually enhances the spinning.[8] The beak is relatively long and slender but variable in length, averaging 8.0–9.5% of body length.[9] The flippers have a strongly curved leading edge and are sharply pointed: they are somewhat longer in the dwarf form (14–18% of body length) than in pelagic spinners (13.5–16.5%). The dorsal fin is subtriangular and slightly hooked or erect and placed centrally. Its height

Length:weight relationship

The length:weight relationship for 166 spinner dolphins (different subspecies combined)[11] is $W = 0.000038L^{2.74}$, where W = weight in kilograms and L = total length in centimetres. This implies that a 1.4 m dolphin would weigh 28.9 kg and a 1.8 m animal 57.4 kg.

Skeletal data

Vertebral formula (from western and central/south Pacific; n = 8–61): 7 cervical, 14–16 thoracic, 16–19 lumbar and 32–38 caudal; total 70–75.[3] One animal from the subregion is within this range (C7, T15, L18, Cd 32 = 72).[12] The dwarf form tends to have fewer vertebrae: 7 cervical, 13–15 thoracic, 16–19 lumbar and 28–32 caudal; total 66–71 (n = 19–35).[3]

The atlas and axis are fused.[13]

Phalangeal formula, including metacarpals (n = 17): I_{0-1}, II_{9-11}, III_{7-8}, IV_{3-4}, V_{1-3}.[13]

Considerable geographical variation in skull morphology occurs, particularly in overall size and relative rostrum length. Nevertheless, the skull of the spinner dolphin can be distinguished from that of the common dolphin by its much shallower palatal grooves, from the striped dolphin by its smaller cranium and narrower facial region, from the clymene dolphin by its longer rostrum, and from the pantropical spotted dolphin by its greater number of teeth.[6]

FIG. 122 The tripartite character of the spinner dolphin's coloration is clearly seen in these photographs from East Africa. In the bottom photograph, the contortions of a dolphin are caught mid-spin.

is slightly greater in the dwarf form (9.5–12.5% of the body length) than in the pelagic spinner dolphin (8–11%).[3] The flukes are lunate in shape and their width is on average 21.0–23.5% of body length.[9] A post-anal hump develops in adult males.[6]

The colour pattern of Gray's spinner dolphin is tripartite: a dark grey dorsal field, a lighter grey lateral field and a white or very light grey belly. The margin between the dorsal and lateral field is distinct and runs back from the side of the melon, dipping above the eye and then below the dorsal fin before terminating on the back halfway between the dorsal fin and the tail. A grey band runs from the flipper to the eye, and a dark ring around the eye extends forward as a streak to the apex of the melon. A median streak from the blowhole runs onto the rostrum as a dark streak that widens to give a black tip to the snout, and both upper and lower lips are edged with black. There may be grey flecking on the belly.[5] Apart, possibly, from a pinker belly, no differences in the colour pattern of the dwarf form have been described.[51]

There are 47–62 (mean 54.1, n = 113) pairs of teeth in the upper jaw and 46–61 (mean 52.1, n = 129) pairs in the lower jaw of the spinner dolphin: the dwarf form has fewer teeth, 42–52 (mean 46.5) in the upper and 41–50 (mean 45.7) in the lower jaw.[3] They are slender and conical in shape, with the largest (in spinner dolphins) averaging 2.5–3.1 mm in diameter.[10]

Distribution and migrations

Spinner dolphins occur around the world in tropical waters, sometimes in the open ocean in relatively deep water but more usually close to shore, with a dwarf subspecies that inhabits shallow water (less than 50 m deep).[3] In the subregion, spinner dolphins (without reference to subspecies) have been recorded from Gabon,[14] Angola,[15] St Helena,[16] South Africa,[4] Mozambique,[17] the Comoro Islands,[18] Madagascar,[17] Réunion,[19] Mauritius,[20] Tanzania,[21] Kenya, and the Seychelles.[22] In the south-west Indian Ocean they have been recorded as far south as Cape Vidal, South Africa (28°08′ S),[23] but in the south-east Atlantic are only known as far as 16° S (St Helena),[6] possibly because of the northward extension of the cold Benguela System.

In most oceans, spinner dolphins tend to be associated with inshore waters, islands or banks, but in the eastern tropical Pacific they also occur pelagically very far from land.[6] At least some of the sightings off South Africa were in water less than 50 m deep,[24] while others off Mauritius were in water between 2 000 and 3 000 m deep.[20] In the northern Gulf of Mexico, spinner dolphins occurred from the upper continental slope into deeper water, or between water depths of 526 and 1 776 m,[25] while in the south-west Atlantic they were found on the outer continental shelf and beyond the slope, in water depths of 170 –2 700 m.[26]

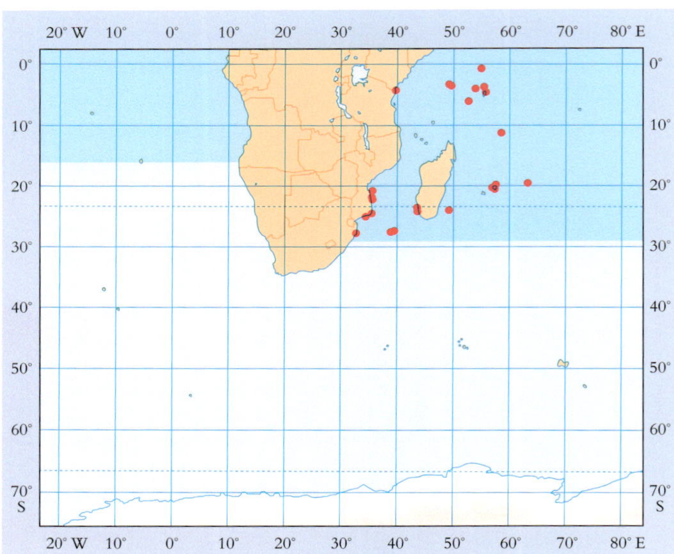

Postulated range ■ Sighting or capture

FIG. 123 Distribution map for spinner dolphin.

In the eastern tropical Pacific, pelagic spinner dolphins have the same habitat preferences as spotted dolphins, occurring primarily in tropical surface water with surface temperatures of more than 25 °C, overlaying a sharp, shallow thermocline.[27] Seasonal changes in distribution occurred between winter and summer, with dolphins seeming to shift away from the equator and further offshore in summer.[28] In the south-west Atlantic, a southward movement in summer under the influence of the Brazil Current has been proposed.[29] In shore-based observations from Cape Vidal, KwaZulu-Natal, between June and October, spinner dolphins were recorded 16 times, four each in June, July, September and October; summer and autumn observations were lacking.[23]

Gray's spinner dolphins inhabiting coastal bays and island groups show a diurnal pattern of movement, tending to enter specific sheltered bays early in the morning and resting there for several hours before moving out in the afternoon or evening. The length of stay varies seasonally with day length.[30]

Swimming speeds for Gray's spinner dolphins while resting inshore averaged 1–1.6 knots for most of the day but increased to 1.6–3.2 knots in the afternoon.[30] Swimming speeds for three schools of mixed spinner and pantropical spotted dolphins responding to the approach of a research vessel averaged 5.9 knots, 8.8 knots and 7.2 knots respectively.[31]

External commensals and parasites

There is no information for the subregion. Elsewhere, stalked barnacles (*Conchoderma auritum*) have been found attached to the teeth,[32] and cyamids (*Syncyamus aequus*) to the body surface of the dwarf form.[33] Whalesuckers (*Remora australis*) are commonly found attached to spinner dolphins: in one study off Brazil, the ratio of whalesuckers to dolphins was estimated as 1–1.8 to 1 throughout the year, and the number of fish per dolphin as 1 to 3. Most frequently attached to the belly or flank, these fish seem to 'clean' dead skin from the dolphins as well as feed on their faeces and independently filter feed.[34] Some whalesuckers may have associations with individual dolphins lasting for up to 87 days.[35]

Food and feeding behaviour

There is no local information. Prey items for pelagic spinner dolphins in the North Pacific were all small (less than 20 cm long) and composed mainly of midwater fish and squid. The dominant fish families were Myctophidae, Photichthyidae, Bregmacerotidae and Melamphidae, the dominant cephalopod families were Onychoteuthidae, Ommastrephidae and Enoploteuthidae, while crustaceans, especially the shrimps *Sergia* and *Pasiphaea*, were also eaten.[36] The diet was similar in the Philippines, consisting mostly of midwater and deep-living fish, midwater squid and crustaceans. Dwarf spinner dolphins fed mostly on shallow-water benthic and reef-dwelling fishes (families Apogonidae, Bregmacerotidae, Carangidae, Congridae, Monocanthidae, Muraenidae and Ophichthidae), unidentified cephalopods and shrimp.[3]

Most prey items of pelagic spinner dolphins are members of the mesopelagic boundary community, many species of which migrate to depths of 400–700 m during the day but at night move inshore and are found from the surface to 400 m. Dolphins track both the horizontal and vertical movements of their prey, and dive to depths corresponding to the highest prey densities. Foraging is cooperative, with dolphins forming pairs in echelon in a diagonal or V-shaped formation, surrounding and apparently altering the characteristics of prey patches.[37]

Spinner dolphins seem to vomit particularly after a meal of squid.[38] Such vomit and faeces produced while resting inshore serve as a source of nutrition for various reef fishes.[39]

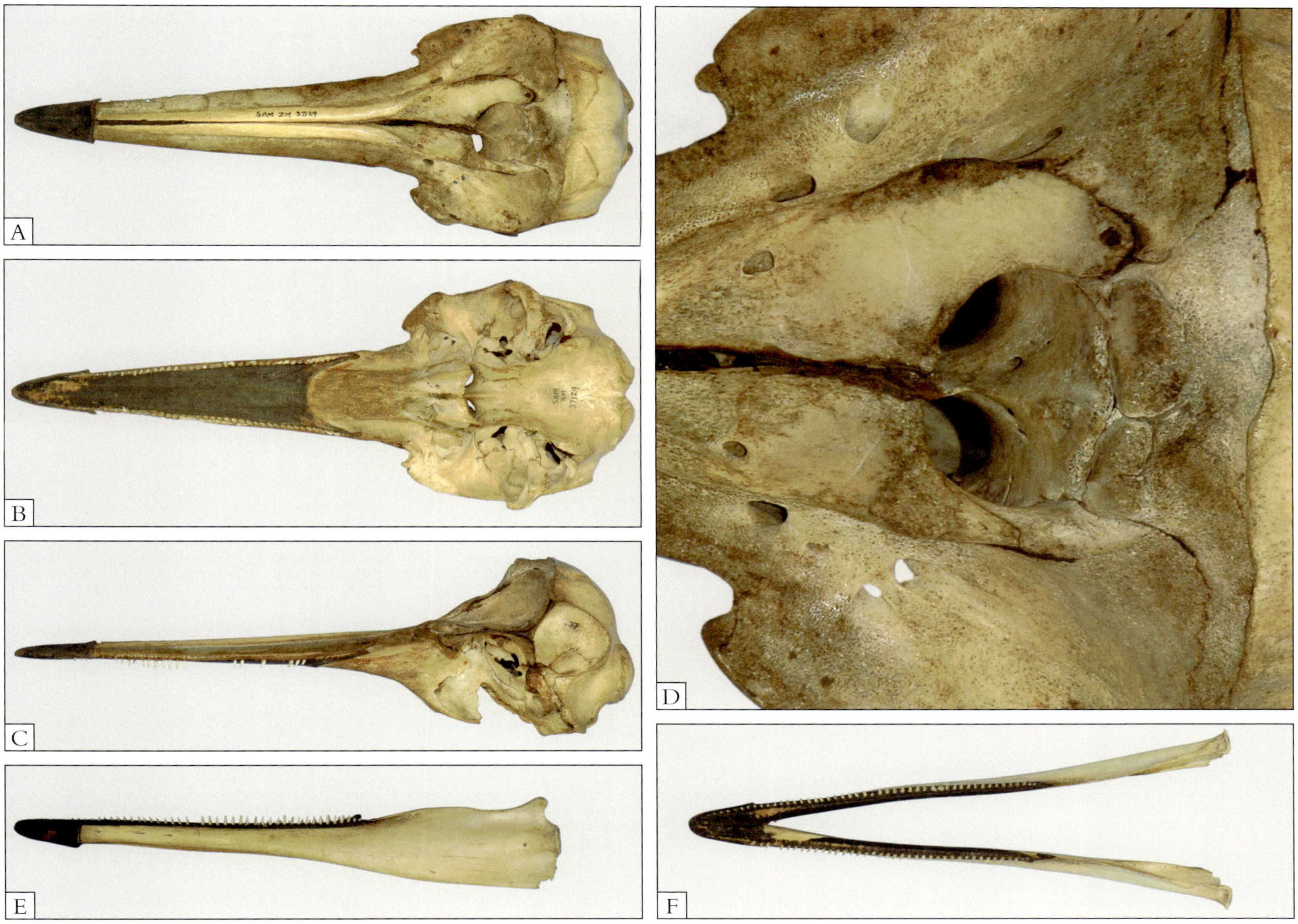

FIG. 124 Dorsal (A), ventral (B) and left lateral (C) views and vertex of skull (D); left lateral (labial) (E) and dorsal (F) views of mandible (ZM 37129).

Reproduction and growth

The estimated length at birth of North Pacific spinner dolphins was 77 cm,[6] although a calf in the subregion that died soon after birth measured 82.5 cm.[12] The seasonality of reproduction seems to vary between populations, but in Gray's spinner dolphin complete mating was only seen in summer and autumn,[40] while serum testosterone levels in a captive male also peaked strongly in summer.[41] Ovulation may be spontaneous.[42] If gestation lasts 10.5 months,[6] births would peak in spring/summer. Suckling takes place alternately from each mammary gland in bouts lasting 5–20 (mean 11.3) sec. Calves up to a length of 1.3 m have been seen suckling.[43] Lactation may last 1–2 years and the average calving interval is about three years. Sexual maturity is attained at about 1.65–1.7 m in females and 1.6–1.8 m in males, at estimated ages of 4–7 and 7–10 years respectively.[6]

The maximum age estimated from tooth readings is about 20 years. About 1% of adult females reach a post-reproductive phase.[6]

Very little information is available for the dwarf form of spinner dolphin, except that females seem to reach sexual maturity at a much smaller size, animals with lengths of 1.29–1.37 m already being mature.[44]

Two instances of apparent hybridisation between female spinner dolphins and male spotted and clymene dolphins have been described.[45]

Behaviour and vocalisations

Spinner dolphins in the subregion have been seen in schools of 1–1 000 animals, with a mean of 56.8 (n = 39). Excluding one sighting of 1 000 animals, schools ranged from one to 200, with a mean of 29 animals.[23] True mixed associations with other species were not noted for any of these sightings, but the species has been recorded with pantropical spotted dolphins off Tanzania.[46] In open waters of the eastern tropical Pacific such associations are frequent, and it has been mooted that in this environment the spinner dolphins are using the spotted dolphin schools as refugia during diurnal resting periods.[6] Such associations are rarer for Gray's spinner dolphin in inshore waters, and are sometimes accompanied by interspecific aggression and copulation.[43, 47]

The nature of society in Gray's spinner dolphin seems to change in accordance with its environment. Around larger islands of the Hawaiian chain, where suitable resting sites are

Getting rid of the suckers?

FIG. 125 Subjected to a force of up to 9 G, a remora clings on for dear life to the chest of a spinning dolphin off the East African coast.

Aerial behaviour in spinner dolphins includes nose-outs, fluke-outs, tail slaps, head and back slaps, arcuate leaps, salmon leaps, tail-over-head leaps and spins. It has been suggested that the most athletic of these (tail-over-head leaps and in particular the spins) could serve to dislodge remoras.[1]

The spinning leaps of *Stenella longirostris* are spectacularly distinctive. The dolphin emerges from the water beak-first, the body rotating as it moves upward, with the rate of rotation increasing once the dolphin is completely airborne. While executing a parabolic trajectory, the dolphin performs 2 to 5.5 (mean 3.7) rotations before re-entering the water, usually on its side.[2] Typically four or five (and up to 14) such spinning manoeuvres are repeated in succession by the same individual, with declining activity level as the series continues. All age classes perform spins, they can take place in any part of the school, and they seem to increase in frequency at dusk (and probably continue on into the night).[1] Even in highly active schools, perhaps only one-third of the animals indulge in spinning.[3]

Application of a mathematical model to the spinning dolphin suggests that during a spin an attached remora might be subjected to a force of 9 G, flinging the fish into a radial orientation so that when the dolphin strikes the water on re-entry the remora would be subjected to a drag force about 700 times its own weight! This contrasts with a normal dolphin leap, in which the remora would presumably maintain its axis longitudinal to that of the dolphin and so be subject to far less drag force on re-entry. Spinning could therefore facilitate the dislodging of remoras, even though that may not be the prime reason for the development of the behaviour.[2]

However, even if a remora cannot be dislodged by such a manoeuvre, it might be persuaded to shift its position on the body. Some of these fish attach themselves onto regions of the dolphin that are disadvantageous (even life-threatening) to the host, such as the throat or genital slit (preventing mating), below the eye (hindering vision) or even on the blowhole, and spinning can result in the fish relocating themselves on the dolphin's body.[4]

But spinning occurs in a number of individuals that do not carry remoras, and shows a clear diurnal pattern of occurrence, even in captive animals with no remoras present. An alternative suggestion is that it is the end of the spin, the re-entry, that is particularly important: the noise of the body striking the water and especially the large plume of bubbles created by the dolphin as it sinks down after the jump, may serve to produce a series of acoustic markers that define the school's shape as it moves along.[1]

References

1 Norris *et al.* 1994
2 Fish *et al.* 2006
3 Johnson and Norris 1994
4 Silva-Jr and Sazima 2006

relatively abundant and closely-spaced, there is a marked fission–fusion system: groups of less than 100 dolphins are found inshore during the day but groups of several hundred offshore at night, so that group size and composition are constantly changing. Under these circumstances there is much fluidity of association, with a group entering a bay on one day being composed of somewhat different individuals from the previous or subsequent days. Off remote atolls of the Hawaiian chain, however, where suitable resting sites may be few and widely spaced, spinner dolphins live in stable societies of long-term associates with strong geographic fidelity. Presumably it is energetically more efficient in this case to remain at home than to travel to other atolls.[48]

Daily bay usage is accompanied by a distinct behavioural sequence. Animals enter the bay rapidly in scattered groups with much aerial behaviour, but then assemble in a tight, slow-moving school, with little above-surface activity and declining frequency of vocalisations – a 'descent into rest'. For the next 4–5 hours the school moves slowly back and forth in the bay as a single unit, usually over a flat sandy bottom, their dives becoming longer and surfacings brief, and with very little vocal activity. Awakening is abrupt, usually marked by a sharp increase in aerial behaviour. The school speeds up and commences zigzag swimming, in which great changes in speed and direction of movement occur, as the school apparently oscillates between moving offshore to feed and returning to their resting area, before it finally leaves for the open sea.[30]

Spinner dolphin behaviour (either echolocation, caressing or aerial) seems to occur cyclically in bouts that last 3–30 min., and could involve a few individuals or a major part of the school. Caressing involves rubbing a partner with flippers or flukes and sometimes making bodily contact in a mating posture. Beak-genital propulsion, in which one animal places the tip of its beak in the genital slit of another and gently propels it forward, is the most unusual caressing behaviour and is associated with the peak of the female's oestrous cycle. During the summer, intromissions become a common feature in caressing groups.[40] Mating can involve a single male-female pair, but more frequently occurs in groups of 2–5 (sometimes more) mature males and a female.[43] Although male coalitions are formed, these do not seem to monopolise access to females.[40] Each male will attempt copulation in turn with the female, although sometimes a particular male will dominate matings. Penetration (achieved by the male inverting himself below the female) lasts 16 sec. on average. Such courting groups will persist for up to 35 min.[43]

As with other oceanic delphinids, sounds produced by spinner dolphins include broadband clicks (up to at least 65 kHz in frequency), pure tone whistles and burst-pulsed sounds. The average frequency range of whistles is from 9.1 kHz to 13.7 kHz, with starting and ending frequencies of 10.4 kHz and 12.4 kHz respectively.[49] Their typical duration is 0.1–1.5 sec.[50] with an average of 0.6 sec. There are an average of 1.9 inflection points and 0.7 steps.[49] Burst-pulsed sounds are the most complicated class, lasting 0.2–4.0 (mean 0.5) sec. and most ranging in frequency between 5 kHz and 60 kHz. All three categories of sounds are produced more frequently at night or when the school is spread out (and to a lesser extent when it is travelling) than when it is entering a bay, descending into rest, resting, socialising or meeting another school.[50]

Conservation status

There are no estimates of population size for spinner dolphins in the subregion. Numbers may be taken as a by-catch in coastal fisheries in the south-west Indian Ocean (44 in Zanzibar in 2000–2003, for instance[46]). The effects of such catches are unknown.

Spinner dolphins are listed as Lower Risk: Conservation Dependent by the IUCN and as Data Deficient in the South African Red Data Book. They are included on Appendix II of CITES, and the populations in the eastern tropical Pacific and south-east Asia are listed on Appendix II of CMS.

References

1 Rice 1998
2 Perrin 1990
3 Perrin *et al.* 1999
4 Ross *et al.* 1985
5 Karczmarski, L., pers. comm.
6 Perrin and Gilpatrick 1994
7 Perrin *et al.* 1981
8 Fish *et al.* 2006
9 Mead *et al.* 1980
10 Schnell *et al.* 1982
11 Perrin *et al.* 2005
12 Ross 1984
13 Perrin 1975
14 Rosenbaum and Collins 2006
15 Weir, C., pers. comm.
16 Perrin 1985
17 Peddemors *et al.* In press
18 Kiszka *et al.* 2006a
19 Boucaud and Rota 2006
20 Corbett 1994
21 Howell and Pearson 1977
22 Leatherwood 1986
23 MRI Whale Unit data
24 Findlay *et al.* 1992
25 Davis *et al.* 1998
26 Moreno *et al.* 2005
27 Au and Perryman 1985
28 Reilly 1990
29 Secchi and Siciliano 1995
30 Würsig *et al.* 1994b
31 Au and Perryman 1982
32 Morris and Mowbray 1966
33 Sedlak-Weinstein 1991
34 Silva-Jr and Sazima 2006
35 Silva-Jr and Sazima 2003
36 Würsig *et al.* 1994a
37 Benoit-Bird and Au 2003
38 Silva-Jr *et al.* 2004
39 Sazima *et al.* 2003
40 Johnson and Norris 1994
41 Wells 1984
42 Benirschke *et al.* 1980
43 Silva-Jr *et al.* 2005a
44 Perrin 1998
45 Silva-Jr *et al.* 2005b
46 Amir *et al.* 2005b
47 Psarakos *et al.* 2003
48 Karczmarski *et al.* 2005
49 Oswald *et al.* 2003
50 Brownlee and Norris 1994
51 Perrin *et al.* 2007

Clymene Dolphin

Stenella clymene

IT HAS BEEN proposed that the name 'clymene' might come from the Greek klymenos, meaning famous or notorious.[1] However, for a long time this species was far from either, and it is still the least known member of the genus *Stenella*. Despite being first described in 1846 from a skull of unknown provenance, it was another 135 years before the species was 'rediscovered' and its habitat and external appearance described.[2] Until recently its presence in the subregion was represented by a single specimen harpooned near Ascension Island in the nineteenth century. It has now been sighted off Angola and the Congo,[3] but most of the information in this account comes from elsewhere.

Appearance

Clymene dolphins are quite similar in appearance to short-beaked common dolphins: the two species are almost identical in body shape and the basic colour pattern is similar. One of the major differences is the shape of the margin between the dark pigmentation on the back and the pale coloration of the chest and flanks. In the clymene dolphin this dips down below the dorsal fin in a smooth curve, whereas in the short-beaked common dolphin it forms a distinct V-shape.[4] (This dip in the margin also distinguishes the clymene dolphin from the spinner dolphin, in which this margin is more or less straight for the length of the animal.[2]) In addition, at close range the black 'moustache' marking on the snout of the clymene dolphin is distinctive, and there is an eye-to-flipper, rather than a chin-to-flipper, stripe as in the short-beaked common dolphin.[4]

Adults reach a length of 176–197 cm,[2] males perhaps growing some 7–8 cm longer than females. The similarity in body shape to the short-beaked common dolphin has already been mentioned. Compared to the spinner dolphin, the clymene dolphin has a shorter beak (4–6% of body length, mean 5.5%, n = 14), shorter flippers (13–17% of body length, mean 14.7%, n = 15), and smaller dorsal fin (8–12% of body length, mean 9.4%, n = 15). A post-anal bump in the outline of the caudal peduncle is present in some adult males.[5]

As in spinner dolphins, the colour pattern of clymene dolphins has three parts: a dark grey cape over the back, light grey sides and a white belly. There is usually a dark line separating the

Length:weight relationship

The maximum recorded weight for a clymene dolphin is 80 kg.[7] The length:weight relationship for 15 females has been recalculated as $\mathbf{W = 9.8322L^{2.9458}}$, where W = weight in kilograms and L = total length in centimetres.[6] This implies that a 135 cm dolphin would weigh 24.2 kg and a 175 cm animal 50.5 kg. Males are likely to be some 8% heavier at a given length.

Skeletal data

Vertebral formula (n = 8–11): 7 cervical, 13–15 thoracic, 17–21 lumbar and 31–35 caudal; total 73–76.[1,5]

The atlas and axis are probably fused, and the other cervical vertebrae may partly fuse in older animals, but published statements to this effect seem to be unavailable.

Phalangeal formula, including metacarpals (n = 10): I_{0-2}, II_{7-9}, III_{5-7}, IV_{2-3}, V_{0-2}.[2]

The skull of the clymene dolphin differs from that of the spotted dolphins in the following ways: it has a relatively small temporal fossa, there are a larger number of teeth (usually more than 40 per row), the posterior sections of the toothrows are almost parallel, the anterior portion of the rostrum is markedly flattened, and there is a grooved palate.[1] It differs from that of the spinner dolphin in being relatively shorter, the upper toothrow not exceeding 215 mm in length, while it differs from the striped dolphin in being smaller, with a preorbital width rarely exceeding 170 mm, and lacking a raised and rounded area on the premaxillae close to the base of the rostrum.[5]

light grey on the side and the white belly. Unlike the pattern on the spinner, the margin between the dark cape and light grey takes a marked downward dip at about the level of the dorsal fin. There is a dark stripe from the flipper to the eye, which narrows as it approaches the eye (in contrast to the spinner dolphin, where the stripe maintains the same width throughout). The black edging to the upper lip is interrupted just behind the tip of the beak by a pale grey to white blaze, which is not seen in the spinner. And most characteristically of all, the black stripe running from the eye towards the apex of the melon dips onto the beak and zigzags several times before merging with the black central stripe of the rostrum. This marking has been referred to as the 'moustache'[1] and its position on the beak varies, from close to the apex of the melon to about halfway along.[5] There are no indications of age- or sex-related variations in the colour pattern.

There are 39–52 (mean 44, n = 56) pairs of teeth in the upper jaw and 39–48 (mean 43, n = 63) pairs in the lower jaw. These counts are from cleaned skulls; field counts (in the flesh) may have 4–5 fewer pairs in each jaw.[5] The teeth are small, slender and conical in shape.

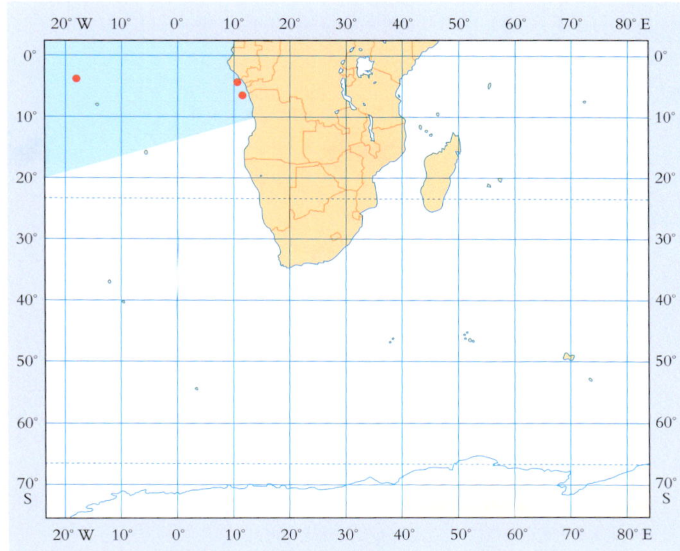

FIG. 126 Distribution map for clymene dolphin.

Distribution and migrations

Clymene dolphins are known only from oceanic waters of the tropical and warm-temperate Atlantic Ocean. The most southerly records on the western side of the Atlantic are at 29°58′ S and in the mid-Atlantic (18°05′ W, near Ascension Island) at 3°40′ S.[4] Until recently, the latter was the only known record for the subregion. Two sightings (one off Angola and the other off the Congo) now establish the presence of the species on the eastern side of the South Atlantic at least as far south as 6°26′ S.[3]

Clymene dolphins have been encountered in water depths of 44–4 500 m, but 94.6% of the 94 records were in water more than 500 m deep.[4] The two sightings in the subregion were in water 466 m and 684 m deep.[3] Their habitat in the Gulf of Mexico has been characterised as deep water beyond the continental shelf.[8]

The species was encountered most often in the Gulf of Mexico in winter and spring, although it was seen in all seasons except autumn.[9] Strandings in that area have apparently occurred throughout the year.[10]

FIG. 127 A leaping clymene dolphin shows how similar it is to, and yet different from, a short-beaked common dolphin.

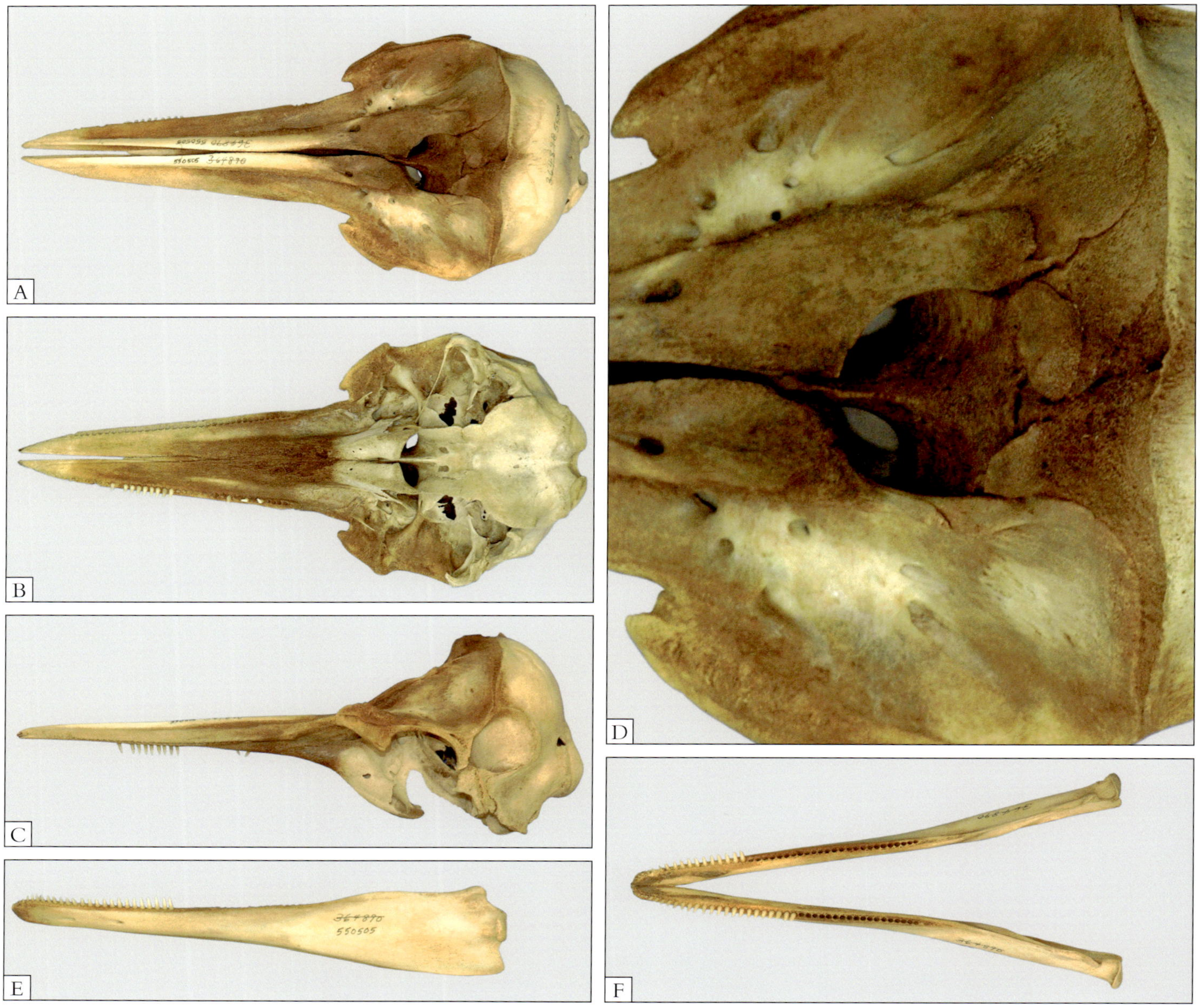

FIG. 128 Dorsal (A), ventral (B) and left lateral (C) views and vertex of skull (D); left lateral (labial) (E) and dorsal (F) views of mandible (USNM 550505).

External commensals and parasites

There are no records from the subregion. In the Gulf of Mexico, the stalked barnacle *Xenobalanus* was found on the dorsal fin, flippers, and flukes of one individual, and 10 cyamids (possibly *Syncyamus pseudorcae*) and one barnacle (on the dorsal fin) of another. Scars resulting from attacks by the cookie-cutter shark *Isistius* were found on all six animals examined, ranging from two to 31 in number.[6]

Food and feeding behaviour

There is no information from the subregion. The only published information is from one stranded animal from the USA: its stomach contained a single pair of cephalopod beaks and over 800 otoliths. Among these, myctophids predominated, including *Ceratoscopelus* sp. (780), *Lampanyctus* (7) and *Symbolophorus* (7), but a possible argentinid (5) and a single *Bregmaceros* sp. were also present. The majority of these are midwater fish that migrate to the surface at night, but *Lampanyctus* usually does not reach the surface, even at night.[1]

Clymene dolphins have been seen feeding cooperatively on small fish (10–15 cm long) at the surface during the day. The fish were kept in a tight ball about 15 m in diameter by about 30 dolphins, with individuals taking turns to pursue the fish.[11]

Reproduction and growth

There is no local information. The size at birth is not known, although calves as small as 1.2 m have been recorded. Seasonality of reproduction is unknown.[2] In a very small sample (seven females, 12 males), the shortest known mature male and female were 176 cm and 171 cm long respectively, while a 168 cm female was immature.[7] The ages of two females were determined to be 15 and 16 years old.[6]

Behaviour and vocalisations

Clymene dolphins have been encountered in schools of 250 off the Congo and in a pod of 12 off Angola.[3] Elsewhere they have been found as single animals (in a school of spinner dolphins) or in groups of up to 1 000. The most usual group size was 50 animals or fewer, and the mean was 76.1 (n = 105).[4] There may be social segregation between groups, with some consisting mostly of males and others being 'nursery' schools consisting mostly of females with dependent young and a few adult males.[6] Calves were observed with 45% of schools in the Gulf of Mexico.[10] Associations with other species (principally spinner dolphins) occur but are uncommon.[7]

The species performs aerial displays similar to those of the spinner dolphin (spins of up to 3–4 revolutions being recorded) and actively rides the bow, stern or beam waves of vessels under way. On one occasion dolphins were seen playing with pieces of floating seaweed (*Sargassum*), picking it up on their flippers or tail-slashing at it.[10]

Clymene dolphins are highly vocal, and whistles can often be heard by observers on the bow of a ship. The structure of these whistles is similar to that of other stenellids, but generally higher in frequency: sounds were between 6.33 kHz and 19.22 kHz, with average starting and ending frequencies of 10.64 kHz and 12.68 kHz respectively. The mean whistle duration was 0.61 sec.[10]

Conservation status

There are no estimates of population size for clymene dolphins in the subregion, and no indications of any substantial human-induced mortality. Their restricted distribution does, however, make them potentially vulnerable to depletion.[4]

Clymene dolphins are listed as Data Deficient by the IUCN and are included on Appendix II of CITES. They are not listed in the South African Red Data Book or by CMS.

References

1 Perrin *et al.* 1981
2 Perrin and Mead 1994
3 Weir 2006a
4 Fertl *et al.* 2003
5 Jefferson 1996
6 Jefferson and Curry 2003
7 Jefferson *et al.* 1995
8 Davis *et al.* 1998
9 Mullin *et al.* 2004
10 Mullin *et al.* 1994a
11 Fertl *et al.* 1997

Striped Dolphin

Stenella coeruleoalba

THIS DOLPHIN HAS had several different common names over the years, including Euphrosyne dolphin and blue-white dolphin, but even by its more prosaic current name it is still undoubtedly one of the more attractive of the small cetaceans. Although the common dolphin is frequently considered to be the classic dolphin of Greek art and literature, striped dolphins are equally as characteristic of the Mediterranean region, and some depictions on Greek pottery dating back to 1500 BC have been attributed to this species.[1]

Appearance

Striped dolphins are most likely to be confused at sea with Fraser's dolphin as the adults of both carry prominent lateral stripes. Adult Fraser's dolphins actually have a thicker stripe and can also be readily distinguished by the much shorter, almost insignificant beak and the highly variable nature of the stripe itself, which is absent in juvenile Fraser's dolphins but present at all ages in the striped dolphin.

Adults in the subregion have been recorded with lengths of up to 2.65 m in males (n = 58) and 2.4 m in females (n = 44).[2,3] The asymptotic lengths from growth curves are 2.46 m and 2.30 m respectively, indicating that males grow slightly larger than females.[2] The body shape is similar to other stenellids, although the beak (at 4.3–6.1% of body length, mean 5.2%, n = 28) is shorter than in the spinner dolphin, and the flippers and dorsal fin are somewhat smaller than in any of the other *Stenella* species in the subregion except the pantropical spotted dolphin (flipper length 11.6–16.2% of body length, mean 13.5%, n = 27; dorsal fin height 6.0–9.7% of body length, mean 8.1%, n = 25). The dorsal fin is also hooked, whereas that of the spinner dolphin tends to be more triangular. The tail flukes have a fairly straight trailing edge but are hooked at the tips, with a width of 17.4–24.6% (mean 21.4, n = 26) of body length.

Length:weight relationship

The length:weight relationship for 28 striped dolphins from the subregion[2–4] is **$W = 8.9436L^{3.1336}$**, where W = weight in kilograms and L = total length in metres. This implies that a 1.5 m dolphin would weigh 31.9 kg and a 2.25 m animal 113.5 kg.

Skeletal data

Vertebral formula from the subregion (n = 25): 7 cervical, 13–16 thoracic, 18–25 lumbar and 31–39 caudal; total 75–81.[3,4]

The atlas and axis are fused, and in some animals the third or even fourth cervical may partly fuse with them. Occasionally there may be separate fusion between a combination of cervicals 4–6.[6]

Phalangeal formula, including metacarpals (n = 17): $I_{2\text{–}3}$, $II_{7\text{–}10}$, $III_{6\text{–}8}$, $IV_{3\text{–}4}$, $V_{2\text{–}3}$.[6]

The skull of the striped dolphin differs from that of the spotted dolphins in having a relatively small temporal fossa, a highly flattened rostrum, and a grooved palate. The mandible may also be slightly S-shaped when seen from above, whereas that of the spotted dolphins is more simply curved.[7] The skull differs from that of the spinner dolphin in being shorter, with the width of the rostrum at its base exceeding one-third of the length of the rostrum itself (as opposed to being less than one-third in spinners).[8] It is more similar to that of the clymene dolphin, but can be distinguished by less well-developed palatal grooves (which only extend for the basal third of the rostrum, as opposed to at least half the rostrum in the clymene dolphin), as well as distinct rounded areas on each premaxilla at the base of the rostrum, which are lacking in the clymene.[9]

A post-anal bump in the outline of the caudal peduncle does not seem to develop in adult males.[3,4]

Although there may be considerable individual variation in the colour pattern of striped dolphins, there is no indication that this is linked to age. The most distinctive feature is a dark line running along the side of the animal, from the eye back along the flank to the anal region. This stripe is not completely straight and tends to dip down the side in the thoracic region; it also broadens as it passes aft, widening especially behind the anus. There is usually a secondary, shorter stripe underlying the main one that also starts at the eye but does not extend back much behind the flipper. A dark stripe connecting the base of the flipper with the eye may be single, double or even treble in nature. Above the dark lateral line, and running from above the eye as far back as the caudal peduncle, there is a prominent blue-grey blaze, usually with a separate 'spinal' fork that curves up towards the dorsal fin. The snout and chin are black, but the throat and belly are white, sometimes tinged with pink.[5]

There are 34–49 (mean 41.8, n = 64) pairs of teeth in the upper jaw and 37–52 (mean 43.1, n = 54) pairs in the lower jaw.[3,4] They are slender and conical in shape, with the largest being 2.5–4.0 mm (mean 3.6 mm, n = 7) in diameter at the gum.[3]

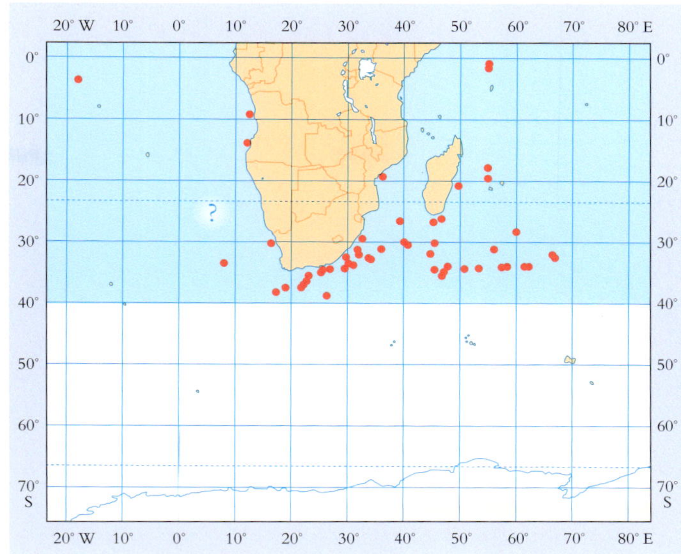

Postulated range · Sighting or capture · ? Questionable origin

FIG. 129 Distribution map for striped dolphin.

Distribution and migrations

Striped dolphins occur in warm-temperate to tropical waters worldwide.[7] In the eastern tropical Pacific they are described as preferring equatorial and subtropical waters with relatively large seasonal changes in surface temperature and thermocline depth, and with seasonal upwelling.[10] In the subregion they have been recorded from Gabon,[11] Angola,[12] South Africa,[13] Mozambique[14] and Kenya[15] (although the origin of the Kenyan record is uncertain). The southern extent of their distribution in the south-east Atlantic is uncertain. Strandings occur up the west coast of South Africa as far as Yzerfontein, but whether these are Indian Ocean animals rounding the Cape or stragglers from a population to the north or further offshore in the South Atlantic is unknown. The most southerly sighting on the west coast of South Africa is a single animal seen at about 30° S in the company of 15 dusky

FIG. 130 This striped dolphin jumping in the Maldives clearly displays all the features of its attractive colour pattern.

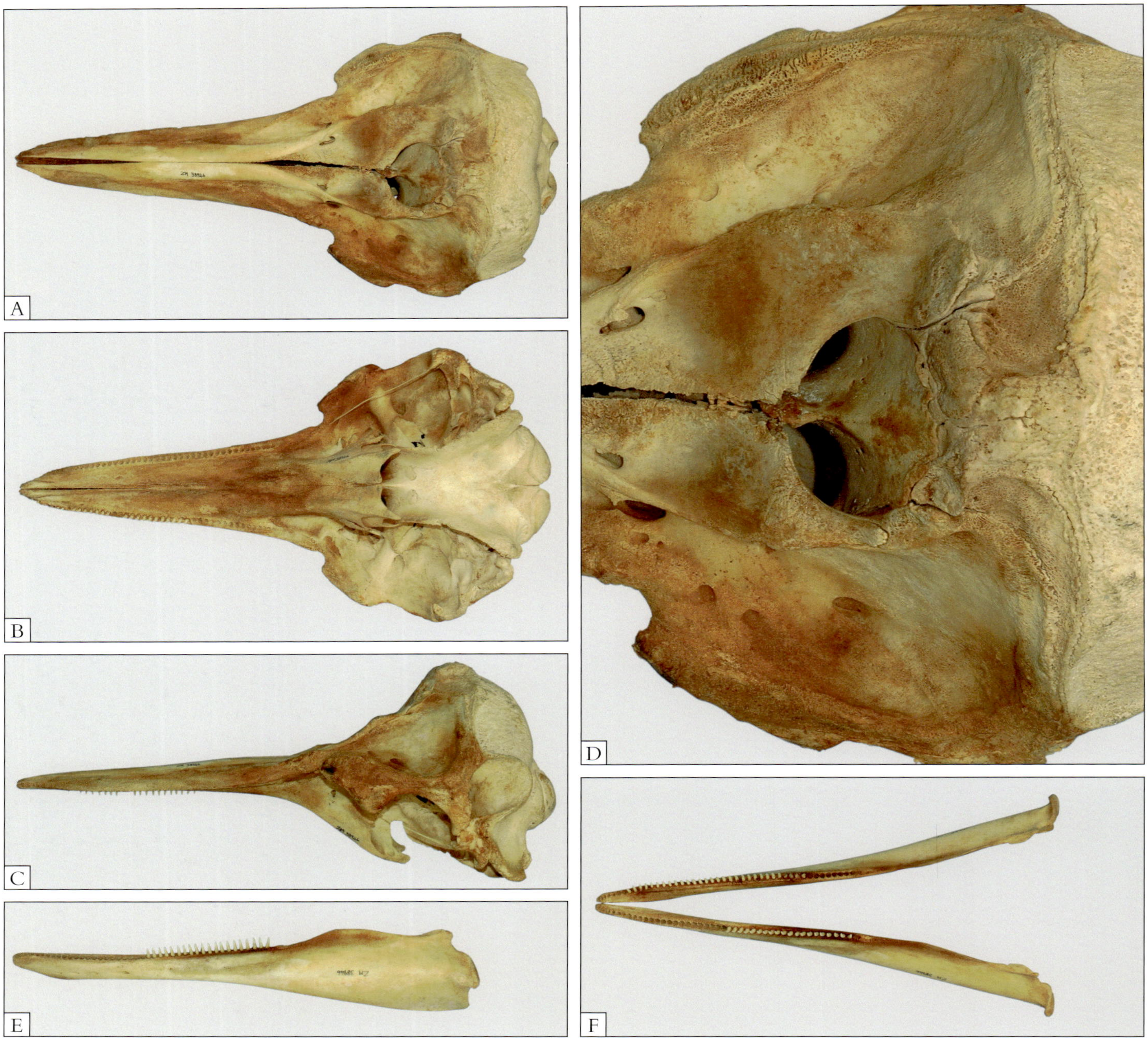

FIG. 131 Dorsal (A), ventral (B) and left lateral (C) views and vertex of skull (D); left lateral (labial) (E) and dorsal (F) views of mandible (ZM 38946).

dolphins, but this was in unusually shallow water (200–220 m) for the species.[13] There are no sightings of the species in the subregion south of 39° S.[3]

Off Angola, the species is reported to occur only in water depths of 1 600–2 400 m,[12] whereas off South Africa all sightings (except that mentioned above) were in water 500 m deep or greater.[13] Elsewhere striped dolphins also show a preference for water that is more than 600 m deep, and with an intermediate or steep slope to the sea floor.[16, 17]

Striped dolphins are apparently found all year round off Angola.[12] Although scientific sightings off South Africa were only made in summer, this seems to have been the result of uneven sighting effort.[13] Strandings in South Africa occur more or less evenly between seasons: spring 18, summer 15, autumn 13, winter 12.[3, 4] A seasonal shift into deeper water has been suggested in the Mediterranean Sea,[18] as well as an offshore morning and in-shore evening movement, associated with nocturnal feeding close to the shelf break.[19] Off Japan, latitudinal shifts in distribution with season seemed to be related to prevailing oceanographic conditions.[20]

External commensals and parasites

Only one of 21 striped dolphins examined in the subregion carried any external parasites: a male with a single stalked barnacle (*Xenobalanus globicipitis*) on the tip of the right flipper, and several cyamids in the blowhole area.[3, 4] These cyamids, and those from two other striped dolphins in the subregion, were identified as

Syncyamus aequus.[21] Studies of dolphins dying in the Mediterranean Sea in an epizootic event (see box below) showed an incidence of 51.5% carrying *Xenobalanus* (average 11.7 per dolphin), 30.5% carrying the parasitic copepod *Pennella* (average 4.6 per dolphin, on the flanks), and 19.1% carrying *Syncyamus aequus* (average 2.9 per dolphin, in the blowhole region). A small number (1–2%) also had goose barnacles (*Lepas*) attached between their teeth and the stalked barnacle *Conchoderma virgatum* attached either to the teeth or a *Pennella*. It is possible that the high incidence of these organisms reflected the poor condition of the animals, leading to a reduced motility.[22]

Food and feeding behaviour

Information is available from 25 stomachs from the subregion. Of 1 421 identified prey items (mainly from trace remains such as fish otoliths or squid beaks), 1 032 (72.6%) were fish and 389 (27.4%) cephalopods. Among the fish, light fishes or myctophids (principally *Symbolophorus*, *Hygophum* and *Diaphus*) were the most abundant (78.2%), followed by hakes (*Merluccius* sp., 10.5%), horse mackerel (*Trachurus trachurus*, 4.6%) and codlets (*Bregmaceros* sp., 3%). In the cephalopods, chokka squid (*Loligo reynaudi*) was numerically dominant (62.5%), followed by the lycoteuthid squid *Oregoniateuthis* (21.9%), ommastrephids (principally *Sthenoteuthis oualensis*) at 4.9% and cuttlefish (*Sepia* sp.) at 4.1%.[3,4]

Many (up to 80%) of these prey carry luminous organs,[4] suggesting that striped dolphins may be feeding at depths of 200–700 m, or at night when many of these organisms migrate closer to the surface.[8] Feeding at dusk or in the early hours of the night was also suggested by the state of digestion of food items in the stomachs of animals taken in a North Atlantic drift-net fishery.[23] An archival tag attached for a week to a striped dolphin in the western North Pacific showed a clear diel pattern of diving, with the average depth reached during the day being 22.6 m and at night 126.7 m. The maximum depth reached was 705 m.[24]

Reproduction and growth

Calves as small as 93 cm and 98.5 cm (n = 11) have been recorded in the subregion,[2,3] suggesting that the length at birth is in the same range as that estimated for the Mediterranean (92.5 cm)[25] and western North Pacific (100 cm).[26,27] Gestation may last for 12–13 months.[26,28] In the subregion, births can occur in summer (January–February) or winter (May–August).[2] Two parturition seasons were also described in the North Pacific, one in May–June and the other in November–December,[26] but only one in the Mediterranean.[25] Lactation can last for 18 months,[26] although in the western North Pacific the calf may start taking solid food as young as three months (and at a length of 133 cm) and be weaned at a size of 176 cm.[27] A milk sample contained 28% fat.[29] In the subregion the smallest mature female was 200 cm long and the largest immature 211.5 cm (n = 22), while the equivalent sizes of males were about 215 cm and 225 cm respectively (n = 24).[2] These ranges probably cover the likely size at sexual maturity in each sex. Average ages at sexual maturity in the western North Pacific are about nine years for both sexes, although full social maturity in the male may only be reached at around 17 years. Adult females calve on average at 3.2 year intervals, but the interval increases with age as more time is spent

Dead in the Med

Mass mortalities of cetaceans can take the form of strandings (in which a social grouping of animals beaches itself in a circumscribed locality over a period of hours – see the species account of the false killer whale) or, more rarely, an epizootic event, in which individual animals die and wash up individually over an extended period of time (usually weeks or months) and over a broad geographical front. The striped dolphin epizootic event in the Mediterranean in the 1990s is one such phenomenon. Between July 1990 and the spring of 1992, at least 1 100 striped dolphins died in three main outbreaks, the first on the coasts of Spain, France, and North Africa, the second (between June and September 1991) in southern Italy, and the last in the region of the Greek Islands in spring 1992. Some dolphins reached the shore alive but many more must have died at sea and were never properly accounted for. Post mortems revealed the proximate cause of the mortalities to be a morbillivirus similar to canine distemper but of unknown origin, the infection producing *inter alia* pneumonia, meningoencephalitis and lymphoid depletion, leading to partial lung collapse, breathing difficulties and nervous and locomotory disorders. However, there may have been other contributory factors. The diseased dolphins were found to carry concentrations of organochlorine pollutants (such as PCBs) two to three times the levels found in the healthy population. These substances are known to be depressors of the immune system and may have increased the susceptibility of individuals to infection. In addition, during the first 70 days of the 1990 outbreak the nutritional state of the dolphins that died was unusually poor, with fat reserves reduced by about 40%, and an abnormally high number of barnacles suggesting that they had restricted mobility. Depressed marine productivity as a result of unusually high temperatures in the winter preceding the outbreak may have been responsible for the poor condition of the animals and hence triggered the outbreak. The impact of the epizootic event on the striped dolphin population in the Mediterranean is not fully known. At the peak of the first outbreak, dolphin schools fell to about one-third of their normal size, but a year later were back to normal numbers, and it was estimated that about 225 000 striped dolphins survived the 1990 outbreak in the western Mediterranean Sea.[1]

Such marine mammal die-offs are known to have occurred 14 times between 1982 and 1996, although mostly (in nine cases) involving seals, sea otters or manatees and all but one (involving Namibian fur seals) occurring in the northern hemisphere. Causes for these mortalities were not always well established, but candidates included biotoxins (four), environmentally-induced starvation (four), morbillivirus (three), accidents (two) or a combination of these (one). Climate change may well influence the rate at which these incidents occur in future.[2]

References

1 Aguilar and Raga 1993
2 Geraci *et al.* 1999

lactating or resting, and the very oldest females may even stop conceiving completely.[28] Physical maturity is attained at ages of 15–20 years in males and 13–18 years in females.[30] Stranding patterns in the western Mediterranean suggest that mortality of neonates and calves in early lactation is much higher than that of calves during mid-lactation and late lactation.[25] Both sexes can live for up to 45–46 years.[28, 31]

Behaviour and vocalisations

Striped dolphins in the subregion occur in schools of 1–1 000 animals, with an average of 126 (n = 48), although 79% of sightings were of schools of 150 animals or fewer.[3] Off Angola the groups were somewhat smaller, usually between eight and 50, with one record of 125 animals.[12] Mixed schools were encountered on four occasions (n = 55), twice with pantropical spotted dolphins, once with dusky dolphins and once (off Angola) with long-beaked common dolphins. Striped dolphins predominated, were equal, or were in the minority on each of three occasions.[3, 12]

Schools in the Japanese drive fishery were of three main types – juvenile, adult and mixed – with the last two further divided into breeding and non-breeding schools. School sizes did not differ significantly between these categories. Juvenile schools were those where 90–100% of animals were immature: most were 3–9 years old and males always predominated. Adult schools contained 60–80% mature individuals and animals of both sexes, although females tended to be in the majority. There were two peaks in the age distribution, at 0–1 year (representing dependent calves) and another at 12–13 years (representing mature adults). Animals of both sexes seem to leave these schools after weaning (presumably to form juvenile schools), though some young individuals may join the school as they reach puberty. Mixed schools contained 50–60% immature animals. Breeding schools had a higher ratio of socially mature males to females in lactating, resting or early pregnant stages than non-breeding schools, and the males tended to be older and with better developed testes. Once most of the available females have been fertilised, such males leave the breeding school, which then reverts to a non-breeding school.[32]

Striped dolphins exhibit aerial behaviour such as breaching, chin slaps and 'roto-tailing'. The last is described as a high arcing jump during which the tail is rapidly and violently rotated several times before re-entering the water.[8] Spinning (as performed by spinner and clymene dolphins) has not been recorded.

Whistles produced by striped dolphins have a frequency range of 8–15 kHz with an average duration of 0.8 sec. Starting and ending frequencies average 10.2 kHz and 12.0 kHz respectively, and each whistle has on average 1.9 inflection points and two steps. In general the whistles are more variable than in some other delphinids, and quite difficult to distinguish accurately.[33]

Conservation status

There are no estimates of population size for striped dolphins in the subregion, and no indications of any substantial human-induced mortality. The potential for their involvement in an expanding drift-net fishery south of Madagascar has been noted.[2]

Striped dolphins are listed as Lower Risk: Conservation Dependent by the IUCN and as Least Concern in the South African Red Data Book. They are included on Appendix II of CITES, and the eastern tropical Pacific and Mediterranean populations are on Appendix II of CMS.

References

1 Pilleri and Pilleri 1982
2 Kroese 1993
3 MRI Whale Unit data
4 Ross 1984
5 Fraser and Noble 1970
6 Ito and Miyazaki 1990
7 Perrin *et al.* 1994c
8 Archer and Perrin 1999
9 Jefferson 1996
10 Au and Perryman 1985
11 Rosenbaum and Collins 2006
12 Weir, C., pers. comm.
13 Findlay *et al.* 1992
14 Peddemors *et al.* In press
15 Wilson *et al.* 1987
16 Cañadas *et al.* 2002
17 Davis *et al.* 1998
18 Gannier 1998
19 Gannier 1999
20 Miyazaki *et al.* 1974
21 Lincoln and Hurley 1981
22 Aznar *et al.* 1994
23 Ringelstein *et al.* 2006
24 Minamikawa *et al.* 2003
25 Aguilar 1991
26 Kasuya 1972
27 Miyazaki 1977
28 Miyazaki 1984
29 Kawai and Fukushima 1981
30 Calzada *et al.* 1997
31 Kasuya 1976
32 Miyazaki and Nishiwaki 1978
33 Oswald *et al.* 2003

Short-beaked Common Dolphin

Delphinus delphis

UNTIL COMPARATIVELY RECENTLY it was generally assumed that there was one species of common dolphin worldwide, with the scientific name *D. delphis*. Now we know the situation is somewhat more complicated. Two species (and one subspecies) of common dolphin are currently recognised: the short-beaked common dolphin, *Delphinus delphis,* and the long-beaked common dolphin, *D. capensis*, with the Indo-Pacific common dolphin, *D. capensis tropicalis*, as a subspecies.[1, 2] The morphological distinctions between the two species in the eastern North Pacific had supporting genetic evidence, from both mitochondrial[3] and nuclear DNA.[4] However these animals exhibit much morphological variation, and in some regions the criteria used to distinguish between short- and long-beaked common dolphins in the eastern North Pacific do not necessarily apply, or are somewhat different.[5, 6] Southern Africa happens to be one of those areas, where additionally the same mitochondrial DNA markers as used in the eastern North Pacific failed to reveal any distinction between the morphotypes.[7] Although the occurrence of the short-beaked dolphin in the Indian Ocean and the South Atlantic has been doubted,[8] the balance of evidence suggests that the species does occur in the subregion,[7] and this has been the assumption followed here.

To make matters worse, much of the old literature (and records) simply refer to 'common dolphins' without reference to species, while the scientific name is given as '*Delphinus delphis*' irrespective of which species it might actually have been. This means that much information has had to be disregarded in compiling these species accounts.

Appearance

Common dolphins (both short-beaked and long-beaked) are relatively easy to distinguish from other small dolphins at sea from the distinctive criss-cross (or 'hourglass') marking on the side, involving a buff or golden-yellow patch on the thorax and a light grey patch on the flanks. The only species with which they might be confused is the clymene dolphin, as the overall body shape and coloration are similar. The most obvious differences are the following: 1) The lower margin of the dark coloration of the back dips down below the dorsal fin in a sharp V-shape in common dolphins but in a more rounded shape in the clymene dolphin. 2) There is a moustache-like marking on the upper surface of the beak on the clymene dolphin, which is absent in common dolphins. 3) There is an eye to flipper stripe in the clymene dolphin, but only a chin to flipper stripe in common dolphins.[9]

Telling short-beaked from long-beaked common dolphins at sea is not easy and requires experience. The characteristics that might help distinguish them at close range or as specimens are included below.

Only three specimens of short-beaked common dolphins from the subregion have been measured, two males at a length of 2.07 m and 2.1 m and an adult female at 1.95 m.[7] In the eastern North Pacific adult males range in length from 172 cm to 223 cm and adult females from 164 cm to 215 cm.[1] The species is thus substantially smaller than the long-beaked common dolphin.

Short-beaked common dolphins from the subregion have (as one might expect) shorter beaks than long-beaked common dolphins: 5.2–7.0% of body length, mean 6.2% ($n = 3$), compared to 6.4–8.4% of body length, mean 7.5% ($n = 20$). They also have longer flippers (15–17.8% of body length, mean 16.3%, compared to 13.5–15.7%, mean 14.4%), and wider flukes (21–26.2% of body length, mean 24.1%, compared to 17.9–24.2%, mean 20.7%).[7] The short-beaked common dolphin is relatively heavier and deeper-bodied anteriorly than the long-beaked, with a more rounded melon that joins the beak at a sharper angle when viewed in profile.[1] Otherwise the overall body shape of the two species is similar, with wide wedge-shaped flippers, an erect recurved dorsal fin and lunate flukes.[7] Adult males develop a distinct anal 'hump'.[10]

The following are some of the 'classic' criteria used for distinguishing between the two common dolphin species from their colour patterns in the eastern North Pacific but which may not always apply in the subregion: 1) The contrast between the yellow-brown colour on the thorax and the dark colour of the back is bright and sharp in the short-beaked, but dull and weak in the long-beaked. 2) A stripe from the flipper towards the anus is absent or weakly developed in the short-beaked, but weak to strongly developed in the long-beaked. 3) The stripe from the flipper to the chin fuses with the lip patch a third to half of the gape length in front of the corner of the mouth in the short-beaked, but at the corner of the mouth or up to a third of the gape length in front of the corner of the mouth in the long-beaked. 4) An extension of the white colour on the belly continues above the flipper stripe at least as far as the eye in the short-beaked, but is rare in the long-beaked.[1]

Anomalously pigmented short-beaked common dolphins, in which the dark grey dorsal cape extends down to replace the buff thoracic patch, have been recorded from New Zealand, Mexico, California and France.[11] Such individuals may comprise as much as 1% of the population in some areas.[12]

Short-beaked common dolphins in the subregion (n = 3) had 48–54 (median 50.5) pairs of teeth in the upper jaw and 48–53 (median 51) pairs in the lower jaw.[7] The teeth are slender and conical, with a maximum diameter at the gum of 2.0–2.5 mm (n = 2).[13] In the eastern North Pacific, tooth counts are 42–54 (upper) and 41–53 (lower),[1] in southern Australia 43–54 (upper),[6] and in the eastern North Atlantic 41–56 (upper).[5]

Distribution and migrations

Although common dolphins occur worldwide in warm-temperate and tropical waters, published records do not always specify which of the two species is involved. Hence the true distribution of short-beaked common dolphins is probably wider than appears from the authenticated records.[1, 2]

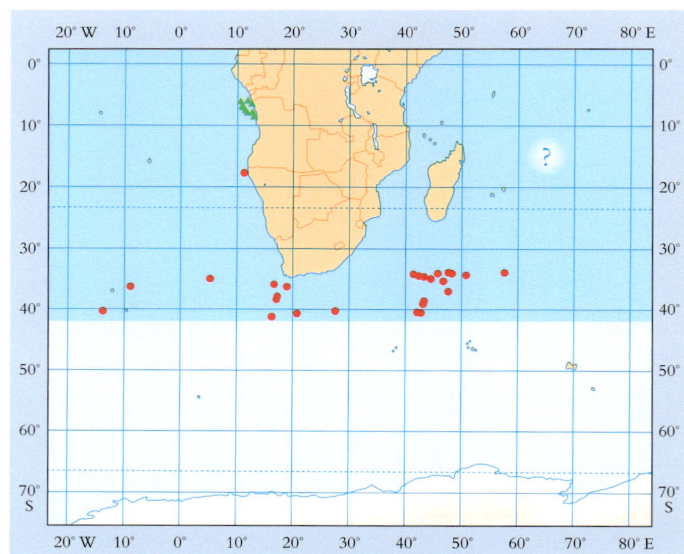

Postulated range ■ Sighting or capture
? Questionable origin ▲ Unidentified common dolphin

FIG. 132 Distribution map for short-beaked common dolphin.

In the eastern tropical Pacific, the habitat preference of offshore (and therefore short-beaked) common dolphins seemed to be equatorial and subtropical waters with relatively large seasonal changes in surface temperature and thermocline depth indicative of upwelling.[16] An association with escarpments and other offshore bathymetric features has also been noted.[17]

Within the subregion, short-beaked common dolphins are known from Gabon, Angola,[18] South Africa, and the south-west Indian Ocean.[7] Parasite and stomach content data suggest that they have a more offshore habitat than the long-beaked common dolphin, and open ocean sightings have been made.[7] In the

Length:weight relationship

There are too few records from the subregion for separate analysis. By combining the two animals from the subregion with eight animals from the eastern tropical Pacific[14] and 20 from the Mediterranean,[15] the following length:weight relationship was obtained: **$W = 12.417L^{2.63}$**, where W = weight in kilograms and L = total length in metres. This implies that a 1.5 m dolphin would weigh 36.1 kg and a 2 m animal 76.9 kg.

Skeletal data

Vertebral formula from the subregion (n = 3): 7 cervical, 14–15 thoracic, 19–20 lumbar and 31–35 caudal; total 73–76.[7] The total counts are somewhat lower than those given for the species in the eastern North Pacific (total 74–80),[1] and unlike the situation in that region, are not very different from the counts for long-beaked common dolphins.[7]

The atlas and axis are fused, and in some animals up to two more cervical vertebrae may be fused.[1]

Phalangeal formula, including metacarpals (n = 29–33): I_{2-3}, II_{8-10}, III_{6-8}, IV_{2-4}, V_{2}.[1]

Morphometric studies of common dolphin skulls from the subregion have effectively classified all the specimens as belonging to the long-beaked species,[2, 7] although some outliers from these analyses suggested the presence of a very small number of skulls with short-beaked characteristics.[7] Hence the reproduction here of the eastern North Pacific criteria for distinguishing between the two, even though they may not be totally applicable to the subregion. Skulls of adult short-beaked common dolphins are shorter overall (44.5 cm or less cf. 44.5 cm or more), with a shorter rostrum (27.5 cm or less cf. 28.1 cm or more), and a lesser distance between the tip of the rostrum and both the internal (31.8 cm or less cf. 32.2 cm or more) and external (32.1 cm or less cf. 32.6 cm or more) nares than those of adult long-beaked common dolphins. The length of the upper toothrow is also shorter (23.2 cm or less cf. 23.2 cm or more) than in long-beaked common dolphins. Proportionally, the rostrum length expressed as a ratio to the zygomatic width is 1.47 to 1 or less in short-beaked but 1.52 to 1 or more in long-beaked common dolphins.[1]

FIG. 133 Coloration of short-beaked dolphins may vary between individuals and populations. This animal was photographed south of South Africa.

Mediterranean, short-beaked common dolphins are found in both shallow (less than 200 m) and deep (more than 1 000 m) water, with some indication of a shift from the first to the second in a year with atypical oceanographic conditions.[19]

In the eastern tropical Pacific, no seasonal distributional changes were noticed,[20] but in New Zealand waters an offshore seasonal movement in autumn (and possibly even further offshore in winter) has been proposed.[21] A marked decline in availability inshore during the day in the same area may reflect a movement out of the survey area (possibly offshore) in the afternoon.[22]

In short, high-speed swimming runs, captive short-beaked common dolphins averaged 23–25 km/hr, with a maximum speed of 27–29 km/hr.[23]

External commensals and parasites

Neither of the two fresh examples of short-beaked common dolphins examined in the subregion carried any external parasites.

One individual from the subregion bore three healed scars, indicating attacks by the cookie-cutter shark *Isistius*.[7]

Food and feeding behaviour

Stomach contents of two short-beaked common dolphins from the subregion contained 1 680 identified prey items, mainly myctophids (97.1% by number) and chokka squid (*Loligo* sp., 2.3% by number).[7]

Elsewhere the species seems to exhibit relatively flexible feeding habits, perhaps associated with their ability to occupy both neritic and pelagic habitats. In the western Mediterranean, the diet is based primarily on shoaling fish such as anchovy and pilchard but also contains cephalopods and crustaceans.[24] In the western North Pacific, the food of short-beaked common dolphins consisted numerically mostly of midwater organisms such as myctophids (98% of the fish eaten) and gonatid squids (78% of the cephalopods eaten), the overall numerical proportion of fish to squid being about 6:1. Most of the prey were species that migrated vertically to shallow depths and even the surface at night, or species that inhabited surface layers of the ocean. Feeding seemed to take place early in the morning or at night, and to a lesser extent during the day.[25] Radio-telemetric and stomach studies off southern California (presumably involving short-beaked common dolphins) indicated that feeding started at dusk and continued throughout the night, with feeding being directed at organisms associated with the deep scattering layer. The diving pattern changed in the evening from shallow dives (less than 8 m) to regular deeper dives (to 50 m). As the scattering layer moved upwards, the dolphin changed its dive profile accordingly. When the layer descended again at first light, the dolphin ceased diving.[26]

Several different feeding strategies have been described for short-beaked common dolphins in New Zealand waters. Individual strategies include high-speed chasing of fish, fish-whacking (in which a fish is struck from the water with the flukes) and 'kerplunking', where the dolphin raises its body almost as far as the dorsal fin out of the water, and brings its flukes down rapidly onto the surface, causing an underwater bubble cloud followed by a vertical splash above the surface. The last behaviour may be more to startle prey than to capture them. Coordinated strategies involved line abreast pursuit, wall formation (where the fish were driven into shallow water or against an obstruction such as another line of dolphins), carouselling (active herding of fish into a ball against the surface) and bubble-blowing (in which a dolphin would position itself below and on the edge of a ball of fish surrounded during a carousel, and release a cloud of large bubbles). As the bubbles reached the school, some of the fish would be startled and, in trying to avoid the cloud, would make themselves more vulnerable to attack. As dolphins other than the bubble-blower often benefited in the short term from such behaviour, it could be construed as a form of reciprocal altruism.[27]

In captive short-beaked common dolphins from New Zealand, daily food intake fell from about 12% of body weight for a 60 kg animal (= 7.2 kg) to about 6% for a 100 kg animal (= 6 kg).[28]

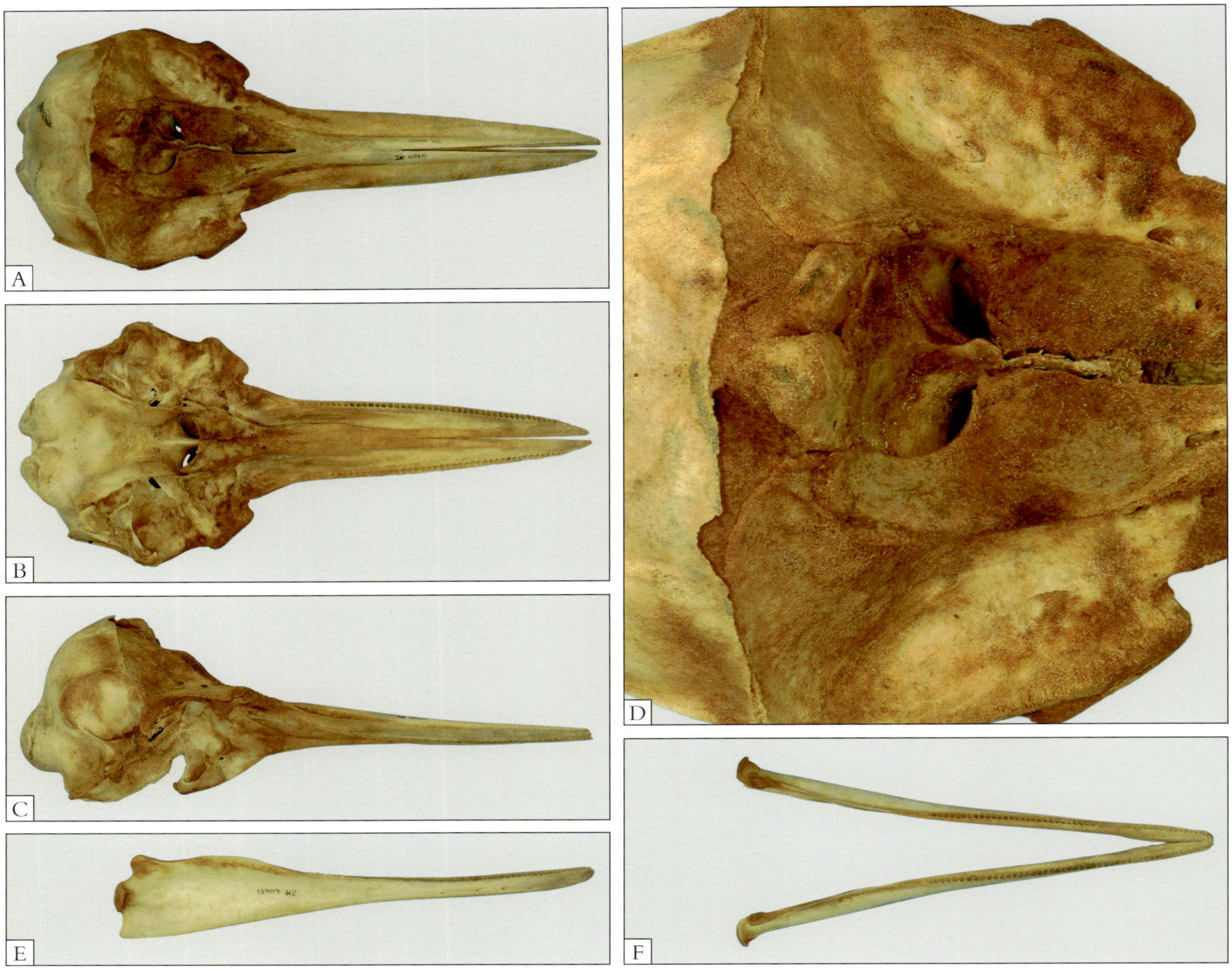

FIG. 134 Dorsal (A), ventral (B) and right lateral (C) views and vertex of skull (D); right lateral (labial) (E) and dorsal (F) views of mandible (ZM 40451).

Reproduction and growth

There is almost no local information. In the eastern tropical Pacific, the average size at birth is estimated as 85.2 cm, and gestation is thought to last 11.3 months. Births appear to be largely non-seasonal in the equatorial, central stock,[29] but vary with latitude to the north and south,[30] so that north of 30° N there is a definite peak in births in May/June.[31] Marked reproductive seasonality is apparent in North Atlantic males, where gonadal changes suggest that the mating period lies between May and September.[32] Lactation is estimated to last just over 16 months on average, with a mean calving interval of 2.15–2.5 years, but the proportion of pregnant animals falls with age while that of lactating animals increases, implying either that younger females wean their calves earlier or do not successfully rear their calves. Despite the fall in fecundity with age, no genuine post-reproductive females were encountered.[29] Sexual maturity is reached at an average age of 10.5–11.9 years in males[31, 32] and 8.1 years in females.[29] The equivalent lengths are about 179–182 cm in males and 170.7 cm in females,[31] although males and females from some stocks are as big as 200 cm and 185.9 cm respectively at maturity.[29, 32] The moderate difference in size between the sexes, together with relatively large testes in the male, suggests that short-beaked common dolphins may have a promiscuous mating system, in which females mate with multiple males and those males that produce larger amounts of sperm may be more successful ('sperm competition').[32] Both sexes can live for at least 26–27 years.[31]

Behaviour and vocalisations

Seven schools of common dolphins seen offshore in the south-west Indian Ocean consisted of 11–200 animals with an average of 72. Two of these schools were positively identified as short-beaked common dolphins.[7] Schools of short-beaked common dolphins in New Zealand waters ranged in size from three to about 300 animals with a mean of 55 (n = 72), with a tendency for groups to be larger in late summer/early autumn than in late spring/early summer. There were between 0 and 15 (mean 1.9) calves per school, of which 1.6 on average were newborns.[22] In the Mediterranean, school size differed with water depth, groups being larger (mean 118, n = 188) in water less than 400 m than in deeper water (mean 40, n = 119). In the same area, 18.8% of sightings (n = 313) included other species such as striped dolphins (16.9%), pilot whales (1.6%) and Risso's dolphins (0.3%).[19]

In New Zealand waters, short-beaked common dolphins spent 54.8% of their daytime travelling, 20.5% milling, 17% feeding, 7.3% socialising and 0.4% resting. Larger groups seemed to spend more time travelling. Fusion or fission of groups was frequently followed or preceded respectively by a change to or from sexual or feeding behaviour.[22]

In the Mediterranean Sea, cooperation between short-beaked common dolphins and a local fishery for skippers (*Scomberesox saurus*) is said to occur, in which the fishermen set nets on fish schools aggregated at the surface by dolphins, and offer fish rewards to the dolphins in return.[33] In Spencer Gulf, Australia, short-beaked common dolphins follow the cod ends of prawn trawls as they are brought to the surface, and both pick protruding fish from the net and feed on discards thrown overboard.[34]

Whistles produced by short-beaked common dolphins have a frequency range of 7–14 kHz with an average duration of 0.8 sec. Starting and ending frequencies average 9.8 kHz and 11.4 kHz respectively, and each whistle has on average1.2 inflection points and one step.[35]

Conservation status

There are no estimates of population size for short-beaked common dolphins in the subregion, and no indications of any substantial human-induced mortality. There is the potential, however, for their involvement in a midwater trawl fishery being conducted off South Africa and especially Namibia.

Short-beaked common dolphins are not specifically recognised by the IUCN: common dolphins per se are listed as Lower Risk: Least Concern. Short-beaked common dolphins are listed as Least Concern in the South African Red Data Book, and will be included under Cetacea spp. on Appendix II of CITES. The Mediterranean population of *D. delphis* is included on Appendix I and the North Sea and Baltic, Black Sea and eastern tropical Pacific populations on Appendix II of CMS.

References

1 Heyning and Perrin 1994
2 Jefferson and Van Waerebeek 2002
3 Rosel *et al.* 1994
4 Kingston and Rosel 2004
5 Murphy *et al.* 2006
6 Bell *et al.* 2002
7 Samaai *et al.* 2005
8 Perrin 2002
9 Fertl *et al.* 2003
10 Neumann *et al.* 2002b
11 Perrin *et al.* 1995
12 Neumann *et al.* 2002a
13 MRI Whale Unit data
14 André *et al.* 1990
15 Gihr and Pilleri 1969
16 Au and Perryman 1985
17 Bearzi 2005
18 Van Waerebeek 1997
19 Cañadas *et al.* 2002
20 Reilly 1990
21 Neumann 2001a
22 Neumann 2001b
23 Rohr *et al.* 2002
24 Bearzi *et al.* 2003
25 Ohizumi *et al.* 1998
26 Evans 1994
27 Neumann and Orams 2003
28 Kastelein *et al.* 2000b
29 Danil 2004
30 Perryman and Lynn 1993
31 Ferrero and Walker 1995
32 Murphy *et al.* 2005
33 Mussi *et al.* 2002
34 Svane 2005
35 Oswald *et al.* 2003

Long-beaked Common Dolphin

Delphinus capensis

THE TAXONOMIC DEBATE about the existence of two species of common dolphin has been touched on in the species account of the short-beaked common dolphin. But there is also uncertainty about the number of forms of long-beaked common dolphin that may exist. A form with an exceptionally long beak that occurs adjacent to the subregion has been described as a separate species (*D. tropicalis*),[1] but is now considered a subspecies of the long-beaked common dolphin (see box on page 222)[2]. Unlike short-beaked common dolphins, long-beaked common dolphins worldwide are highly differentiated genetically, and may have evolved independently in several different coastal locations from adjacent oceanic common dolphin populations. South Africa is actually the type locality for *Delphinus capensis*, and long-beaked common dolphins from there were some of the most genetically differentiated of those that were examined.[3]

Although most of the published (and unpublished) work to date on common dolphins in the subregion has referred to the species as *D. delphis*, it has been assumed here that unless specifically mentioned, these results refer to long-beaked dolphins, as these are the ones that are believed to be distributed closer inshore off South Africa and so more accessible to researchers.[4]

Appearance

Common dolphins (both short-beaked and long-beaked) are relatively easy to distinguish from other small dolphins at sea from the distinctive criss-cross (or 'hourglass') marking on the side, involving a buff or golden-yellow patch on the thorax and a light grey patch on the flanks. Over the continental shelf where they normally occur, long-beaked common dolphins are unlikely to be confused with any other small delphinid in the subregion (although if their distribution does overlap, some confusion with short-beaked common dolphins is possible).

Telling long-beaked from short-beaked common dolphins at sea is not so easy and requires experience. Characteristics that might help distinguish them at close range or as specimens are included below.

Length:weight relationship

There are enough records from the subregion for a separate analysis of stranded and captured animals. For 37 captured animals (that were presumably healthy), $W = 11.145L^{2.8616}$, where W = weight in kilograms and L = total length in metres. This implies that a 1.5 m dolphin would weigh 35.6 kg and a 2.2 m animal 106.4 kg. Equivalent weights for stranded animals (n = 62) are 8–10% lighter, or 32.3 kg and 97.7 kg respectively, presumably reflecting malnutrition or post-mortem dehydration.[5,7,8]

Skeletal data

Vertebral formula from the subregion (n = 57): 7 cervical, 13–16 thoracic, 18–23 lumbar and 30–35 caudal; total 72–77.[7] The total counts are very similar to those for short-beaked common dolphins from the same area, but lower than those given for short-beaked common dolphins in the eastern North Pacific.[6]

The atlas and axis are fused.[6]

Phalangeal formula, including metacarpals (n = 9–10): I_{2-4}, II_{9-11}, III_{7-8}, IV_{2-4}, V_{2}.[6]

The following features of the skull have been used to distinguish long-beaked from short-beaked common dolphins in the eastern North Pacific, but may not always apply in the subregion (for details see the species account for *D. delphis*).

Skulls of adult long-beaked common dolphins have a condylobasal length equal to or exceeding 44.5 cm, with a rostrum exceeding 28 cm. The length of the rostrum as a ratio of the zygomatic width is 1.5 or more. The distances between the tip of the rostrum and the internal and external nares exceed 32 cm and 32.5 cm respectively, while the length of the upper toothrow is 23.2 cm or greater.[6]

FIG. 135 Differences in coloration between long-beaked and short-beaked common dolphins are subtle and may not be consistent between ocean basins. This long-beaked common dolphin was photographed off Plettenberg Bay, South Africa.

Long-beaked common dolphins in the subregion reach maximum lengths of 2.54 m in males and 2.22 m in females, and there is evidence that adult males may be some 20–30 cm longer than adult females.[5] Both sexes are apparently substantially larger than their short-beaked common dolphin equivalents, but in conformation are less heavily built, especially in the front half of the body.[6] Adult dolphins have beaks averaging 7.5% of body length (range 6.4–8.4%, n = 20),[4] with a flatter melon than the short-beaked common dolphin in profile so that there is a less acute angle where it meets the beak.[6] The flippers are relatively smaller, averaging 14.4% of body length (range 13.5–15.7%, n = 20), and the flukes narrower, their width averaging 20.7% of body length (range 17.9–24.2%).[4] Adult males develop a distinct anal hump.[7]

The following features of the colour pattern have been used to distinguish long-beaked from short-beaked common dolphins in the eastern North Pacific, but may not always apply in the subregion (for details see the *D. delphis* species account): 1) the weak contrast between the yellow-brown colour on the thorax and the dark colour of the back; 2) the presence of a stripe from the flipper towards the anus; 3) the position of fusion of the stripe from the flipper to the chin with the lip patch; and 4) the presence of an extension of the white colour on the belly above the flipper stripe.[6]

There are 46–58 (median 52, n = 33) pairs of teeth in the upper jaw and 47–56 (median 50.5, n = 28) pairs in the lower jaw of long-beaked dolphins from the subregion.[4] The teeth are slender and conical, with a maximum diameter at the gum of 2–4 mm (mean 3.2, n = 38) in the adult.[7]

Distribution and migrations

Common dolphins per se occur worldwide in warm-temperate and tropical waters, although long-beaked common dolphins seem to be distributed as a series of disjunct populations in near-shore waters off California (USA), Mexico, Peru, Venezuela south to Argentina, West Africa, South Africa, Korea, southern Japan and Taiwan.[6]

Within the subregion, long-beaked dolphins are known from Gabon, Congo-Brazzaville, Angola,[9] and South Africa.[10] Their reported occurrence in Madagascar is unsubstantiated[2] and the identity of the dolphins seen on the Mozambique and Kenyan coasts is unknown. Similarly, common dolphins occur off Namibia but it is not known which species.[10]

Parasite and stomach content data suggest that long-beaked common dolphins off South Africa have a more inshore habitat than short-beaked common dolphins, and the allopatric nature of the distribution of these two species may account for the fact that they do not seem so well differentiated in this region as in the North Pacific.[4] If all the inshore sightings off South Africa are

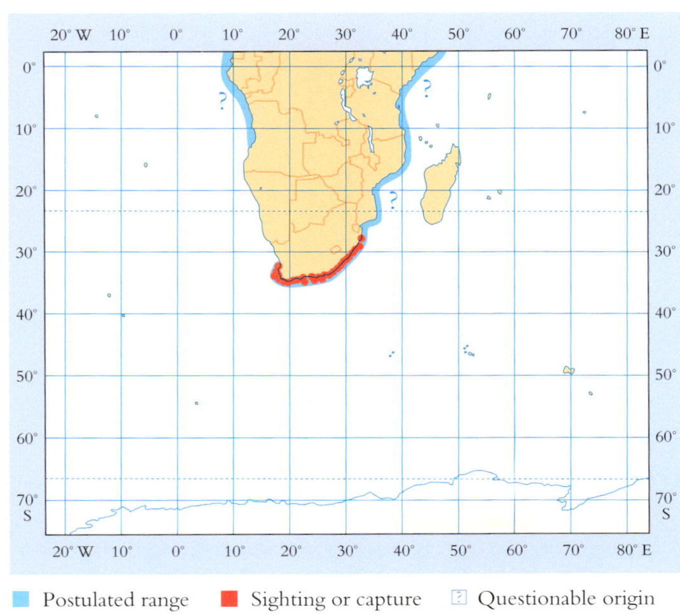

FIG. 136 Distribution map for long-beaked common dolphin.

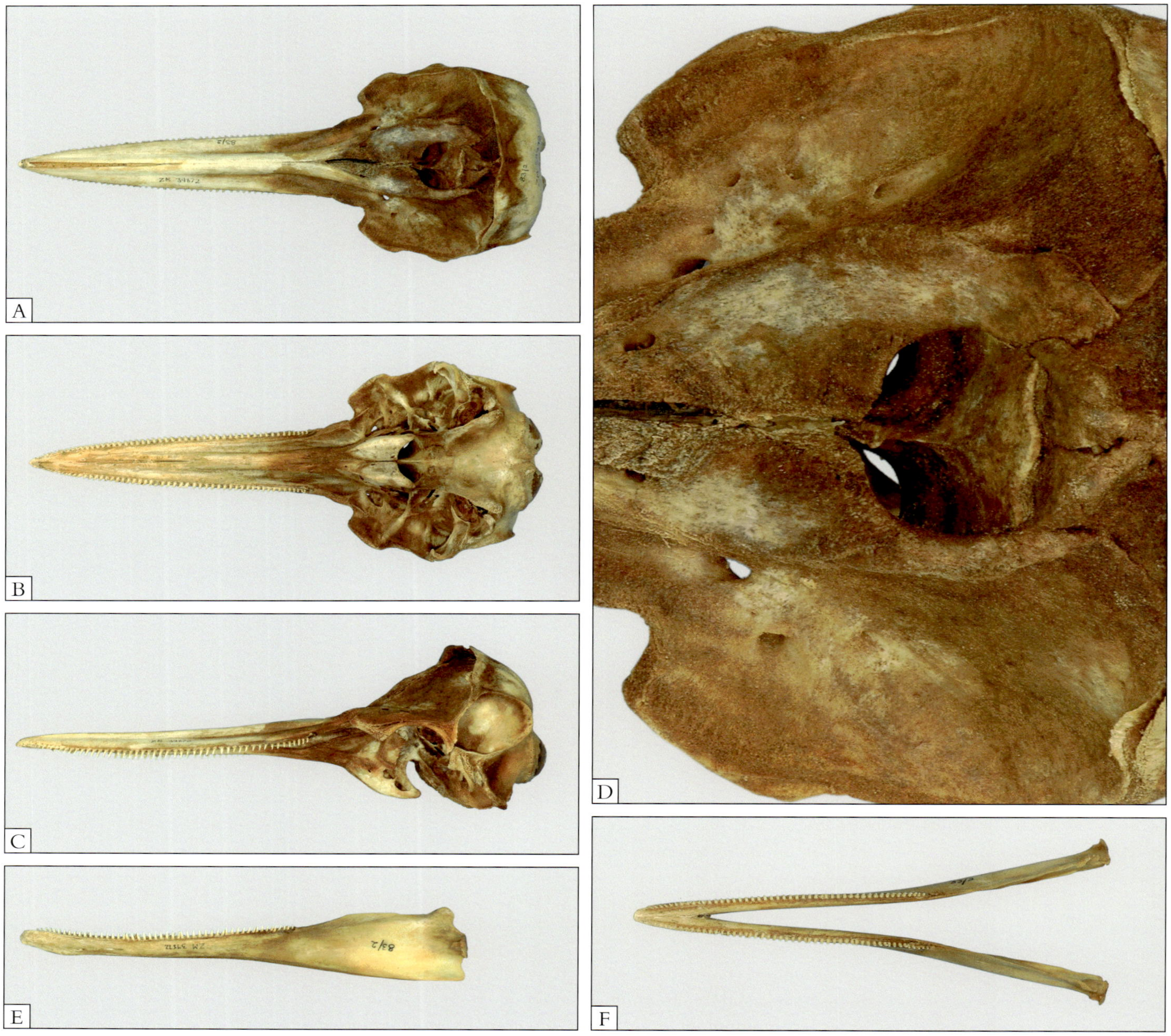

FIG. 137 Dorsal (A), ventral (B) and left lateral (C) views and vertex of skull (D); left lateral (labial) (E) and dorsal (F) views of mandible (ZM 39872).

attributed to the long-beaked common dolphin, then it is distributed principally along the south coast between St Helena Bay in the west and Richards Bay in the east, and in water less than 500 m deep.[10]

At the eastern end of their range in South African waters, common dolphins may migrate in winter from the Eastern Cape into KwaZulu-Natal waters, following shoals of pilchards,[11] although sightings have been made in KwaZulu-Natal waters in both summer and winter.[10] Similar excursions up the west coast of South Africa may also be made sporadically.[7]

Maximum swimming speeds of long-beaked common dolphins off California in reaction to a low-flying aircraft were measured at 20.5–32 km/hr. These measurements were only obtained over periods of one to two sec., and from non-porpoising animals.[12]

External commensals and parasites

In 33 long-beaked common dolphins captured at sea in the subregion, one carried cyamids, four had stalked barnacles (*Xenobalanus* sp.), and one had both cyamids and *Xenobalanus*. Cyamids from a stranded specimen were identified as *Syncyamus aequus*. Of the 28 cyamids found on four dolphins (both stranded and captured), all were located in the head region: nine in the region of the blowhole, five around the eyes, five on the throat, two on the upper lips, one on the lower jaw, and six on the side of the head. Numbers per dolphin ranged from one to 21 with all but one animal having three or fewer. The *Xenobalanus* were either located on the tip of a flipper (one animal) or on the trailing edge of the flukes (six animals). The numbers involved ranged from one to five, with a mean of 2.1 per dolphin.[7]

FIG. 138 Common dolphins are compulsive bow-riders of even some of the largest vessels.

Food and feeding behaviour

A comparison between the food consumed by short-beaked and other (presumably long-beaked) common dolphins in the subregion suggested that long-beaked common dolphins may feed more on surface shoaling fish.[4] Common dolphins caught in winter in the nets set along beaches in KwaZulu-Natal to protect bathers from sharks are presumably long-beaked common dolphins and consumed 34 different fish and four different cephalopod species. Their prey was dominated by pilchard (*Sardinops ocellatus*, 48.1% by mass, n = 297), mackerel (*Scomber japonicus*, 13.5%), elf (*Pomatomus saltatrix*, 8.7%), squid (*Loligo* spp., 8.4%), and strepie (*Sarpa salpa*, 8.1%), although there were significant inter-annual variations in the importance of these (and other) species. Most items were between 50 g and 120 g in weight.[13] In contrast, the stomachs of common dolphins stranded in the Eastern Cape (and assumed to be mostly if not completely long-beaked, judging by the lack of short-beaked specimens in Eastern Cape museums), were dominated by cephalopods. Although there were temporal variations in the diet, overall squid (*Loligo* spp.) comprised 40.3% by weight, with the pilchard forming 21.3% and the redeye roundherring (*Etrumeus whiteheadi*) 17.5% (n = 28). The difference in diet from KwaZulu-Natal was taken to be a reflection of a seasonal change in feeding habits driven largely by prey availability.[14] Feeding habits in the Western Cape were similar to those off the Eastern Cape, with *Loligo v. reynaudi* predominating (42.7% by modified volume, n = 42), followed by pelagic fish, of which anchovy (*Engraulis japonicus*) was the most important (17%), and hake (*Merluccius* sp.) comprising 9.8%. Most of the major prey items were between 11 and 19 cm long.[15]

Calves and adolescents in both the Eastern Cape and KwaZulu-Natal consumed fewer prey species than older individuals, while lactating females in both areas ate a higher proportion of cephalopods than other classes.[13, 14]

Feeding associations between long-beaked common dolphins and a variety of other vertebrate predators, including seabirds, sharks, fur seals and Bryde's whales, are a frequent phenomenon in South Africa's inshore waters,[16] particularly during the annual 'sardine run' up the east coast in winter. Interspecific interactions have not been systematically studied, but the common dolphins have been visualised as herding and concentrating the prey into 'balls' and forcing it to the surface, making it more available to the birds and sharks.[17] The reported release of bubble clouds (similar to that recorded for short-beaked dolphins[18]) may play some part in this process.[19]

Reproduction and growth

The lengths of 13 neonates from the subregion varied from 0.86 m to 1.12 m, with an average of 0.98 m.[5, 7] Births may occur throughout the year,[5] but there is a strong peak in summer, with 11 of 16 neonates being recorded in February/March.[5, 7] Foetal length data suggest that most conceptions occur in March/April, indicating a gestation period of about 11 months.[8] Suckling continues at least until the calf is 1.75 m long,[13] although solid food may be taken earlier.[5] The duration of lactation is not known but is at least six months and may persist until after the female has fallen pregnant again.[8] A milk sample from an unknown stage in lactation contained 29.2% fat, 11.3% protein and 1.41% carbohydrate.[5] Testes weights increase greatly after males have attained 2.2 m in length when they are about 10–12 years old. Females first ovulate at an average length of 213 cm and an age of 8–9 years. Although the ovulation rate in mature females was calculated as 0.53 a year, 36.5% (n = 93) of mature females were pregnant, suggesting an average calving interval of just under three years, but there was a marked decline in the proportion of pregnant individuals in females of greater reproductive age. Females may live for up to 26 years.[8]

Indo-Pacific Common Dolphin

Delphinus capensis tropicalis

Its relegation to a box does not mean that its validity is in doubt, only that there is no certainty that this subspecies occurs in the subregion. To date it has been recorded from southern China and western Indonesia in the east, around the rim of the northern Indian Ocean, to Somalia and the Red Sea in the west. Throughout this range it seems to be confined to waters of the continental shelf and slope. Although there are sightings of common dolphins along the coast of Kenya and Mozambique, it is not known to which subspecies they belong (and there may even be interbreeding or an intergradation between Indo-Pacific and long-beaked common dolphins over this range). The only documented difference externally between Indo-Pacific and long-beaked common dolphins is the longer beak of the former (9.4–9.7% cf. 6.9–7.6% of body length), and there appear to be no major colour pattern differences. Internally, the skull differs in being narrower (rostral length/zygomatic width ratio 1.6–2.06 cf. 1.46–1.77) with a higher tooth count (upper counts 54–67 cf. 47–60). Very little is known about the biology of this subspecies. It is found in schools of variable size (2–1 700), there is little obvious seasonality of reproduction, and food consists of squid and fish.[1]

Reference

1 Jefferson and Van Waerebeek 2002

Behaviour and vocalisations

Schools of common dolphins seen inshore on the South African coast (and therefore probably long-beaked common dolphins) ranged in size from one to 10 000, with an average of 619 (n = 119), but if information from aerial surveys (in which smaller schools may have been missed) is excluded, the average school size falls to 302 animals (n = 65).[7, 11] Mixed schools with other delphinid species are rarely seen. The composition of schools has not been specifically investigated. A mass stranding of 15 animals consisted of 13 mature females (11 of which were pregnant) and two males,[14] and in shark-net catches off KwaZulu-Natal there were significantly more mature females than males, with the majority of calves taken being weaned.[20]

Long-beaked common dolphins in the subregion seem to be compulsive bow-riders.[7] Evidence from the eastern North Pacific suggests that this may involve only certain individuals from each school – in the case of females, mainly young, sexually immature animals.[21]

Common dolphins in general are highly vocal, producing clicks, whistles, squeaks and creaks. While travelling, large schools may produce almost synchronous whistles, timed for each group that is porpoising. In the eastern North Pacific, whistles and screams predominated in daylight, pulse trains at night. The clicks are of very short duration (20–50 μsec) with most energy between 15 kHz and 100 kHz.[21]

Whistles produced by long-beaked common dolphins have a frequency range of 8–16 kHz with an average duration of 0.7 sec. Starting and ending frequencies average 10.1 kHz and 14.1 kHz respectively, and each whistle has on average 1.3 inflection points and 1.5 steps.[22]

Conservation status

The population of common dolphins inshore on the south-east coast of South Africa (Port Elizabeth to Richards Bay) was estimated as 15 000–20 000 animals in 1988/89.[11] These are presumably all long-beaked common dolphins. As the estimate only covered part of the range of this population, and included a maximum count on any one survey of 15 800, it is probably a lower bound to the population size. No other estimates exist for this species in the subregion.

Between 1980 and 2005, a total of 1 023 common dolphins died as a result of entanglement in shark nets off Durban, South Africa (see the account of the Indo-Pacific bottlenose dolphin), or an annual average of 39. Most if not all of these are believed to have been long-beaked common dolphins. In addition, some common dolphins die annually as a by-catch in the South African purse-seine fishery for pelagic fish, mostly in the Western Cape. Unsubstantiated estimates suggest the total dolphin catch might be about 100 animals, of which the majority would probably be long-beaked common dolphins.[23] Although all these mortalities are unfortunate, given the size of the population it is likely that they are sustainable.

Long-beaked common dolphins are not specifically recognised by the IUCN: common dolphins per se are listed as Lower Risk: Least Concern. Long-beaked common dolphins are listed as Least Concern in the South African Red Data Book, and will automatically be included under Cetacea spp. on Appendix II of CITES. The species is not listed by CMS.

References

1 Van Bree and Gallagher 1978
2 Jefferson and Van Waerebeek 2002
3 Natoli *et al.* 2005
4 Samaai *et al.* 2005
5 Ross 1984
6 Heyning and Perrin 1994
7 MRI Whale Unit data
8 Mendolia 1989
9 Van Waerebeek 1997
10 Findlay *et al.* 1992
11 Cockcroft and Peddemors 1990
12 Rohr *et al.* 2002
13 Young and Cockcroft 1994
14 Young and Cockcroft 1995
15 Sekiguchi *et al.* 1992b
16 Best *et al.* 1984a
17 Peschak 2005
18 Neumann and Orams 2003
19 Maxwell 2002
20 Cockcroft 1990
21 Evans 1994
22 Oswald *et al.* 2003
23 Best and Ross 1977

Fraser's Dolphin

Lagenodelphis hosei

♂

♀

Juvenile

THE STORY OF how this dolphin was discovered and then disappeared from view for 76 years is a fascinating one (see box on page 225), and illustrates how easy it is for even a relatively large mammal to go undetected in the open ocean.

Although the subregion is one in which Fraser's dolphin was 'rediscovered',[1] comparatively little has been published here on the species subsequently, so this account has had to be supplemented by information from other parts of the world.

Appearance

Fraser's dolphins are most likely to be confused with striped dolphins at sea (at least, that was our experience when we first saw them). However, the extreme variability in the presence and thickness of the lateral stripe (compared to its universal presence and more or less uniform thickness in striped dolphins) should be sufficient to make the distinction. The dorsal fin is also smaller, and assumes a pointed, triangular shape in adult males.

In the north-west Pacific males can achieve a maximum length of 2.68 m and females 2.58 m, with equivalent asymptotic lengths of 2.3–2.6 m and 2.2–2.5 m respectively.[2] The largest male (2.64 m) and female (2.36 m) from the subregion are within these ranges.[3] It has been described as a small stocky dolphin, with a short well-defined beak, but the latter is so small (1.1–2.9% of body length, mean 1.7%, n = 31) that it is barely distinguishable at sea. The flippers are curved along both leading and trailing edges, and their length is 9.7–13.8% (mean 10.6%, n = 31) of body length. The dorsal fin is small, with a height of 4.3–9.2% (mean 6.7%, n = 31) of body length,[3–6] and although it is recurved in most individuals, in mature males the point becomes erect and makes the whole fin almost triangular in shape.[5,7] The span of the flukes is 19.7–24.2% of body length, with a mean of 21.8% (n = 20),[3,4,6] and there may also be age- or sex-related changes in their shape. Most mature males also develop a moderate to large post-anal hump[7] as well as a deepening of the caudal peduncle.[5]

Length:weight relationship

Combining data from five animals from the subregion with four from other areas,[1,4,8,10,11] $W = 13.642L^{2.7123}$, where W = weight in kilograms and L = total length in metres. This implies that a 1.5 m dolphin would weigh 41 kg and a 2.2 m animal 115.8 kg.

Skeletal data

Vertebral formula (n = 8–12): 7 cervical, 15–16 thoracic, 20–24 lumbar and 34–39 caudal; total 78–81.[12]

The atlas and axis are fused, and the third and/or fourth cervicals may also be partly fused to the axis.[6,9,11]

Phalangeal formula, including metacarpals (n = 6): I_1, II_9, III_{6-7}, IV_{3-4}, V_2.[3,11]

The characteristics of the skull are a broad rostrum, deep longitudinal grooves in the palate (almost as deep as those in the common dolphin) and small temporal fossae.[12]

There are size and probably sex-related differences in the colour pattern of Fraser's dolphins. The most obvious feature is a dark grey lateral stripe from the region of the eye to the anus. In the calf, the stripe is narrow and starts above the eye, but fades out before reaching the genital slit, while in adults the stripe is broader, darker, and extends the whole way from the side of the head to the anus. There is some evidence that the stripe is more intense in adult males than in adult females. The back above the stripe, as well as the dorsal fin, flippers and flukes, are dark greyish-blue. The belly is white tinged with pink. There is a dark eye patch with a stripe running forward to the apex of the melon, where it joins a stripe from the melon to the blowhole. Apart from calves, all Fraser's dolphins have a light grey blaze on either side of the beak, leaving a thin strip of dark pigment in the mid-line that connects with a wedge-shaped patch at the tip of the beak and a narrow line around the edge of the gape. A dark stripe from the flipper meets the dark lower lips just in front of the angle of the gape.[1,7]

There are 35–42 (mean 37.9, n = 12) pairs of teeth in the upper jaw and 36–42 (mean 38, n = 12) pairs in the lower jaw of Fraser's dolphins from the subregion.[3,8] The teeth are slender, conical, and slightly oval in cross-section, with their maximum greater diameter (at right angles to the longitudinal axis of the jaw) 3.9–4.6 mm and their maximum lesser diameter 2.4–3.7 mm.[9]

Distribution and migrations

Fraser's dolphin occurs worldwide in tropical waters, and strandings in temperate waters (such as the North Atlantic,[4] even as far north as Scotland[13]) may involve extralimital forays in response to unusual oceanographic events.[12] In the subregion it has been recorded from Gabon,[14] South Africa[8] and Réunion.[15] The records shown for southern Madagascar[12] cannot be substantiated. Sightings indicate that in the south-western Indian Ocean, at least, distribution ranges from the equator to 35° S.

The species is normally considered to be an open ocean species, and all sightings in the subregion have been made in water more than 1 000 m deep.[16] Close to oceanic islands, however, it has been encountered in water as shallow as 150–200 m.[17] In the eastern tropical Pacific it has been characterised as occurring in equatorial-southern subtropical surface-water or other areas where variable oceanographic conditions such as upwelling exist.[12]

Seasonal movements have not been described.

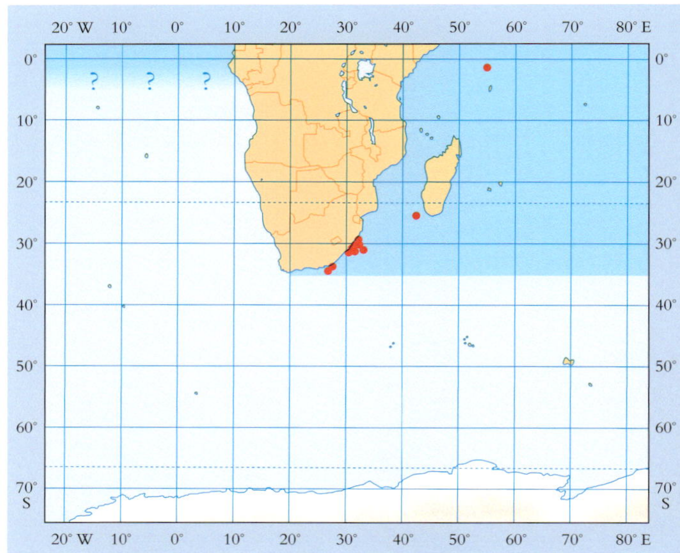

Postulated range ■ Sighting or capture [?] Questionable origin

FIG. 139 Distribution map for Fraser's dolphin.

External commensals and parasites

None were seen on any of the five animals examined from the subregion, and none seem to have been recorded elsewhere.

Food and feeding behaviour

Identified remains from the stomachs of two animals from the subregion included the following: 16 viper fish (*Chauliodus sloani*), 11 light fish (10 *Photichthys argenteus*, one *Ichthyococcus*), six bigscale fish (*Melamphaes* sp.), three frost fish (probably the cutlass fish, *Trichiurus lepturus*), one loose jaw (*Malacosteus niger*), and seven possible scaly dragonfishes (*Stomiidae*), nine cephalopods (six *Chiroteuthis capensis*, two *Histioteuthis* sp. and one *Taningia danae*) and two mysid prawns.[3] These prey items are mainly midwater or deepwater organisms, some of which may migrate up to or closer to the surface at night. Similar prey has been recorded for Fraser's dolphins from the eastern tropical Pacific,[18] the western North Pacific,[11] the Sulu Sea,[19] and the North Atlantic,[4] although the relative importance of fish and squid differed between areas.

From the depth distribution of several of the prey items it has been concluded that Fraser's dolphins must feed at depths of 250–500 m[18] or above 200 m as well as down to 600 m.[19] However, apparent surface feeding during the day has also been noted.[17]

Quantities of food in the stomachs of individuals caught incidentally in a tuna drift-net fishery in the Philippines suggested that feeding occurred in the afternoon and at night.[19]

Cooperative feeding on fish (*Elagatis bipinnulatus*) by Fraser's dolphins at the surface involved chasing in line abreast,

Lost – and found

FIG. 140 Francis Charles Fraser, 1903–1978.

It is not often that a cetacean story makes it to the cover of the prestigious scientific journal *Nature*, but in 1973 it announced the 'rediscovery' of a dolphin species.

In 1895 the Natural History Museum in London had purchased a batch of cetacean specimens from Charles Hose in Borneo. During a reorganisation of the museum's collections 60 years later, one of Hose's skeletons from Sarawak caught the attention of Francis Fraser, Keeper of Zoology, who noticed its unique characteristics. Possessing some of the features of two different dolphin genera (*Lagenorhynchus* and *Delphinus*), the skeleton was finally described by Fraser in 1956 as a new genus *Lagenodelphis* and species *hosei* (after the collector).[1] For another 15 years it remained as the sole documented representative of its species. The fact that the external appearance of this dolphin was virtually unknown must have helped it to keep its anonymity. Some sceptics even doubted that the species was genuine, opining that it might have been a hybrid. Then, in one of those weird twists of fate, in 1971 over the space of 34 days examples of an unusual-looking dolphin turned up in three widely separated places in the world – as a by-catch in the tuna fishery in the eastern tropical Pacific, as specimens collected off the eastern seaboard of South Africa, and as a stranding in New South Wales, Australia. Perhaps not so surprisingly, given his reputation as the leading guru in cetacean taxonomy at the time, photographs of the dolphins from all three localities ended up independently on Fraser's desk in London. Generously acting as a link between all three groups of researchers, and providing a cast of the characteristic rostrum of the type skull for comparison with the new specimens, Fraser must have felt thoroughly vindicated when, in the year of his retirement, the existence of the mysterious 'Sarawak dolphin' was confirmed and its external appearance described for the first time. It was only appropriate that the authors should give the animal the common name of 'Fraser's dolphin'.[2]

As a postscript, an entire specimen of Fraser's dolphin was shipped frozen from South Africa to the Natural History Museum in London in 1974, partly so that some of the sceptics could establish for themselves the validity of the species. When an enquiry was made 30 years later about what the stomach of the specimen might have contained, we were informed that the animal was still frozen!

References

1 Fraser 1956
2 Perrin *et al.* 1973

FIG. 141 Although this juvenile Fraser's dolphin has not yet developed a dark lateral stripe, other distinguishing features (the diminutive beak, flippers and dorsal fin) are clearly seen.

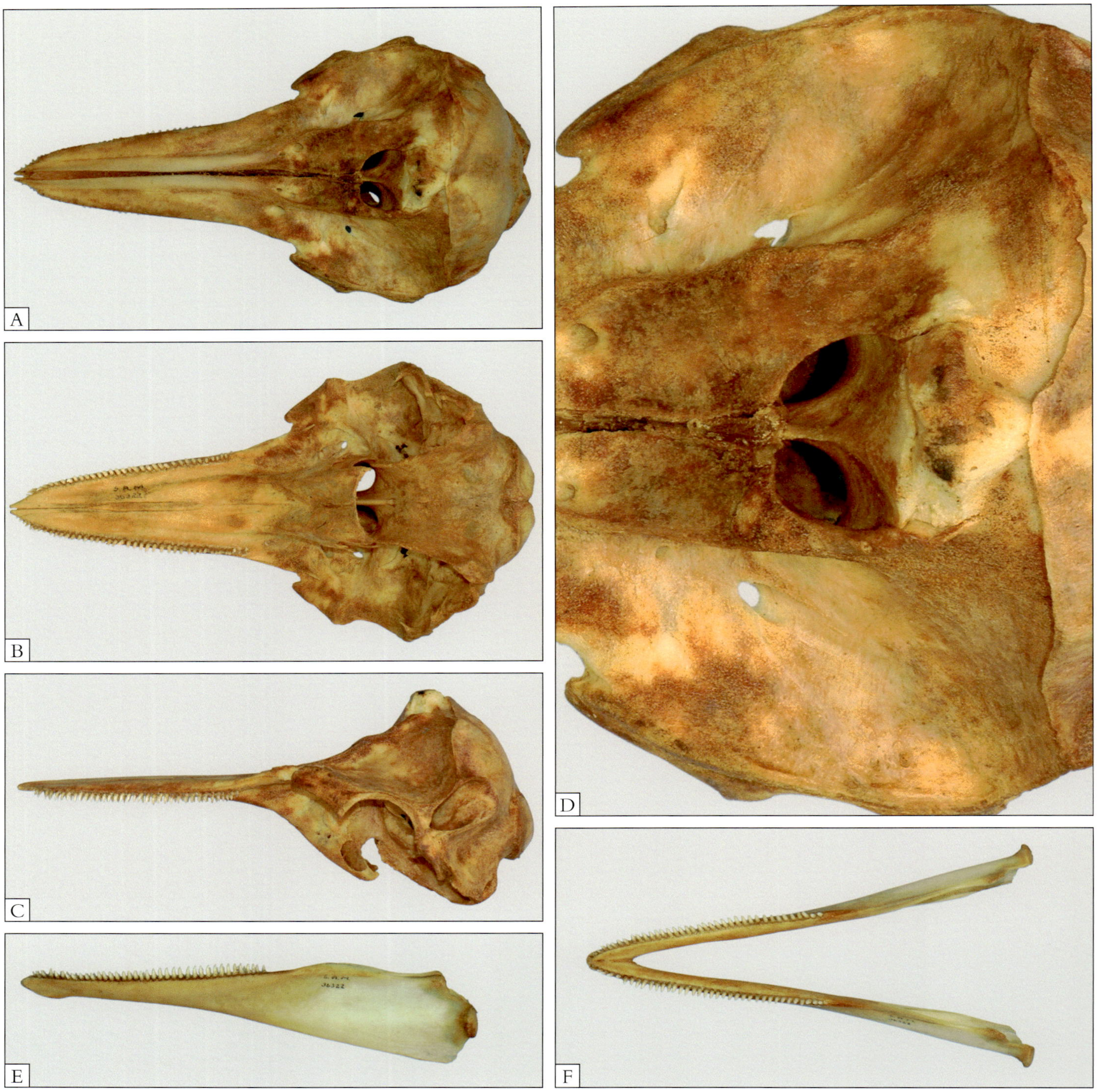

FIG. 142 Dorsal (A), ventral (B) and left lateral (C) views and vertex of skull (D); left lateral (labial) (E) and dorsal (F) views of mandible (ZM 36322).

surrounding the school and compacting it, and then attacking from above and below.[17]

Reproduction and growth

Very little information is available for the subregion, so the following is based on data from the North Pacific. Calves are born at a length of about 110 cm after an estimated gestation period of 12.4–12.6 months; births apparently occur in two periods, spring and autumn. The length of lactation is unknown. Sexual maturity is attained at a length of 2.2–2.4 m in males and 2.1–2.2 m in females, and at the ages of 7–10 years and 5–8 years respectively. Mature females ovulate at a rate of 0.49/year, and the mean calving interval was estimated to be 1.96–2.26 years. The oldest individuals aged in both sexes were estimated to be 17.5 years old. Although these conclusions are based on a relatively small sample (57 males and 51 females), they suggest that Fraser's dolphin may mature slightly earlier, have a shorter calving interval and a lower longevity than most other small delphinids.[2]

Behaviour and vocalisations

Fraser's dolphins in the subregion have been encountered in groups of 7–1 000 animals, with a mean of 167 (n = 16).[3] Mixed groups are not uncommon: in the Pacific they often occur with melon-headed whales,[12] and sightings in the subregion involved melon-headed whales and common bottlenose dolphins in one case, sperm whales in another and false killer whales in the third.[3] Other species with which it has been seen include spotted, spinner and striped dolphins.[12]

Although individuals in the subregion and the western Pacific will bow-ride vessels, Fraser's dolphins show ship avoidance in the eastern tropical Pacific. The little information available on school composition suggests that the species may have a social organisation similar to that of striped and pantropical spotted dolphins, and the existence of non-mating schools[2] and schools with and without calves has been noted.[17] School cohesion can be particularly tight.[12]

The few individuals kept in captivity have not fared well owing to their generally nervous disposition and 'fragility'.[20]

Fraser's dolphins produce both whistles and clicks similar to those of other small delphinids. During at-sea encounters the clicks were described as short, broadband sounds produced in long sequences at a wide variety of repetition rates. Whistles were frequency-modulated narrow-band tones with one or more harmonics. Starting frequencies averaged 11.9 kHz and ending frequencies 13.1 kHz, while the highest and lowest frequencies averaged 16.9 kHz and 9.4 kHz respectively. The whistles ranged in duration from less than 0.1 sec. to 2.15 sec. with an average of 0.77 sec. The intensity of vocalisation seemed to be associated with bursts of surface activity, while during travel only occasional whistles and very little clicking was heard.[17]

Conservation status

There are no estimates of population size in the subregion, but the impression is that Fraser's dolphin is relatively rare and perhaps localised in distribution. At least one individual was taken in shark nets off KwaZulu-Natal between 1981 and 2005, but there is no evidence of significant fisheries involvement in the subregion. Fraser's dolphins are listed as Data Deficient by the IUCN and in the South African Red Data Book. They are included on Appendix II of CITES and the south-east Asian population on Appendix II of CMS.

References

1 Perrin *et al.* 1973
2 Amano *et al.* 1996
3 MRI Whale Unit data
4 Van Bree *et al.* 1986
5 Azevedo *et al.* 2003
6 Miyazaki and Wada 1978b
7 Jefferson *et al.* 1997
8 Ross 1984
9 Fraser 1956
10 Mignucci-Giannoni *et al.* 1999a
11 Tobayama *et al.* 1973
12 Perrin *et al.* 1994b
13 Bones *et al.* 1998
14 Rosenbaum and Collins 2006
15 Boucaud and Rota 2006
16 Findlay *et al.* 1992
17 Watkins *et al.* 1994
18 Robison and Craddock 1983
19 Dolar *et al.* 2003
20 Hammond and Leatherwood 1984

Dusky Dolphin

Lagenorhynchus obscurus

THIS ANIMAL LOOKS very similar to the Pacific white-sided dolphin (*Lagenorhynchus obliquidens*) of the northern hemisphere, and the two are generally considered sister species.[1] Whether either should be in the genus *Lagenorhynchus*, however, is another matter, as this is often considered an artificial grouping. If this view prevails, the dusky dolphin might find itself in the genus *Sagmatias*, along with its sister species, Peale's dolphin and the hourglass dolphin.[2]

First described from the 'Cape of Good Hope', surprisingly little has been published subsequently on the dusky dolphin in its type locality. This account has made use of what unpublished information is available from the subregion, but has had to borrow data from other, better-studied populations where necessary.

Appearance

Dusky dolphins are unlikely to be confused with any other dolphins occurring in the subregion: they are partly sympatric with Heaviside's dolphins but are easily distinguished from them by their recurved (rather than triangular) dorsal fin. The diminutive beak and white flank patch also differentiate them from all other dolphins in the area except their congener, the hourglass dolphin. Although they are unlikely to be found in the same habitat, dusky dolphins differ from hourglass dolphins externally by their smaller dorsal fin and the light grey colour of the thorax above the flipper: in the hourglass dolphin the body immediately above the flipper is black.

The largest individuals among 45 male and 46 female dusky dolphins examined in the subregion were 1.90 m and 1.905 m long respectively, indicating little difference in size between the sexes. The head slopes gently to a small snout, which is well differentiated laterally from the head by shallow grooves but almost imperceptibly in the mid-line. Its length (not easily measured) was 0.6–3.5% (mean 2%, n = 49) of body length. The flippers are long, narrow and gently tapered, ranging in length (in non-neonates) from 14.9–21.3% (mean 18.4%, n = 60) of body length. The centrally-placed dorsal fin is erect and hooked, its height (in non-neonates) being 7.2–14.3% (mean 10.7%, n = 59) of body length.[3] In adult males the fin is more curved, has a broader base and a greater surface area than in adult females, but because of the extent of individual variation these differences in size and shape cannot be used as reliable field

Length:weight relationship

Using data from 63 animals from the subregion weighed whole,[3] **$W = 12.909L^{2.9918}$**, where W = weight in kilograms and L = total length in metres. This implies that a 1.2 m dolphin would weigh 22.3 kg and a 1.8 m animal 74.9 kg.

Skeletal data

Vertebral formula (n = 49): 7 cervical, 12–15 thoracic, 16–26 lumbar and 28–36 caudal; total 71–75.[3]

The atlas and axis are fused.[5]

Phalangeal formula, including metacarpals (n = 3): I_{2-3}, II_{9-10}, III_{6-7}, IV_{2-3}, V_{1-2}.[3]

The characteristics of the skull that distinguish *L. obscurus* from *L. cruciger* are: 1) a long, dome-shaped or lanceolate rostrum, as opposed to a relatively short, triangular rostrum; and 2) a somewhat elongated, rectangular (as opposed to square-shaped) neurocranium.[6] The zygomatic width of the skull averages about 45% of its total (condylobasal) length in the dusky dolphin, compared to about 55% in the hourglass dolphin.[7]

criteria of gender.[4] The width of flukes in non-neonates is 15–28.3% (mean 24.1%, n = 52) of body length.[3]

The basic colour pattern is black above, grey laterally and white below. The snout and chin are black, and a dark band may connect the flipper to a patch around the eye. A grey 'cape' runs from the side of the head, above the flipper-eye band, and along the side of the thorax as far as the mid-length of the animal. Above the cape the body is dark, and this extends ventrally to cover the whole of the caudal peduncle behind the anus. Within this dark area, and positioned mostly behind the dorsal fin, there is a characteristic white flank patch, variable in size, shape and position, but usually with a dorsal extension that runs forward as a thin streak below the dorsal fin and towards the head. The belly from throat to anus is white. Most individuals carry a grey blaze on the rear half of the dorsal fin, and a greyish tone to the upper surface of the flippers. This flipper tone and the chin are less pigmented in juveniles than adults, but otherwise the colour pattern is common to all ages and size classes. Colour variants ('piebalds'), however, have been recorded in the subregion and elsewhere.[4]

There are 26–32 (mean 28.6, n = 106) pairs of erupted teeth in the upper jaw and 24–31 (mean 27.2, n = 106) pairs in the lower jaw of dusky dolphins from the subregion. The teeth are slender and conical, with their maximum diameter at the gum being 2.3–5.0 mm (mean 3.5 mm, n = 22).[3]

Distribution and migrations

Dusky dolphins are found in coastal and continental shelf waters off parts of South America, the west coast of southern Africa, New Zealand and southern Australia, as well at several subantarctic islands.[8, 9] These populations are physically discontinuous, and the rate of transoceanic genetic interchange seems low.[10]

In the subregion, the species has been recorded from Angola, Namibia, South Africa,[11] Gough Island (South Atlantic), Prince Edward Islands (Indian Ocean) and Amsterdam Island and St Paul Island (Indian Ocean).[6] There are also reports of dusky dolphins between Kerguelen and the Crozet Islands in the Indian Ocean,[12] as well as north of the Crozets, but they have been considered not well substantiated.[6]

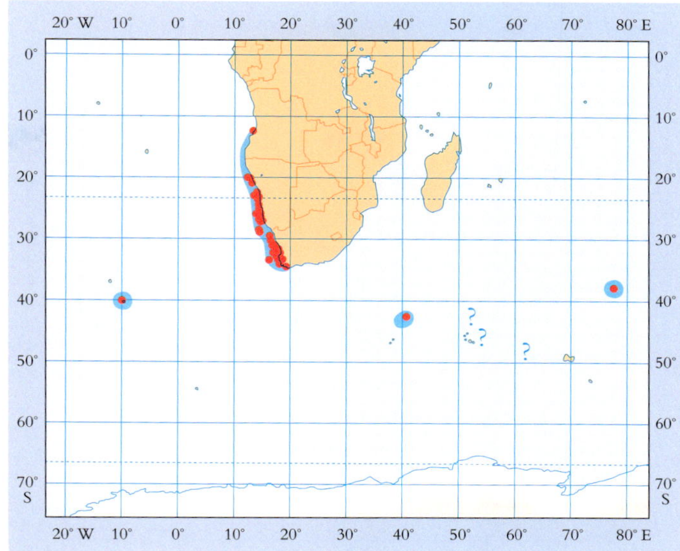

Postulated range · Sighting or capture · Questionable origin

FIG. 143 Distribution map for dusky dolphin.

In southern African waters, dusky dolphins are distributed from at least 12° S on the Angolan coast to the vicinity of Cape Town.[11] The most easterly sighting on the South African south coast has been at 19°21.36′ E.[3] An apparent hiatus in distribution between 27° S and 30° S (or around the Orange River mouth) has been noted,[11] but a comparison between Namibian and South African dolphins failed to reveal any significant genetic differentiation.[10] (For possible tropical sightings, see box on page 232.)

FIG. 144 Dusky dolphins possess an imposing repertoire of aerial manoeuvres, some of which are associated with feeding bouts, while others can be produced in several contexts, such as in this apparent response to the presence of our research boat off Cape Town.

On the southern African coast, sightings of dusky dolphins have been made mainly in water shallower than 500 m, but include encounters in water over 2 000 m deep.[11] Even in the vicinity of oceanic islands, the species may be restricted to water less than 2 000 m deep.[6]

In summer, dusky dolphins in New Zealand waters are usually found close to shore in the morning, move offshore in late afternoon and return inshore around sunrise. This movement is possibly related to feeding on organisms associated with the deep scattering layer.[13] Judging by the nature of the stomach contents of some dusky dolphins taken close to the shore in the morning, such an onshore-offshore movement may also occur in South African waters.[14]

Seasonal movements are unrecorded in the subregion, but off New Zealand dusky dolphins make regular 500 km round-trip migrations between localities on South Island, as well as longer excursions from the east to the west coast.[15] Long-range movements (of up to 780 km) have also been recorded for dusky dolphins off Argentina.[16]

The speed of movement of dusky dolphins off Argentina normally averaged 6.3 km/hr, but increased to about 15 km/hr around feeding bouts.[17]

External commensals and parasites

Cyamids were found on two of 50 dusky dolphins examined in the subregion: one animal carried 47 *Scutocyamus antipodensis* on the body surface and appendages and one in the region of the mouth, while a second had a single cyamid on its dorsal fin.[3]

Stalked barnacles (*Xenobalanus globicipitis*) occurred on the flukes, flippers and dorsal fin of dusky dolphins in Peruvian waters: prevalence overall was 38.9%, but was significantly higher in winter (June) and summer (December).[18]

Food and feeding behaviour

Food remains from the stomachs of 36 dusky dolphins from the subregion contained 42 600 identified prey items. Fish predominated, making up 99.4% by number: the principal species consumed were myctophids (47.5%, of which *Lampanyctus hectoris* formed 91.8%), hatchet fish (*Maurolicus muelleri*, 26.3%), maasbanker (*Trachurus trachurus*, 18.8%), anchovy (*Engraulis japonicus*, 2.6%) and shallow-water hake (*Merluccius capensis*, 1.6%). Of the few cephalopods consumed, arrow squid (*Todarodes sagittatus*) and ommastrephid squid (especially *Sthenoteuthis oualaniensis*) were the most frequent. The incidence of pelagic fish (maasbanker, anchovy and pilchard) seemed to be reciprocally related to that of mesopelagic fish (myctophids and hatchet fish). In 30 out of 32 stomachs, either pelagic fish or mesopelagic fish formed more than 90% of the two groups combined. This suggests that the dolphins were exploiting two alternative resources. The presence (and even dominance) of southern mullet (*Liza richardsonii*) in some stomachs also indicates occasional exploitation of extreme nearshore resources.[3]

Separate utilisation of surface shoaling fish and those associated with the deep scattering layer has also been described in New Zealand. In shallow waters, dusky dolphins tended to feed during the day, either in a coordinated fashion, herding schooling fish close to the surface,[19] or in small subgroups at low prey densities. In deeper waters dusky dolphins fed at night when the deep scattering layer migrated up to within 130 m of the surface.[20]

Coordinated surface feeding by dusky dolphins in Argentina began with long dives, interspersed with clean, noiseless leaps, as dolphins breathed rapidly and dived forcefully. This was followed by the appearance of tightly bunched fish at the surface, after which long dives decreased. Noisy leaps, in which the dolphin re-entered the water on its side or back, began before the fish surfaced but continued throughout feeding, especially at the periphery of the school, presumably helping to keep the fish tightly bunched. As the feeding bout continued, clean leaps and steep dives still took place, the dolphins resuming their attacks on the school from below, while other individuals moved rapidly through the school and appeared on the other side with fish in their mouths.[17] Many aspects of this behaviour pattern have been seen in South African nearshore waters.[3]

Annual food consumption by a captive female dusky dolphin at Sea World, Durban, varied between 1 870 kg (at estimated age three) and 3 100 kg (at estimated age 11), and thereafter stabilised at about 2 600 kg. This represented 9–12% of body weight a day.[21]

Severe interruptions in feeding rate, such as that caused by El Niño events, may be reflected by atypical layers in the dentine of dusky dolphin teeth.[22]

Reproduction and growth

Ten live or freshly dead neonates (each with a raw unhealed umbilicus) from the subregion ranged from 81.5 cm to 91 cm in length (mean 86.4 cm), and occurred between 15 January and 8 March, with a mean date of 2 February. A regression of the sizes of these neonates and six calves, 92.1–120.3 cm long, against date of death indicates a rate of growth of 0.27 cm/day,[3] very similar to the rate of foetal growth in dusky dolphins from Peru (0.261 cm/day), in which gestation is estimated to last 12.9 months. In the same region, lactation is calculated as 12 months, followed by a resting period of 3.7 months (possibly less), giving an average calving interval of 28.6 months, or perhaps closer to two years. Such frequent calving might be a response to the large take of dolphins in the area, or to the extreme unpredictability of the Peruvian upwelling system.[23] The smallest mature female from the subregion was 1.68 m long ($n = 14$) and the largest immature 1.76 m ($n = 20$), suggesting puberty was reached at an intermediate length. Using criteria (from Peru) of a combined testis weight of 610 g and testis length of 24 cm at maturity, the smallest mature male from the subregion was 1.66 m long ($n = 9$) and the largest immature 1.76 m ($n = 24$), suggesting that maturity was reached at a similar size in both sexes.[3] There were insufficient data to test for the seasonal increase in testis size seen in Peruvian dusky dolphins.[23]

Very little has been published on age and growth in dusky dolphins. Sexual maturity in both sexes may be reached at about seven years of age, and the oldest of 32 specimens from New Zealand was 35–36 years old.[8] In a sample of 18 females from off Patagonia, the youngest mature and oldest immature females were six and seven years respectively,[24] suggesting that sexual maturity is attained in that age range.

Possible cases of hybridisation in the wild between dusky dolphins and southern right whale dolphins[25] and between dusky dolphins and long-beaked common dolphins[26] have been noted off South America.

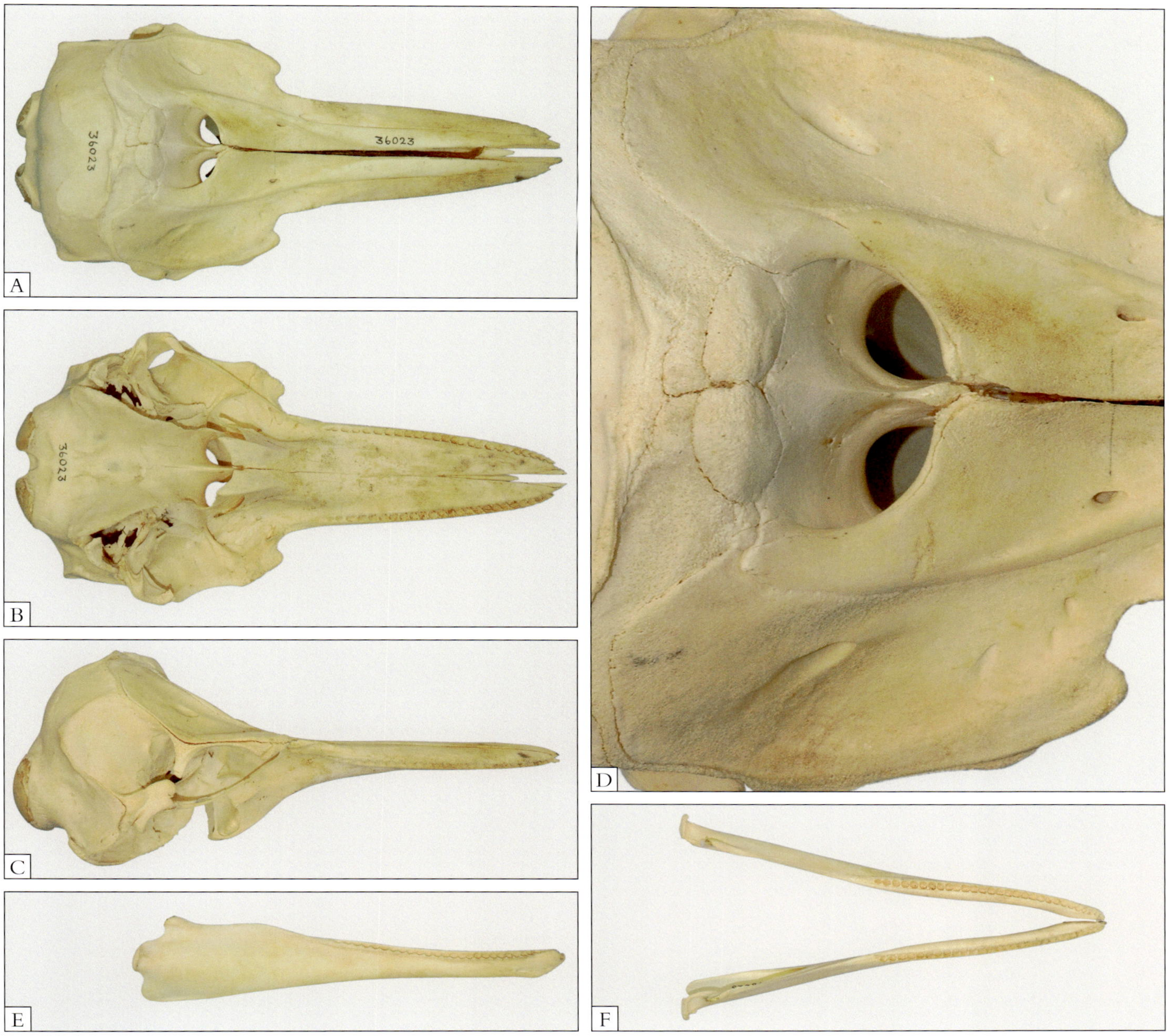

FIG. 145 Dorsal (A), ventral (B) and right lateral (C) views and vertex of skull (D); right lateral (labial) (E) and dorsal (F) views of mandible (ZM 36023).

Behaviour and vocalisations

In southern African waters, dusky dolphins occur in groups of 2–800 with a mean of 35.3 (n = 100). There is no significant difference between the size of schools in summer and winter.[11] The few sightings made in deeper water in the subregion are of larger groupings of 50–300 with a mean of 212.5 (n = 4).[3]

In New Zealand waters, four groupings of dusky dolphins have been recognised: large groups (50 or more individuals), nursery groups (less than 50 individuals, including calves), mating groups (less than 50 individuals exhibiting sexual activity), and adult groups (less than 50 individuals with neither calves nor sexual activity). Large groups comprised 50–1 000 animals, and tended to be largest with fewest calves and juveniles in winter and the reverse in summer. Mating groups consisted of 3–35 adult animals and 0–6 juveniles, and most of the small groups had less than 10 males and a single female. Calves made up about 25% of nursery groups, suggesting that there were three older animals for each calf. Non-mating adult groups consisted on average of nine adults and one juvenile.[27] In South African waters, nursery groups seem to form mainly in late summer, shortly after the postulated calving season, and are quite infrequent by the end of the year. Exactly when they rejoin the rest of the population is unknown.[3]

Dusky dolphins exhibit some of the most spectacular aerial behaviour of any cetacean in the subregion. Off Argentina simple jumps, head-over-tail leaps, backslaps, headslaps, tailslaps, spins, nose-outs, head-first re-entry leaps, and humping have all been described. The first six of these usually occurred in bouts, from three to about 20 times in succession, sometimes at least performed by the same individual. Such leaps resulted in noise as the animal re-entered the water that could be detected on hydrophones up to 0.5 km away ('noisy leaps'). The last two behaviours did not occur in sequences, and were usually only performed once in about 30 sec., with the animal diving deeply thereafter, creating little underwater sound ('clean leaps'). Both noisy and clean leaps occurred more frequently during surface feeding than at other times. After feeding, most leaps were the acrobatic head-over-tail and spins.[17]

In mating groups off New Zealand, periods of sexual activity ranged from 4 to 34 min., during which both males and females had multiple sexual partners, although usually single females mated with multiple males. Inverted swimming and chasing occurred most frequently in such contexts, and the inverted animals were always male. Mothers with calves avoided such groups.[27]

The strength of associations between individuals off New Zealand varied in different locations: in Marlborough Sound in winter, when surface shoaling fish were the principal prey, dusky dolphins associated with preferred long-term hunting companions (more than 1 000 days), while off Kaikoura in summer, when feeding was directed at midwater fish at night, dolphins associated with non-random casual acquaintances (200 days).[27] Strong inter-individual fidelity is also suggested by two dolphins tagged together off Argentina and re-sighted together almost eight years later.[16]

Dusky dolphins produce whistles, click trains, and burst pulses, with the majority of sounds being the last-named.[28] Whistles ranged in frequency from 1 kHz to 27 kHz, with the start and end frequencies averaging 13.4 kHz and 11.6 kHz

Is there a tropical *Lagenorhynchus*?

FIG. 146 Peale's dolphin *Lagenorhynchus australis*

Dusky dolphins (and other members of their genus) are considered to be creatures of temperate or even cold-temperate waters of less than 18 °C.[1] It was therefore something of a surprise when in 1988 at 18°10′ S 163°20′ W (near the Cook Islands, South Pacific), bow-riding dolphins were encountered that from their description and photographs most closely resembled a *Lagenorhynchus* species, either the dusky dolphin or (more likely) Peale's dolphin (*L. australis*).[2] However, this is some 9 000 km from the nearest record of Peale's dolphin (from the coast of Chile), and in an entirely different habitat (tropical, open-ocean, in a depth of over 3 600 m, as opposed to temperate, nearshore, in depths of less than 300 m).[3] Was this just an anomalous range extension, perhaps related to an El Niño type event, or is there an undescribed tropical form of *Lagenorhynchus* out there?

Subsequently other, less well-documented sighting records of *Lagenorhynchus*-type dolphins in tropical regions have surfaced. In 1986 six unusual dolphins that most closely resembled the hourglass dolphin were seen at about 12° N in the Gulf of Aden. In 1988 trained observers from the tuna-dolphin programme in the eastern tropical Pacific described and sketched 40 dolphins at 6°46′ S 89°10′ W that strongly resembled a *Lagenorhynchus* species. In the same year, observers on a whale survey vessel south of Madagascar at 31°46′ S 49°18′ E encountered three dolphins which they tentatively identified as dusky dolphins. And in 1996 a mixed group of small cetaceans was encountered by a whale survey vessel south-west of Madagascar (at 26°52.9′ S 40°36.5′ E) that included 10 dolphins identified by an experienced Japanese observer as Pacific white-sided dolphins (*L. obliquidens*). These sightings were in sea-surface temperatures of 25–32 °C, or 7–14 °C warmer than 'normal' for a *Lagenorhynchus* dolphin habitat.[1]

Nearly all of these sightings were made either during dedicated cetacean surveys or during surveys on which dedicated cetacean observers were on board, so the uncertainty over identification cannot be ascribed to inexperience. Given the way Fraser's dolphin kept itself out of sight for so long (see the species account), it is always possible that this is an unknown dolphin species (or even genus), and observers on vessels traversing tropical regions of the Indo-Pacific should be aware of this possibility. On the other hand, the dusky dolphin that lived for 13 years at Sea World, Durban, in tank temperatures of 22–27 °C[4] should alert us to the adaptability of these animals.

References

1 Van Waerebeek *et al.* 1997
2 Leatherwood *et al.* 1991a
3 Brownell *et al.* 1999
4 Kastelein *et al.* 2000d

respectively. Each whistle lasted an average of 1 sec. and contained 1.97 inflections.[29] Most clicks had a bimodal frequency spectrum, with a low-frequency peak at 40–50 kHz and a high-frequency peak at 80–110 kHz. The centre frequencies of these broadband echolocation signals were among the highest for dolphins measured in the field.[30]

Conservation status

There are no estimates of population size in the subregion. Dusky dolphins on the southern African coast are involved as a by-catch in several fisheries: between 1963 and 2005, at least 11 are known to have died as a result of entanglement in midwater trawls, three (possibly five) in set nets, three (possibly four) in purse seines, one (possibly two) in beach seines and one after entanglement in a hand line. In addition, two dusky dolphins were harpooned from fishing boats, and three washed ashore after being shot in the head (also probably from fishing boats).[3] Although no reliable estimates of this by-catch exist, it has been estimated that about 100 dolphins (of all species) may die annually in the purse-seine and set-net fisheries alone.[31] In addition, 37 dusky dolphins were captured live between 1961 and 1981, of which 13 were immediately released.[32]

Dusky dolphins are listed as Data Deficient by the IUCN and in the South African Red Data Book. They are included on Appendix II of both CITES and CMS.

References

1 Cipriano 1997
2 LeDuc *et al.* 1999
3 MRI Whale Unit data
4 Van Waerebeek 1993a
5 True 1889
6 Van Waerebeek *et al.* 1995
7 Fraser 1966
8 Brownell and Cipriano 1999
9 Gill *et al.* 2000
10 Cassens *et al.* 2005
11 Findlay *et al.* 1992
12 Stahl 1982
13 Cipriano *et al.* 1989
14 Sekiguchi 1994
15 Harlin *et al.* 2003
16 Würsig and Bastida 1986
17 Würsig and Würsig 1980
18 Van Waerebeek *et al.* 1993
19 Markowitz *et al.* 2004
20 Benoit-Bird *et al.* 2004
21 Kastelein *et al.* 2000d
22 Manzanilla 1989
23 Van Waerebeek and Read 1994
24 Dans *et al.* 1997
25 Yazdi 2002
26 Reyes 1996
27 Markowitz 2004
28 Yin 1999
29 Wang *et al.* 1995a
30 Au and Würsig 2004
31 Best and Ross 1977
32 Best and Ross 1984

Hourglass Dolphin

Lagenorhynchus cruciger

CETOLOGISTS MAY DISAGREE on which cetacean species is the most attractive, but the hourglass dolphin must surely be one of the strongest candidates. Unfortunately, though, the chances of most people ever seeing this animal in the wild are slim indeed, as it lives in one of the least inhabited and stormiest places on our planet, the roaring forties and furious fifties of the southern hemisphere.

Because of the remote nature of its habitat, little is known about the hourglass dolphin (up to 1997, only 20 specimens had been examined),[1] and what is known is mostly derived from outside the subregion.

Appearance

Hourglass dolphins are unlikely to be confused with any other dolphins in the subregion apart from the dusky dolphin (though their distributions rarely overlap). As the animal surfaces, the startling white of the successive patches on the thorax and flank should distinguish the species from the dusky dolphin,[1] as should the dark 'shoulder' region.

The largest of nine male and six female hourglass dolphins examined were 1.87 m and 1.83 m long respectively.[1–4] The animal is distinctly more chunky-looking than the dusky dolphin (this is borne out by the relative weights of the animals – see Length:weight relationship below), accentuated perhaps by strong keeling of the caudal peduncle in both sexes.[5] The head is of a similar shape, with a 'ploughshare' snout, but the profile of the head is less tapered than in the dusky dolphin. The snout is barely distinguishable, and forms only 1.2–2.3% (mean 1.7%, n = 3) of body length. The flippers, dorsal fin and flukes are all of a similar size to those of dusky dolphins (16.1–20.8%, mean 18.5%, n = 3; 9.1–12.4%, mean 10.1%, n = 4; and 18.3–28.5%, mean 24.3%, n = 4, respectively). There is noticeable variation in the size and particularly the shape of the dorsal fin, some individuals having quite erect fins while others incline back and some are strongly hooked, with the tip pointing almost backwards.[1] It is possible that there is similar sexual dimorphism in the shape of the dorsal fin to that found in the dusky dolphin, but the sample is too small to demonstrate this. The strongest hooked fins seen so far in sexed animals have been in adult males.[1, 3] Judging by a photograph of one of these animals, there may also be some development of a post-anal hump in adult males.

The basic theme of the colour pattern of the animals examined to date has been similar, but the details have been highly variable. A white blaze extends back from the side of the melon above the eye and tapers out at about the level of the anterior insertion of the dorsal fin. Starting at about the level of the posterior insertion of the dorsal fin is a second white blaze that

Length:weight relationship

Weights are only available for four animals; a 1.635 m female weighed 73.5 kg, a 1.74 m male 94 kg, a 1.8 m male 93 kg and a 1.83 m female 88.2 kg.[1, 3]

Equivalent weights for dusky dolphins of these sizes (using the length:weight relationship developed for that species) would be 56.2 kg, 67.7 kg, 74.9 kg and 78.7 kg respectively.

Skeletal data

Vertebral formula (n = 5–8): 7 cervical, 12–13 thoracic, 18–22 lumbar and 29–33 caudal; total 69–71.[1, 3]

The atlas and axis are fused.[1]

Phalangeal formula, including metacarpals (n = 4): I_{2-3}, II_{9-11}, III_{6-8}, IV_{2-4}, V_{1-2}.[1]

The characteristics of the skull that distinguish *L. cruciger* from *L. obscurus* are listed in the latter's species account.

rapidly expands to cover most of the lower half of the caudal peduncle, before tapering out on the upper half of the caudal peduncle at the base of the flukes. The belly from the throat to the anus is also white, but the rest of the body is mostly black (sometimes described as brownish, but this may partly be the effect of diatom infestation). There is a black band around the eye, extending forward as a streak towards the tip of the snout, and the snout and chin are black. Individual variation mainly involves the pattern of pigmentation below the flank patch (including the presence of brownish-grey pigment), the extent of black around the gape, and the patterning and coloration (including light brown or ochre pigmentation) of the stripe from flipper to eye. There is also considerable variation in the pattern surrounding the ano-genital region: in both sexes the white of the belly bypasses the anus as a narrow tongue on either side, but in some animals these tongues extend further back and leave a narrower, more well-defined isthmus of black around the anus. This difference is not necessarily gender-related.[1, 3, 6] The colour pattern of calves has not yet been investigated.

There are 26–34 (mean 29.6, n = 14) pairs of erupted teeth in the upper jaw and 27–35 (mean 29.4, n = 14) pairs in the lower jaw.[1, 3] The teeth are slender and conical, with their maximum diameter at the gum being 2–3 mm.[3, 7]

Distribution and migrations

Hourglass dolphins are only found in the southern hemisphere, in an area north and south of the Antarctic Convergence.[5] The most northerly record is not clear. A sighting at 33°40′ S may well not be of this species,[1] while the identification of one made at 36°14′ S does not seem very definite.[8] The sighting at 42°24′ S in 1912 described by the ornithologist Robert Cushman Murphy may be the best candidate, despite its vintage.[9] Within the subregion, sightings on the IDCR/SOWER cruise programme have occurred as far north as 43°04′ S.[7] The most southerly sighting is at 67°38′ S,[10] but within the subregion the species has not been recorded south of 63°54.4′ S.[7]

The hourglass dolphin has been recorded from Kerguelen[11] and Heard Island,[12] but because of its pelagic habit it probably occurs close to other subantarctic islands in the subregion such as the Prince Edward and Crozet Islands. A sighting of six dolphins resembling *L. cruciger* at about 12° N in the Gulf of Aden remains unexplained (see the box on page 232).[13]

Although no analysis of their depth distribution has been published, hourglass dolphins are essentially pelagic, deep-sea animals. Most sightings have been in sea surface temperatures of 0–13 °C, with peak numbers around 3–5 °C.[14]

A movement south in summer to latitudes greater than 60° S as water temperatures increase has been postulated, but this interpretation may be confounded by incomplete seasonal effort coverage.[10] Winter sightings of hourglass dolphins have been made between 55° S and 58° S.[15]

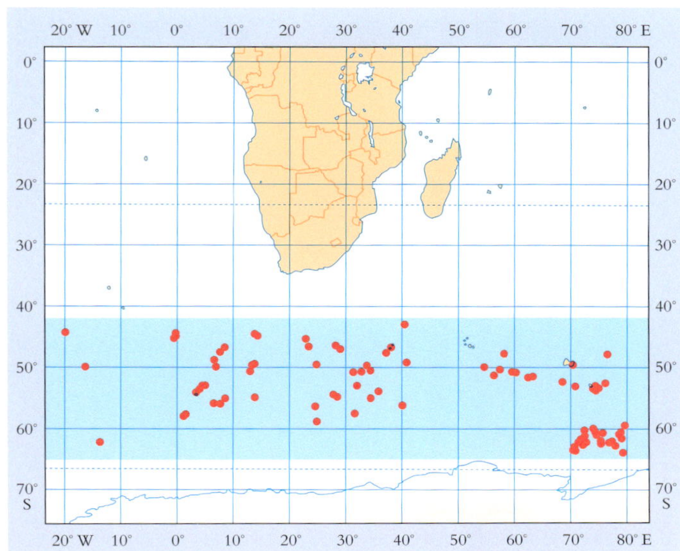

FIG. 147 Distribution map for hourglass dolphin.

FIG. 148 This image clearly demonstrates the difference in dorsal fin size and shape (that may correlate with gender) between individual hourglass dolphins.

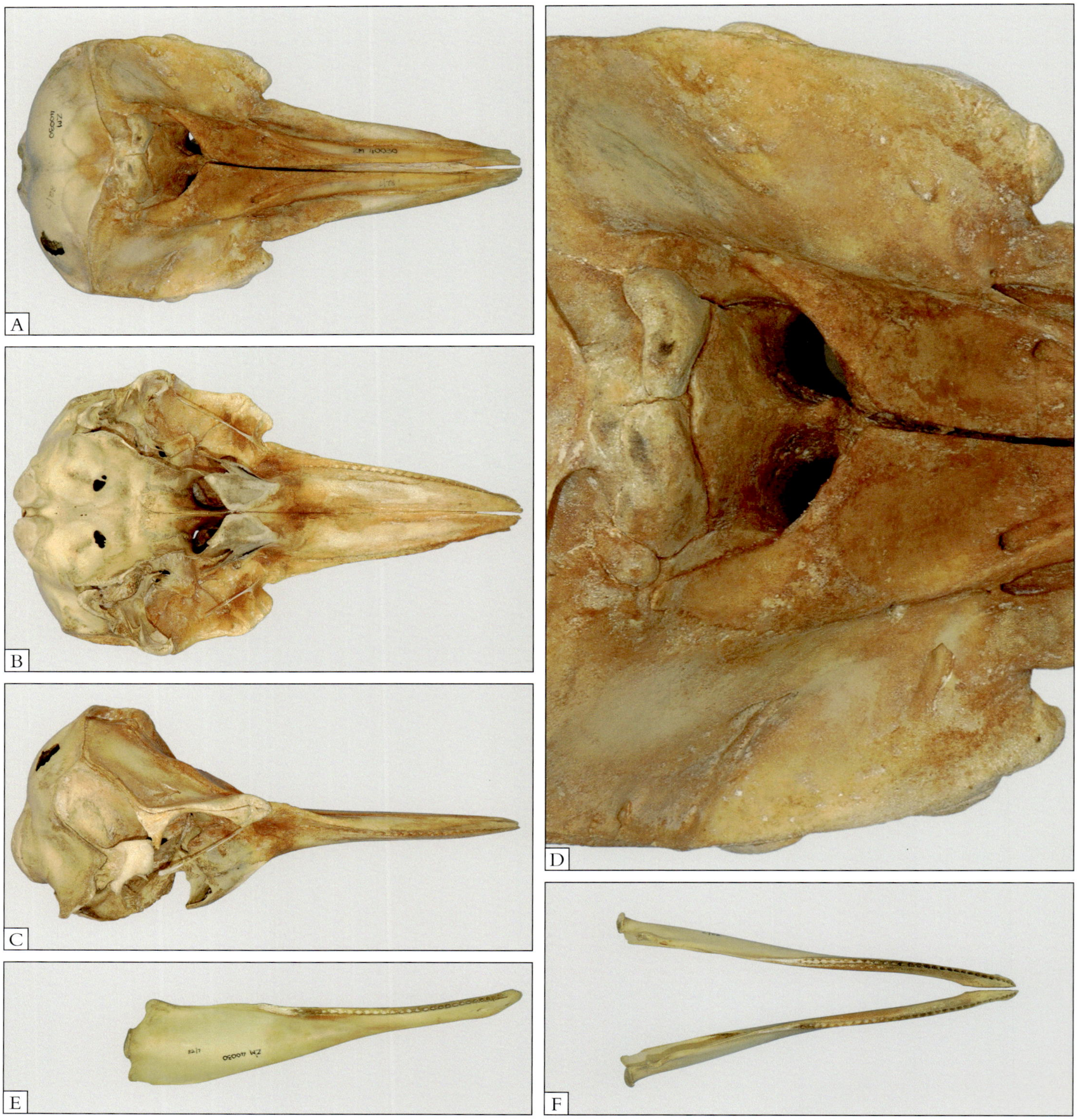

FIG. 149 Dorsal (A), ventral (B) and right lateral (C) views and vertex of skull (D); right lateral (labial) (E) and dorsal (F) views of mandible (ZM 40030).

Estimates of swimming speed range from 4 to 16 knots (mean 10.5 knots, n = 22).[14]

External commensals and parasites

A specimen taken at sea carried small circular patches of possible diatoms on the back and sides.[7] No other external parasites or commensals have been recorded.

Food and feeding behaviour

The stomach contents of four specimens have been identified. The only animal from the subregion contained the beaks of 73 cephalopods, 45 of which were identified as Onychoteuthidae and 20 as Enoploteuthidae; the remainder were unidentified.[16] A specimen taken south-east of Cape Horn contained 591 otoliths from the myctophid fish *Krefftichthys andersonii*, the beaks from nine unidentified cephalopods and a very small amount of unidentified crustacean remains.[7] In one stranded dolphin from Patagonia, the bobtail squid (*Semirossia tenera*) was the most abundant prey item, followed by the Patagonian squid (*Loligo gahi*) and juvenile Argentine hake (*Merluccius hubbsi*). Other prey included several crustaceans (*Heterosquilla* sp.) and the benthic cephalopod *Eledone massyae*. A second stranded Patagonian dolphin contained mainly myctophid fish of the genus *Protomyctophum*, as well as the cephalopods *Illex argentinus* and *L. gahi*.[2]

Although the nature of some of the prey would suggest that feeding must take place at night on midwater species that migrate closer to the surface, daytime feeding has apparently been witnessed several times.[14]

Reproduction and growth

Almost nothing is known. Animals accompanied by individuals small enough to be termed calves have been seen in high latitudes of the Antarctic in summer, but their low overall incidence (only 0.2% of 1 634 individuals seen)[14] suggests that they may occur mainly north of 58° S.[17] Of the seven males examined above 1.64 m in length, six have been described as (or can, from the size of their testes, be assumed to be) sexually mature:[1–4] two had combined testis weights of 3.75–4.0 kg[4, 5] while there is no information about the seventh (which was 1.78 m long). Of the two females examined, one at 1.635 m was sexually immature[1] and the other at 1.83 m was approaching maturity.[5]

Males 1.64 m, 1.78 m and 1.8 m long had, respectively, eight, nine and eight dentinal growth layers in the teeth.[2, 3]

Behaviour and vocalisations

Hourglass dolphin groups sighted in the subregion have ranged in size from one to 31 (mean 5.6, n = 83), while 89% have contained less than 10 animals.[7] They have been reported in the company of a number of other cetaceans, including fin whales, sei whales, minke whales, beaked whales, pilot whales and southern right whale dolphins.[14]

Schools of the species are known for their fast, plunging swimming movement, likened by some observers to a flock of fast-moving penguins. They are often attracted to the bows of moving ships,[14] possibly as an extension of their behaviour in accompanying large whales.

Clicks have been recorded in the vicinity of hourglass dolphins,[18] but no analysis of their vocalisations is available.

Conservation status

There are no estimates of population size in the subregion, but an estimate of 144 300 (CV = 0.17) hourglass dolphins has been made for the entire area south of the Antarctic Convergence in summer.[10] Given the remote and inhospitable nature of its environment, and its lack of commercial value, the species seems to be faced by few anthropogenic threats.

Hourglass dolphins are listed as Lower Risk: Least Concern by the IUCN and included on Appendix II of CITES. They are not included in the South African Red Data Book and are not listed by CMS.

References

1 Goodall *et al.* 1997
2 Fernández *et al.* 2003
3 Gazitúa *et al.* 1999
4 Goodall, R.N.P., pers. comm.
5 Brownell and Donahue 1999
6 Fraser 1966
7 MRI Whale Unit data
8 Nichols 1908
9 Murphy 1947
10 Kasamatsu and Joyce 1995
11 Robineau 1989
12 Anon. 1993
13 Small and Small 1991
14 Goodall 1997
15 Thiele and Gill 1999
16 Clarke 1986
17 Kasamatsu *et al.* 1988
18 Leaper *et al.* 2000

Southern Right Whale Dolphin

Lissodelphis peronii

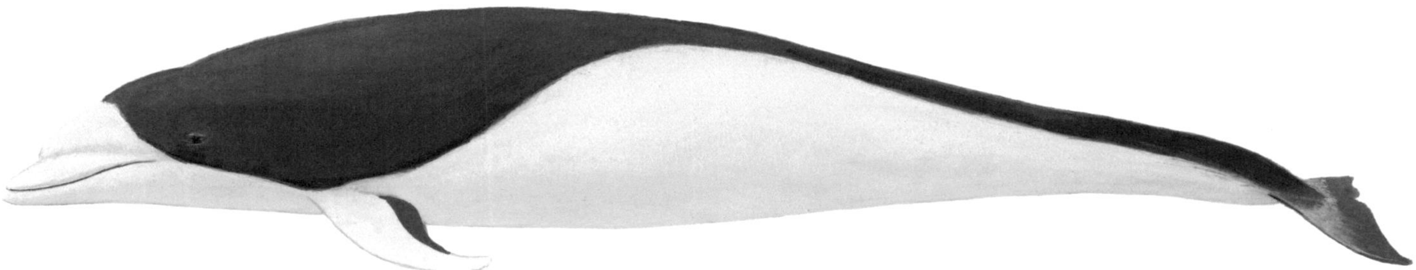

Dorsal view

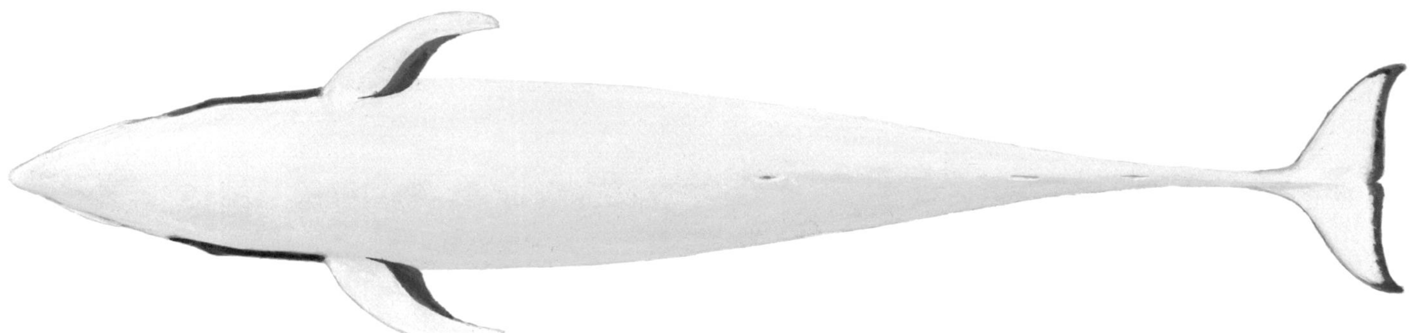

Ventral view

IT IS OFTEN thought that the rather clumsy common name of this dolphin originated because, like a right whale, it lacks a dorsal fin, but according to Herman Melville the name arose from the dolphin being chiefly found in the vicinity of right whales. Melville actually preferred to call it the mealy-mouthed porpoise, since he thought that the white marking on its mouth and head looked 'as if he had just escaped from a felonious visit to a meal-bag'.[1] (The confusing construction of the common name can lead to some interesting typological errors – a paper in a prestigious journal carried the running head 'Hybrid between the dusky dolphin and the southern right whale'.)

Appearance

Southern right whale dolphins are very unlikely to be confused with any other dolphins in the subregion, given their striking two-tone coloration and lack of dorsal fin.

The largest male reported was 2.97 m long[2] and the largest female was 2.3–2.4 m.[3] It has been suggested that this represents a real difference in size between the sexes,[4] although the numbers measured have been small (about 11 males and six females). The body shape of right whale dolphins is unique: apart from the lack of a dorsal fin, the body is distinctly broader than deep in the thoracic region, and the caudal peduncle is very slender. The beak is short and well differentiated from the rest of the head by a prominent groove, but in profile is hardly distinguishable. The flippers are large (15–16% of body length), with a markedly convex leading edge and an S-shaped trailing edge, which is strongly concave near the tip and then convex nearer the body. The flukes are moderately wide (17.5–21% of body length) but strongly curved backwards with a prominent notch.[3]

The colour pattern of southern right whale dolphins is one of the most striking of any cetacean. The body is basically black above and white below, with a sharply defined border between the two running from a widow's peak in the centre of the melon down above the angle of the gape to just above the flipper, sweeping back up the side to form a plimsoll line along the side of the body as far as the insertion of the flukes. The flippers are white with a black trailing edge, both above and below. The flukes are black above and white below with a black trailing edge. Juveniles have a similar coloration to adults, although calves appeared grey above when submerged[3] (and grey pigmentation seen on calves suggests some age-related change in coloration: see Fig. 152).

There are many small variations on this pattern, and some major ones. There may be a small white patch on the melon; the plimsoll line on either side may extend high enough to form a saddle across the middle of the back; the amount of black on the flippers may vary;[5] and the dorsal surface of the flukes may be greyish or even white.[6] Major variants include individuals that are completely black,[7] more melanistic than normal (with both sides of flukes and flippers black, and extensive darkening of the flanks),[3] and completely white.[8]

There are 37–49 (n = 11) pairs of slender and conical teeth in the upper jaw and 37–48 (n = 10) pairs in the lower jaw.[3, 9]

Length:weight relationship

Very few individuals have been weighed. A 0.86 m neonate weighed 5.1 kg,[2] a 1.73 m individual was 65 kg,[10] a 1.86 m male was 60 kg,[11] and a 2.51 m male was 116 kg.[4] These have been used to calculate the following relationship: $W = 9.0358L^{3.0167}$, where W = weight in kilograms and L = total length in metres. This should be viewed with caution given the small sample size.

Skeletal data

Vertebral formula (n = 1): 7 cervical, 14 thoracic, 32 lumbar and 31 caudal; total 84.[4]

The atlas and axis are fused.[3]

The phalangeal formula seems undocumented, but that for the northern right whale dolphin is (n = 4): I_{1-2}, II_{7-8}, III_{5-6}, IV_{3}, V_{2-3}.[12]

The skull bears a strong resemblance to that of the striped dolphin, but can be distinguished by the shorter and more pointed rostrum, and the flatter, fully separated maxillae on the rostrum.[3] The skull can also be separated from that of the dusky dolphin (with which it has sometimes been confused) by the greater extension of the prenarial triangle (almost to the mid-length of the rostrum), the prominent maxillary foramina, and the sinusoidal ventral border and greater maximum height of the lower jaw.[9]

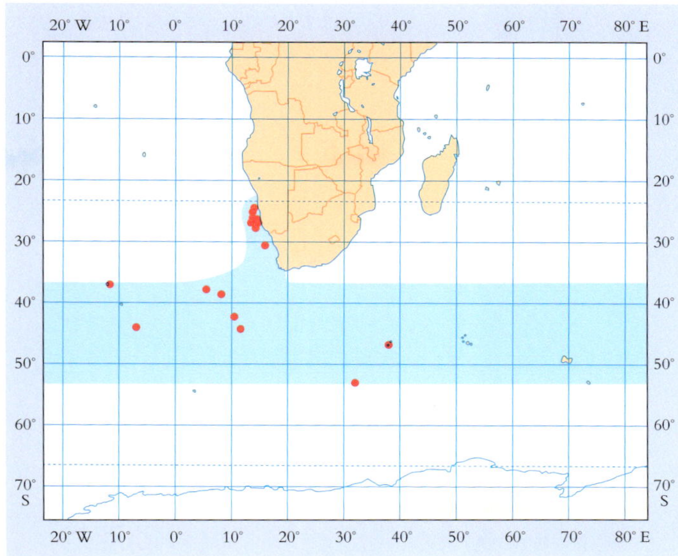

Postulated range ■ Sighting or capture

FIG. 150 Distribution map for southern right whale dolphin.

Distribution and migrations

Southern right whale dolphins are usually only found in subantarctic waters, but extend their distribution significantly further north in cold boundary current systems such as the Peru Coastal Current and the Benguela Current.[4]

Sightings are relatively infrequent in the subregion. Away from the influence of the Benguela System, southern right whale dolphins have been found between 37° S and 53° S. The rarity of sightings in the south-west Indian Ocean may be more apparent than real, as the species has been seen at 82° E, just outside the eastern limit of the study area.[13] On the west coast of southern Africa, sightings with accurate locations extend as far north as 24°26′ S,[14] but there is an additional record only described as 'north of Walvis Bay' (22°59′ S).[15] A skull originating from the island of St Helena is believed to have been an importation from elsewhere.[16] Most of the records on this coast occur in a relatively restricted area between 26° S and 28° S, or off Lüderitz, Namibia, possibly in association with an upwelling cell.[17] There are two nineteenth-century museum specimens that may have originated from South Africa,[5] but otherwise there is only a single sighting just south of the Orange River to testify to its presence in South African waters.[17]

Seasonal movements are not well documented. Although most records of southern right whale dolphins off Namibia are in summer/autumn, leading to suggestions that the population is migratory,[5] most of the cruise effort was at that time of year and there are at least some winter sightings in the area, which would not exclude some residency.[17] Off Chile, a partial northward movement in winter and spring has been proposed,[2] but such a phenomenon would not be compatible with the pattern of availability seen off Namibia.

Sightings off Namibia mainly occurred in water depths of 100–200 m to 1 000–2 000 m,[14] although some have obviously been seen in water less than 100 m deep.[17] Generally, however, the species seems to avoid shallow water (less than 200 m deep).[2]

All accounts indicate that the species is a fast swimmer, sustaining speeds of up to 12–12.5 knots while accompanying ships.[5, 17]

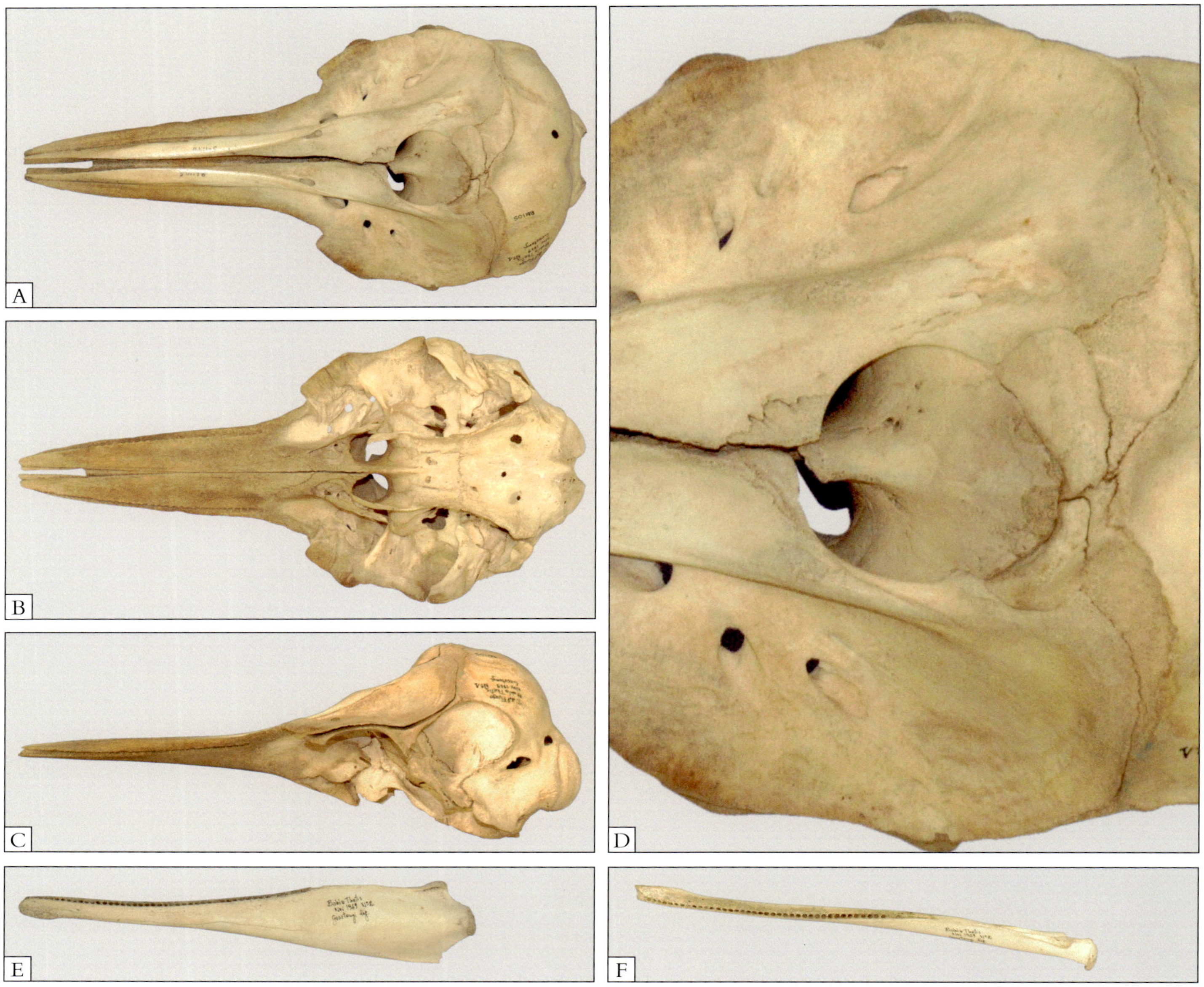

FIG. 151 Dorsal (A), ventral (B) and left lateral (C) views and vertex of skull (D); left lateral (labial) (E) and dorsal (F) views of mandible (USNM 501198).

External commensals and parasites

None seem to have been recorded, although both the stalked barnacle *Xenobalanus* and the parasitic copepod *Pennella* have been found on northern right whale dolphins.[18]

Food and feeding behaviour

There is no information from the subregion, although it has been speculated that individuals off Namibia might be feeding on lantern fish (*Lampanyctodes hectoris*), thumbnail squid (*Lolliguncula* sp.) or the euphausiid *Nyctiphanes capensis*.[5] Elsewhere the stomachs of only four southern right whale dolphins seem to have been examined in any detail. Two specimens off New Zealand were reported as feeding mainly on 'adolescent' common arrow squid (*Nototodaris sloani*) and juvenile hake (*Macruronus novaezelandiae*).[3] Off the Pacific coast of South America one individual had been feeding mainly on myctophids (especially *Hygophum hanseni*) and the squids *Gonatus antarcticus* and *Mastigoteuthis* sp.,[19] while a second individual had eaten anchoveta (*Engraulis ringens*), silversides (*Odontesthes ragia*) and horse mackerel (*Trachurus murphyi*).[18]

The nature of the prey reported indicates that the dolphins could be feeding in a variety of modes: close to the surface at night on species that vertically migrate,[19] diving to at least 150 m to access midwater species that do not migrate closer to the surface,[3] or at the surface on epipelagic shoaling fish.[18]

Reproduction and growth

Very little is known. Near-term foetuses have been recorded from November (six) and April (one) and a neonate from October. The largest foetus measured 1.02 m and the neonate (possibly premature) 0.86 m.[4] A 2.18 m female was pregnant[2]

FIG. 152 These pictures capture the exuberance of southern right whale dolphins on the move. The extent of grey pigmentation on the calf compared to its mother suggests some age-related change in coloration.

and a 2.283 m female was classified as 'adult',[3] while males of 2.505 m and 2.71 m were sexually mature.[2]

Possible hybridisation between a southern right whale dolphin and a dusky dolphin has been noted off Argentina.[20]

Behaviour and vocalisations

Schools of southern right whale dolphins in the subregion varied in size from one to 500, with a mean of 56.6 (n = 28), and only four of these exceeded 100 animals. As found elsewhere, the species is frequently seen in association with other cetaceans. In the subregion, 45.8% (n = 24) of schools encountered were of mixed species. Of those on the southern African coast, dusky dolphins were included in 77.8%, pilot whales in 11.1% and bottlenose dolphins with long-finned pilot whales in 11.1%. In higher latitudes one of the mixed sightings involved pilot whales and the other pilot whales and hourglass dolphins.[21]

The swimming behaviour of southern right whale dolphins has been classified as slow-moving (exposing just the head to breathe, when the animals must be hard to spot without dorsal fins), rapid subsurface swimming, fast surface swimming (with much shallow leaping), belly flop/side slap/lobtail on re-entering, sounding and bow-riding. At very high speed the animals give the impression of hardly re-entering the water between leaps, and the progress of a school engaged in such activity could be monitored for half an hour by means of the spray crest it generated.[5] A specific behaviour pattern involved a large school milling at the surface, from which a single dolphin commenced a series of jumps around the periphery of the group, making 20 jumps and two complete circles of the school in under 2 min. About 30 sec. after it had stopped jumping, the whole school exploded outwards in an ever-increasing circle. After they had progressed like this for a nautical mile or more, the dolphins stopped and re-grouped before moving off. Although the species has been termed a reluctant bow-rider, on occasion very large numbers of animals ride the bow. Submergence times on such occasions were 10–14 sec. Other features of group behaviour involve simultaneous jumping (occurring every few minutes), and sudden explosions into high-speed activity for a minute or two.[17]

No vocalisations have been described for the species, but the northern right whale dolphin produces far fewer whistles than most pelagic dolphins, uttering mainly pulsed sounds (clicks) at a high repetition rate (700–1 000/sec.). Some of these click trains resemble frequency-modulated whelps or moans to the human ear, while at least some of the whistles (extending above 16 kHz in frequency) actually blend with the click trains.[22]

Conservation status

No estimates of population size exist for this species. There is no indication of any by-catch in the subregion, though concerns have been expressed over possible involvement of the species in a drift-net fishery for swordfish off Chile,[2] and the vulnerability of northern right whale dolphins to drift-net fishing has been highlighted.[23]

Southern right whale dolphins are listed as Data Deficient by the IUCN and included on Appendix II of CITES. They are not included in the South African Red Data Book and are not listed by CMS.

References

1 Melville 1851
2 Van Waerebeek *et al.* 1991
3 Baker 1981
4 Newcomer *et al.* 1996
5 Cruickshank and Brown 1981
6 Fraser 1955
7 Visser *et al.* 2004
8 Brown 1973
9 Van Waerebeek 1993b
10 Philippi 1893
11 Martuscelli *et al.* 1995
12 Tobayama *et al.* 1969
13 Anon. 1969
14 Findlay *et al.* 1992
15 Brown 1982
16 Perrin 1985
17 Rose and Payne 1991
18 Jefferson *et al.* 1994
19 Torres and Aguayo 1979
20 Yazdi 2002
21 MRI Whale Unit data
22 Leatherwood and Walker 1979
23 Mangel 1993

Risso's Dolphin

Grampus griseus

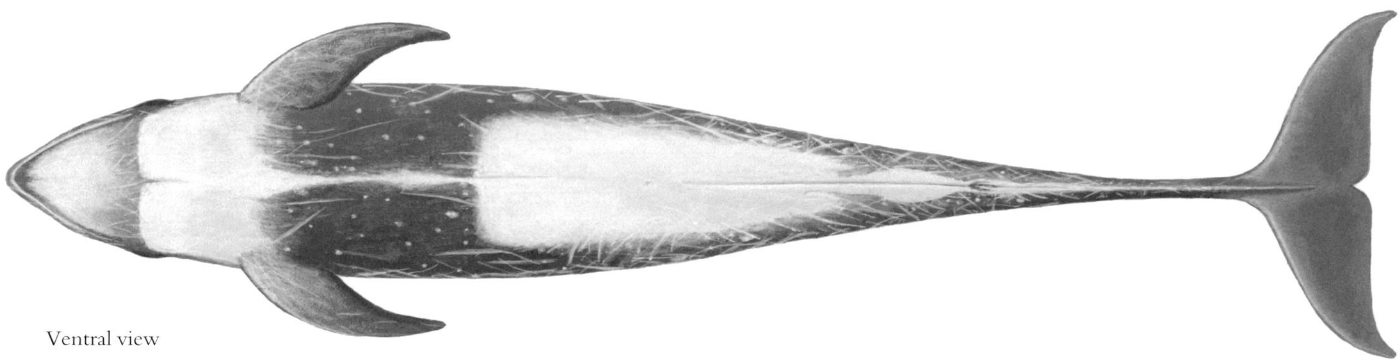

Ventral view

Calf

THE LARGEST DOLPHIN not to be called a whale, Risso's dolphin is something of a maverick. Taxonomically it has roamed between the family Delphinidae, its own family Grampidae, and has joined pilot whales and other larger dolphins in a subfamily Globicephalinae.[1] Anatomically this is perhaps not surprising, as its 'top-heavy' front half, large dorsal fin and teeth confined to the tip of the lower jaw make it look as though it was designed by a committee.

Largely owing to a number of mass strandings in South Africa between 1983 and 1991, a reasonable amount of information on the biology of Risso's dolphins has accumulated in the subregion, although little has yet been published. This account has drawn on much of this information where necessary.

Appearance

With their unique combination of a blunt head, prominent centrally-placed erect dorsal fin and (in adults) almost pure white coloration, Risso's dolphins are unlikely to be confused with any other cetacean in the area.

The largest individuals among 50 male and 43 female Risso's dolphins measured in the subregion were 3.41 m and 3.18 m long respectively, while 20% of males exceeded the biggest female in length.[2,3] Although age-length relationships are not yet available, there seems little doubt that the species is sexually dimorphic in the region. The body shape is distinctive, with a blunt forehead carrying a vertical furrow in its anterior face and projecting beyond the tip of the lower jaw. There is no trace of a beak. The line of the gape is a flattened S-shape, and forms a distinct angle with the main longitudinal axis of the body. The front half of the animal is noticeably 'heavier' than the body behind the dorsal fin. The flippers are large, in adults ranging in length from 19.2% to 24.4% (mean 21.1%, n = 45) of body length, and sickle-shaped. The dorsal fin is also large (its height in adults 10–15.3% of body length, mean 12.4%, n = 45), erect and hooked, and placed in the centre of the back.[3]

The coloration of Risso's dolphin undergoes a marked change with age. At birth the animals are a brownish-grey above with white below, the margin between the two being well defined and running from the angle of the gape to the base of the flipper and from there back along the side, terminating at the anus. The upper lip and the front surface of the head, chin and throat are yellowish-white, while both surfaces of the flipper and flukes are the same colour as the back. As the animal ages it first darkens overall, but retains a whitish-grey anchor-shaped mark on the chest connected by a thin mid-ventral streak with an irregular lozenge-shaped mark extending from the umbilical region to the anus. With increasing age the head, abdomen and flanks lighten, and the body surface begins to accumulate a motley assembly of spots and scars, many of the latter linear in nature and almost certainly caused by the teeth of conspecifics (and even other small cetaceans[4]). Hardly any part of the body is immune from such scarring, although most seems to be targeted at the back and flanks. The combination of gradually lightening pigmentation and the accumulation of scars make the larger and presumably older Risso's dolphins become almost completely white around the head and back (although the dorsal fin usually survives much of this process and tends to form a striking contrast in colour). The anchor- and lozenge-shaped markings are usually still discernible on the belly of even the whitest animals. The aetiology and development of this scarring still have to be studied in detail, since it is unclear whether the accumulation of scars is a combination of a lack of repigmentation and a slower healing process than in other cetaceans,[1] or whether the process differs between the sexes.[5] The use of such scarring to indicate 'quality' of individuals during social interactions has been proposed.[6]

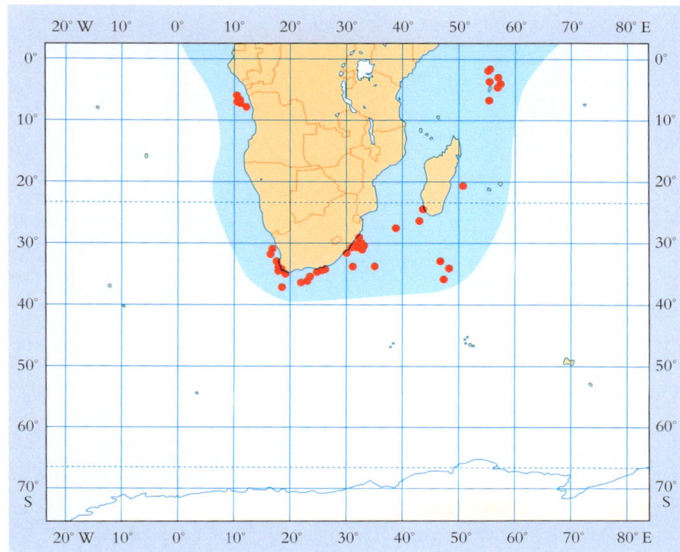

FIG. 153 Distribution map for Risso's dolphin.

Length:weight relationship

Data are available for 32 animals from the subregion weighed whole, and five weighed only in pieces.[2,3] Using an adjustment factor of 1.0468 (developed from 10 adults weighed whole and in parts), the following length:weight relationship is obtained for all 37 animals: **$W = 6.8786L^{3.4979}$**, where W = weight in kilograms and L = total length in metres. This implies that a 2 m dolphin would weigh 77.7 kg and a 3 m animal 320.9 kg.

Skeletal data

Vertebral formula (n = 32–33): 7 cervical, 12–13 thoracic, 17–23 lumbar and 26–33 caudal; total 67–71.[2,3]

All seven cervical vertebrae are fused together, although the centra of the sixth and seventh may not be completely fused.[8]

Phalangeal formula, including metacarpals (n = 3): I_2, II_{9-10}, III_{7-8}, IV_{3-4}, V_{1-3}.[3,8]

The characteristics of the skull that distinguish Risso's dolphin from the pilot whales (the only other cetaceans in the subregion with which it might be confused) are: 1) the absence of any alveoli for teeth in the upper jaw; 2) the width of the premaxillae in the middle of the rostrum being less than at its base; and 3) the surface of the premaxillae in front of the nares being convex.[8]

Erupted teeth have only been found in the lower jaw of Risso's dolphins from the subregion, although elsewhere they have been described in the upper jaw.[7] The teeth are confined to the very front of the jaw, and may erupt by a body length of 1.88 m (although they have been seen still unerupted in an animal 2.26 m long). In adults (more than 2.8 m long) there are between two and six pairs of erupted teeth (mean 3.8, n = 72), but in the oldest and largest animals these become worn down and may even fall out, leaving some individuals with no functional teeth at all. Teeth when present are conical and robust, their maximum diameter at the gum being 6.5–12.0 mm in adults (mean 9.7 mm, n = 17).[3]

Distribution and migrations

Risso's dolphins are distributed worldwide in tropical and temperate seas.[1] In the subregion they have been recorded from Gabon,[9] Angola,[10] Namibia, South Africa,[11] the Comoro Islands,[12] Madagascar[13] and the Seychelles,[14] as well as pelagically. The most southerly record in the subregion is at 37°14′ S.[3]

In several areas this species is strongly associated with the upper continental slope, particularly between depths of 350–600 m and 975–1 500 m,[15,16] although it also occurs more rarely in deeper water.[15,17] A preference for regions of high bathymetric relief has been demonstrated in more than one region.[1,15,17] Off South Africa Risso's dolphins are associated with the shelf edge and pelagic waters.[11] In the Seychelles all sightings were in water deeper than 3 200 m.[14]

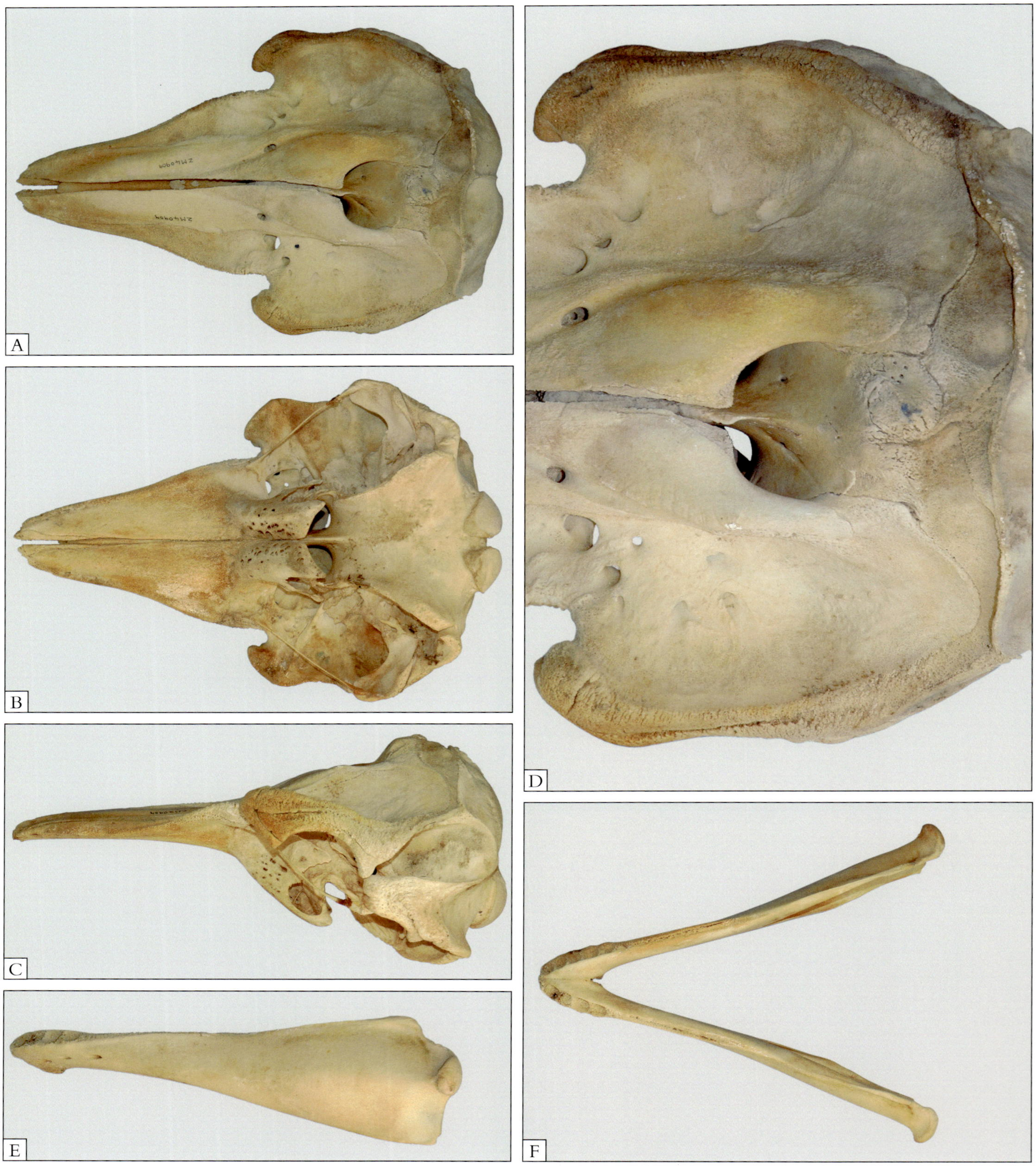

FIG. 154 Dorsal (A), ventral (B) and left lateral (C) views and vertex of skull (D); left lateral (labial) (E) and dorsal (F) views of mandible (ZM 40909).

Seasonal movements in the subregion are unknown. In the eastern tropical Pacific, movement onto the continental shelf may occur in response to changes in sea temperature,[17] and in the western Mediterranean Risso's dolphins show a similar movement onto the upper portions of the continental slope in spring and summer.[18] Off California, however, winter abundance within 100–150 nautical miles of the coast was greater than in summer, although the distribution of the dolphins was the same: a seasonal migration north in spring and summer was one possible explanation.[19] Longer-term shifts in distribution, probably related to environmental changes, have been reported in the North Pacific.[17,20] In one case there was a simultaneous reduction in the abundance of short-finned pilot whales, which were possibly competitively displaced by Risso's dolphins.[20]

There are no accurate measurements of swimming speed, but Risso's dolphins have been estimated to reach 28 km/hr (and up to 31.5 km/hr for short periods).[21] At these times they can create such a 'rooster-tail' bow-wave that it is hard to identify the species.[3] Normal travelling speeds, however, are much slower, around 5.5 km/hr or less.[22]

External commensals and parasites

Stalked barnacles (*Xenobalanus* sp.) occurred on four of 33 Risso's dolphins examined in South Africa, ranging from one to eight individuals per animal with 75% on the trailing edge of the tail flukes and the remainder on the dorsal fin. Cyamids were detected on two of these dolphins, identified as *Scutocyamus* sp. in one case, where 134 individuals were scattered over the body surface (45.5%), in the axillary region of both flippers (43.3%), on the dorsal fin (7.5%) and on the tail (3.7%).[3]

Food and feeding behaviour

Only cephalopods were found in the stomachs of 32 Risso's dolphins stranded in the Eastern Cape, South Africa. Among the 1 341 items and 17 species identified, the five most important were the squids *Loligo vulgaris reynaudii*, *Lycoteuthis diadema*, and *Ancistrocheirus lesueri* and the octopods *Argonauta nodosa* and *Octopus magnificus*. Overall, however, *Loligo v. reynaudii* and *Lycoteuthis diadema* were clearly dominant in terms of numbers eaten and frequency of occurrence in stomachs. There was a tendency for larger cephalopod species (such as *Todarodes angolensis*) to be more important for larger dolphins, while smaller species (such as *Sepia*) that were important for smaller dolphins were not taken by larger dolphins. Males tended to eat larger prey items than females.

From the species composition of their food, Risso's dolphins seemed to be feeding primarily over the continental shelf, although slope species were also consumed.[23] In 17 Risso's dolphins stranded in the Western Cape, cephalopods accounted for 98% of the 352 identified prey items, the remainder being pelagic fish (anchovy and horse mackerel) that might have been secondary prey of the squid. Of the 16 cephalopod species, *Todaropsis eblanae*, *Loligo vulgaris*, *Argonauta nodosa* and *A. argo* predominated numerically, followed by *Moroteuthis robsoni* and *Taningia danae*.[3] This diet suggested that the dolphins were feeding on the shelf edge as well as over the continental shelf.

Studies elsewhere have shown similar feeding habits, with Risso's dolphins consuming both oceanic and neritic cephalopods, as well as sepiids and octopods.[24] One problem with all these studies, however, is that they are heavily dependent on stranded animals, whose feeding habits may not be typical for the species since they might, for instance, include a greater number of nearshore prey items.

One animal stranded in Hawaii contained the remains of 1 349 cephalopods, all oceanic species, many of which are midwater in habit and migrate to the upper 200 m at night.[25] From the rarity of observed feeding behaviour during the day, it has been proposed that Risso's dolphins may be predominantly nocturnal feeders.[22]

A captive 2.75 m female consumed up to 25 kg of fish a day, but maintenance requirements were met by feeding 15 681 kcal daily.[26]

Reproduction and growth

The length at birth is not well established. A live calf 1.293 m long with an unhealed umbilicus was caught in a fishing net on the South African coast in January, suggesting birth may occur at a length of about 1.3 m. Regressing the lengths of 11 foetuses and six calves (with unerupted teeth) against date of death produces an overall rate of growth of 0.46 cm/day. Given a size at birth of 1.3 m, this implies a period of linear foetal growth of 9–10 months, and a total gestation period of 10–11 months. Births would then peak in late spring/early summer.[3] The narrow spread of foetal size at any one time suggests that the season of births would be restricted in time.[5]

All but one of 18 females 2.8 m or more in length were sexually mature, and all nine females smaller than this were immature (although only one between 2.6 and 2.8 m was examined).[3] Using maturation criteria based on testes weights,[27] only two out of 10 males less than 3 m long, 57% of seven males between 3 m and 3.1 m, and all seven males over 3.2 m long were fully mature.[3] The age at female maturity is 8–10 years, and possibly 10–12 years in males. The length of lactation is unknown, but calves may be born on average at intervals of at least 2.4 years. Females may live for up to an estimated 34.5 years[27] and males for at least 30 years.[1]

Hybridisation has occurred with bottlenose dolphins, both in captivity[28] and apparently in the wild.[29]

Behaviour and vocalisations

The size of 94 schools of Risso's dolphins encountered in the subregion varied from one to 80 with a mean of 11.6, but 92.6% comprised 25 animals or fewer.[3] Such groups mostly travel as tight ovals, or in mixed geometry, or occasionally in line abreast.[22] Mixed associations with other cetacean species are frequently seen elsewhere,[1] but rarely in the subregion.[3]

Four mass strandings in South Africa consisted of: (May 1983) four males, four females; (April 1989) 10 males, 11 females; (February 1990) one male, six females; and (August 1991) nine males, two females. Some of these were released alive but their lengths were measured. From these (and the examination of animals that died) it can be deduced that the last two incidents involved six mature females and one juvenile male, and nine mature males and two pubertal females respectively.[3] Schools of Risso's dolphins may reflect at least some sex and age segregation. The young may stay in their natal school until they are about five years old, mature females in a similar reproductive

FIG. 155 Risso's dolphins might well be dubbed 'graffiti' dolphins from the extent of scribbling that their skin incurs with linear and other scarring. A spy-hopping dolphin at the Maldives (*below*) shows the vertical furrow to the front of its blunt head.

condition may school together, and mature males may move between schools.[27] Some of these subgroups exhibit individual cohesion over periods of at least 15 months.[1]

The reaction of Risso's dolphins to ships varies: off South Africa they rarely bow-ride powered vessels, preferring to ignore or actively avoid them,[3] but elsewhere they have been reported bow-riding vessels, including the famous 'Pelorus Jack' that piloted ships in New Zealand waters between 1888 and 1912.[30] Aerial activity, including side-breaching and lob-tailing, is frequently seen.[1]

Vocalisations produced by Risso's dolphins have been divided into seven types: broadband clicks, barks, buzzes, grunts, chirps, whistles and simultaneous whistle/burst-pulse sounds.[31] On average, the whistles range in frequency from 8.83 kHz to 13.44 kHz, with starting and ending frequencies of 12.1 kHz and 10.83 kHz respectively; they have 1.37 inflections and last about 0.53 sec.[32] The broadband clicks have peak frequencies around 50 kHz and centroid frequencies of 60–90 kHz.[33] The use of the clicks in echolocation has been demonstrated in a captive Risso's dolphin tasked with discriminating between objects of different shapes and materials while blindfolded with eye-cups.[34] A comparison of the hearing abilities of a juvenile 1.47 m long with an adult female Risso's dolphin estimated to be over 30 years old suggested (despite differences in experimental procedures) that the older animal had suffered significant hearing loss at higher frequencies (above 100 kHz).[35]

Conservation status

There are no estimates of population size in the subregion, and no indications that the species is subject to any significant fisheries-related mortality.

Risso's dolphins are listed as Data Deficient by the IUCN and in the South African Red Data Book. They are included on Appendix II of CITES, and the North and Baltic Sea populations on Appendix II of CMS.

References

1 Kruse *et al.* 1999
2 Ross 1984
3 MRI Whale Unit data
4 Frantzis and Herzing 2002
5 Mizue and Yoshida 1962
6 MacLeod 1998
7 Harmer 1927
8 Flower 1872a
9 Rosenbaum and Collins 2006
10 Weir, C., pers. comm.
11 Findlay *et al.* 1992
12 Kiszka *et al.* 2006a
13 Peddemors *et al.* In press
14 Keller *et al.* 1982
15 Baumgartner 1997
16 Cañadas *et al.* 2002
17 Leatherwood *et al.* 1980
18 Gannier 1998
19 Forney and Barlow 1998
20 Shane 1995b
21 Mörzer Bruyns 1971
22 Shane 1995a
23 Cockcroft *et al.* 1993
24 Clarke 1996
25 Clarke and Young 1998
26 Nachtigall *et al.* 1990
27 Amano and Miyazaki 2004
28 Sylvestre and Tasaka 1985
29 Fraser 1940
30 Baker 1974
31 Corkeron and Van Parijs 2001
32 Rendell *et al.* 1999
33 Madsen *et al.* 2004a
34 Philips *et al.* 2003
35 Nachtigall *et al.* 2005

Melon-headed Whale

Peponocephala electra

Ventral view

FOR A LONG time assigned to the genus *Lagenorhynchus*, this animal now resides in its own genus, *Peponocephala*, in recognition of its closer similarity to pygmy and false killer whales than to the smaller delphinids. The common name is supposedly an English translation of the generic name, but other interpretations are that *pepo* is Latin for pumpkin rather than melon, so the name 'Punkinhead whale' has been suggested.[1] Mercifully this suggestion, which is only marginally better than another alternative, 'many-toothed blackfish', has not been taken seriously.

Little is known about this species in the subregion, so much of the information has been adopted from studies elsewhere.

Appearance

Melon-headed whales are most likely to be confused with pygmy killer whales, which they closely resemble in size, body shape and general coloration. At sea, one clue is the size of the school: melon-headed whales are usually found in gatherings of more than 100 (and more often several hundred) individuals, whereas pygmy killer whale schools are much smaller. Melon-headed whales are often found together with Fraser's dolphin, whereas pygmy killer whales are not. And the species behave differently: melon-headed whales form tightly-packed schools that travel rapidly with frequent course changes, whereas pygmy killer whales tend to raft at the surface and will move away in ranked files when disturbed.[2] Other differences require close examination of a specimen.

The largest individuals in 68 males and 41 females stranded in Japan were 2.64 m and 2.565 m respectively,[3] and among 42 males and 73 females stranded in Brazil the largest were 2.61 m and 2.75 m respectively.[4] Growth curves for the Japanese animals indicate asymptotic lengths of 2.524 m for males and 2.427 m for females, thus showing slight sexual dimorphism in size.[3]

The snout is rounded and projects beyond the tip of the lower jaw. The head shape is distinctive, being tapered in profile but more pointed when seen from above, as the sides of the melon are 'pressed in'.[2] The flippers differ from those of the pygmy killer whale in being more pointed at the tip and with a straighter trailing edge. The dorsal fin is large, erect, falcate and centrally placed, while the flukes have an almost straight trailing edge.

Adult males develop larger flippers (17.5–20.4% cf. 16.2–17% of body length), taller dorsal fins (8.4–10.9% cf. 7.3–8.0%) and broader flukes (notch to leading edge is 6.1–7.2% cf. 5.1–5.5%) than females.[5] In addition, the girth at the anus is greater,

Length:weight relationship

Data from 25 animals (including five foetuses) from Japan (all presumably weighed whole, although the weight of the foetuses was subtracted from that of the pregnant animals) gave the following relationship: $W = 0.0000192L^{2.91}$, where W = weight in kilograms and L = total length in centimetres.[3] This implies that a 1.5 m dolphin would weigh 41.3 kg and a 2.5 m animal 182.5 kg. A 2.48 m male from South Africa weighed 206 kg.[5]

Skeletal data

Vertebral formula (n = 20): 7 cervical, 13–16 thoracic, 15–18 lumbar and 39–45 caudal; total 78–83.[3, 5]

The first 3–4 cervical vertebrae are fused together, as are the fifth and sixth, and the sixth and seventh on occasion.[3, 5]

Phalangeal formula, including metacarpals (n = 3): I_{2-3}, II_{8-9}, III_{6-7}, IV_{3-4}, V_{2-3}.[5–7]

The characteristics of the skull that distinguish melon-headed whales from pygmy killer whales are as follows: 1) it has a greater number of teeth; 2) the external nares are bordered rostrally by the maxillae, which emerge as a triangle between the premaxillae; 3) the pair of premaxillary foramina are anterior to the most anterior pair of maxillary foramina (sometimes the right premaxillary and maxillary foramina are almost at the same level); and 4) the left external naris in dorsal view shows a markedly larger aperture than the right naris.[8]

presumably as a result of the development of a distinct post-anal hump in adult males (2.31 m or more), and the snout becomes more rounded in profile.[3]

Melon-headed whales are a dark greyish-black overall, but in some living animals a dark dorsal cape is apparent (but not as distinct as in pygmy killer whales), narrow over the head and thorax but widening near the dorsal fin before narrowing again on the rear half of the caudal peduncle. The grey pigment on the side responsible for this marking fades rapidly after death (or is not present in some individuals/populations) so that the animal appears almost completely black above. Also present is a dark eye spot that extends forward onto the side of the melon, where it meets a pale stripe from the blowhole to the tip of the snout. The lips may also be edged with white.[2]

The most conspicuous markings, however, are pale grey throat chevrons, a midventral streak and a genital patch, very similar in conformation to those seen on the belly of a Risso's dolphin. The pale genital patch extends back as far as the anus in males but only to the ends of the mammary slits in females. These pale ventral markings seem to be present in all age classes.[3]

Tooth counts have usually been given as 20–26 pairs in the upper jaw and 22–25 in the lower jaw,[5] but in a mass stranding from Japan the counts ranged lower: 14–25 (mean 20.7, n = 14) in the upper and 18–25 (mean 21.4, n = 14) in the lower jaw.[3]

The teeth are conical and more robust than in many smaller delphinids, their maximum diameter at the gum being 6.3–6.6 mm, but the tips can show considerable wear in older individuals. Unworn teeth are shorter (7.6–8.8 mm) than those of pygmy killer whales (19.2–23.2 mm).[5]

Distribution and migrations

Melon-headed whales are distributed worldwide in tropical and subtropical seas, with most records occurring between latitudes 20° N and 20° S. Records from more temperate regions (such as the stranding that occurred near Cape Town in 1976[5]) are likely to be strays.[9]

In the subregion they have been recorded from Gabon,[10] Angola,[11] South Africa,[5] Mozambique,[12] the Comoro Islands,[13] Réunion,[14] Tanzania[15] and the Seychelles,[5] as well as pelagically. The most southerly sighting in the subregion is at 25°36′ S.[2]

The species is described as oceanic, being usually found in water depths in excess of 500 m,[16, 17] although it can be found in close proximity to land where the continental shelf is very narrow.[18]

There is no reliable information on seasonal movements anywhere in its range.

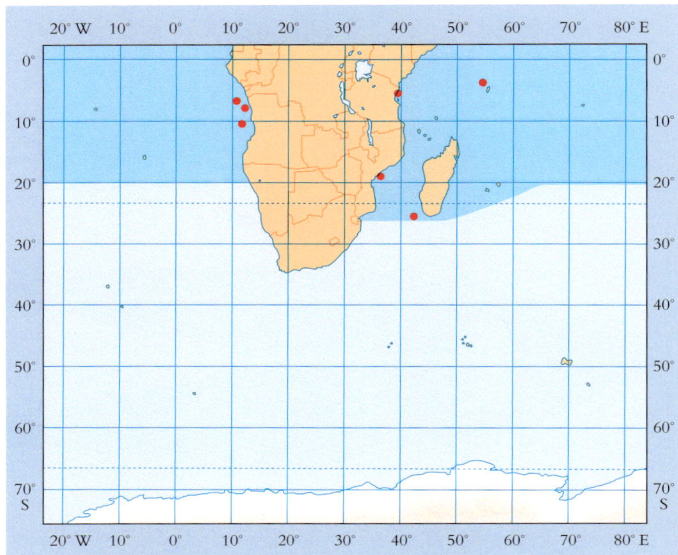

Postulated range · Sighting or capture

FIG. 156 Distribution map for melon-headed whale.

External commensals and parasites

Scars on the trailing edge of the flukes of an animal stranded in Australia suggested prior attachment of the stalked barnacle *Tubicinella*.[7] Cyamids were recorded between the flippers, around the mammary gland and umbilicus, and under the right eye of an immature female.[19] Twelve cyamids collected from cookie-cutter shark lesions on a stranded individual proved to be *Isocyamus delphinii*.[20] Wounds supposedly caused by this shark (*Isistius* sp.) have been reported on several melon-headed whales,[5, 21] and are especially prevalent on the belly rather than the back or sides.[3] Remoras have been seen attached to these whales in the eastern tropical Pacific.[2]

Food and feeding behaviour

The only stomach examined in the subregion (from a stranded whale) contained remains of two hake (*Merluccius* sp.), two chokka squid (*Loligo reynaudi*) and one ommastrephid sp.[5] Elsewhere there is a similar scarcity of information: an animal from Brazil contained unidentified fish remains and the remains of 11 cephalopods (nine *Ornithoteuthis antillarum*, two *Abralia veranyi*);[22] one from Hawaii had unidentified fish remains and the remains of six cephalopods (one *Enoplcteuthis*, one possible *Teuthowenia*, two *Abraliopsis*, one probable *Abralia* sp., and one *Bathyteuthis abyssicola*);[23] and one from the West Indies had unidentified fish and cephalopod remains.[24]

Unpublished data indicate feeding on fish of the families Myctophidae, Paralepididae and Scopelarchidae, and on cephalopods of the families Ommastrephidae, Onycoteuthidae, Chiroteuthidae, Mastigoteuthidae, Cranchiidae, Enoploteuthidae and Histioteuthidae. Most of these families consist of midwater species, found at depths of up to 1 500 m, indicating that feeding may take place at depth.[9] The sizes of the squid (mainly *Dosidicus gigas*) eaten by individuals stranded in Costa Rica suggested that these whales fed on larger cephalopods than did the spotted or spinner dolphins in the same area.[2]

Several of the prey types are apparently similar to those consumed by Fraser's dolphin, a species with which melon-headed whales are frequently associated.[9]

Reproduction and growth

The largest foetus examined was 1.01 m long,[3] while four calves measured 1.06 m, 1.10 m,[4] 1.12 m[25] and 1.15 m. The last animal, which had a partly closed umbilical slit, was estimated to be about a week old.[26] The length of gestation is not known. The two largest calves (both from the North Pacific) were encountered in May and July, while 10 foetuses in a mass stranding in Japan in January ranged from 0.293 m to 1.01 m with a slight peak around 0.8–0.85 m.[3] This suggests that at least some births may occur in late spring through summer, but that the calving season is quite protracted (as one might expect in a largely tropical animal).

The 1.06–1.1 m calves from Brazil were from a stranding in April, which suggests either that there is little difference in the timing of reproduction between hemispheres, or that the season is extremely protracted. The length of lactation is unknown, but individuals less than 2 m in length had higher nitrogen isotope ratios than larger animals, suggesting that they were still suckling.[27]

Sexual maturity is attained in males at about age 15, when they are 2.45–2.5 m long, while females mature at around 11.5 years and at a length of 2.35 m. The average ovulation rate is estimated at 0.28, implying a calving interval of 3.6 years. No post-reproductive females have yet been found.[3]

The maximum estimated ages of individuals are 38.5 years for males and 45.5 years for females.[3]

Behaviour and vocalisations

Nine sightings of melon-headed whales in the subregion have involved schools ranging in size from one to 600, with a mean of 239, but if the single whale is omitted the range and mean become 100–600 and 269 respectively.[28] Such large groups are often subdivided into tightly-bunched subgroups.[16]

Sex ratios in mass strandings have varied widely, from 64.8%[4] and 68.6%[29] female to 39.5% female, suggesting some behavioural segregation of the sexes. The ratio of sexually active mature males to non-pregnant mature females in one stranding was similar to that expected in a breeding school of striped dolphins.[3]

Elsewhere schools of melon-headed whales are often seen mixed with other cetacean species, especially Fraser's dolphin.[9]

FIG. 157 The grey band running forward from the blowhole onto the front of the head is a good field character for distinguishing melon-headed whales from pygmy killer whales, which they closely resemble.

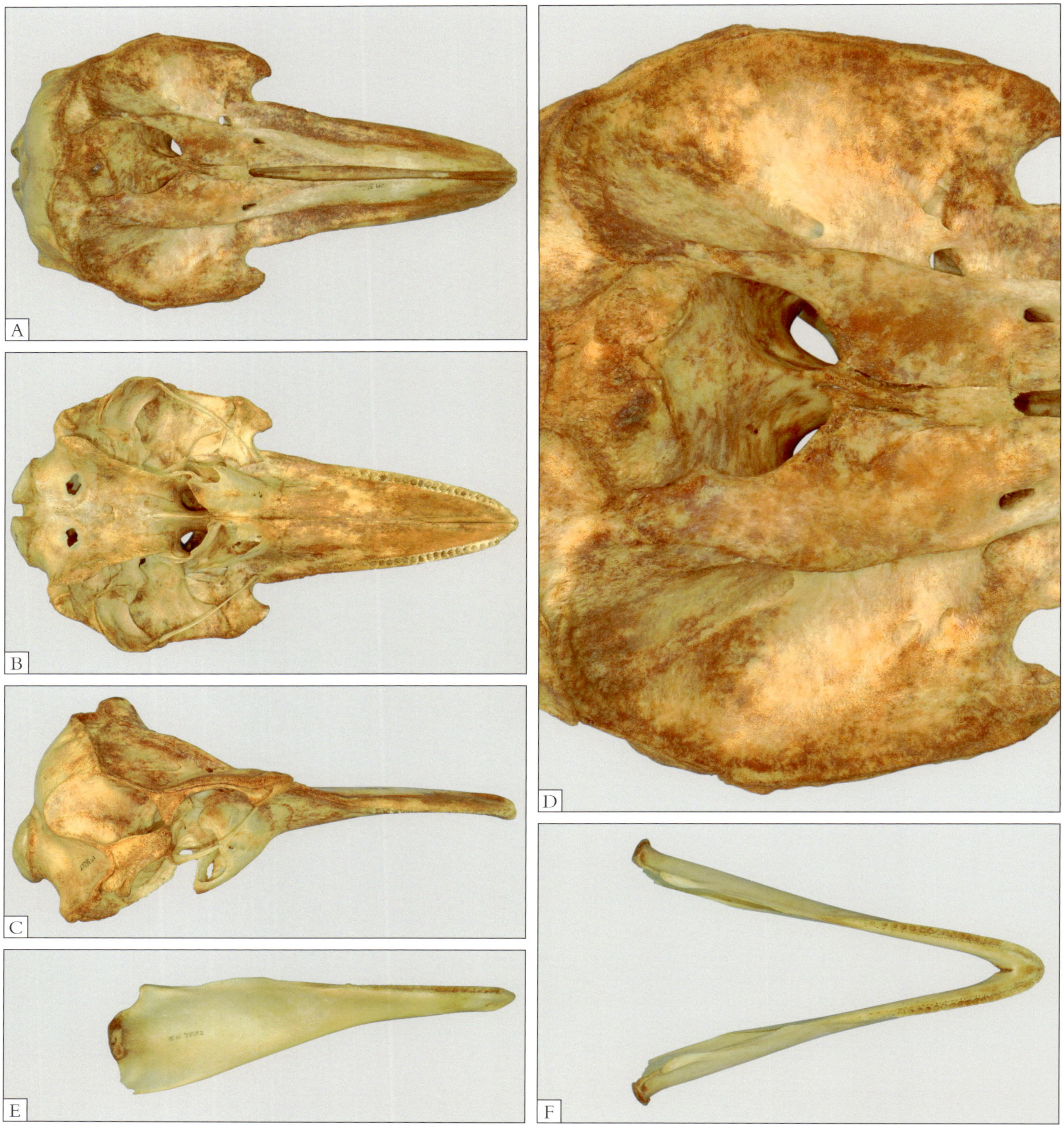

FIG. 158 Dorsal (A), ventral (B) and right lateral (C) views and vertex of skull (D); right lateral (labial) (E) and dorsal (F) views of mandible (ZM 38245).

In the subregion two of the nine sightings involved other species: one an unidentified cetacean and the other a mixed group of 175 Fraser's dolphins and 75 bottlenose dolphins.[28]

The species is said to be particularly aggressive towards humans when in captivity.[2]

When swimming rapidly, melon-headed whales tend to group together and break the surface with a flat trajectory so that they rarely clear the water completely. Their head is said to push a crescent of water ahead of it that breaks over their back as they surface.[2] A similarly distinctive but smaller disturbance (possibly caused by their slight upper lip extension) was seen ahead of surfacing animals even in calm water.[18]

Vocalisations recorded from melon-headed whales at sea included whistles and click sequences. The whistles were short (less than 0.1–0.9 sec.) with dominant frequencies around 8–12 kHz. Starting frequencies were 10.5–18.0 kHz for both upsweep and downsweep whistles, but ending frequencies were 12.0–24.5 kHz for upsweeps and 5.5–12.5 kHz for downsweeps. Click frequencies were mainly between 20 kHz and 40 kHz, most often in bursts of 40 or more at rates of up to 1 200/sec. and lasting 0.1–0.2 sec. The incidence and level of sounds produced seemed to be positively related to the amount of physical activity by the whales.[18]

Conservation status

There are no estimates of population size in the subregion, and no indications that the species is subject to any significant fisheries-related mortality.

Melon-headed whales are listed as Lower Risk: Least Concern by the IUCN and as Least Concern in the South African Red Data Book. They are listed on Appendix II of CITES, but not by CMS.

References

1 Rice 1998
2 Perryman *et al.* 1994
3 Miyazaki *et al.* 1998
4 Lodi *et al.* 1990
5 Best and Shaughnessy 1981
6 Nakajima and Nishiwaki 1965
7 Bryden *et al.* 1977a
8 Van Waerebeek *et al.* 1999
9 Jefferson and Barros 1997
10 Rosenbaum and Collins 2006
11 Weir, C., pers. comm.
12 Peddemors *et al.* In press
13 Kiszka *et al.* 2006a
14 Boucaud and Rota 2006
15 Peddemors and Ross 1988
16 Mullin *et al.* 1994c
17 Mullin *et al.* 2004
18 Watkins *et al.* 1997
19 Miyazaki and Wada 1978a
20 Wardle *et al.* 2000
21 Gasparini and Sazima 1996
22 De Gurjão *et al.* 2004
23 Clarke and Young 1998
24 Caldwell *et al.* 1976
25 Perrin 1976
26 Southall *et al.* 2006
27 Umezaki *et al.* 2003
28 MRI Whale Unit data
29 Bryden *et al.* 1977b

Pygmy Killer Whale

Feresa attenuata

Ventral view

THIS SPECIES WAS first described in 1827 and a second specimen in 1874, but it was another 80 years before a third specimen surfaced and its external appearance was described for the first time. Since then more information has accumulated, but the species is still far from well known.

The common name was suggested because the skull (and especially the dentition) was more reminiscent of a killer whale than a melon-headed whale, to which it was previously thought to be related.[1] The appropriateness of the common name seemed to be borne out by subsequent reports of the behaviour of individuals in the wild and captivity, although the feeding habits of the species are still poorly known.

Appearance

Pygmy killer whales are easily confused with melon-headed whales when seen at sea, being of similar size, body shape and general coloration. Clues to distinguishing between the two species in the field are given under the melon-headed whale species account, and include school size, species associations and school behaviour.[2] Other differences require close examination of a specimen.

The largest individuals in 20 males and 23 females were 2.59 m and 2.45 m long respectively.[3–12] Interestingly, both these individuals originated from the southern hemisphere, and were 15 cm and 18 cm longer respectively than their counterparts from the northern hemisphere (despite the southern sample of measured animals being much smaller in either case). Whether this implies some population differences between hemispheres (and whether there is a size difference between sexes) can only be decided after a fuller investigation of age and growth.

The overall body shape is very similar to that of the melon-headed whale, but the melon is more rounded in the front, giving the animal a less tapered snout in profile. Nevertheless the head is shorter, the distance from tip of snout to centre of blowhole being 6.7–11.9% of body length (mean 9.8%, n = 19) compared to 12.9–14.9% (mean 13.9%, n = 10) in melon-headed whales.

The flippers are also noticeably rounded at the tip (compared to a pointed tip in melon-headed whales) and their leading edge more curved.[13] The dorsal fin and flukes are similar in size and shape to those of melon-headed whales, the height of the fin being 9.4–12.3% (mean 10.5%, n = 20) and the width of the flukes 23.8–28.4% (mean 25.5%) of body length.[9] There does not appear to be the sexual dimorphism in the size of the dorsal fin, flippers and flukes shown by adult melon-headed whales,[8] nor has a post-anal hump been described in mature males. Several individuals seem to have irregular indentations on the trailing edges of the dorsal fin, flippers and flukes.[3, 8]

Pygmy killer whales are generally dark bluish-black over most of the back, sides and belly, but with a paler grey lateral marking and white or grey markings on the belly. The upper margin of the lateral marking commences above the eye, runs back along the upper thorax and then dips down the side at about the level of the anterior insertion of the dorsal fin, before curving up again onto the caudal peduncle. This grey coloration is less visible when the skin is dry and fades rapidly after death.

On the chest there is a grey marking that resembles the anchor marking of the Risso's dolphin and melon-headed whale and extends towards the rear as a narrow streak. Behind it there is a roughly elliptical white marking enclosing the anus and genital slits. Both these markings are clearly visible after death, before sun-darkening takes place.

The lips are edged with white, forming a node at the tip of the snout (and occasionally extending over a large part of the throat[14]), and there may be irregular white patches in the axillary region of the flippers. The development of this pattern with age is unknown, although at birth the snout and chin appear lighter in colour than the rest of the body and this must intensify with age to form the white lips.[3, 8, 10]

There are 8–11 (mean 9.8, n = 27) pairs of teeth in the upper jaw and 10–13 (mean 12.2, n = 27) pairs in the lower jaw.[9] They are conical and pointed, with a maximum diameter of 6–8 mm (mean 7.4 mm, n = 4),[8, 15] but may become worn or chipped in older animals.

Distribution and migrations

Pygmy killer whales are found worldwide in tropical and subtropical waters, although most of the records outside the tropics have been associated with warm western boundary current systems that may extend the species' range.[9]

Within the subregion it has been recorded from Gabon,[16] Namibia, South Africa,[17] the Comoro Islands[18] and the Seychelles,[3] as well as pelagically. Several other records are available in which it was not possible to be certain whether the species involved was the pygmy killer or melon-headed whale. It has been encountered in water as shallow as 47 m but is more commonly found in oceanic water.[19]

There is little information available from which to assess whether seasonal movements occur, although the pygmy killer whale is apparently found year-round off Sri Lanka and in the West Indies.[9]

A swimming speed of about 30 km/hr has been estimated.[19]

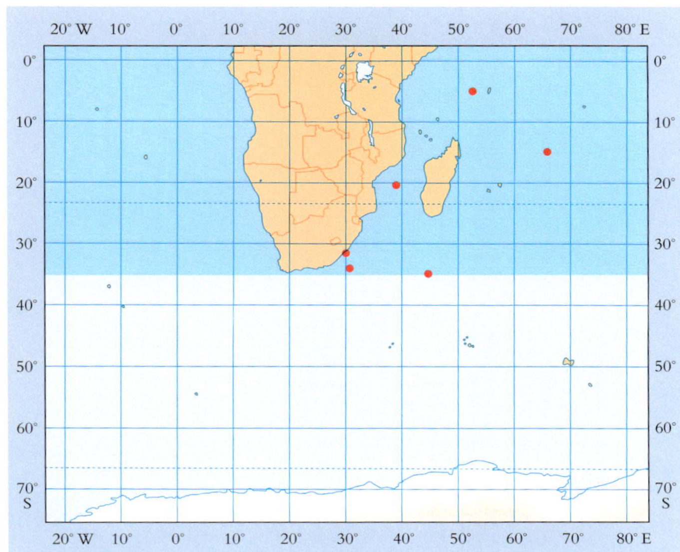

FIG. 159 Distribution map for pygmy killer whale.

> ### Length:weight relationship
>
> Data from 11 animals (all presumably weighed whole) gave the following relationship: $W = 0.00002502L^{2.882}$, where W = weight in kilograms and L = total length in centimetres.[9] (This regression does not include two individuals from South Africa, a 2.235 m female weighing 131.1 kg in pieces and a 2.33 m male weighing 166 kg whole.) The regression implies that a 1.5 m animal would weigh 46.8 kg and a 2 m animal 107.1 kg.
>
> ### Skeletal data
>
> Vertebral formula (n = 24): 7 cervical, 12–13 thoracic, 15–17 lumbar and 32–34 caudal; total 67–70.[9]
>
> The first 3–6 cervical vertebrae are fused in physically mature animals.[9]
>
> Phalangeal formula, including metacarpals (n = 17): I_{2-3}, II_{7-10}, III_{6-8}, IV_{3-5}, V_{2-3}.[9]
>
> For the skull characteristics that distinguish pygmy killer whales from melon-headed whales, see the latter's species account.

External commensals and parasites

Stalked barnacles (*Conchoderma* sp.) were found attached to the teeth of an adult male (three on left lower 3; one on left lower 4), presumably because a number of teeth were broken or worn. Otherwise no external parasites were found on two other individuals examined in the subregion.[15] A young individual stranded in Australia carried numerous stalked barnacles, *Xenobalanus* (misidentified as *Conchoderma*) on the trailing edges of the dorsal fin and tail flukes,[10] possibly because it was in poor condition, and 'barnacles' (unidentified) were found attached to the lips, flippers and flukes of animals from Sri Lanka.[9] Cyamids (unidentified) were described from the body surface of a stranded male in the Caribbean.[20]

Food and feeding behaviour

Very little is known. One stomach examined in South Africa contained 'squid beaks',[4] and another the remains of 17 hake (*Merluccius* sp.) and one bearded goby (*Sufflogobius bibarbatus*).[15] An individual stranded in Brazil had eaten five cephalopods (two *Loligo plei*, two *Illex argentinus* and one *Ornithoteuthis antillarum*) and fish eye lenses were present.[21] Unidentified squid beaks and

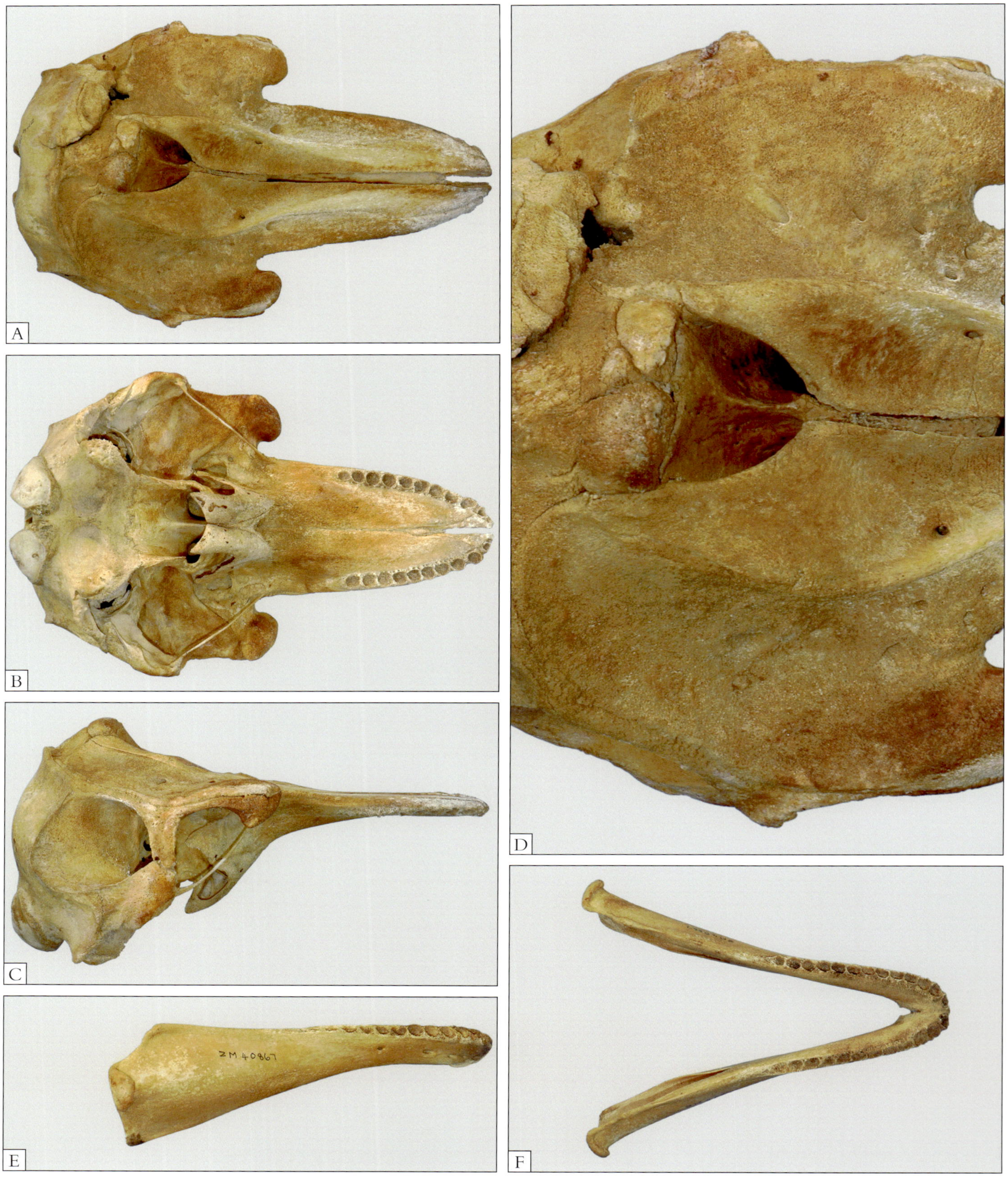

FIG. 160 Dorsal (A), ventral (B) and right lateral (C) views and vertex of skull (D); right lateral (labial) (E) and dorsal (F) views of mandible (ZM 40867). (Damage from gun shot to left maxillary crest.)

fish otoliths were found in two individuals from Sri Lanka[9] and three from the Caribbean,[6, 20] while the stomach of an animal stranded in Argentina contained an unidentified eye lens.[5]

Despite these seemingly innocuous stomach contents, the pygmy killer whale has gained something of a sinister reputation. Besides being responsible for the deaths of pilot whale[7] and dusky dolphin[3] tank-mates in captivity (see box alongside), these little whales have possibly been implicated in attacks on dolphins during tuna purse-seining operations,[22] and have been seen skulking around cetaceans with calves[23] (though whether with malicious intent is not known).

The high concentrations of myoglobin found in the muscle of a pygmy killer whale suggest that this may be a relatively deep-diving species.[24]

Two captive adult males consumed up to 5.4 kg[7] and 8 kg[8] of food a day.

FIG. 161 A pygmy killer whale leaping in the Philippines reveals the rounded tip to the flipper that is one of the features distinguishing it from the melon-headed whale.

Sea wolf

In 1963 the collecting team from Sea Life Park, Hawaii, brought back a strange male 'porpoise' for the oceanarium. Not only was its appearance unusual, but its behaviour gave some cause for alarm, as it was snapping at its captors and emitting blatting or growling noises. Still, staff at the Park were unprepared for the events that took place once the pygmy killer whale got to the oceanarium. Unlike most newly captured cetaceans, it made almost no attempt to avoid observers but acted as if it expected them to move. Swimming rapidly around the tank it leaped out of the water and snapped three times at onlookers, 'like a seagoing wolf'.

Although its attitude to people calmed down over subsequent days, it often growled and had the disturbing habit of lying on its side in the centre of the tank, watching passers-by with one eye and patting the water surface with its flipper in a manner that resembled 'an irritable man drumming his fingers on a tabletop'.

Because the animal seemed bored and lonely it was transferred to a tank holding two pilot whales. Occasionally it took to charging at the smaller pilot whale at right angles, and although the pilot whale easily dodged these, one morning it was found dead. A post-mortem revealed it had died from a sharp blow at the base of the skull, a technique sometimes practised by killer whales. After a period of isolation, the *Feresa* was introduced to a little spinner dolphin: it responded by playing cat and mouse with it, stationing itself in the centre of the tank and making rushes at the spinner, which circled the tank at ever-increasing speed with the pygmy killer pivoting in the centre 'no doubt laughing sadistically'. The spinner was rescued.[1]

With these experiences in mind, it was with some trepidation that an adult female pygmy killer whale that stranded near Cape Town in 1970 was placed in a tidal pool with bathers celebrating their New Year holiday, while awaiting personnel from a local zoo. Fortunately nobody was hurt, but once the whale got into captivity it revealed another side to its nature, killing the smallest of four dusky dolphins housed with it during the night, and tooth-raking another before dying after six days.[2]

Neither whale attempted to feed on the cetaceans it killed, and given its somewhat weak dentition and small size it is difficult to visualise how it could effectively dismember such large 'prey', at least by itself.

Other recently-captured pygmy killer whales were kept for some days in an enclosure in Japan with dolphins without any (reported) interactions.[3] Nevertheless, Sea Life Park's experiences at least were sufficient to dissuade them from attempting to collect any further individuals.[1]

References

1 Pryor 1975
2 Best 1970b
3 Nishiwaki *et al.* 1965

Reproduction and growth

Very little is known. An individual that may have been newly born was 0.822 m long,[11] while a 1.105 m female had been alive long enough to acquire a heavy infestation of barnacles.[10] It is believed that births take place mainly in summer[9] and mating has been seen in spring.[25] A 1.65 m female stranded in the Caribbean in September was still nursing.[6] Eight females 2.05–2.45 m (mean 2.21 m) in length were judged to be mature, as they were pregnant or lactating.[3, 5, 6, 8, 9] Three males 2.07–2.33 m long (the largest from South Africa) were judged to be mature from testis size or the presence of spermatozoa,[7, 14, 15] while two at 1.70 m and 2.2 m were immature.[9, 20]

Behaviour and vocalisations

Six groups sighted in the subregion ranged in size from 10 to 27, with a mean of 15.2.[15] Similar relatively small schools have been seen elsewhere, generally consisting of 70 individuals or fewer.[9]

Schools are frequently organised in ranks,[2] and when swimming slowly tend to surface synchronously.[23] When travelling fast, pygmy killer whales clear the water with clean leaps, entering either head first or on their side. Individuals will also bow-ride.[19] Vertical spy-hopping (when they appear to inspect objects or people) has also been reported.[8]

Pygmy killer whales have been reported producing whistles,[14, 19] 'blats and growls'[7] and click sounds similar to those recorded for other toothed cetaceans of a similar size. The clicks were produced at intervals of 50–120 ms (or an instantaneous repetition rate of 8–20/sec.), had peak frequencies between 45 kHz and 117 kHz, and centroid frequencies between 70 kHz and 85 kHz. Assuming that the pygmy killer whale can detect and process the incoming signals with the same capability as other dolphins, it should be able to detect prey at ranges of 50–200 m.[26]

Conservation status

There are no estimates of population size in the subregion, and no indications that the species is subject to any significant fisheries-related mortality.

Pygmy killer whales are listed as Data Deficient by the IUCN and in the South African Red Data Book. They are included on Appendix II of CITES, but not listed by CMS.

References

1 Yamada 1954b
2 Perryman *et al.* 1994
3 Best 1970b
4 Ross 1984
5 Lichter *et al.* 1990
6 Mignucci-Giannoni *et al.* 1999b
7 Pryor *et al.* 1965
8 Nishiwaki *et al.* 1965
9 Ross and Leatherwood 1994
10 Bryden 1976
11 Perrin and Hubbs 1969
12 Forrester *et al.* 1980
13 Best and Shaughnessy 1981
14 White 1976
15 MRI Whale Unit data
16 Rosenbaum and Collins 2006
17 Findlay *et al.* 1992
18 Kiszka *et al.* 2006a
19 Castro 2004
20 Rodríguez-López and Mignucci-Giannoni 1999
21 Dos Santos and Haimovici 2001
22 Perryman and Foster 1980
23 Williams *et al.* 2002
24 Dolar *et al.* 1999
25 Würsig *et al.* 2000
26 Madsen *et al.* 2004b

False Killer Whale

Pseudorca crassidens

♂

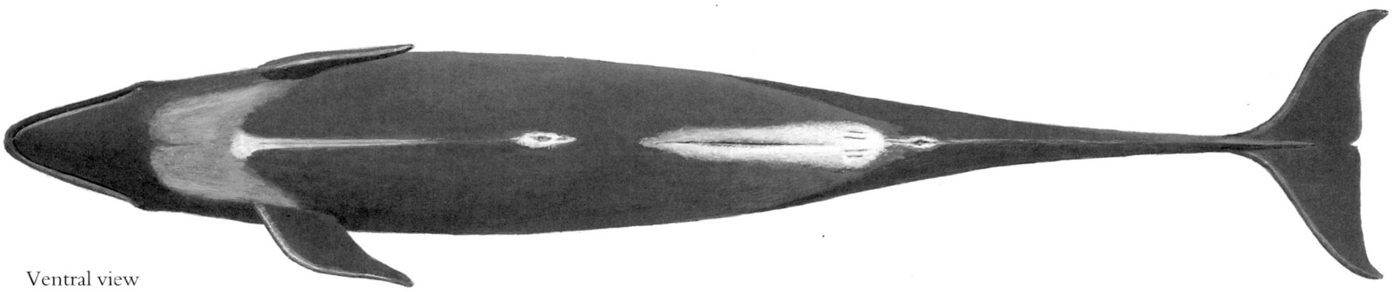

Ventral view

♀

IN YET ANOTHER example of the bizarre ways in which cetacean species have been 'rediscovered', the false killer whale was first described as a sub-fossil from a fen in England. It was only 15 years later, when a school of the whales stranded in Germany, that it was realised that the species was still alive and well.[1]

Although generally recognised as one species worldwide, analyses of skull morphology (and growth rate) indicated substantial differences between populations off Scotland, South Africa and Australia.[2] Furthermore, a preliminary genetic study of seven animals from China suggested two distinct clades, between which the divergence was greater than between some cetacean species.[3] It would be premature to suggest any revision of the taxonomy, but it is clear that false killer whale populations in different parts of the world may differ substantially. This account therefore attempts to make as much use as possible of information from the subregion.

Appearance

False killer whales are most likely to be confused with melon-headed or pygmy killer whales. Apart from their greater size, false killer whales can be distinguished by their relatively smaller dorsal fin, which is narrower and with a more rounded and backward-pointing tip. At close range the distinctive S-shape of the leading edge of the flipper (sometimes referred to as a hump on the forward margin) may be detected.[4] Although the coloration is often said to be dark all over without any trace of the cape marking seen in pygmy killer and to some extent melon-headed whales, some live individuals photographed in the subregion have a distinct cape marking on the thoracic region.

The largest individuals among 28 males and 43 females measured in the subregion recently were 5.14 m and 4.57 m long respectively.[5] A male 5.36 m long was recorded from a 1936 mass stranding in South Africa, but it is not clear how the animals were measured: 47.3% of the 19 females at that stranding exceeded 4.1 m in length, whereas in more recent data only 4.7% of females have been in this size category.[6] Asymptotic lengths are 4.64–4.79 m for males and 3.85–3.92 m for females, depending on the growth model used,[7] but obviously the species is strongly dimorphic in size by sex.

The body shape is slim, with the snout rounded and tapered but more swollen at the tip than in pygmy killer and melon-headed whales, coming to project well beyond the tip of the lower jaw, especially in adult males. The distinctively S-shaped flippers are 9.4–14.5% (mean 12%, n = 108) of the body in length, while the height of the dorsal fin is 6.3–10.5% (mean 8.2%, n = 87) of body length.[8] There does not appear to be the development of a post-anal hump in adult males, but a peculiar feature is the presence of non-functional 'false mammary slits', even in males.[9]

Photographs of live false killer whales in St Helena Bay, South Africa, taken from the air, show that at least some individuals carry a dark central cape marking on the front part of the back, which is narrowest about halfway between the blowhole and the dorsal fin and then rapidly widens. The melon may be pale. Unfortunately details of the sides of the animals cannot be distinguished. The cape marking does not seem to be present in a freshly killed animal from the east coast of South Africa, and may in any case fade rapidly after death. The remainder of the body is a dark greyish-black, apart from a light grey anchor-shaped marking on the chest that extends back as a narrow streak in the mid-line towards the umbilicus. The ano-genital region is also enclosed in a more diffuse pale grey area. There is no white edging to the lips as seen in adult pygmy killer and melon-headed whales.

There are 7–9 (mean 8.4, n = 52) pairs of erupted teeth in the upper jaw and 8–10 (mean 9.2, n = 53) pairs in the lower jaw.[5, 6] They are conical, roughly circular in cross-section (unlike those of killer whales) and the largest may measure 1.6–1.9 cm in diameter.[5, 10] Wear on the teeth may include some undercutting, suggesting that the animal is capable of making small sideways and backward movements of the jaw as well as vertical movements, possibly associated with the breaking up of large prey, and in this respect resembles the killer whale.[11]

Length:weight relationship

Published information is available for five animals[8] and unpublished data for four, the largest of which was a 5.14 m male weighing 1 302.79 kg in pieces.[5] A value of 5% has been used to estimate weight loss for this individual, and the following relationship calculated for all nine animals: **$W = 8.9509L^{2.9247}$**, where W = weight in kilograms and L = total length in metres. The regression implies that a 3 m whale would weigh 222.5 kg and a 4.5 m animal 728.3 kg. Data from captive animals[12] have not been used as they seem to lie mainly above this regression, possibly indicating a higher level of nutrition.

Skeletal data

Vertebral formula (n = 103): 7 cervical, 9–11 thoracic, 9–13 lumbar and 16–26 caudal; total 47–52.[13] Four animals from the subregion fell within these limits, with total vertebral counts of 47–51.[5, 11]

The degree of fusion of the cervical vertebrae is variable. The first five vertebrae are fused in 23.8% of cases, the first six in 67% and all seven in the remaining 12.2% (n = 82). There is a possibility that the extent of fusion increases with increasing age.[14]

Phalangeal formula, including metacarpals (n = 93): I_1, II_{5-9}, III_{5-7}, IV_{2-4}, V_{0-2}.[15] One individual from the subregion fell within these limits.[5]

The skull of the false killer whale can be readily distinguished from that of the killer whale by the greater relative width of the premaxillae, which occupy more than half the width of the rostrum just in front of the antorbital notches.[9] In addition, the pterygoid bones at their base are closely apposed in the mid-line, and the alveoli are completely separated from each other throughout their whole extent.[16]

Distribution and migrations

False killer whales are found worldwide from tropical to temperate waters, with their latitudinal limits generally being between 50° N and 50° S.[8] In the subregion they have been recorded from Gabon,[17] Angola,[18] Namibia, South Africa,[19] Mozambique,[20] Tanzania (Zanzibar),[21] Madagascar,[22] the Comoro Islands[23] and the Seychelles,[21] as well as pelagically, with the most southerly record being at 35°31′ S.[5]

Off South Africa most sightings were made in water more than 1 000 m deep, but occasionally the species is found closer inshore in much shallower water.[19] Elsewhere it has been described as an oceanic species that makes occasional forays into inshore waters.[24]

There is no real evidence of seasonal movements. Resightings of individuals in the same vicinity after intervals of 11–24 months suggest some fidelity to particular areas.[24]

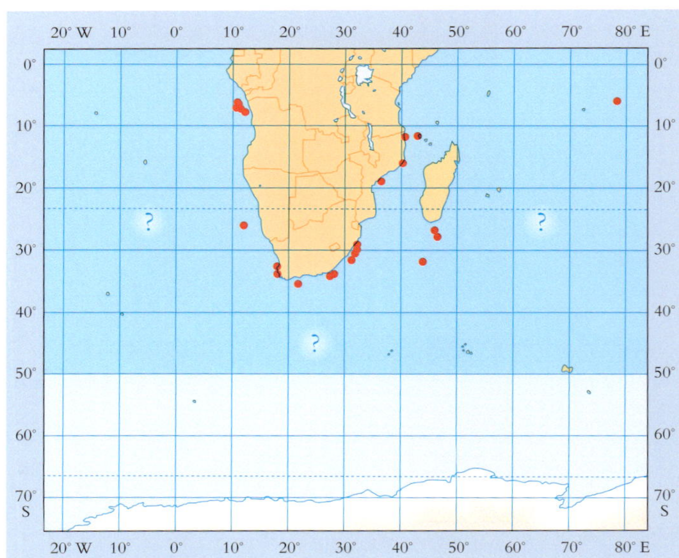

Postulated range ■ Sighting or capture ? Questionable origin

FIG. 162 Distribution map for false killer whale.

Mass strandings

FIG. 163 Alison and Robert Best inspect some of the false killer whales that stranded in St Helena Bay, South Africa, in 1981.

Multiple beachings of whales or dolphins juxtaposed in time and space are normally called mass strandings, and differ from sequential beachings of animals over a protracted period and an extended geographical region, as typically occur in an epizootic event, or mass mortality associated with disease. In southern Africa, false killer whales are the cetacean most likely to strand *en masse*. Since 1928 eight such incidents have been recorded: 24 December 1928, Long Beach, Kommetjie, more than100 animals; 2 December 1933, Mtoni Beach, Zanzibar, 54 animals; 19 November 1935, Seaspray, Mamre district, 200–300 animals; 27 December 1936, Varkvlei, St Helena Bay, 58 animals; 19 August 1958, Die Kelders, Gansbaai, four animals; 12 June 1973, Berg River mouth, St Helena Bay, four animals; 19 August 1981, Varkvlei, St Helena Bay, 65 animals; 29 September 1982, Elizabeth Bay, Namibia, five animals. The fact that three of these incidents occurred over much the same stretch of coastline is a feature common to many mass stranding events. Typically the species involved are also oceanic, gregarious animals, such as pilot, false killer, melon-headed or sperm whales.

Several ingenious hypotheses to explain this phenomenon have been put forward over the years. A gently shelving sea floor (together with micro-bubble formation in the nearshore region) could interfere with echolocation.[1,2] If cetaceans navigate using the earth's geomagnetic field, then a geomagnetic disturbance could interfere with their navigation and cause them to follow the field inshore, mass stranding sites frequently occurring where geomagnetic contours intersect the coast.[3,4] One or more individuals in the school may be debilitated (through parasitic infestation of the middle ear, for instance[5]), causing them to become disoriented or sick enough to seek out land, other animals following through social bonding. Or the phenomenon could simply be a density-dependent response, in which as a population approaches its carrying capacity it suffers increased natural mortality, manifested in the case of these social animals by the loss of whole schools.[6] It is very likely that no single cause is responsible for each and every event, and elements of some of the above hypotheses may have to act in combination to produce the circumstances necessary for a mass stranding to occur. In the case of beaked whales it now seems that human intervention, in the form of mid-range sonar, may have to be added to the list of probable causes (see Cuvier's beaked whale account).

References

1 Dudik van Heel 1962
2 Chambers and James 2005
3 Klinowska 1985a
4 Klinowska 1985b
5 Morimitsu *et al.* 1987
6 Sergeant 1982

In captivity false killer whales have achieved maximum swimming speeds of 28.8 km/hr in short bursts.[25] In the wild they have been estimated to swim at a speed of at least 18.5 km/hr.[26]

External commensals and parasites

No external parasites were found on five false killer whales examined in the subregion.[5] Elsewhere stalked barnacles (*Conchoderma auritum*) have been reported attached to the teeth,[27] *Xenobalanus globicipitis* on appendages[9] and two species of cyamids (*Syncyamus pseudorcae*[28] and *Isocyamus delphinii*[29]) on the body surface.

Food and feeding behaviour

All 48 food items identified from the stomachs of 13 false killer whales from the subregion were cephalopods, of which *Lycoteuthis diadema* (43.8%) and *Todarodes sagittatus* (37.5%) were numerically the most important. The diamondback squid (*Thysanoteuthis rhombus*) was only represented by a single specimen, but with an estimated weight of just under 29 kg it was by far the biggest item consumed.[5] All but two of these stomachs came from mass stranded animals that (as found elsewhere[30]) contained very little in the way of food, so may not be representative of the diet. In addition, if the species does break up larger food items (see below) then using trace remains such as beaks or otoliths to determine diet may severely underestimate their contribution. The stomach contents as described suggest that false killer whales in the subregion were foraging at the edge of the continental shelf.[31]

Elsewhere the species has been recorded feeding on a variety of fish, including salmon (*Oncorhynchus* sp.), sciaenids, carangids, bonito (*Sarda* sp.), dorado (*Coryphaena* sp.), yellowfin tuna (*Thunnus albacares*), yellowtail (*Pseudosciana*), barracuda, perch (*Lateolabrax japonicus*), amberjack (*Seriola alandi*), catfish (*Tachysurus* spp.),[8] and Patagonian grenadier (*Macruronus magellanicus*).[30] Cephalopods included *Gonatopsis borealis* or *Berryteuthis magister*,[8] *Ommastrephes bartrami*, *Argonauta* sp.,[27] *Martialia hyadesi*, *Illex argentinus*, *Todarodes fillipovae*, and *Moroteuthis ingens*.[30]

Although no marine mammal remains have yet been found in false killer whale stomachs, there are several anecdotal accounts that suggest an apparently unhealthy (for the victim) interest in other cetaceans. The pursuit, capture and killing of dolphins has been recorded,[24, 32] but not their actual consumption, although the breaking up of large prey items such as dorado (*Coryphaena*) has been seen in both captivity and the wild.[26] False killer whales have also been seen harassing humpback whales[5] and even attacking sperm whales, removing pieces of flippers or flukes. Although the latter incident was considered a physical attack, its motive was unclear and could have represented either a predation attempt, competition for food, or food robbery (in which

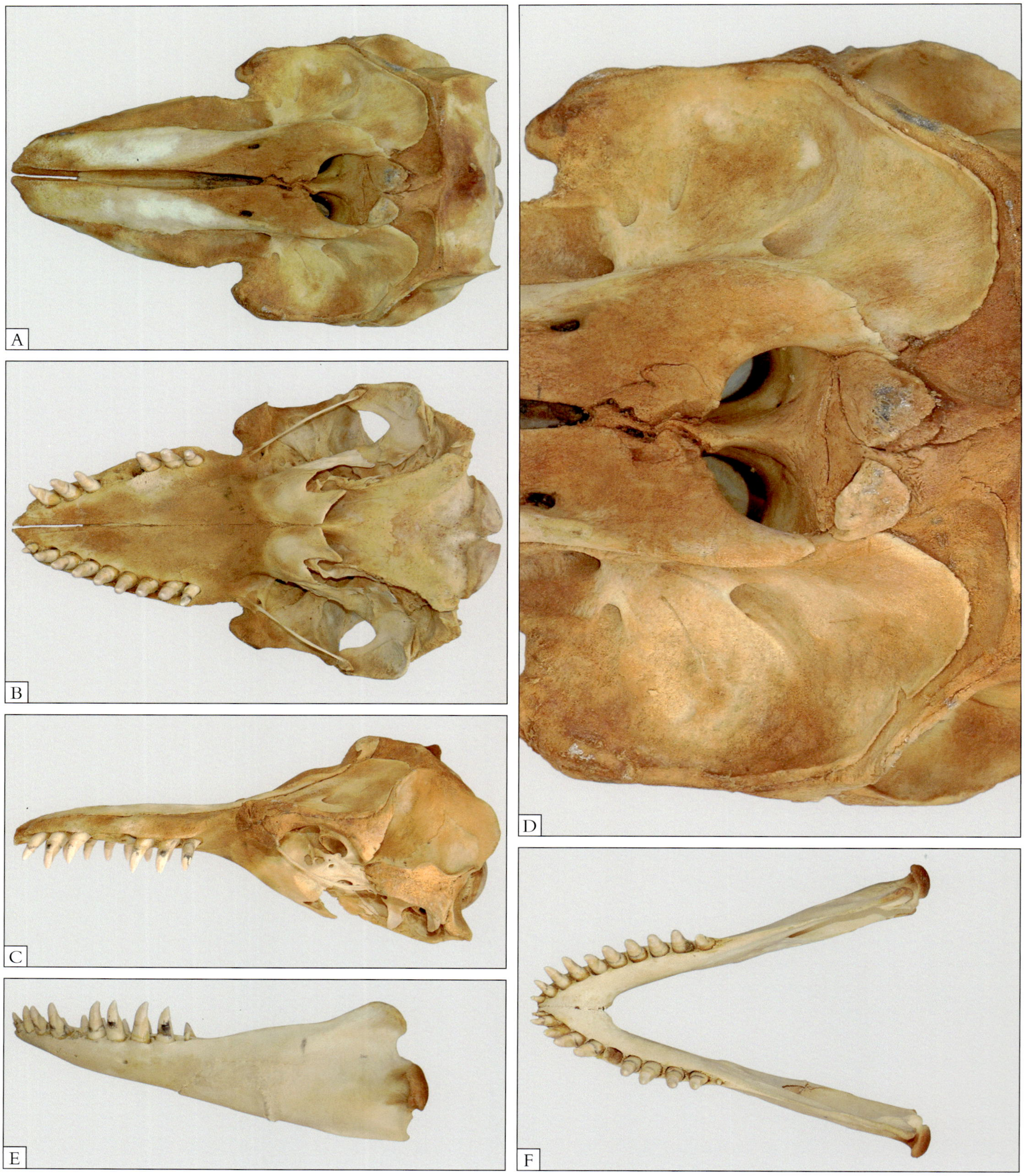

FIG. 164 Dorsal (A), ventral (B) and left lateral (C) views and vertex of skull (D); left lateral (labial) (E) and dorsal (F) views of mandible (ZM 36811).

sperm whales could be induced to vomit food that would be consumed by the false killer whales).[33]

Cooperative feeding was demonstrated by a false killer whale that allowed two other individuals to tear chunks of flesh from a dorado that it was holding in its mouth.[34]

Three captive false killer whales weighing 350–380 kg consumed food equivalent to 4.1–4.3% of their body weight (and a 450 kg female 3.4% of its body weight) per day.[12]

Reproduction and growth

In the subregion the largest foetus found was 1.48 m long (in August), whereas calves of 1.58 m (in December), 1.606 m (in February) and 1.755 m (in August) have been measured.[5, 6] Length at birth is therefore about 1.55 m and the season of birth probably protracted. Captive mature females ovulated spontaneously and were reproductively active for most of the year, but with elevated progesterone levels in summer.[35] Pregnancy lasted 14 months in a captive female[8] and is estimated at about 15 months in wild animals.[36] Suckling is particularly intense in the first week of life and then gradually decreases over time. Nursing may continue in captivity for over two years, although the calf begins to take solid food between six and eight months of age.[37] A 2.25 m male stranded in Chile had both milk and squid remains in the stomach.[30] Growth is rapid in the first year of life, when both sexes may increase in length by 40–50%: thereafter males begin to grow more rapidly than females.[36] Off South Africa females reach sexual maturity at a length of about 3.2 m and an age of 9.25–10.5 years. The situation in males is less clear because of a scarcity of material but they may reach puberty at about 3.6–3.8 m and full maturity (when testis size increases markedly) at around 4.3 m.[5] Fecundity seems low, with the average calving interval being about seven years. In Japanese waters the proportion of pregnant animals declines with increasing age to zero in females over 45 years old, while the proportion of lactating females increases with age but falls to zero in females over 53.5 years. It is concluded that both the interval between calves and the length of lactation increase with age, and that reproduction ceases in females over 45 years old.[36]

Hybridisation with bottlenose dolphins has occurred in captivity on several occasions.[38]

Ages of up to 57 years have been estimated for male false killer whales and up to 62 years for females. Adults of both sexes may have annual mortality rates of 5–6%.[36]

Behaviour and vocalisations

Groups of false killer whales sighted in the subregion varied in size from one to 100, with a mean of 20.2 (n = 26), but only three schools contained more than 50 animals. Two schools were accompanied by common bottlenose dolphins,[5] an association frequently seen in schools driven ashore in Japan.[36]

Stranded schools tended to be bigger than those sighted (4–200, mean 63.6 animals, n = 8; see box on page 259), possibly representing an amalgamation of social groupings. Although it is difficult to know how representative the animals examined were, both strandings and schools driven ashore tended to contain animals of both sexes (although females usually predominated) and of most reproductive classes. Maturing males, however, were largely absent, suggesting that young males temporarily segregate themselves from breeding schools.[7, 36] Re-sightings of several individuals together in the same school after intervals of 11 months to two years support the concept that stable associations are formed within some of these schools.[24] The strong social nature of these groupings was apparent during a birth observed at sea.[39]

When travelling, schools of false killer whales tend to be in tight subgroups that dive synchronously and move in the same direction. During foraging, the subgroups are further apart and surfacing becomes more asynchronous and accompanied by leaps. Seabirds are also more likely to be associated with such groups. Feeding groups can either be directional (travelling in one direction) or non-directional, and feeding can take place at the surface (when there is much high-speed activity and leaping) or at depth (when larger subgroups dive together).[24]

False killer whales are among the few species that habitually strand *en masse* (see box on page 259). While the reasons for this behaviour are unknown, some incidents are revealing. Repeat strandings (after an interval of three days and at a distance of 235 km from the original site) were recorded for some members of one group of about 30 whales,[40] and in another case there was apparent succourant behaviour shown towards an adult male by other members of the group.[41]

FIG. 165 When on the move, subgroups of false killer whales tend to surface synchronously.

Although the false killer whale has the typical odontocete repertoire of vocalisations that can be classified as whistles, burst-pulse sounds and click trains, in reality these lie along a continuum from widely-spaced pulses to whistles.[42] Whistles recorded at sea ranged in frequency from 1.2 kHz to 17.89 kHz and in duration from 0.03 sec. to 2.07 sec., and could be classified by their contour into eight types. They appeared to be somewhat different when produced by feeding and travelling whales.[43] The clicks produced by wild whales are short duration (30 µsec) broadband signals with peak frequencies around 40 kHz and centroid frequencies of 30–70 kHz. Click repetition rates ranged from 5 to 40 clicks/sec., rising to more than 300 clicks/sec. during burst pulses or buzzes.[44] Echolocation capabilities determined experimentally indicate that the whales can detect a 7.62 cm diameter sphere with 75% success rate at about 98 m,[45] while modelling using source levels determined in wild animals suggest that large fish such as tuna might be detected at ranges of up to 210 m and small squid (20 cm) up to 80–85 m.[44]

Conservation status

There are no estimates of population size in the subregion, and no indications that the species is subject to any significant fisheries-related mortality. However, false killer whales have been implicated in interactions with long-line fisheries in the Indian Ocean, and in the Seychelles region were held responsible for damaging 14–27% of the catch annually between 1996 and 1999.[46] Entanglement while purloining fish in such circumstances may result in disfigurement and occasionally death.[47]

False killer whales are listed as Lower Risk: Least Concern by the IUCN and as Least Concern in the South African Red Data Book. They are included on Appendix II of CITES, but are not listed by CMS.

References

1 Rice 1998
2 Kitchener *et al.* 1990
3 Zhang *et al.* 2005
4 Perryman *et al.* 1994
5 MRI Whale Unit data
6 Smithers 1938
7 Ferreira, M. Sc. thesis, University of Pretoria
8 Odell and McClune 1999
9 Stacey *et al.* 1994
10 Fraser 1936
11 Ross 1984
12 Kastelein *et al.* 2000c
13 Purves and Pilleri 1978
14 Yamada 1956
15 Gihr *et al.* 1982
16 Reinhardt 1866
17 Van Waerebeek and De Smet 1996
18 Weir, C., pers. comm.
19 Findlay *et al.* 1992
20 Peddemors *et al.* In press
21 Leatherwood *et al.* 1991b
22 Rosenbaum 2003
23 Kiszka *et al.* 2006b
24 Acevedo-Gutiérrez *et al.* 1997
25 Rohr *et al.* 2002
26 Brown *et al.* 1966
27 Hernández-Garcia 2002
28 Bowman 1955
29 Leung 1967
30 Alonso *et al.* 1999
31 Sekiguchi *et al.* 1992b
32 Perryman and Foster 1980
33 Palacios and Mate 1996
34 Connor and Norris 1982
35 Atkinson *et al.* 1999
36 Kasuya 1986
37 Clark and Odell 1999b
38 Sylvestre and Tasaka 1985
39 Di Sciara *et al.* 1997
40 Odell *et al.* 1980
41 Porter 1977
42 Murray *et al.* 1998
43 Nester *et al.* 2001
44 Madsen *et al.* 2004a
45 Thomas and Turl 1990
46 Bargain *et al.* 2000
47 Baird and Gorgone 2005

Killer Whale

Orcinus orca

Type A

♂

Ventral view

♀

'EXCEPTION MIGHT BE taken to the name bestowed upon this whale, on the ground of its indistinctness. For we are all killers, on land and at sea; Bonapartes and sharks included.'[1] Melville's logic may be part of the rationale for the popularity of the name orca instead of killer whale. But as the word orca is derived from Orcus, the Roman god of the underworld who tormented evil-doers in the afterlife, the name is hardly any more complimentary than being called a killer whale. I prefer the latter term, as it acknowledges the role of the species as the top predator in the marine ecosystem, with a diet that includes just about anything in the sea (and a few things out of it) that it considers large enough to be worth consuming.

However, not all killer whales are equal. In both northern and southern hemispheres the community of killer whales is sub-structured by dietary specialisation, some subsisting largely on marine mammals, others on fish and some apparently on a mixture of the two. These dietary preferences are associated with a number of behavioural and even morphological differences that border on being sufficient for taxonomic recognition.

In the 1980s, two new species of killer whales were described from the Antarctic, *Orcinus nanus*[2] and *O. glacialis*.[3] Both were considerably smaller than the 'standard' killer whale, and (in the case of *glacialis* at least) differed also in skull conformation, tooth size and external morphometrics.[3] It is not clear whether the two descriptions referred to the same or different animals, and in any case either no type material was assigned (*nanus*) or it has subsequently been lost (*glacialis*). More recently, photographs and observations have revealed the existence of three forms of killer whale in high latitudes of the Antarctic, known as Types A, B and C.

- Type A is the 'standard' killer whale, with no sign of a dorsal cape marking (the back behind the grey saddle being as dark as in front of it), and with a moderately-sized white patch above the eye, whose long axis is roughly parallel to the body axis.
- Type B has a distinct cape marking, with its margin running back and down from above the eye patch to join the grey saddle at its lowest point, but then curving up again so that the whole back behind the dorsal fin is grey rather than black. The eye patch itself is about twice as large as in Type A whales, but with the same orientation relative to the body axis.
- Type C also has a distinct cape marking, but the eye patch is much smaller than in Type B, and angled downwards in the front.

Whether all three types constitute subspecies or distinct species, and how Types B and C might relate to either *O. nanus* or *O. glacialis*, is unclear, but all three have been recorded from the subregion.[4] Most sightings of killer whales in latitudes outside the Antarctic are believed to belong to Type A, so although they have not been formally identified, most of the animals for which data are presented here are assumed to have belonged to this form.

Appearance

It would seem difficult to confuse this species with any other in the subregion, given the large, sexually-dimorphic dorsal fin, prominent white eye patch and grey post-dorsal fin saddle, but in practice there has been confusion with long-finned pilot whales, which qualitatively at least share all these same characters. The two can easily be distinguished by the shape of the dorsal fin, which in pilot whales has an extremely convex leading edge (especially in adult males) so that its tip points almost straight back, whereas the killer whale's dorsal is more erect and in adult males the tip points almost straight up in a large isosceles triangle.

The largest individuals among 27 males and 20 females measured in the subregion were 8.81 m and 7.9 m long respectively, but only 10% of females exceeded 7 m in length, compared to 51.9% of males. Most of the whales measured were landed at whaling stations at Durban or Donkergat, and are believed to be referable to Type A killer whales.[5] Male *O. glacialis* do not seem to exceed 8.1 m, and females 7.1 m in length,[3] while female *O. nanus* are said to be 1.0–1.5 m shorter than female *O. orca*.[2] In Type C killer whales, adult females average 5.2 m (n = 33) and adult males 5.6 m (n = 65), with the largest reaching 6.1 m.[6]

Killer whales are heavy-bodied animals. The head in profile has traces of a snout, but this is only weakly demarcated from the melon. The upper jaw typically protrudes some distance in front of the tip of the lower jaw. The dorsal fin in juveniles (3 m long or less) of both sexes is typically delphinid in shape, with a convex leading edge and a concave trailing edge, and its height is 7.3–8.3% of body length (mean 7.8%, n = 4). The fin continues to grow in both sexes, its height in adult females (more than 5 m long) reaching 8.9–12.2% (mean 10.4%, n = 9) of body length, but with much the same shape as in juveniles. In males, however, the fin not only enlarges but also changes shape, both leading and

Length:weight relationship

The length:weight relationship for 16 males and 16 females up to 7 m in length from the North Pacific was $W = 0.000208L^{2.577}$, where W = weight in kilograms and L = total length in centimetres.[15] Data from seven other animals, including five from the subregion (four of which were weighed in pieces and have been adjusted upwards by 5%) gave a rather different regression: $W = 8.6791L^{3.2769}$, where W = weight in kilograms and L = total length in metres.[5, 16] The latter equation implies that a 6 m whale weighs 3 079 kg and an 8 m whale 7 903 kg.

Skeletal data

Vertebral formula (n = 2): 7 cervical, 11–12 thoracic, 10–11 lumbar and 21–25 caudal; total 50–54.[5, 10]

The first three cervical vertebrae are fused, to which the fourth may be partially joined (and numbers 5 and 6 may also be partially fused together). Some of these variations may be age related.[12]

Phalangeal formula, including metacarpals (n = 6): I_2, II_{5-7}, III_{4-5}, IV_{3-4}, V_3.[12]

The skulls of killer whales can be distinguished from those of false killer whales by the narrower premaxillae (their width in front of the antorbital notches being less than 50% of the width of the rostrum), the wide separation of the pterygoid bones at their base by a protrusion of the palatines, and the compressed roots of the teeth.[16]

FIG. 166 Type A killer whales, the more widespread of the ecotypes, lack the grey cape of Types B and C, and are generally larger animals.

trailing edges straightening so that the fin becomes more triangular, and the height much greater than that in adult females, attaining as much as 1.78 m. Such large fins actually wobble when the whale surfaces, and because they are unsupported by any bone or cartilage it is not surprising that they can eventually cant or curl over to one side. This fin enlargement in males seems more related to the maturation stage of the individual than to its absolute size: individuals with a combined testis weight of less than 6 kg have dorsal fins of a similar relative height to that of adult females, but from a testes weight of 10 kg upwards the fin height becomes 15.6–23.0% of body length (mean 18.7%, n = 12).[5] In captive northern hemisphere animals this acceleration in fin height commences at a body length of about 5.1 m.[7] The flippers are quite unlike those of any other delphinid, being roughly oval or paddle-shaped, and are also sexually dimorphic in size, adult females having flipper lengths of 10.6–14.4% of body length (mean 13.3%, n = 13), while adult males (combined testis weight more than 6 kg) have flippers of 17.7–24.1% of body length (mean 20.7%, n = 13). A similar proportional difference in flipper width occurs between sexes.[5] This acceleration in flipper growth commences at a body length of about 4.35 m in captive northern hemisphere males.[7] The width of the flukes also differs between sexes in adults: males have proportionally wider flukes (31.9–36.5% of body length, mean 33.2%, n = 6) than females (22.9–28.8% of body length, mean 25.7%, n = 10).[5] The flukes of the nominal species *O. glacialis* are some 40% smaller than those of *O. orca*,[3] but no such difference was apparent in *O. nanus*.[2]

The basic colour pattern of all killer whales is similar: there is a distinctive white lower jaw and throat connected to a broad white belly patch extending back between the flippers and trifurcating just behind the umbilicus, into lobes that extend up each side and a central tongue that encloses the genital slit and anus. The underside of the flukes is also white, with a narrow black trailing edge. Above the eye there is a roughly elliptical white patch that runs back to about the level of the anterior insertion of the flipper, and is highly variable in size and shape, even within Type A whales.[8] The size and orientation of the eye-patch are also characters that are used to distinguish between Types A, B and C killer whales in the subregion (see above). The rest of the body is black, apart from a grey saddle behind the dorsal fin. The size and shape of this saddle are also highly variable between individuals, even within Type A whales[9] (including some where it may be absent[10]), but in Types B and C the saddle is connected to a grey cape marking, as described above, and the back behind the saddle is distinctly greyer than in front of it. Finally, the overall coloration of Types B and C whales may be affected by diatom infestation so that black and white areas appear brownish and yellowish respectively.[4]

Tooth eruption starts at an age of 9.5 weeks (upper) and 13 weeks (lower).[11] In the adult there are 11–12 (mean 11.6, n = 14) pairs of erupted teeth in the upper jaw and 10–13 (mean 11.6, n = 15) pairs in the lower jaw.[5, 10] The teeth are characteristic in shape, with a conical curved tip but a base that is somewhat compressed, so that in cross-section it is more box-shaped than oval, with the inner margin being concave and the outer margin convex.[12] In Type A killer whales the largest whole teeth may be up to 11.0–13.3 cm in length, and with a maximum transverse diameter of 3.9–4.5 cm and an anterior-posterior diameter of 2.7–3.0 cm: at gum level the respective diameters may be 1.9–3.1 cm and 1.4–1.9 cm.[13] The teeth of *O. glacialis* are described as being about half the height and diameter of the teeth of *O. orca*.[3] The situation in *O. nanus* or Type B and C killer whales is unknown. Wear in killer whale teeth is frequently seen, including a peculiar under-cutting of the crown, and may result from the greater flexibility of the lower jaw in this species.

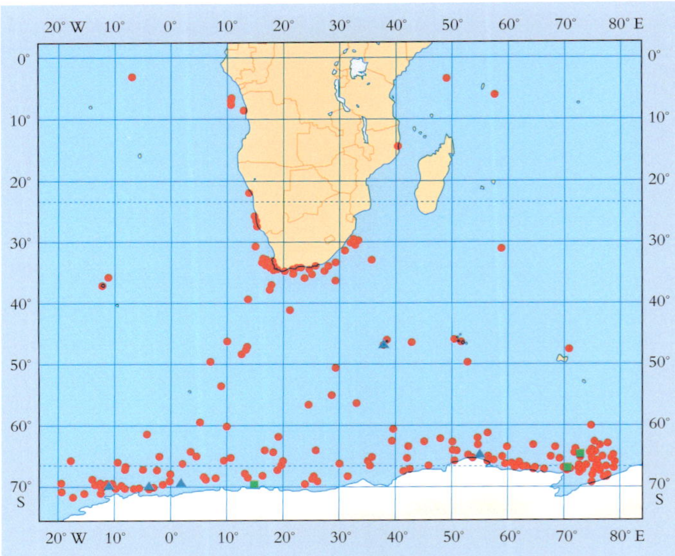

■ Postulated range ■ Sighting or capture
▲ Type B killer whale ■ Type C killer whale

FIG. 167 Distribution map for killer whale.

In extreme cases all the teeth may become flattened virtually down to the gum.[5, 14]

Distribution and migrations

Killer whales are said to be the world's most widely spread mammal, occurring from the tropics to the high latitudes of both hemispheres, and including both oceanic and coastal waters. Abundances may be higher in coastal waters and at high latitudes than elsewhere.[16] In the subregion they have been recorded from Gabon,[17] Angola,[18] Tristan da Cunha,[5] Namibia, South Africa,[19] Mozambique,[20] the Seychelles,[21] Madagascar,[22] Amsterdam and St Paul[23] and subantarctic islands such as Prince Edward,[24] Crozet[25] and Kerguelen,[26] as well as pelagically, with sightings extending into the pack ice in the Antarctic.[27] In most of these cases the killer whales have not been assigned to Types A, B and C as this requires either photographic/descriptive evidence or prior experience with identifying all three types. Nevertheless, the overall impression is that most records outside high latitudes of the Antarctic (with the exception of Marion Island) have been of Type A whales. At Marion Island, photographs indicate that these are Type B whales.[5] The only other confirmed sightings of Types B and C in the subregion have been south of 64° S,[5] although elsewhere in the southern hemisphere both of these types have been seen as far north as 35–36° S (and one of the two at 28°31′ S).[4]

Off South Africa, killer whales were seen in all water depths, from 0–50 m to over 3 000 m.[19] At Marion Island, more than 50% of sightings were within 5 m of the shore, and no more than 20% in the other distance intervals monitored (5–100 m, 100–500 m and more than 500 m from the shore).[28] The distribution around the island was noticeably non-uniform, with most sightings occurring on the east coast, coinciding with the majority of breeding and moulting sites for elephant seals.[29] Within the pack ice, killer whales were seen in more open water (three-tenths ice or less) than minke whales (up to nine-tenths ice).[27]

Sightings of killer whales were made year-round off southern Africa,[19, 20] although on the east and west coast whaling grounds off South Africa their abundance seemed to be correlated with that of baleen whales, especially sei whales on their migration south.[30] There was a definite cycle of seasonality of occurrence at Marion Island and Crozet Islands (both at 46–47° S), with a peak between October and December and a virtual absence in winter,[24, 31] and at Amsterdam/St Paul (at 38–39° S), where abundance peaked in February and March.[23] These local cycles of abundance seem to be roughly consistent with an overall pattern of migration proposed in the southern hemisphere, with most killer whales being in warm waters in winter and in high latitudes in summer. The southern migration is supposed to begin in September/October and be completed by January, while the return migration begins at the end of February and by the end of April most animals have left higher latitudes.[2] Some killer whales have been seen south of 60° S in midwinter, well within the pack ice,[32, 33] but these have all been Types B and C whales,[4] so it is possible that only Type A whales make the migrations proposed above.

Individual killer whales have been reported making large latitudinal movements, three animals moving 2 660 km from Alaska to California,[34] and an adult male apparently making a 5 535 km transequatorial movement from Peru to Mexico (although the quality of one of the images used in this match leaves a lot to be desired).[35] The recovery of the remains of at least three elephant seals from the stomach of a killer whale taken at Durban, about 1 800 km from the nearest colony, provides indirect evidence of long distance movement in the subregion.[19] Conversely, no exchange of known individuals between Marion Island and the Crozet Islands (about 1 000 km apart) has yet been recorded.[36]

Swimming speeds around Marion Island varied between 11.9 km/hr and 14.4 km/hr,[37] although a maximum speed of 28.5 km/hr has been recorded in captive killer whales. Higher speeds in the literature may include some of the effect of bow-riding.[38]

External commensals and parasites

In eight killer whales examined in the subregion, stalked barnacles (*Xenobalanus*) were found on two, one individual carrying 30 along the trailing edge of its flukes, while the second had five and two attached to the right and left tips of the flukes respectively, and nine and five attached to the tips of the right and left flippers respectively. Whale lice (cyamids) were also found on four whales, three of which were from the series of eight whales: those from two whales have been identified, proving to be *Isocyamus orcini* in one case, and *I. orcini* (in the angle of the gape) and *Cyamus antarcticensis* (in scratches on the body surface) in the other.[5] Skin-film diatoms occur frequently on killer whales in high latitudes, causing their skin to appear yellowish,[4] and although the species does not seem to have been identified, it could be *Bennettella* (=*Cocconeis*) *ceticola*. Diatoms found on a killer whale in South African waters originally identified as *Stauroneis olympica* have subsequently been described as a new species, *Epiphalaina radiata*.[39]

Food and feeding behaviour

Killer whales worldwide have accumulated an impressive list of prey. Marine mammals seen to be harassed, chased, attacked or

FIG. 168 Types B (*above*) and C (*below*) killer whales tend to be more prevalent than Type A at higher latitudes. The yellowish stains on Type C are probably films of skin diatoms.

killed include at least 20 species of cetaceans (from blue whale to harbour porpoise), 14 pinniped species, dugongs and sea otters,[40] while other prey items documented include at least 11 species of teleost fish,[41] 12 species of elasmobranch fish,[42] birds,[41] turtles,[43] cephalopods[44] and even deer and moose.[45] The diversity of this list conceals a considerable degree of dietary specialisation: in the north-east Pacific, communities of killer whales are organised around either a fish-eating or mammal-eating 'culture' that has resulted in inter-community differences in schooling, social and acoustic behaviour and even to some extent in morphology.[46] It is possible that such a system has also developed among killer whales in the southern hemisphere. Within the Antarctic, 98.5% of the stomachs of the proposed species *O. glacialis* contained fish, 0.4% marine mammals and 1.1% squid (n = 629), while 89.7% of those attributed to *O. orca* contained marine mammals, 7.1% squid and 3.2% fish (n = 156).[3] Of the three southern killer whale 'types' recognised, A is believed to feed mainly on minke whales, B on seals (and possibly whales and penguins), and C on fish (Antarctic toothfish) but may occasionally take penguins or seals.[4]

The feeding habits of killer whales in South African waters remain poorly known. Instances of predation on dolphins, fur seals, penguins and cormorants have been reported by the public,[20] although some of the reported attacks on seabirds seem more like 'play' than actual predation, as victims may be maimed or killed but not necessarily consumed.[47] Numerous reports of killer whales stealing hooked tuna from sport and commercial fishermen have been received, and at least two 'attacks' on southern right whales have been reported.[5] The stomachs examined came mainly from animals killed in whaling operations, so their contents may not be fully representative of prey because of the opportunity provided to scavenge from dead whales and the greater likelihood therefore of mammal-feeding killer whales being sampled. Of 24 stomachs examined, 20 contained mammal remains, consisting of the tongue and/or meat of large whales (six), blubber or other remains of small whales (four, including one identified as pygmy sperm whale), baleen plates of juvenile mysticetes (two, one identified as pygmy right whale), blubber or other remains of dolphins (three), and nails/vibrissae of elephant seals (one). Two stomachs contained fish remains, one of which was tentatively identified as tuna, and the contents of the other were accompanied by pieces of fishing line and thin rope, suggesting that the whale had been scavenging. Omitted from this list is a stranded male with extreme tooth wear, whose stomach contained a heavily patched piece of a diving suit, a plastic milk bottle, a 75 cm square piece of plastic sheeting, a cardboard soft drink carton, two egg cases of the elephantfish

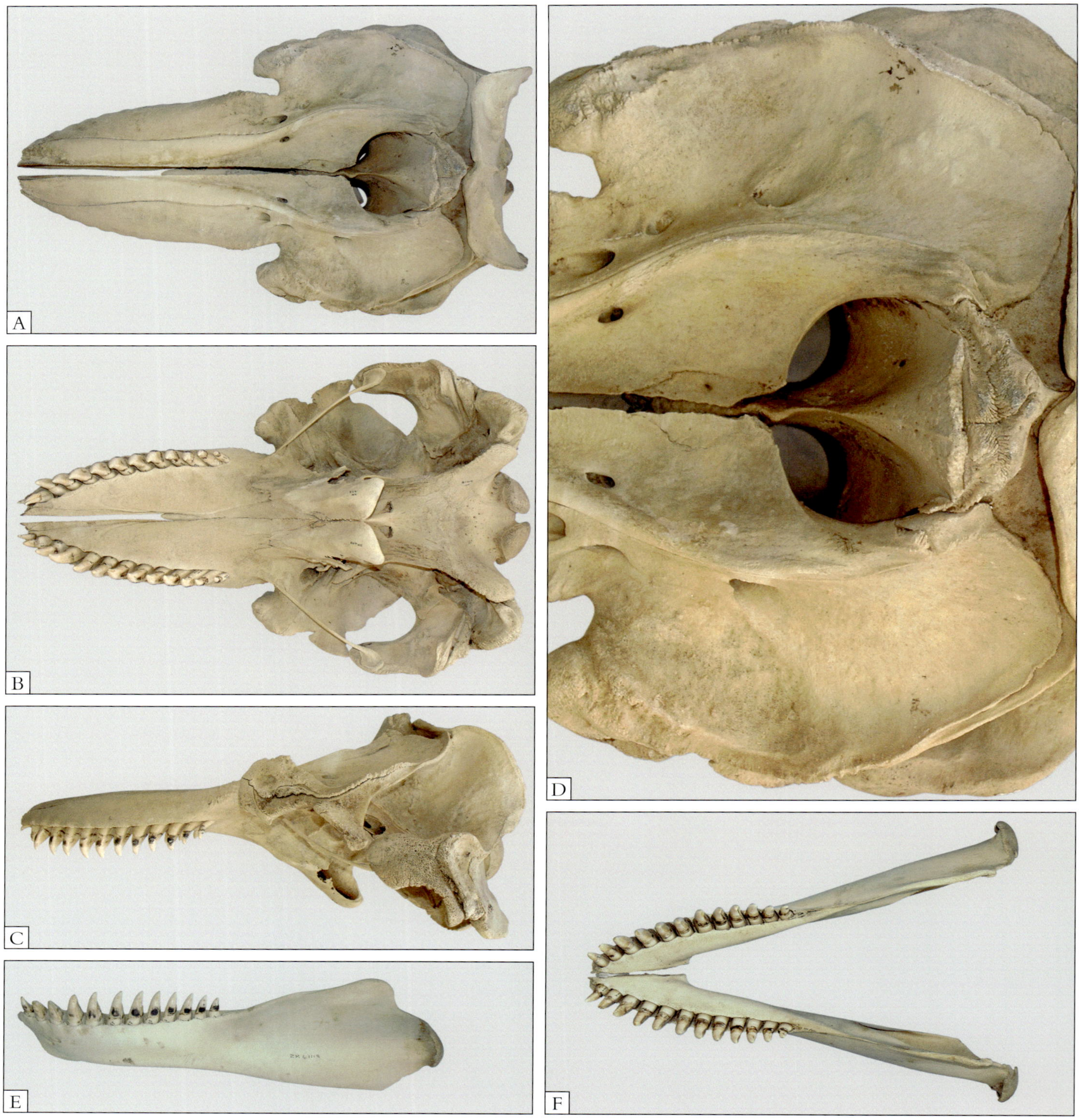

FIG. 169 Dorsal (A), ventral (B) and left lateral (C) views and vertex of skull (D); left lateral (labial) (E) and dorsal (F) views of mandible (ZM 41118).

(*Callorhincus capensis*), two albatross skins (either black-browed or yellow-nosed), and 7.5 kg of kelp.[5]

At Marion Island, consumption of elephant seals (*Mirounga leonina*), king penguins (*Aptenodytes patagonica*) and rockhopper penguins (*Eudyptes chrysocome*) has been seen.[24] At the Crozet Islands, predation on elephant seals and king penguins was most frequently observed, although fish was also consumed and large whales were occasionally attacked.[48]

Killer whales exhibit a variety of techniques to capture prey.[16] Within the Antarctic, a concerted charge by several individual whales is used to create a wave that rocks an ice floe and dislodges a seal thereon.[49] Alternatively, the floe may be physically tilted by a whale to spill the seal into the water.[4] Judging by associated damage and scarring, the flippers of Antarctic minke whales may be grasped by pursuing killer whales in an attempt to slow down their victim.[50] In the Crozet Islands, killer whales routinely used different techniques at specific locations chosen according to topography and climate to hunt penguins and elephant seals. Penguins were hunted in open water or against the coast and within kelp beds, and were usually consumed individually. Elephant seals were taken at river mouths (especially at times of heavy rain and a corresponding increase in outflow), when swimming along the shore, or on the beach by intentional stranding. The seals were usually dispatched by powerful blows of the tail or head and in most cases their consumption was shared between individuals.[48] The technique of intentional stranding is potentially hazardous (six individuals being found dead on the Crozet archipelago over a 30-year period), and young undergo an extensive apprenticeship during which simultaneous stranding occurs with the mother or another adult female. The first solo attempts occurred at an estimated age of 4–5 years, and the first attempt to capture an elephant seal on the beach at age 5–6 years (although the assistance of an adult female was still required in order to take the seal off the beach).[51, 52] Similar predation of southern sea lions and elephant seals through intentional stranding has been recorded in Argentina (although, unlike the Crozets, involving adult male killer whales)[53] and apparently occurred historically at Tristan da Cunha on elephant seals,[54] but does not occur at Marion Island.[28]

Food consumption in captive killer whales averaged 3.6% of body weight per day in an 18-year-old female,[55] and can be estimated at 2.2% of body weight per day in an 11-year-old female.[56] Generally, daily food consumption declined from about 6% of body weight in an animal of 450 kg to about 1.5% in a 2 900 kg whale, but with obvious variations depending on circumstances. For instance, a female increased its daily intake from about 55 kg to 80 kg after giving birth.[57]

Reproduction and growth

The longest of six foetuses from whales landed at the Durban and Donkergat whaling stations were 2.18 m (14 August) and 2.22 m (23 May),[5] and the smallest of five calves from the subregion was 2.27 m long (5 September),[5, 10] suggesting that birth takes place at a length of about 2.25 m. Ovulation is spontaneous, and oestrous cycle activity in captive animals can occur year-round but peaks in spring.[58] Successful live births have occurred in captive females after gestation periods ranging from 15 to 18 months, with a mean of 517 days (n = 10).[59] All indicators of nursing intensity (number of suckles per day, bouts per day and suckle duration) peaked within the first two days of birth in captive calves, and then declined exponentially over time. For instance, the number of suckles per day declined from over 400 in the first two days to about 30 by day 90, while the time spent nursing per day dropped from about 2 100 sec. to about 200 sec. over the same time period. Although there were statistical differences in the rate of these changes between sexes and between the calves of naïve and experienced mothers, by 42 days of age the cumulative values for all indicators of nursing intensity were similar.[60] The age and length at weaning are not well established (and may be highly variable). A captive calf first took solid food at 11 weeks of age but only occasionally from 3.5 to 5.5 months. After 14 months, nursing decreased and ceased at about 18 months.[11] Other captive-born calves have taken fish as early as six weeks to two months of age, and regularly consumed solid food from six to seven months.[57] Wild calves as small as 2.6 m (and judged to be 'a few weeks old') have been found with solid food in their stomachs.[61] The smallest of 15 sexually mature females from South Africa was 5.28 m long, and the largest immature was 3 m.[5] In the Antarctic all 91 females over 5 m in length were mature and the only immature was 3.7 m long.[2] This suggests that female sexual maturity is reached at about 5 m. If killer whales mature when their combined testis weight is more than 6 kg,[16] then in 21 males from South Africa all but one less than 6 m long were immature and all but one over 6 m were mature,[5, 10] indicating a length at sexual maturity of about 6 m. Captive females have given birth for the first time at between eight years and an estimated 17 years,[59] while wild females from the resident population in the north-east Pacific were seen with their first viable calves at 12–16 years of age. Using the growth of the dorsal fin as an indicator of maturity, male north-east Pacific killer whales were judged to reach sexual maturity at 10.5–17.5 years, 50% of them reaching maturity at about age 15. The mean age at physical maturity (when the fin had essentially stopped increasing in height) was estimated at 21 years.[62] Out of 15 mature females examined in South Africa, six (40%) were pregnant,[5] while among 91 mature females from the Antarctic, 27.5% were pregnant.[2] Reciprocals of these (and allowing for a 17-month gestation) would suggest average calving intervals of 3.5–5.2 years. Intervals between the first and second calves of captive animals were 32–38 months (average three years) and increased to 58 months for a third calf.[59] Intervals between observations of viable births among the resident north-east Pacific killer whales were 2–12 years with a mean of 5.2 years. The fertility of females in this population declined with age until about age 30, with the last calf being born at about age 39. Maximum longevity, however, was about 80–90 years, indicating that the population was likely to contain a substantial proportion of post-reproductive females. Male survival was lower than that of females, with a maximum estimated longevity of 50–60 years.[62]

Behaviour and vocalisations

The size of killer whale schools in the subregion varied by latitude. Sightings at 0–20° S, 20–40° S, 40–60° S and south of 60° S averaged 4.8 (n = 8), 4.5 (n = 71), 5.8 (n = 17) and 12.3 animals (n = 196) respectively, and while there were no schools larger than 23 sighted north of 60° S, south of that latitude schools of 25 or larger formed 9.2% of all sightings and included aggregations as large as 200.[5] Probably part of the reason for the larger schools in higher latitudes was the presence of Type C killer whales that tend to form larger social groupings.[4] At Marion Island a school as large as 25 was only seen in one month, and most consisted of 13 animals or fewer, with a mean of 4.2 (n = 281). The composition of this population in the period 1973–1976 was described as 28.7% adult males, 21.4% adult females, 7.8% calves, 24.7% subadults and 17.4% unidentified (n = 954),[24] but later observations suggested a different composition, with fewer adult males (16.4%) and more adult females (28.3%) and calves (14.5%), although such a comparison may have been affected by differing interpretations of age/sex classes.[28] At the Crozet archipelago the size of groups was 2–7 with a mean of 4.2 animals, 21% of which were adult males, 34% adult females and 45% immatures (although the latter figure could have included some reproductively inactive adult females).[31] Among 27 killer whales landed at the Durban whaling station

Swimming with killer whales – reckless recreation?

In February 1991, Keltie Byrne, a 20-year-old part-time trainer at Sealand of the Pacific in Canada, slipped and fell into a pool with three killer whales. One whale took her in its mouth and dragged her around the pool. Despite Keltie being a champion swimmer, and at one stage managing to reach the side, she was pulled back in and eventually drowned. Her body was only recovered several hours later. Although Keltie was the first human known to have been killed by killer whales, her accident was by no means the first negative interaction between captive killer whales and trainers. In 1987 alone there were over a dozen cases, some resulting in fractured bones, surgery and permanent disability for the human. The causes of these incidents are largely unknown, and might be attributed to the stress of captivity, inappropriate training procedures or simply a miscalculation on the part of the whale.[1] Nevertheless, they indicate the potential danger of working with a captive carnivore weighing several tonnes.

Although there have as yet been no documented fatal attacks on humans by wild killer whales, some incidents bear a distinct similarity to hunting strategies developed in securing marine mammal prey such as seals. A killer whale in the Antarctic lunged out of a lead in the ice and pinned a man against the ice briefly before sliding back into the water.[2] Three killer whales charged an inflatable carrying five observers in the Antarctic, causing a prominent bow wave that rocked the boat in a tactic resembling that used to dislodge seals from ice floes.[3] And in perhaps the most famous incident, during the British Antarctic expedition of 1910–12, six or seven killer whales surfaced under a floe carrying two huskies and the expedition's photographer, Herbert Ponting, splitting the floe, and spy-hopping through the cracks so created.[4] Although some solace might be sought in the fact that humans are terrestrial organisms, and so presumably not part of the normal 'search image' of a killer whale, the capture of moose and deer, and an apparent attack on a barking dog, suggest that one should not be too complacent. As recommended by leading killer whale researchers, 'it is advisable [for a swimmer] to leave the water should [mammal-eating killer whales] appear in the vicinity'.[5] It may be wise not to wait around to determine whether or not you are listed on their menu.

References

1 Hoyt 1992
2 Pitman and Ensor 2003
3 Pierpoint *et al.* 1995
4 Huxley 1947
5 Ford and Ellis 1999

between 1971 and 1975, 51.8% were mature males, 40.7% mature females and 7.4% immature males, but this composition was clearly affected by selection for larger animals.[5] In the Antarctic, most schools contained one or more adult males, either with one or more adult females and their calves (44%), or with other unspecified animals (33.3%), but rarely alone (5.6%).[63] Mixed species associations are rare, but have been recorded with minke whales, where killer whales predominated in half the cases and minke whales in the other half (n = 8).[64]

In the north-east Pacific, the social organisation of fish-eating and mammal-eating killer whales (or so-called residents and transients respectively) differs greatly. Residents occur in a wide variety of group sizes (3–59), whereas transients occur in much smaller groups (1–4, occasionally up to 15), apparently because remaining in a group larger than about three would result in a reduced rate of individual energy intake. Because of these differing pressures on group size, the dispersal of animals from groups in which they were born (maternal pods) follows quite different patterns in residents and transients. Calves of both sexes born in resident groups tend to stay in that group for life. Mating seems likely to occur between pods rather than within, so that dispersal from the natal group to avoid inbreeding is not essential. Calves born in transient groups, on the other hand, appear to exercise a number of options. Female calves seem to stay with their maternal pod until they reach sexual maturity, when they may leave their mother to associate with several other pods, either succeeding in establishing themselves reproductively or, if unsuccessful, possibly returning to their maternal pod. The dispersal of male calves depends on their social status: first-born males probably stay with their maternal pods, whereas later born males (being unable to compete with their older sibling) leave their mother at sexual maturity to become largely solitary, though if a group lacks an adult male it may be accepted into that pod for a longer period of time. The occasional larger groupings of transients may represent opportunities for mating and developing other social behaviours and skills.[65] It is not known to what extent these social patterns apply to killer whales in other parts of the world, more especially to Types A, B and C in the southern hemisphere, but the contrast in group sizes between at least Type A and Type C (believed to be largely mammal- and fish-eating respectively[4]) is reminiscent of that between transient and resident killer whales in the North Pacific.

Intraspecific aggression between killer whales is rarely reported. Resident and transient whales in the North Pacific either ignore each other or the transients tend to avoid the residents. A case of active pursuit (and apparent injury) of transients by a group of residents has been recorded, possibly because they considered the mammal-eating transients a threat to a newborn calf.[45] Extensive scarring on two males in New Zealand waters was attributed to tooth raking by conspecifics, but the context in which the injuries were received is unknown.[66] In one (so far unique) instance of apparent cannibalism, the remains of an 8.2 m killer whale were found in the stomachs of two male killer whales taken from the same group between 30° S and 50° S:[67] whether the victim was killed or scavenged is not known.

The diving patterns of telemetered resident killer whales in the north-east Pacific were extremely variable. While they regularly dived to depths below 150 m, reaching a maximum of 264 m (in water less than 330 m deep), more than 70% of their time was spent in the top 20 m of the water column. Adult males tended to dive deeper more frequently than adult females during daylight hours, while diving rates and swimming speeds for all individuals were also higher during the day than at night. Over a period of 10 years, the rate of diving to deeper depths during the day decreased, possibly because of shifts in prey distribution.[68] On four occasions killer whales in the North Atlantic dived to depths of 160–180 m to force herring shoals nearer to the surface.[69] A telemetered transient killer whale made extremely regular dives, with all long-duration dives being to depths of

20–60 m, a stratum in which it spent more than two-thirds of its time. It has been speculated that swimming at these depths allows the whale to see its prey silhouetted against surface light.[65] A group of killer whales tracked on asdic from a Durban whale catcher dived to a depth of about 120 m for 3 min.[5]

Killer whales produce underwater sounds of three basic types, clicks, burst-pulsed sounds and whistles. Clicks are used for echolocation, while whistles and burst-pulsed sounds are used primarily for communication. Clicks range in duration from 0.8 ms to 13 ms and cover a broad frequency range (up to 22 kHz) but with most energy below 2 000 Hz. Click trains are usually short (less than 20 clicks) with an average interval of 63 ms, but some have click repetition rates as high as 200/sec. Burst-pulsed sounds (cries) are highly variable in nature but contain a fundamental tone (600–6 000 Hz) that is pulsed at rates of 60–2 200 Hz and is rich in harmonics. Whistles are tonal calls between 500 Hz and 8 900 Hz, with durations of 0.05–1.85 sec.[70] Each pod of resident killer whales in the north-east Pacific has a repertoire of 7–17 discrete calls that can vary considerably between groups. These consistent variations form the group's 'dialect' that can be used to distinguish each pod acoustically. It is possible that these calls are passed down from generation to generation through learning, as the composition of the schools is fairly stable. The vocal repertoire of transient killer whales is much smaller, with only 4–6 discrete calls, and although none of these are shared with resident whales, there is little variation between pods. This lack of a well-developed dialect may reflect their more fluid social structure than that of resident whales.[45] Resident whales are generally more vociferous than transient whales, producing trains of clicks 27 times more often, with the average duration of trains being more than twice as long as those from transient whales, while transients produced isolated single or paired clicks about four times as often as residents. These dissimilarities presumably reflect different foraging strategies, in that mammalian prey is more likely to detect clicks than fish, so that passive listening may be the better strategy to use when killer whales are hunting marine mammals.[71] While much less is known about the acoustic behaviour of killer whales in the southern hemisphere, the same three basic call-types (clicks, cries and whistles) seem to be produced.[72] At the Crozet Islands, hunting of both penguins and elephant seals was conducted mainly in silence, with isolated clicks being the most frequent signals emitted, possibly to coordinate group movements. Attacks on elephant seals were sometimes preceded by a particular short-distance contact call that caused convergence of group members on the individual that had detected the prey, while successful captures were frequently accompanied by excited long-distance calls that could bring other killer whales to the scene at high speed from up to several kilometres away.[48]

Conservation status

The abundance of killer whales in the Antarctic in summer from line transect surveys between 1976/77 and 1987/88 was estimated at 80 400 animals, of which about 42% occurred in the eastern half of Area II and Area III (roughly equivalent to the high latitudinal sector of the subregion).[73] The proportions of these that might be Types A, B or C are unknown. At Marion Island it was estimated that there were 25–30 killer whales present at their peak abundance in December 2000.[37] This compares with a photo-identification catalogue of 26 individuals compiled between 1975 and 1989 (although 42% of photographs taken did not contain fins with any identifiable markings).[28] At the Crozet Islands, mark-recapture estimates indicated a decline from 85 individuals in 1988/89 to 52 animals in 1998–2000, possibly as a result of fishery-induced mortality or a decline in prey abundance.[36] Between 1971 and 1975, 36 killer whales were taken at the Durban whaling station,[74] and from 1947 to 1980 a total of 1 287 were taken by the Antarctic whaling expeditions of the Soviet Union, mostly in high latitudes, of which 916 were taken in 1979/80.[75, 76] Between 1969/70 and 1978/79, 70 (or 24.2%) of the Soviet takes were in the subregion,[2] but it is not known if this proportion refers to the catches in other seasons. Although direct takes have ceased, killer whales are frequently implicated in scavenging fish from long-line and other hook-and-line fisheries. In the subregion such incidents have been reported from South Africa, and from the vicinity of the Prince Edward, Crozet and Kerguelen islands in the Southern Ocean. The extent of such depredations can be considerable: 53.8% of 26 long-line cruises to the region of the Prince Edward Islands between 1999 and 2001 reported interactions with killer whales, and in the South Georgia region more than 50% (and up to 80–90%) of the catch could be lost to this species. Certain pods may be repeat offenders, and some species of fish may be preferentially selected. While few killer whales ever get entangled,[77] unconfirmed reports suggest that retaliatory actions such as shooting do occur. Killer whales are listed as Lower Risk: Conservation Dependent by the IUCN and as Data Deficient in the South African Red Data Book. They are included on Appendix II of CITES and CMS.

References

1 Melville 1851
2 Mikhalev *et al.* 1981
3 Berzin and Vladimirov 1983
4 Pitman and Ensor 2003
5 MRI Whale Unit data
6 Pitman *et al.* 2007
7 Clark and Odell 1999a
8 Visser and Mäkeläinen 2000
9 Baird and Stacey 1988
10 Ross 1984
11 Asper *et al.* 1988
12 Eschricht 1866
13 Tomilin 1957
14 Caldwell and Brown 1964
15 Bigg and Wolman 1975
16 Dahlheim and Heyning 1999
17 Rosenbaum and Collins 2006
18 Weir, C., pers. comm.
19 Findlay *et al.* 1992
20 Rice and Saayman 1987
21 Keller *et al.* 1982
22 Rosenbaum 2003
23 Roux 1986
24 Condy *et al.* 1978
25 Voisin 1972
26 Robineau 1989
27 Condy 1977
28 Keith *et al.* 2001
29 Bester *et al.* 2002
30 Findlay 1989
31 Guinet 1991b
32 Taylor 1957
33 Gill and Thiele 1997
34 Goley and Straley 1994
35 Guerrero-Ruiz *et al.* 2005
36 Poncelet *et al.* In press
37 Pistorius *et al.* 2002
38 Fish 1998
39 Holmes *et al.* 1993
40 Jefferson *et al.* 1991
41 Martinez and Klinghammer 1970
42 Fertl *et al.* 1996
43 Pitman and Dutton 2004
44 Baird *et al.* 2006
45 Ford and Ellis 1999
46 Ford *et al.* 1998
47 Williams *et al.* 1990
48 Guinet 1992
49 Smith *et al.* 1981
50 Best 1982
51 Guinet 1991a
52 Guinet and Bouvier 1995
53 Hoelzel 1991
54 Lockerby 1925
55 Kastelein *et al.* 2000e
56 Kastelein and Vaughan 1989
57 Kastelein *et al.* 2003a
58 Robeck *et al.* 1993
59 Duffield *et al.* 1995
60 Clark and Odell 1999c
61 Heyning 1988b
62 Olesiuk *et al.* 1990
63 Miyazaki 1989
64 Miyazaki and Kato 1988
65 Baird 2000
66 Visser 1998
67 Shevchenko 1975
68 Baird *et al.* 2005
69 Nøttestad *et al.* 2002
70 Steiner *et al.* 1979
71 Barrett-Lennard *et al.* 1996
72 Awbrey *et al.* 1982
73 Kasamatsu and Joyce 1995
74 Best and Ross 1977
75 Secretariat of the International Whaling Commission 1988
76 Zemsky *et al.* 1995
77 Kock *et al.* 2006

Long-finned Pilot Whale

Globicephala melas

♂

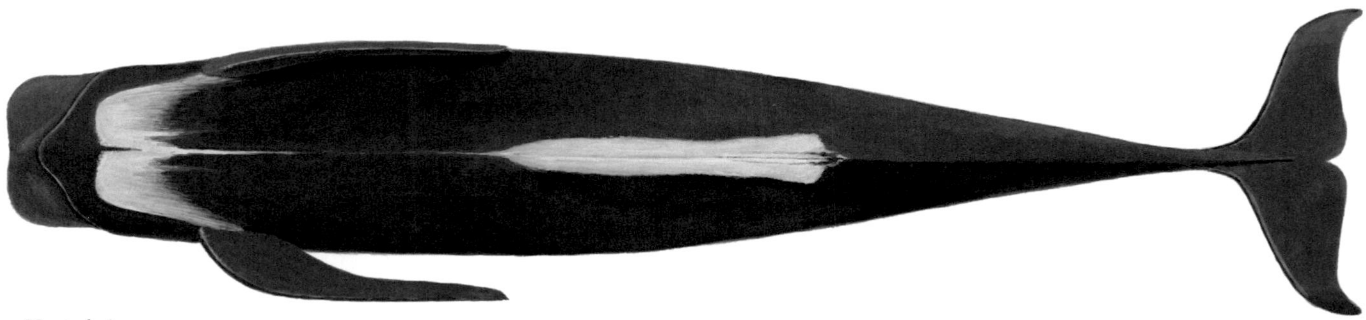

Ventral view

♀

THERE IS GENERAL agreement that there are two principal types of pilot whale at large, the long-finned pilot whale (*Globicephala melas*) and the short-finned pilot whale (*G. macrorhynchus*),[1] 'finned' referring to the flipper rather than the dorsal fin. The two are readily distinguishable on body proportions and skull characters, but pigmentation patterns seem to vary geographically and can be confusing. To complicate matters further, there may be within-ocean population structure in both species,[2, 3] although little research on this has been carried out in the southern hemisphere.

Long-finned pilot whales occur in the North Atlantic and southern hemisphere, and while there are no recent records from the North Pacific, archaeological remains suggest that that they also occurred there up to 6 400 years ago.[1] Southern populations have been recognised as distinct enough externally from those in the North Atlantic to be made a separate subspecies, *G. m. edwardii*,[4] the original specimen of which hailed from Slangkop, South Africa.[5]

Appearance

The possibility of confusing long-finned pilot whales with killer whales – at least on the west coast of South Africa, where an earlier publication (misleadingly) referred to pilot whales as 'very rare in Cape waters'[6] – has already been discussed in the killer whale species account.

Otherwise the principal problem is distinguishing between the two pilot whale species at sea. Adult long-finned pilot whales throughout the subregion are characterised by a wedge-shaped grey saddle behind the dorsal fin and a slender diagonal grey streak running back from the region of the eye. Juveniles may have neither marking or only faint traces of them. In the Indian Ocean, even adult short-finned pilot whales do not have obvious grey saddles or eye streaks, and appear completely dark dorsally,[7] so should be readily distinguished from long-finned pilot whales. The appearance of short-finned pilot whales in the tropical South Atlantic is not well established but seems to resemble that in the Indian Ocean.[8]

Relatively few long-finned pilot whales have been examined in the subregion. The largest of four males was 5.56 m long, and the largest of eight females was 4.75 m.[7,9] In the North Atlantic, the mean lengths of males and females more than 25 years old were 5.57–5.7 m and 4.46–4.49 m respectively,[10] indicating a similar size and degree of sexual dimorphism.

The head is bulbous even in a calf, and with age grows to protrude beyond the short and barely noticeable beak, especially in adult males, where it can become almost square in outline when viewed from above or below. The line of the mouth slopes diagonally up towards the eye. The length of the dorsal fin is far greater than its height: the base measured 15.4–21.2% (mean 18.5%) and the vertical height 4.8–7.6% (mean 6.5%) of body length in 11 animals from the subregion.[9] Its shape is distinctive (for pilot whales) in that the leading edge is strongly convex and the trailing edge concave, so that the tip points backwards rather than up. It is also situated nearer the head than the tail, making

FIG. 170 Variation in the colour pattern of long-finned pilot whales is captured in this image from the Southern Ocean, with the calf lacking an eye streak and the two adults displaying the eye streak and grey saddle.

the front half of the body appear particularly robust. Although the fin is sometimes said to be markedly sexually dimorphic in size and shape, proportional measurements suggest that this is not so, and adult males simply have a bigger fin because they are bigger than adult females.[11]

The flippers are long and tapered with a definite 'elbow': in 10 individuals from the subregion their length was 17.7–23.7% of body length with a mean of 20.4%.[9] Elsewhere it has been shown that flippers grow rapidly at puberty and that mature males have relatively longer flippers than mature females.[11] Because of the forward placement of the dorsal fin, the tail section of the body appears particularly long, although it is laterally compressed and relatively deep.[12] In 10 individuals from the subregion the width of the flukes was 18.0–25.6% of body length with a mean of 22.4%.[9] Elsewhere mature males have been shown to have relatively wider flukes on average than any other age or sex class.[11]

All long-finned pilot whales from the subregion are basically a dark greyish-black on the back and sides with an anchor-shaped pale grey to white marking on the throat, the central portion of which extends back in the mid-line as a tapering streak, usually ending just behind the flippers. A separate pale grey marking runs back from the umbilical region as a narrow streak, expanding laterally just before terminating at the level of the genital slit. Adults of both sexes also carry a white streak that runs diagonally upwards from above the eye, and a light grey to white wedge-shaped saddle behind the dorsal fin. Typically the saddle has irregular edges and contains dark spots or flecks, and does not extend much, if at all, in front of the trailing edge of the dorsal fin.[9] Photographs suggest that smaller and presumably younger individuals may not carry either the eye streak or the saddle. In the Faroe Islands the youngest animals with either marking were three and five years old respectively, while overall only 22% of whales exhibited an eye streak, 43% a saddle and 15% both.[11]

In adults there are 9–12 (mean 10.6, n = 12) pairs of erupted teeth in the upper jaw and 9–11 (mean 10.1, n = 11) pairs in the lower jaw. The teeth are conical and peg-like, with a

Length:weight relationship

Published data for a very large sample of animals from the North Atlantic indicate that the following relationship is appropriate, regardless of sex, reproductive status or season: $W = 0.00023L^{2.501}$, where W = weight in kilograms and L = total length in centimetres.[14] A 2.56 m male and a 3.77 m male from South Africa weighed 148.5 kg and 520 kg respectively,[9] these weights being somewhat less than predicted from the above regression (242.5 kg and 638.5 kg respectively), possibly because they were both stranded animals.

Skeletal data

Vertebral formula (n = 3): 7 cervical, 11 thoracic, 12–14 lumbar and 27–29 caudal; total 59–60.[9]

The first 5–6 cervical vertebrae are fused.[13]

The phalangeal formula, including metacarpals, has been given as: I_{3-4}, II_{9-14}, III_{9-11}, IV_{2-3}, V_{1-2},[13] but this is for pilot whales in general, and it is not clear whether the flippers of the two species differ in this respect (as one might suspect).

The skull of the long-finned pilot whale can be readily distinguished from that of the short-finned pilot whale by its more elongate rostrum and rather narrow premaxillae, which do not expand towards the front of the skull and so leave at least 1 cm of the maxillae exposed on each side of the rostrum for most of its length. Usually there are more teeth (9–12 pairs as opposed to 7–9), but there can be considerable overlap in this character.[12]

maximum diameter at the gum of 5.5–11.0 mm (mean 7.9 mm, n = 5).[9] The teeth tend to become worn down or fall out in older individuals.[13]

Distribution and migrations

Depicting the distribution of either species of pilot whale is fraught by the problem of possible at-sea confusion between them. Long-finned pilot whales are generally regarded as inhabiting more temperate regions than short-finned, avoiding the tropics (although they may extend further towards the equator in cold current systems such as the Benguela).[12] In the subregion they have been recorded from Namibia, South Africa,[15] Tristan da Cunha,[9] Kerguelen[16] and the Southern Ocean.[17] On the South African coast, long-finned pilot whales have stranded as far east as East London (28° E), but all sightings on the Durban whaling ground were referable to short-finned pilot whales,[5] suggesting that the stranding record may represent an animal at the extreme limit of its range or an extralimital animal.

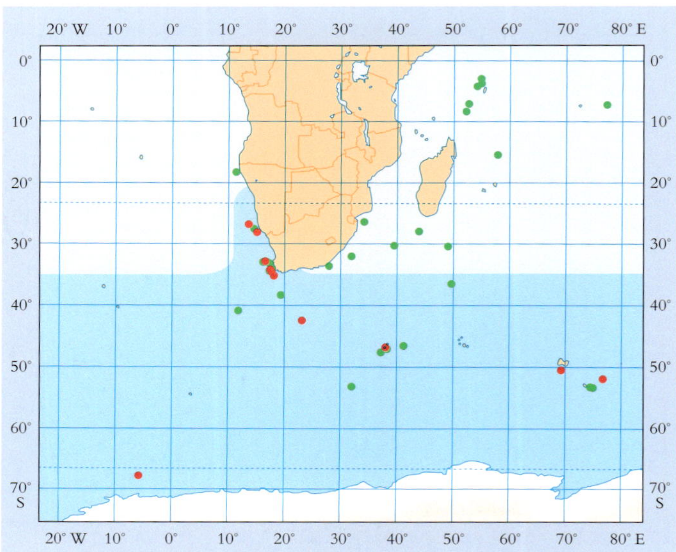

FIG. 171 Distribution map for long-finned pilot whale.

On the South African coast all pilot whale sightings were associated with the shelf edge or in water over 1 000 m deep.[15] The importance of the shelf edge/slope to pilot whales has also been demonstrated in the North Atlantic,[18] and the movements of satellite-tagged long-finned pilot whales have demonstrated an affinity for the continental shelf or deep water adjacent to it.[19, 20]

Sightings of pilot whales off southern Africa occurred in both summer and winter, at least on the west coast of South Africa.[15] In the North Atlantic the long-finned pilot whale is believed to move northward along the shelf edge/slope interface in spring, onto the shelf and into nearshore waters from late spring to early summer, and presumably back south-westwards into slope waters during autumn and winter to complete a clockwise pattern of movement.[18]

Satellite-tagged whales had swimming speeds of 0.2–14.5 km/hr,[19] or median monthly speeds of 1.0–2.7 km/hr.[20]

External commensals and parasites

Of eight long-finned pilot whales examined in South Africa, three carried the stalked barnacle *Xenobalanus* and two also carried cyamids (unidentified). One female had 13 barnacles attached to the tip of the right flipper and three on the trailing edge of the left fluke, a male had a total of 11 attached to its appendages and another female had one on the tip of the dorsal fin and another on the trailing edge of the left flipper. The male carried 13 cyamids dispersed in wounds and scars on the body, and the second female had two cyamids on its left flank.[9]

In a large sample of long-finned pilot whales examined in the North Atlantic, 29.6% overall carried the cyamid *Isocyamus delphinii*, but mature males (77.3%) were much more heavily infested than other classes, possibly because of the greater incidence of wounds and scars arising from inter-male interactions. In general, however, the favoured site of attachment was in the region of the genital slit.[21] In the same region, *Xenobalanus* was found attached to 30.3% of individuals.[22]

Food and feeding behaviour

All 551 food items identified from the stomachs of five long-finned pilot whales from South Africa were cephalopods, of which *Lycoteuthis diadema* (43.4%) and *Todarodes sagittatus* (27.8%) were numerically the most important, with *Histioteuthis* sp. comprising 14.2%.[9] This suggests that the whales were foraging over the continental slope.[23]

Larger samples of long-finned pilot whales from the North Atlantic indicated a greater diversity of prey but still a predominance of cephalopods. Stomachs examined from the drive fishery in the Faroes contained 12 genera of cephalopods and 15 genera of fish, as well as crustaceans, but by number 93% of prey items were cephalopods and 7% fish, with *Todarodes sagittatus* and *Gonatus* sp. (including *G. fabricii*) forming 31.1% and 48.3% respectively (n = 8 870 items). The occurrence of crustaceans (shrimps) was higher in whales less than 4 m long, while fish remains were more prevalent in immature females, lactating females and mature males.[24] The stomach contents of 30 long-finned pilot whales taken incidentally in a mackerel fishery in the western North Atlantic were also dominated by cephalopods, especially long-finned squid (*Loligo pealei*), which numerically formed 70% of the trace remains and 34.6% of the non-trace remains, while Atlantic mackerel (*Scomber scombrus*) was the dominant fish species (3% of trace remains and 44.3% of non-trace remains). The predominance of mackerel in non-trace remains was probably an artefact of the sampled whales being taken in the mackerel fishery.[25]

A theoretical annual consumption of 11.3 tonnes of squid for a 1 000 kg long-finned pilot whale[14] would indicate an average daily food intake of about 3.1% of body mass.

Long-finned pilot whales carrying time-depth recorders in the Ligurian Sea spent daylight hours in the top 15 m of the water column, but shortly after sunset made a few dives as deep as 648 m, possibly intercepting shoals of squid ascending towards the surface as part of their nocturnal vertical migration.[26] Deeper and longer dives at night were also seen in two satellite-tagged *G. melas* in the north-west Atlantic,[27] while off the Faroe Islands three others fitted with satellite tags exhibited maximum daily dive depths averaging around 600 m, with the deepest dive being

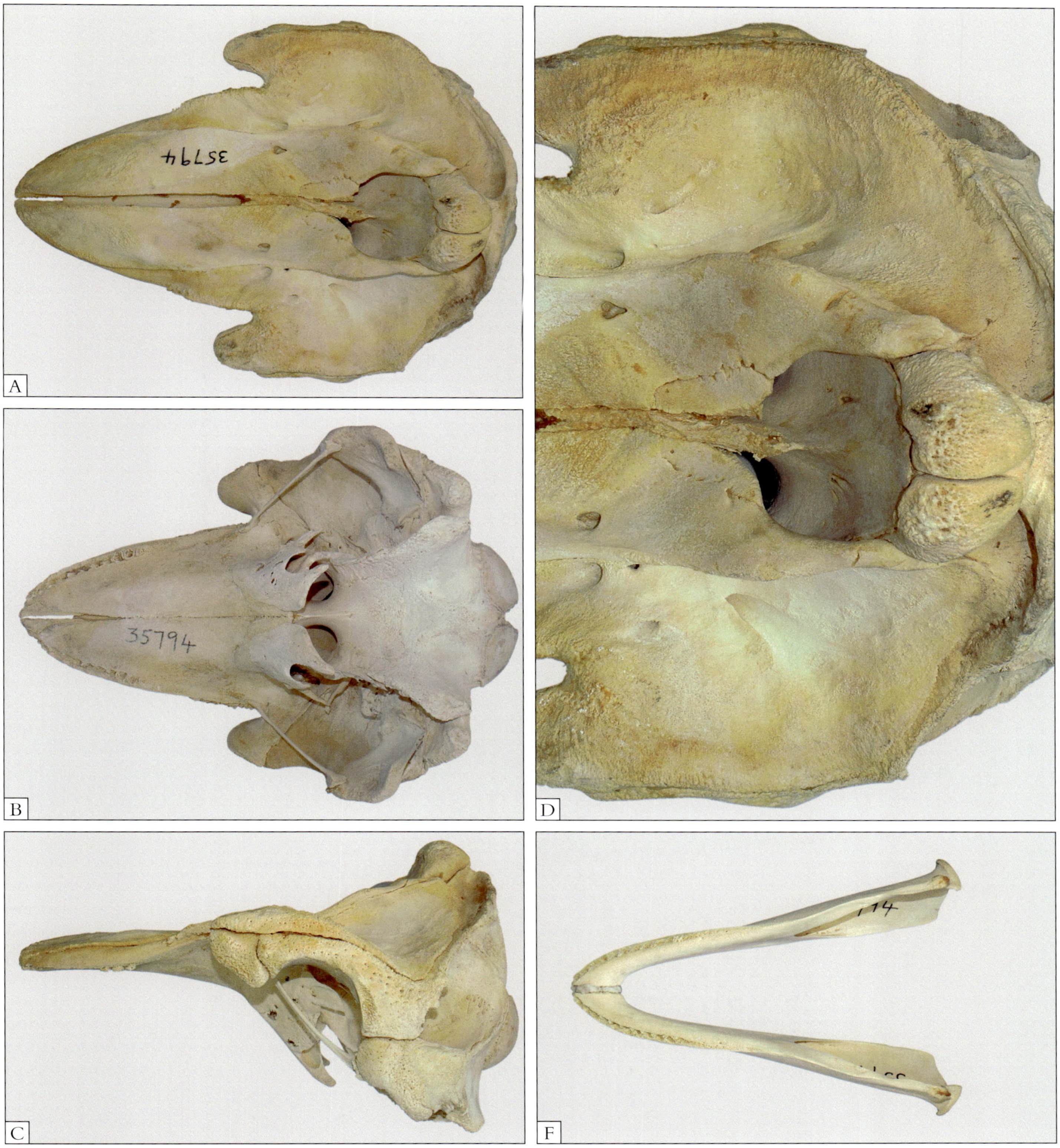

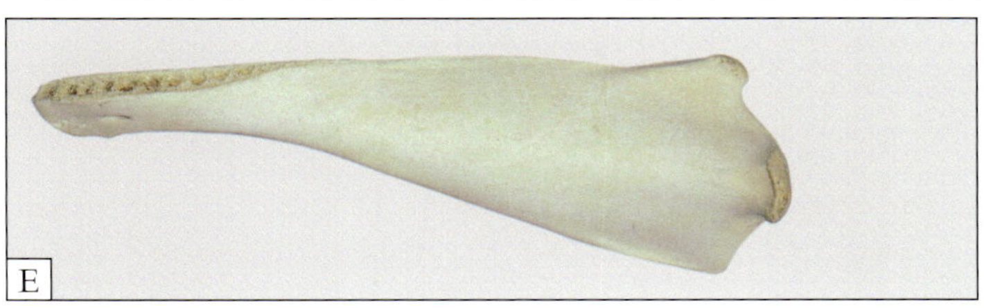

FIG. 172 Dorsal (A), ventral (B) and left lateral (C) views and vertex of skull (D); left lateral (labial) (E) and dorsal (F) views of mandible (ZM 35794).

to 828 m.[28] Captive juvenile long-finned pilot whales used suction pressure generated within the mouth to ingest prey from a distance of 10–20 cm.[29]

Reproduction and growth

Very little information is available for the subregion, so most of the following is taken from populations in the North Atlantic. The size at birth varies between 1.63 m and 1.91 m, with 50% of births occurring at 1.77 m. The length of gestation has been variously given as 12 months[30] or 15.5–16.0 months.[31] The former estimate, with its year-round coverage, is probably the more accurate. While breeding is clearly seasonal, the timing of births is not well established, but most (75%) seem to occur in the six months between mid-spring and mid-autumn,[30] which is consistent with a 1.5 times increase in the testis weight of mature males between March and September.[32]

The average duration of lactation has been calculated as 3.35 years, but varies with the age of the female, from about three years in females 21–39 years old to nine years in older females.[30] Weaning begins at around 6.5 months of age but milk can still be detected in the stomachs of some whales as old as seven years (male) or 12 years (female).[24] Sexual maturity in females is reached on average at eight years of age, and at a body length of 3.74 m.[30] All five females with a length of 3.71 m or more from the subregion were mature.[9] Male long-finned pilot whales can produce spermatozoa when they are as small as 4.38 m and as young as eight years, but puberty seems to occur at lengths of between about 4.6 m and 5.2 m, when the animals are 11–18 years old, and sexual maturity at an average age of 16.8 years and a length of 5.16 m.[32] Calves are produced at an average interval of 5.1 years, but this declines significantly with age, so that while the first four calves might be born at intervals of 2.7 years, the next four would be born at average intervals of 5.8 years, and the last two at about 11.8 year intervals. However, few females become reproductively senescent before they die and pregnancy can occur even in the oldest individuals.[30]

Males suffer a higher natural mortality than females,[31] which may be a contributory reason for the marked disparity in the sex ratio in favour of females in most mixed schools examined.[32] The oldest male aged in a sample from the Faroe Islands was 46 years old (n = 967) and the oldest female 59 years (n = 1 482); 15.1% of males and 24.2% of females were over the age of 25.[10]

Behaviour and vocalisations

Groups of pilot whales sighted in the subregion that were positively identified as long-finned varied in size from two to 120, with an average of 56 (n = 9). Groups of pilot whales sighted, where the species was undetermined (but many of which from their location were probably long-finned), ranged from one to 275 with a mean of 41.5 (n = 31).[9] Although no associations with other species have been recorded in the subregion for groups identified as long-finned pilot whales, 'pilot whale' sightings in the area have been recorded with bottlenose dolphins, striped dolphins, spinner dolphins, southern right whale dolphins and hourglass dolphins, and some of these (from their location) probably involved long-finned pilot whales.[9, 33]

On average, schools in the Faroe Islands (n = 38) contained at most 12% mature males, while 12% were pregnant females, 24% were lactating females and 46% were immature.[34] From their relative absence in most schools it has been suggested that maturing and young mature males emigrate from their natal schools and form separate aggregations, at least temporarily (although such groups are rarely encountered). This would also contribute to the low overall proportion of males in mixed schools noted above. The proportion of mature males also varied between schools, from 0% to 19%, and their numbers together with those of males over 22 years old were weakly correlated with the number of ovulating females, suggesting some temporary aggregation.[32] Genetic analysis of whole schools driven ashore in the Faroe Islands has shown that multigenerational sets of maternally related males and females are captured together, but that breeding within the school is rare or non-existent.[35] This, together with other evidence (such as that from parasites[22] and pollutant loads[36]) that differences exist between some schools, has led to the interpretation that such groupings are permanent entities, that there is little or no dispersal of either sex from its natal group, even by mature males, and that mating must take place reciprocally between pods.[35] Behavioural studies of long-finned pilot whales off Nova Scotia, however, indicated a much greater fluidity of association between individuals, with units disassociating within hours or days. Whether this represents a real difference between populations, or is a reflection of the different research methodologies used, is not clear.[37]

The behaviour of southern long-finned pilot whales has been described as nearly stationary, milling or travelling. Surfing down waves and spy-hopping are frequently noted, but leaping not often. Flukes are sometimes exposed.[38] Support of newborn young by adults, and a possible instance of the death of a mature male through intraspecific aggression, occurred during observations of a school in nearshore waters.[39]

Long-finned pilot whales in the southern hemisphere have been recorded as producing whistles, screams and pulsed sounds.[40] The whistle repertoire of North Atlantic long-finned pilot whales was shown to be a continuum with no clear mutually exclusive whistle types. Besides the shape or contour of the whistles, their frequency, duration and rate of production could all vary.[41] The average minimum and maximum frequencies were 3.48 kHz and 5.78 kHz; the average starting and ending frequencies were 4.43 kHz and 4.23 kHz; and the average duration was 0.72 sec. and 0.98 inflections. Differences in some of the characteristics between populations in the North Atlantic were also apparent.[42] Within a herd, simple whistles (with no inflections) were recorded more frequently when whales were milling; complex whistles and pulsed sounds occurred more frequently when the animals were engaged in energetic, coordinated behaviour at the surface, possibly feeding; and greater numbers of most whistle types were produced when the whales were widely dispersed and in more subgroups. Vocalisations possibly served to maintain contact and coordinate activities.[43]

Conservation status

The abundance of long-finned pilot whales in the entire Southern Ocean south of the Antarctic Convergence (about 50° S) in summer has been estimated as 200 000, but how this relates to pilot whales found further north is not clear. Furthermore, there appeared to be generally low sighting rates in the South Atlantic/western Indian Ocean sector, suggesting a relatively low abundance in higher latitudes of the subregion.[17] There are no indications of any significant fisheries-related mortality, and the species has never been the subject of any directed take in the subregion.

Long-finned pilot whales are listed as Lower Risk: Least Concern by the IUCN and as Least Concern in the South African Red Data Book. The species is included on Appendix II of CITES, while only the North and Baltic Sea populations are listed on Appendix II of CMS.

References

1 Rice 1998
2 Fullard *et al.* 2000
3 Kasuya *et al.* 1988
4 Davies 1960
5 Van Bree *et al.* 1978
6 Barnard 1954
7 Ross 1984
8 Weir, C., pers. comm.
9 MRI Whale Unit data
10 Bloch *et al.* 1993b
11 Bloch *et al.* 1993c
12 Bernard and Reilly 1999
13 Tomilin 1957
14 Lockyer 1993
15 Findlay *et al.* 1992
16 Robineau 1989
17 Kasamatsu and Joyce 1995
18 Payne and Heinemann 1993
19 Bloch *et al.* 2003
20 Mate *et al.* 2005
21 Balbuena and Raga 1991
22 Raga and Balbuena 1993
23 Sekiguchi *et al.* 1992b
24 Desportes and Mouritsen 1993
25 Gannon *et al.* 1997
26 Baird *et al.* 2002
27 Nawojchik *et al.* 2003
28 Heide-Jørgensen *et al.* 2002
29 Werth 2000
30 Martin and Rothery 1993
31 Sergeant 1962
32 Desportes *et al.* 1993
33 Ballance and Pitman 1998
34 Bloch *et al.* 1993a
35 Amos *et al.* 1993
36 Aguilar *et al.* 1993
37 Ottensmeyer and Whitehead 2003
38 Goodall and Macnie 1998b
39 Goodall and Macnie 1998a
40 Fisher *et al.* 1983
41 Taruski 1979
42 Rendell *et al.* 1999
43 Weilgart and Whitehead 1990

Short-finned Pilot Whale

Globicephala macrorhynchus

THE PILOT WHALE'S taxonomic situation has been discussed in the species account of the long-finned pilot whale. The occurrence of the short-finned pilot whale in the subregion is beyond doubt, but the relationship between the populations of this species in the Indian Ocean and the Atlantic is unclear. It would seem from the small amount of available data that they must be largely or completely separate, but whether this geographical separation is accompanied by morphological or genetic distinctions is unknown.

Appearance

The difficulty in distinguishing between the two pilot whale species at sea, discussed under the long-finned pilot whale, is complicated by a scarcity of knowledge about the external appearance of the short-finned pilot whale in the South Atlantic. If a body can be examined, the following characters should distinguish the majority of short-finned from the majority of long-finned pilot whales in the subregion: 1) The tooth counts on each side of the upper jaw are 7–10 (cf. 9–12 in the long-finned pilot whale). 2) The flipper lengths (anterior insertion to tip) are 14.8–17.8% of body length (cf. 17.7–23.7% in the long-finned pilot whale).

Few short-finned pilot whales have been examined in the subregion. The largest of 11 individuals accurately measured was an unsexed animal 5.89 m long. The largest of four known males was 5.2 m and the largest of three known females 4.05 m long.[1] In the North Pacific, males of the southern form reach an asymptotic length of 4.74 m and females 3.64 m (the largest male was 5.25 m and the largest female 4.05 m),[2] while males of the northern form attain asymptotic lengths of 6.5 m and females 4.67 m.[3] Given the small number of animals accurately measured in the subregion, it seems more likely that at least those off the east coast of South Africa are closer in size to the northern form from the North Pacific.

The overall body shape is similar to that of the long-finned pilot whale, with a bulbous head (and little trace of a beak in adults) that is almost square in outline when viewed from above or below. The dorsal fin is also similar in shape, longer than it is high with a tip pointing backwards rather than vertically. In height it may be somewhat larger (7.2–9.8% of body length, mean 8%, n = 8)[1] than that of the long-finned pilot whale, and situated slightly further forward. Although the forward placement of the dorsal fin makes the tail section of the body appear long, in fact it is relatively shorter than in the long-finned pilot whale.[4] The flippers are long and sickle-shaped but shorter than those of the long-finned pilot whale, being 14.8–17.8% (mean 16.3%, n = 7) of the body in length. The width of the tail flukes is 23.6–28.8% (mean 26.1%, n = 7)[1] of body length, and is therefore relatively greater than that of the long-finned pilot whale.[4]

Length:weight relationship

No mass information is available from the subregion. For the North Pacific the weights of 10 post-natal females and one near-term foetus 1.44 m long[2] have been used to produce the following relationship: $\mathbf{W = 14.719L^{2.8345}}$, where W = weight in kilograms and L = total length in metres. The weights of seven of another eight animals that died after being in captivity for up to 18 months fell above this regression line, presumably because they had been on an unnaturally high plane of nutrition, so justifying their exclusion.[2] From this relationship a 2.5 m short-finned pilot whale would weigh 197.6 kg and a 3.5 m animal 512.9 kg.

Skeletal data

Vertebral formula (n = 3–4): 7 cervical, 10–11 thoracic, 11–14 lumbar and 25–28 caudal; total 56–57.[10, 11]

The first 6–7 cervical vertebrae are fused.[11]

The phalangeal formula for pilot whales in general, including metacarpals, has been given as I_{3-4}, II_{9-14}, III_{9-11}, IV_{2-3}, V_{1-2},[12] but it is not clear whether the flippers of the two species differ in this respect (as one might suspect). One short-finned pilot whale had a phalangeal formula of I_3, II_{11}, III_8, IV_3, V_2.[6]

The skull of the short-finned pilot whale can be distinguished from that of the long-finned whale by the lateral expansion of the premaxillae, which in dorsal view almost completely cover the maxillae in the front half of the rostrum. Usually there are fewer teeth (see above), but there can be some overlap in this character.[13]

The pigmentation of short-finned pilot whales in the subregion is only known in some detail for animals from the Indian Ocean, where individuals have been described as almost entirely black except for a dark grey anchor-shaped mark on the chest between the flippers, similar in shape to that of the long-finned pilot whale, and extending back in the midline as far as the ano-genital region. There is no grey saddle behind the dorsal fin, and apparently no trace of a grey or white eye streak.[1] It is not clear whether short-finned pilot whales in the South Atlantic share these characteristics. At least some individuals from the tropical North Atlantic seem to resemble those from the Indian Ocean. A group of pilot whales near the Cape Verde Islands (about 16° N) were described as black with an ill-defined obscure greyish area in front of the dorsal fin and no trace of a saddle patch or eye streak,[5] while a photograph of a young individual from the Ivory Coast (about 5° N) also fails to show any pale dorsal pigmentation.[6] Further north in the North Atlantic, however, short-finned pilot whales from the Canary Islands (about 28° N) have a distinctive pigmentation pattern, with a prominent grey-white saddle behind the dorsal fin that extends forward on each side almost as far as the leading edge of the dorsal fin, and is virtually contiguous with a prominent white oblique eye streak.[7] This situation may be analogous to that in the North Pacific, where the northern form has a brighter, more prominent saddle and a slightly brighter eye streak than the southern form, in which the saddle is much darker and may become indiscernible after death.[8] One might speculate therefore that at least the short-finned pilot whales in the tropical South Atlantic would be of the darker form, similar to that in the Indian Ocean (and recent sightings and photographs off Angola confirm that these animals appear to be uniformly dark dorsally[7]).

In adults there are 7–9 (mean 7.4, n = 18) pairs of teeth in the upper jaw and 7–10 (mean 8.4, n = 16) pairs in the lower jaw.[1] The teeth are conical and peg-like, and of a similar size to those of long-finned pilot whales. They start to erupt in the upper jaw at an age of 3–9 months and slightly later in the lower jaw. By the age of nine months (and a body length of 2.11 m in the southern form in the North Pacific), all individuals have some erupted teeth in both jaws.[9]

Distribution and migrations

As explained for long-finned pilot whales, describing the distribution of short-finned pilot whales in the subregion is complicated by the problem of possible at-sea confusion between the two species. Short-finned pilot whales are generally regarded as more tropical in distribution than long-finned, but the poleward extension of the warm Agulhas Current on the east coast of southern Africa, and the equatorward extension of the cold Benguela System on the west coast, makes it unlikely that there is a common latitudinal limit to the southern distribution of the short-finned pilot whale throughout the subregion. They have been recorded from Gabon,[14] Angola,[7] possibly Namibia,[15] South Africa,[16] Madagascar,[17] Mauritius,[18] Réunion,[19] the Comoro

Islands[20] and the Seychelles.[21] The Namibian record is based on a single skull seen in the Lüderitz Museum in 1976 but which is no longer traceable.[22] On the South African coast the most westerly record is from Mossel Bay (22° E),[23] but this was a skull found in the local museum with no information about its provenance. The next most westerly record is from Jeffrey's Bay (25° E).[1]

Off Angola all short-finned pilot whale sightings were in water depths of 1 206–3 400 m (mean 2 014 m, n = 22).[24] Elsewhere the species has been noted as occurring most commonly along the mid-to-upper continental slope, and in areas with a steep bottom gradient[25] or high relief.[26] Short-finned pilot whales were seen off Angola in all four seasons, although 72.7% of sightings (n = 22) occurred in spring and summer (September–February).[24] In the Gulf of Mexico sightings were also made in all four seasons.[27] Off Japan, both southern and northern forms of short-finned pilot whale were considered likely to shift their distribution towards the equator in winter.[8]

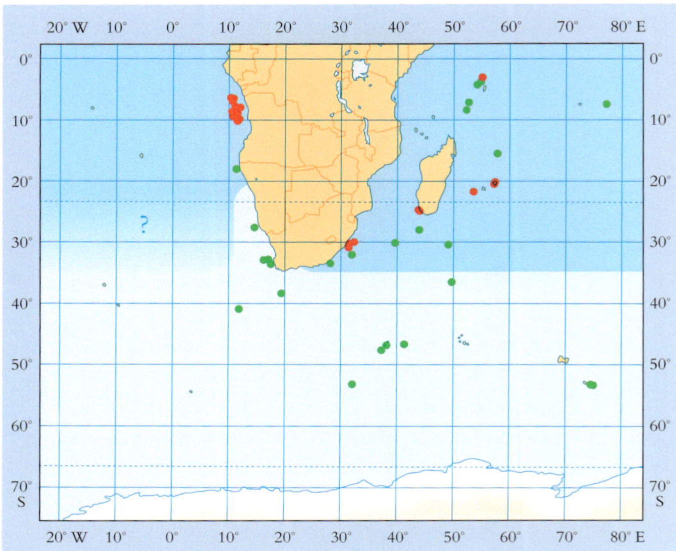

Postulated range Sighting or capture
Unidentified pilot whale ? Questionable origin

FIG. 173 Distribution map for short-finned pilot whale.

External commensals and parasites

There is no data for the subregion. Elsewhere, short-finned pilot whales have played host to three species of cyamids: *Isocyamus deltobranchium* (from healing wounds and scars),[28] *I. delphinii* (from around the anus and blowhole), and *Syncyamus ilheusensis* (from the mucosa of the eye).[29]

Food and feeding behaviour

Very little information is available locally: cephalopod remains were found in four stomachs, 24 beaks from one being identified as *Loligo reynaudi* and 10 as *Oregoniateuthis*.[1] The only other information from the southern hemisphere (Western Australia) indicates a predominance of ommastrephids among the cephalopods eaten.[30] In the eastern North Pacific, four short-finned pilot whales had also been feeding exclusively on cephalopods, consuming at least seven species, of which *Loligo opalescens* predominated (95% by number),[31] while in the western North Pacific animals from the northern population had fed mostly on cephalopods (of which the major species were *Todarodes pacificus*, *Eucleoteuthis luminosa* and *Ommastrephes bartrami*) but also on some fish (e.g. *Trachypterus*).[3]

In captivity, three short-finned pilot whales weighing 229 kg, 617 kg and 1 361–1 588 kg consumed on average 7.9–8.9%, 4.4–5.9% and 2.9–3.3% respectively of their body weight daily.[32]

Newly introduced captive animals seemed to be largely nocturnal in behaviour, spending most daylight hours in slumber.[33] Observations of diving behaviour in a telemetered individual showed that, as darkness approached, the pattern of respiration changed and the frequency of long dives increased, suggesting nocturnal feeding.[13] Behavioural patterns during the day also point to the likelihood that the species feeds at night.[34] A trained individual made many dives deeper than 300 m in the open ocean, including one to 504 m and another possibly as deep as 604 m. The maximum submergence time was 14 min. and 49 sec.[13]

Reproduction and growth

Very little information is available for the subregion, and most of the following is taken from populations in the North Pacific. Birth occurs at a length of about 1.4 m in the southern form and 1.85 m in the northern form. Breeding is diffusely seasonal, with a peak in parturition in midsummer in the southern form but in midwinter in the northern form. Pregnancy is estimated to occupy 14–16 months,[3,9] although this conclusion has been challenged.[35] Testosterone concentrations in the testis and serum of the southern form were significantly higher in spring than in autumn,[36] which is not inconsistent with summer births and a gestation period of a year or slightly more. The length for which the calf is suckled seems to vary with the age of its mother, being as short as two years in some precocious individuals and as long as 13–15 years in others, but usually occupies 3.5–5.5 years. Solid food is first taken at an age of 0.5–1 year. Sexual maturity in females is reached at an average age of 8.5–9.0 years, when they are 3.9–4.0 m in length (northern form)[9] or 3.16 m (southern form).[3] In males physiological maturity is reached at 17–18 years of age,[3,9] when they are on average 4.22 m long in the southern form[9] (and presumably substantially larger in the northern form). The limited information available for the subregion suggests that females reach sexual maturity at 3.8–3.9 m (n = 3), while all four males 4.37–5.2 m long were mature.[1] It therefore seems that females, at least, are more similar in size to the northern form in the North Pacific.

The mean interval between calves in the North Pacific has been estimated as 7.8 years for the southern population and 5.1–7.1 years for the northern population, but in the southern population the incidence of pregnant females falls rapidly from 30.9% at age 10 to 4.7% at age 36, and then to zero in older females. This means that a female produces only 4–5 calves during its entire reproductive life.[9] Females of the northern population are believed to show a similar reproductive pattern. Males in both populations lived to 44–46 years, and females to 61.5–62.5 years.[3] This not only implies that males have a higher mortality rate, but also that mature females must experience a significant post-reproductive life (see box on page 283).[9]

FIG. 174 Unlike long-finned pilot whales, their short-finned relatives (in the western Indian Ocean at least) are a plain dark colour above, although the grey anchor-shaped mark on the chest can be seen in the animal spy-hopping off the Maldives shown in the lower photograph.

Behaviour and vocalisations

Groups of pilot whales sighted in the subregion, and which were positively identified as short-finned, varied in size from two to 200, with a mean of 31.7 (n = 31).[37] Mixed species associations with bottlenose dolphins occurred in six (19.4%) of the sightings.

Solitary schools driven ashore in Japan are considered less subject to selection, and they contained on average 7.8% mature males, 51.9% mature females, 22% immature males, 3.2% maturing males and 14.9% immature females. Some males may leave their natal school at puberty to join other males, and adult males may also move between schools occasionally to form schools dominated by adult males, but there is no evidence of social or geographic segregation by sex. Females probably stay in their natal school at least until they reach sexual maturity, and associations between females probably last more than one calving cycle and possibly for life.[9]

Short-finned pilot whales encountered off Santa Catalina Island in the North Pacific were travelling 73% of the time, while 17% were feeding, and the remainder were either resting or milling. They were never observed socialising (possibly because all observations were made in winter and outside the probable breeding season). Travelling whales formed a tight oval most (62%) of the time, or moved in a tight line abreast (15%) or in mixed geometry (14%). When feeding, the whales were

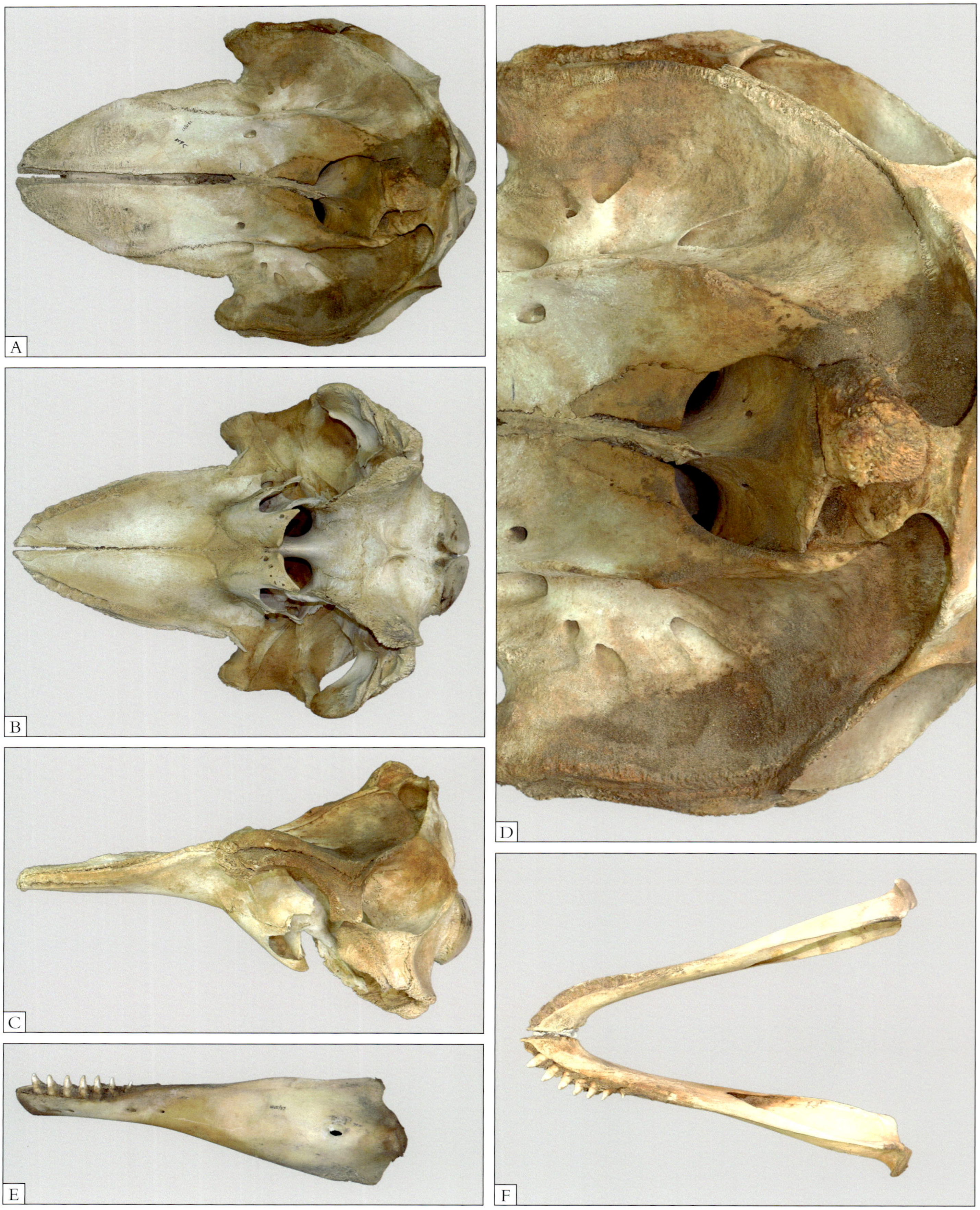

FIG. 175 Dorsal (A), ventral (B) and left lateral (C) views and vertex of skull (D); left lateral (labial) (E) and dorsal (F) views of mandible (PEM 95).

Do grandmothers have a role?

In the great majority of mammals, including cetaceans, females continue to be reproductively active throughout life. Although the tempo of reproduction may slow with age – and declines in pregnancy rate in older females are seen in several toothed whale species – reproductively active individuals can usually be found among even the oldest animals. Short-finned pilot whales, on the other hand, display a protracted period of reproductive inactivity in old age. Despite 26% of the mature females examined being over 35 (and up to 63) years old, the oldest pregnant female was only 34.5 years, the oldest ovulating animal 39.5 years, and the oldest lactating female 50.5 years. Some females may have a post-reproductive lifespan of 20–30 years, although the mean post-reproductive lifespan is about 14 years. Declines in the abundance of egg-bearing follicles, increases in the atresia or degeneration of the remaining follicles, and other anatomical changes in the ovaries confirm that this reproductive failure is the result of true senescence.[1,2] Such a phenomenon is a parallel of the human menopause and is highly unusual among marine mammals, possibly only being also encountered in some populations of killer and false killer whales.[3]

The role that these grandmothers might play in short-finned pilot whale society is uncertain. About one-sixth of them were found to be lactating, suggesting either that they were nursing for extended periods (which would be consistent with the trend shown with age in reproductively active females) or that they were nursing the calves of younger females. The latter behaviour could theoretically benefit the school as a whole by allowing younger females to invest more heavily (in energetic terms) in bearing rather than rearing calves, but there is no behavioural evidence to support it (and the majority of the grandmothers were resting).[3] Assistance with foraging or with defence against predators are possibilities, but are either not seen or occur rarely. On the assumption that females tend to remain in their natal school for life, it is possible that these older individuals improve the overall fitness of the school by acting as repositories of information.[4] When unusual events such as environmental anomalies occur, for example, the experience of the grandmothers in being able to locate alternative resources could be highly advantageous to the school as a whole.

References

1 Kasuya and Marsh 1984
2 Marsh and Kasuya 1984
3 Marsh and Kasuya 1986
4 McAuliffe and Whitehead 2005

usually widely dispersed (56%) or in mixed geometry (40%).[34]

Short-finned pilot whales have been observed harassing sperm whales sufficiently for them to adopt defensive behaviour patterns, including marguerite formation and tight huddling. No combat was seen and no injury recorded to either party. In one instance the remains of two apparently regurgitated squid were seen floating in the water at the end of the encounter,[38] and it is possible that this behaviour represents a form of kleptoparasitism in which the pilot whales are attempting to induce the sperm whales to vomit so that they can purloin the food. Other interactions include the persistent carrying of the bodies of dead conspecifics,[39] other dolphins[40] and sea lions.[41] Some of these corpses were 'defended', while mating was attempted with a dead conspecific, and at least three stillborn or very young dolphins were ultimately consumed by captive pilot whales,[42] leaving the motivation for this behaviour unclear. In an even more inexplicable move, a wild short-finned pilot whale seized a human swimmer who had been petting it and dragged her 12–15 m underwater in a potentially life-threatening action[43] (her companion, who filmed the event, was fined $10 000 for harassing whales).

Courtship behaviour witnessed in captive short-finned pilot whales seemed to be initiated by the females rubbing themselves against the belly of the male. This could be followed by forceful head-butting, in which contact was made from a distance of up to 9 m with such force that sometimes a whip-like crack could be heard through the wall of the tank and the smaller female would be thrust backward some metres. Arousal of the male and attempted intromission would follow either the body-rubbing or butting, and the female would appear to play the major role in pre-copulatory positioning.[39]

Vocalisations of short-finned pilot whales include both whistles and clicks, the latter frequently being emitted in pairs.[44] The whistles produced are generally shorter in duration (mean 0.56 sec.), higher in frequency (mean 7.87 kHz), cover a wider frequency range (5.43–9.6 kHz) and have fewer inflections (0.69) than those of long-finned pilot whales. There are also significant differences in pitch, duration, frequency range and number of inflections between two populations of short-finned pilot whales either side of the North Atlantic.[45]

Conservation status

There are no estimates of abundance for short-finned pilot whales in the subregion.

There are no indications of any significant fisheries-related mortality, and the species has never been the subject of any directed take in the subregion.

Short-finned pilot whales are listed as Lower Risk: Least Concern by the IUCN and as Data Deficient in the South African Red Data Book. The species is included on Appendix II of CITES, but is not listed by CMS.

References

1 Ross 1984
2 Kasuya and Matsui 1984
3 Kasuya and Tai 1993
4 Yonekura *et al.* 1980
5 Fraser 1950b
6 Cadenat 1959
7 Caroline Weir, pers. comm.
8 Kasuya *et al.* 1988
9 Kasuya and Marsh 1984
10 True 1889
11 Dammerman 1924
12 Tomilin 1957
13 Bernard and Reilly 1999
14 Rosenbaum and Collins 2006
15 Griffin and Coetzee 2005
16 Findlay *et al.* 1992
17 Peddemors *et al.* In press
18 Corbett 1994
19 Boucaud and Rota 2006
20 Kiszka *et al.* 2006a
21 Keller *et al.* 1982
22 Mike Griffin, pers. comm.
23 Van Bree *et al.* 1978
24 Weir. In press
25 Davis *et al.* 1998
26 Hui 1985
27 Mullin *et al.* 2004
28 Sedlak-Weinstein 1992
29 Haney *et al.* 2004
30 Clarke 1996
31 Sinclair 1992
32 Gilmore 1962
33 Kritzler 1952
34 Shane 1995a
35 Martin and Rothery 1993
36 Kita *et al.* 1999
37 MRI Whale Unit data
38 Weller *et al.* 1996
39 Brown 1962
40 Brown *et al.* 1966
41 Shane 1994
42 Perryman and Foster 1980
43 Shane *et al.* 1993
44 Higashi *et al.* 1992
45 Rendell *et al.* 1999

Family PHOCOENIDAE

Spectacled Porpoise

Phocoena dioptrica

♂

♀

THIS HIGHLY DISTINCTIVE and very attractive little animal has the distinction of being the only porpoise to occur in the subregion (see box on page 288). Although the type specimen of the finless porpoise, *Neophocaena phocaenoides*, is supposed to have come from the Cape of Good Hope, it is now considered more likely that the skull originated further east, possibly in India,[1] while sightings off South Africa attributed to this species[2] are now believed more likely to have been Cape fur seals. Consequently, the spectacled porpoise has this field to itself.

Appearance

There should be no difficulty in recognising this little animal at sea. Its completely dark back and unhooked dorsal fin (disproportionately large in adult males) distinguish it from all other small cetaceans in the subregion.

No spectacled porpoises from the subregion have been examined. Elsewhere the largest of nine males was 2.24 m long and the largest of seven females 2.035 m.[3] Like all porpoises, the head is rounded in front with no beak. The gape of the mouth is particularly small ('less than half the length of the flipper'), and the upper jaw projects slightly beyond the lower. The flippers themselves are not large (8.3–14.9% of body length, mean 11%, n = 6) and are situated relatively far forward and low down on the body. They are fairly blunt in shape compared to those of most dolphins. Apart from the animal's coloration, the dorsal fin is perhaps the most distinctive characteristic of the species. In females and juvenile males it is broadly triangular, with the leading edge somewhat longer than the trailing edge and both edges slightly convex.[4] It is situated in the middle of the back and its height is 7.5–8.6% of body length. In adult males, however, the fin enlarges massively, both at the base and in height, where it reaches about 12.5% of body length.[5] Such a sexually dimorphic fin is only paralleled by that of the killer whale, but unlike the impressive appearance of the latter's triangular fin, that of the adult male spectacled porpoise somehow seems out of proportion and makes the animal look top heavy, like a capsized yacht. In two individuals examined so far (both female), a single row of up to 26 small tubercles has been described along the leading edge of the dorsal fin.[6,7] The tail flukes have rounded tips and their span is about 23–24% of body length.[5]

In pigmentation, spectacled porpoises are strikingly black above and white below, the border between the two being sharply defined and extending as a plimsoll line from the front of the snout through the eye, above the flipper and along the side of the body before curving up behind the level of the anus onto the caudal peduncle. There may or may not be a narrow ridge of dark pigment connecting the dark back to the dorsal surface of the flukes. Its face looks almost clown-like, with a dark ring thinly outlined in white or grey around each eye, and both lips thickly painted with black. The upper surface of the flippers varies in colour, sometimes being white with or without a grey rear edge, in other individuals being mottled grey to black. The flukes also vary in colour, but are normally dark above and lighter (white, or light grey with a white veination pattern) below. Calves are not as brightly pigmented as adults.[4] In some sightings of adults a pale saddle area around the dorsal fin was evident at close range.[8]

In adults there are 13–22 (mean 18, n = 11) pairs of erupted teeth in the upper jaw and 15–21 (mean 18, n = 9) pairs in the lower jaw. Even in large animals, however, there can be as many as six more pairs of upper teeth concealed in the gum at the rear end of the toothrow. The teeth are small and spade-shaped, although there is considerable variation in their size and shape.[4] The smallest individual with erupted teeth was 1.27 m long[7] and the largest with unerupted teeth was 1.36 m.[4]

Distribution and migrations

The distribution of the spectacled porpoise is circumpolar in temperate and cold waters of the southern hemisphere. Most of the sightings to date have been south of 45° S,[4] up to 64°34′ S,[8] but strandings have occurred on the east coast of South America as far north as 33°05′ S[11] and in South Australia at 35°32′ S.[7] There appears to be a tendency for sightings to be grouped around islands in the West Wind Drift, but whether this is real or an artefact of search effort is not clear. In the subregion the species has been recorded from Kerguelen[12] and Heard Island,[13] as well as about 400 nautical miles east of the South Sandwich Islands.[8]

Length:weight relationship

No mass data are available from the subregion. Only three specimens (including one foetus) have been weighed whole,[4] with the following relationship: **$W = 13.671L^{2.9569}$**, where W = weight in kilograms and L = total length in metres. A fourth eviscerated specimen 1.09 m long weighed 14.8 kg.[4] The regression predicts a mass of 17.6 kg, implying a weight loss of 16%.

Skeletal data

Vertebral formula (n = 14–40): 7 cervical, 13–14 thoracic, 13–16 lumbar and 29–34 caudal; total 66–70.[9]

The first 5–7 cervical vertebrae are fused.[9]

The phalangeal formula, including metacarpals (n = 2), is: I_2, II_7, III_{4-6}, IV_{3-4}, V_{2-4}.[6,7]

The skull of the spectacled porpoise shows all of the characteristics that distinguish porpoises from dolphins: a short rostrum, a large and rounded brain case, raised knob-like protuberances towards the base of the premaxillae, delayed closure of skull sutures, and spade-shaped teeth.[10]

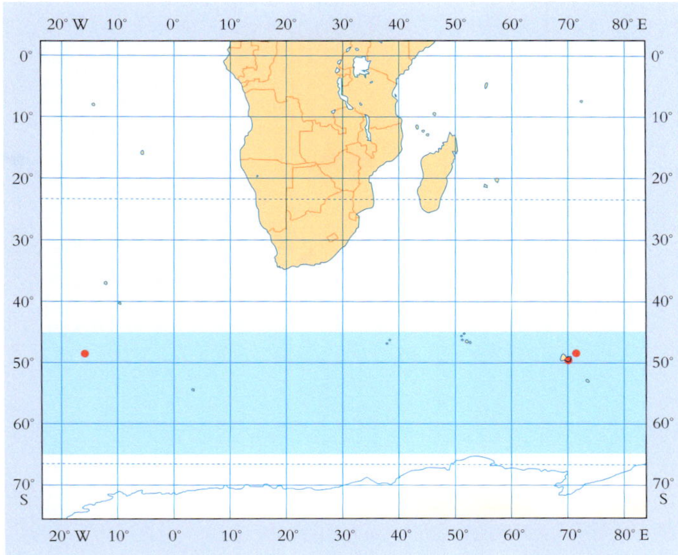

Postulated range ■ Sighting or capture

FIG. 176 Distribution map for spectacled porpoise.

Although many sightings have been made in deep water far from land, some individuals enter estuaries and channels,[3] or even swim among kelp.[12]

There is no information on whether seasonal movements occur.

External commensals and parasites

No information is available for the species.[14]

Food and feeding behaviour

No information is available for the subregion, and very little for the species. In Argentina, one stomach contained a small piece of algae, and another anchovy (*Engraulis* sp.) and small crustaceans, possibly stomatopods.[4] The stomach of a stranded animal in Brazil contained some tunicate remains,[11] and one from Australia contained both some milky fluid and two sets of cuttlefish (*Sepia* sp.) beaks, as well as having a partially digested ornate cowfish (*Aracana ornate*) in the oesophagus.[7] Feeding behaviour has not been observed.[14]

Reproduction and growth

No information is available for the subregion – nearly all of what is known comes from Argentina. Young are born at a length of 90–100 cm, probably in late spring or summer. The length of gestation is unknown, as is the duration of lactation. A female 1.83 m long and aged 2–3 years was taken in a net with a calf. A change in the pattern of layering in the teeth after the second or third group of layers indicates early maturation (as found in some other porpoises). Two other females 1.86 m long were sexually mature, as was a 2.05 m male, and the limited information available suggests that animals may reach sexual maturity at three years or younger.[4]

The species does not appear long-lived: the oldest animal found in a small sample (n = 28) was eight years old, and some males at least are physically mature by four years of age.[4] Calves formed 10.5% of animals seen in the Antarctic (n = 57).[8]

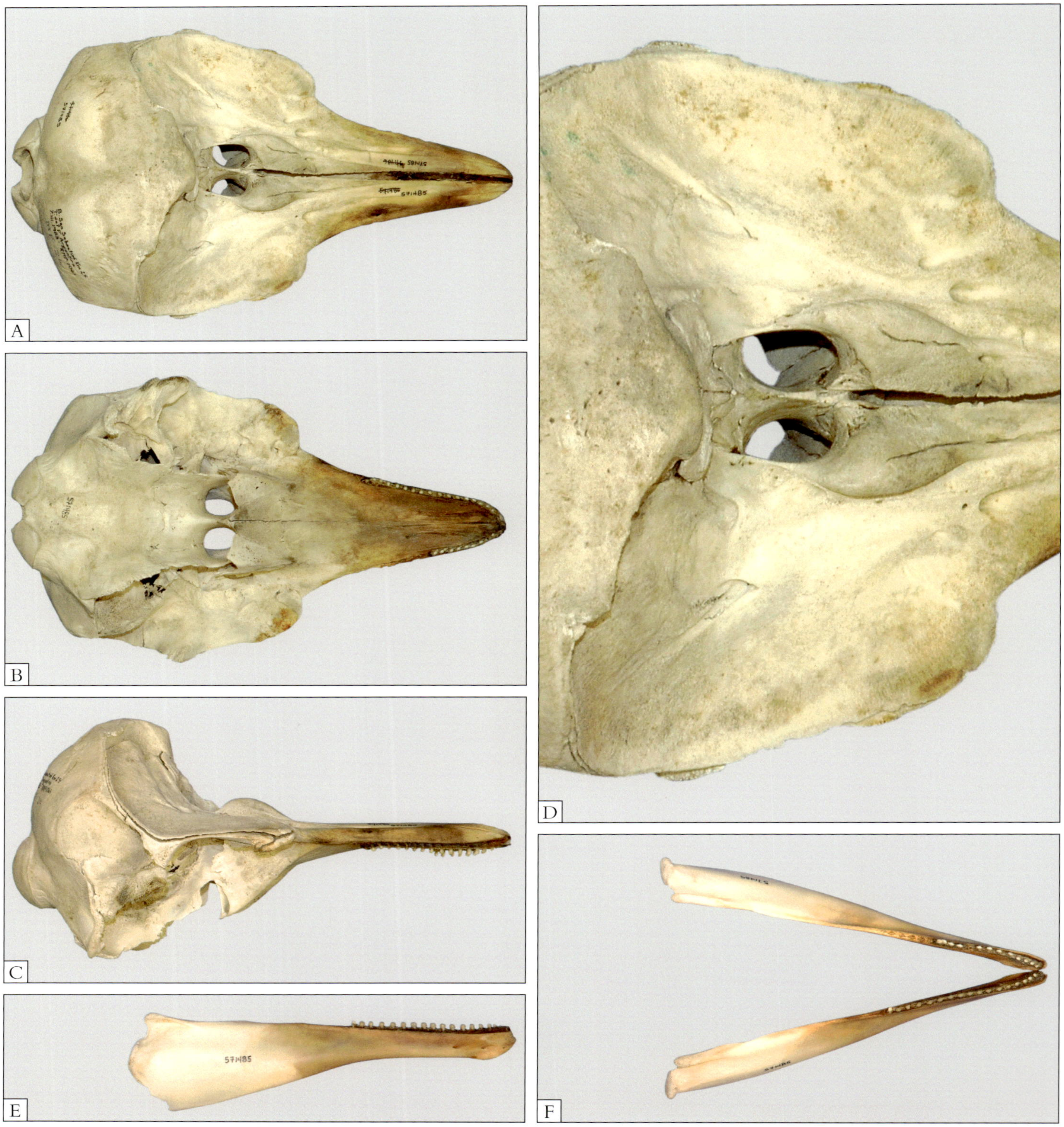

FIG. 177 Dorsal (A), ventral (B) and right lateral (C) views and vertex of skull (D); right lateral (labial) (E) and dorsal (F) views of mandible (USNM 571485).

FIG. 178 The spectacled porpoise is one of the most infrequently sighted cetaceans, and this subsurface photograph (*above*) provides an unusually good opportunity to see the distinctive colour pattern. Also distinctive is the large and rather bizarre dorsal fin of an adult male (*below*).

Behaviour and vocalisations

Spectacled porpoises have so far only been found in small groups of 1–4 individuals, with an average group size of two (n = 28). Calves were seen in 21.4% of the sightings, and apart from their mother were always accompanied by an adult male and on one occasion by a second female. The male tended to leave the cow-calf pair during or towards the end of protracted (16–41 min.) periods of inspection from a vessel.[8]

When undisturbed, spectacled porpoises roll slowly at the surface without making a splash. This plus their small individual and group size makes them difficult to spot at sea except in excellent sea conditions (mainly Beaufort 2 or less) and when relatively close to the vessel. When approached by a vessel they try to avoid it by accelerated swimming. This takes place in a very characteristic manner, with the porpoise remaining submerged for 100–200 m but so close to the surface that its fluke-prints can easily be followed. The animal then reduces speed, surfaces slowly at right angles to its flight path, as if to check whether it is being followed. After one or more slow rolls it submerges again and resumes its rapid flight from the vessel. The species apparently does not readily bow-ride.[8]

No information is available on vocalisations produced by this species.

Conservation status

There are no estimates of abundance for spectacled porpoises in any part of their range. Sightings are relatively few, but this may be due partly to the fact that they are not easy to spot at sea. Throughout most of its range this species probably rarely comes into contact with humans, but close to the South American continent it has suffered some mortality as a by-catch in coastal gill nets and bottom and midwater trawls.[14]

The spectacled porpoise is listed as Data Deficient by the IUCN, but is not included in the South African Red Data Book. The species is included on Appendix II of CITES and on Appendix II of CMS.

References

1 Allen 1923
2 Gibson-Hill 1950
3 Goodall 2002
4 Goodall and Schiavini 1995
5 Fraser 1968
6 Lahille 1912
7 Evans *et al.* 2001
8 Sekiguchi *et al.* In press
9 Perrin *et al.* 2000
10 Read 2002
11 Pinedo *et al.* 2002
12 De Boer *et al.* 2003
13 Brownell *et al.* 1989
14 Brownell and Clapham 1999

Dolphins, porpoises – what's the difference?

A good question, and one which frequently confuses people because for a long time some parts of the world insisted on calling 'genuine' dolphins porpoises to avoid confusion with the dolphin fish (a.k.a. dorado or mahi-mahi, *Coryphaena hippurus*). In fact, porpoises diverged from dolphins at least 11 million years ago and are now placed in a separate family, Phocoenidae. Porpoises are characteristically small, stocky animals without the beak distinctive of dolphins. Their flippers and flukes are relatively small, the dorsal fin is not hooked (as in most dolphins) but usually triangular or rounded, and their teeth are spade-shaped at the tip rather than conical as in dolphins. Behaviourally they are unlike most dolphins in tending to occur alone or in small temporary groupings and are believed to forage independently rather than cooperatively. They also rarely if ever engage in the aerial behaviour that is so typical of dolphins. Vocally they differ in not producing pure-tone signals or whistles, but rather a range of pulsed sounds. And in life-style they appear to be fast-growing, early-maturing and short-lived animals compared to most dolphins: some reach maturity in their third year, and calve annually (so that they are simultaneously pregnant and lactating), and few live beyond 20 years.[1]

But such distinctions are not clear-cut. In particular, the dolphin genus *Cephalorhynchus*, of which we have two representatives in the subregion (Heaviside's and Commerson's dolphins), seems to be trying very hard to blur the distinction with porpoises. They too are small and stocky, without a beak, have small rounded flippers and triangular or lobed dorsal fins, swim around in small schools and even produce pulsed sounds rather than whistles. This seems to be an example of convergent evolution, in which different species develop similar groups of characters in response to selective pressures associated with occupying similar environments.[2] However, an inspection of their teeth (conical, not spade-shaped) reveals that both *Cephalorhynchus* species are in reality dolphins – and one might have guessed as much from the delight that Heaviside's dolphins, in particular, take in leaping out of the backline, giving the lie to the idea that they might be porpoises with an aversion to aerial activity.

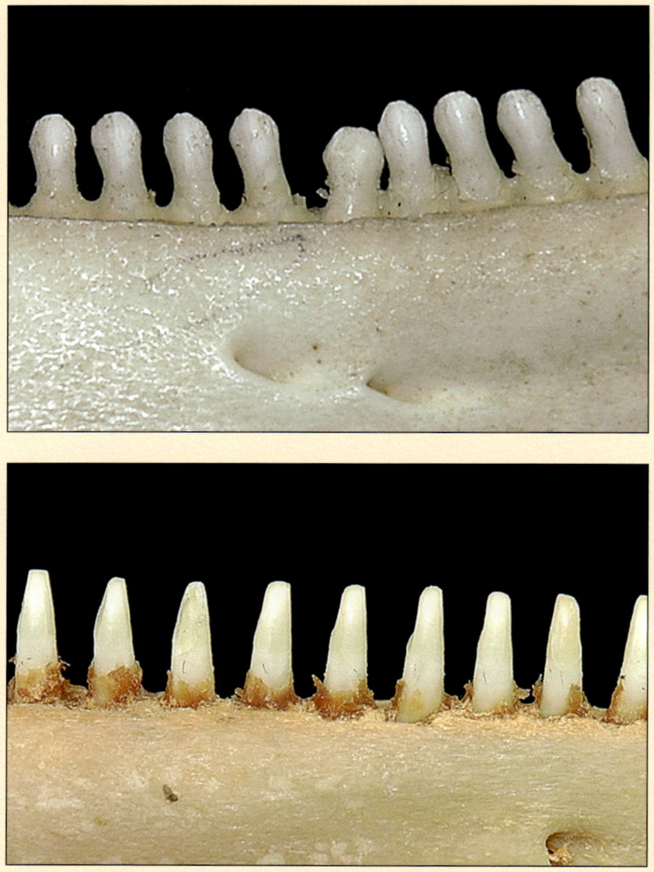

FIG. 179 The spade-shaped teeth of a harbour porpoise (*above*) compared with those of a Heaviside's dolphin (*below*).

References

1 Read 2002

2 Buchholtz 2002

Southern right whale dolphins bowride a seismic exploration vessel in transit between the Orange River and Walvis Bay, Namibia.

BIBLIOGRAPHY

A

Acebes, J.M.V., Bautista, A.L., Yamada, T.K., Dolar, L., and Perrin, W.F. 2005. Stranding of *Indopacetus pacificus* in Davao, Philippines. Abstract submitted to 16th Biennial Meeting of the Society for Marine Mammalogy, San Diego, USA.

Acevedo-Gutiérrez, A., Brennan, B., Rodriguez, P., and Thomas, M. 1997. Resightings and behaviour of false killer whales (*Pseudorca crassidens*) in Costa Rica. *Marine Mammal Science* **13**(2): 307–314.

Acevedo-Gutiérrez, A., Croll, D.A., and Tershy, B.R. 2002. High feeding costs limit dive time in the largest whales. *Journal of Experimental Biology* **205**: 1 747–1 753.

Addink, M.J., and Smeenk, C. 2001. Opportunistic feeding behaviour of rough-toothed dolphins *Steno bredanensis* off Mauritania. *Zoologische Verhandelingen, Leiden* **334**: 37–48.

Agenbag, J.J. 1996. Pacific ENSO events reflected in meteorological and oceanographic perturbations in the southern Benguela system. *South African Journal of Science* **92**: 243–247.

Agenbag, J.J., and Shannon, L.V. 1988. A suggested physical explanation for the existence of a biological boundary at 24°30′ S in the Benguela system. *South African Journal of Marine Science* **6**: 119–132.

Aguilar, A. 1985. Aboriginal whaling off Pagalu (Equatorial Guinea). *Report of the International Whaling Commission* **35**: 385–386.

Aguilar, A. 1991. Calving and early mortality in the western Mediterranean striped dolphin, *Stenella coeruleoalba*. *Canadian Journal of Zoology* **69**: 1 408–1 412.

Aguilar, A., Jover, L., and Borrell, A. 1993. Heterogeneities in organochlorine profiles of Faroese long-finned pilot whales: indication of segregation between pods? *Report of the International Whaling Commission* (Special Issue) **14**: 359–367.

Aguilar, A., and Raga, J.A. 1993. The striped dolphin epizootic in the Mediterranean Sea. *Ambio* **22**(8): 524–528.

Allen, G.M. 1923. The black finless porpoise, *Meomeris*. *Bulletin of the Museum of Comparative Zoology* **65**(7): 233–256 + 3 pl.

Allen, G.M. 1941. Pygmy sperm whale in the Atlantic. *Publications. Field Museum of Natural History (Zoological Series)* **27**: 17–36.

Allen, K.R. 1977. Updated estimates of fin whale stocks. *Report of the International Whaling Commission* **27**: 221.

Allison, C., and Smith, T.D. 2004. Progress on the construction of a comprehensive database of twentieth century whaling catches. Paper SC/56/O27 submitted to the Scientific Committee of the International Whaling Commission, Sorrento, Italy, June 2004. [Catches prior to 1930 obtained from BIWS data.]

Alonso, M.K., Pedraza, S.N., Schiavini, C.M., Goodall, R.N.P., and Crespo, E.A. 1999. Stomach contents of false killer whales (*Pseudorca crassidens*) stranded on the coasts of the Strait of Magellan, Tierra del Fuego. *Marine Mammal Science* **15**(3): 712–724.

Amano, M., and Miyazaki, N. 2004. Composition of a school of Risso's dolphins, *Grampus griseus*. *Marine Mammal Science* **20**(1): 152–160.

Amano, M., Miyazaki, N., and Yanagisawa, F. 1996. Life history of Fraser's dolphin, *Lagenodelphis hosei*, based on a school captured off the Pacific coast of Japan. *Marine Mammal Science* **12**(2): 199–214.

Amir, O.A., and Berggren, P. 2001. A note on recent sightings and strandings of humpback whales (*Megaptera novaeangliae*) around Zanzibar and in the coastal waters of Tanzania. Paper presented to International Workshop on Cooperative Research on humpback whales in the southwest Indian Ocean, Cape Town, April 2001.

Amir, O.A., Berggren, P., and Jiddawi, N.S. 2002. The incidental catch of dolphins in gillnet fisheries in Zanzibar, Tanzania. *Western Indian Ocean Journal of Marine Science* **1**(2): 155–162.

Amir, O.A., Berggren, P., Ndaro, S.G.M., and Jiddawi, N.S. 2005a. Feeding ecology of the Indo-Pacific bottlenose dolphin (*Tursiops aduncus*) incidentally caught in the gillnet fisheries off Zanzibar, Tanzania. *Estuarine, Coastal and Shelf Science* **63**: 429–437.

Amir, O.A., Jiddawi, N.S., and Berggren, P. 2005b. The occurrence and distribution of dolphins in Zanzibar, Tanzania, with comments on the differences between two species of *Tursiops*. *Western Indian Ocean Journal of Marine Science* **4**(1): 85–93.

Amos, B., Bloch, D., Desportes, G., Majerus, T.M.O., Bancroft, D.R., Barrett, J.A., and Dover, G.A. 1993. A review of molecular evidence relating to social organisation and breeding system in the long-finned pilot whale. *Report of the International Whaling Commission* (Special Issue) **14**: 209–231.

Anderson, R.C., Shaan, A., and Waheed, Z. 1999. Records of cetacean 'strandings' from the Maldives. *Journal of South Asian Natural History* **4**(2): 187–202.

André, J.M., Ribeyre, F., and Boudou, A. 1990. Relations entre l'âge, la longeur totale, le poids total et le poids des organes chez trois espèces de dauphins de la zone tropicale Est de l'océan Pacifique. *Mammalia* **54**(3): 479–488.

Andrews, R.C. 1908. Description of a new species of *Mesoplodon* from Canterbury Province, New Zealand. *Bulletin of the American Museum of Natural History* **24**: 203–215 + 1 pl.

Andrews, R.C. 1916. Monographs of the Pacific Cetacea. II. The sei whale (*Balaenoptera borealis*) 1. History, habits, external anatomy, osteology and relationships. *Memoirs of the American Museum of Natural History* n.s. **1**(6): 289–388.

Angot, M. 1951. Rapport scientifique sur les expeditions baleinieres autour de Madagascar (saisons 1949 et 1950). *Mémoires de l'Institut Scientifique de Madagascar (Series A)* **6**(2): 439–486.

Anon. 1920. Report of the interdepartmental committee on research and development in the dependencies of the Falkland Islands. London, H.M.S.O. 164 pp. [Annexure V]

Anon. 1947. International Whaling Conference in London June 14th–June 24th 1938. *International Whaling Statistics* **17**: 35–45.

Anon. 1969. A sighting of southern right whale dolphins. *The Marine Observer* **39**(225): 128 + 1 pl.

Anon. 1993. Australia. Progress report on cetacean research, May 1991 to June 1992. *Report of the International Whaling Commission* **43**: 263–269.

Anon. 1998. Initial alternative hypotheses for the distribution of humpback breeding stocks on the feeding grounds. *Report of the International Whaling Commission* **48**: 181.

Anon. 2006. Minke whale project – Research update – Information Sheet No. 9, 15 December 2006, 4 pp. James Cook University, the Museum of Tropical Queensland and Undersea Explorer, Australia.

Archer (II), F.I. 2002. Striped dolphin *Stenella coeruleoalba*. Pp 1 201–1 203 in (Perrin, W.F., Würsig, B., and Thewissen, J.G.M., eds) *Encyclopedia of Marine Mammals*. Academic Press, San Diego.

Archer (II), F.I., and Perrin, W.F. 1999. *Stenella coeruleoalba. Mammalian Species* **603**: 1–9.

Archer, F.I., and Robertson, K.M. 2004. Age and length at weaning and development of diet of pantropical spotted dolphins, *Stenella attenuata*, from the eastern tropical Pacific. *Marine Mammal Science* **20**(2): 232–245.

Armstrong, A.J., and Siegfried, W.R. 1991. Consumption of Antarctic krill by minke whales. *Antarctic Science* **3**(1): 13–18.

Árnason, Ú., Gullberg, A., and Janke, A. 2004. Mitogenomic analyses provide new insights into cetacean origin and evolution. *Gene* **333**: 27–34.

Árnason, Ú., Gullberg, A., and Widegren, B. 1993. Cetacean mitochondrial DNA control region: sequences of all extant baleen whales and two sperm whale species. *Molecular Biology and Evolution* **10**: 960–970.

Árnason, Ú., Spillaert, R., Pálsdóttir, Á., and Árnason, A. 1991. Molecular identification of hybrids between the two largest whale species, the blue whale (*Balaenoptera musculus*) and the fin whale (*B. physalus*). *Hereditas* **115**: 183–189.

Arnold, A. 1987. Portland's chance encounter with a pygmy right whale. *Australian Natural History* **22**(6): 266–270.

Arnold, P., Marsh, H., and Heinsohn, G. 1987. The occurrence of two forms of minke whales in east Australian waters with a description of external characters and skeleton of the diminutive or dwarf form. *Scientific Reports of the Whales Research Institute, Tokyo* **38**: 1–46.

Arnold, P.W. 1997. Occurrence of dwarf minke whales (*Balaenoptera acutorostrata*) on the northern Great Barrier Reef, Australia. *Report of the International Whaling Commission* **47**: 419–424.

Aron, W., Burke, W., and Freeman, M. 2002. Scientists versus whaling: science, advocacy, and errors of judgment. *BioScience* **52**(12): 1 137–1 140.

Ash, C.E. 1953. Weights of Antarctic humpback whales. *Norsk Hvalfangsttidende* **42**(7): 387–391.

Ashford, J.R., Rubilar, P.S., and Martin, A.R. 1996. Interactions between cetaceans and longline fishery operations around South Georgia. *Marine Mammal Science* **12**: 452–456.

Asper, E.D., Andrews, B.F., Antrim, J.E., and Young, W.G. 1992. Establishing and maintaining successful breeding programs for whales and dolphins in a zoological environment. *International Marine Biological Research Institute Report* **3**: 71–84.

Asper, E.D., Young, W.G., and Walsh, M.T. 1988. Observations on the birth and development of a captive-born killer whale. *International Zoo Yearbook* **27**: 295–304.

Atkins, S., Pillay, N., and Peddemors, V.M. 2004. Spatial distribution of Indo-Pacific humpback dolphins (*Sousa chinensis*) at Richards Bay, South Africa: environmental influences and behavioural patterns. *Aquatic Mammals* **30**(1): 84–93.

Atkinson, S., Combelles, C., Vincent, D., Nachtigall, P., Pawloski, J., and Breese, M. 1999. Monitoring of progesterone in captive female false killer whales, *Pseudorca crassidens*. *General and Comparative Endocrinology* **115**: 323–332.

Au, D., and Perryman, W. 1982. Movement and speed of dolphin schools responding to an approaching ship. *Fishery Bulletin, US* **80**(2): 371–379.

Au, D.W.K., and Perryman, W.L. 1985. Dolphin habitats in the eastern tropical Pacific. *Fishery Bulletin, US* **83**(4): 623–643.

Au, W.W.L., and Würsig, B. 2004. Echolocation signals of dusky dolphins (*Lagenorhynchus obscurus*) in Kaikoura, New Zealand. *Journal of the Acoustical Society of America* **115**(1): 2 307–2 313.

Au, W.W.L. 2002. Echolocation. Pp 358–367 in (Perrin, W.F., Würsig, B., and Thewissen, J.G.M., eds) *Encyclopedia of Marine Mammals*. Academic Press, San Diego.

Avolio, M., Ersts, P., Pomilla, C. *et al.* 2002. Humpback whale distribution and marine mammal diversity in the waters of Mayotte (Comores Archipelago) in the Mozambique Channel. Paper SC54/H18 submitted to the Scientific Committee of the International Whaling Commission, Shimonoseki, Japan, April–May 2002.

Awbrey, F.T., Thomas, J.A., Evans, W.E., and Leatherwood, S. 1982. Ross Sea killer whale vocalizations: preliminary description and comparison with those of some Northern Hemisphere killer whales. *Report of the International Whaling Commission* **32**: 667–670.

Azevedo, A.F., Lailson-Brito Jr, J., Siciliano, S., Cunha, H.A., and Fragoso, A.B.L. 2003. Colour pattern and external morphology of the Fraser's dolphin (*Lagenodelphis hosei*) in the southwestern Atlantic. *Aquatic Mammals* **29**(3): 411–416.

Aznar, F.J., Balbuena, J.A., and Raga, J.A. 1994. Are epizoites biological indicators of a western Mediterranean striped dolphin die-off? *Diseases of Aquatic Organisms* **18**(3): 159–163.

B

Baird, R.W. 2000. The killer whale. Foraging specializations and group hunting. Pp 127–153 in (Mann, J., Connor, R.C., Tyack, P.L., and Whitehead, H., eds) *Cetacean societies. Field studies of dolphins and whales*. University of Chicago Press, Chicago.

Baird, R.W. 2005. Sightings of dwarf (*Kogia sima*) and pygmy (*K. breviceps*) sperm whales from the main Hawaiian Islands. *Pacific Science* **59**(3): 461–466.

Baird, R.W., Borsani, J.F., Hanson, M.B., and Tyack, P.L. 2002. Diving and night-time behaviour of long-finned pilot whales in the Ligurian Sea. *Marine Ecology Progress Series* **237**: 301–305.

Baird, R.W., and Gorgone, A.M. 2005. False killer whale dorsal fin disfigurements as a possible indicator of long-line fishery interactions in Hawaiian waters. *Pacific Science* **59**(4): 593–601.

Baird, R.W., Hanson, M.B., and Dill, L.M. 2005. Factors influencing the diving behaviour of fish-eating killer whales: sex differences and diel and interannual variation in diving rates. *Canadian Journal of Zoology* **83**: 257–267.

Baird, R.W., Ligon, A.D., Hooker, S.K., and Gorgone, A.M. 2001. Subsurface and nighttime behaviour of pantropical spotted dolphins in Hawai'i. *Canadian Journal of Zoology* **79**: 988–996.

Baird, R.W., McSweeney, D.J., Bane, C. *et al.* 2006. Killer whales in Hawaiian waters: information on population identity and feeding habits. *Pacific Science* **60**(4): 523–530.

Baird, R.W., McSweeney, D.J., Ligon, A.D., and Webster, D.L. 2004. Tagging feasibility and diving of Cuvier's beaked whales (*Ziphius cavirostris*) and Blainville's beaked whales (*Mesoplodon densirostris*) in Hawai'i. Report prepared under order no. QAB133F-03-SE-0986 from National Marine Fisheries Service (NMFS) SWFSC, USA.

Baird, R.W., Nelson, D., Lien, J., and Nagorsen, D.W. 1996. The status of the pygmy sperm whale, *Kogia breviceps*, in Canada. *Canadian Field-Naturalist* **110**(3): 525–532.

Baird, R.W., and Stacey, P.J. 1988. Variation in saddle patch pigmentation in populations of killer whales (*Orcinus orca*) from British Columbia, Alaska, and Washington State. *Canadian Journal of Zoology* **66**: 2 582–2 585.

Baker, A.N. 1974. Risso's dolphin in New Zealand waters, and the identity of 'Pelorus Jack'. *Records of the Dominion Museum* **8**(16): 267–276.

Baker, A.N. 1981. The southern right whale dolphin *Lissodelphis peronii* (Lacépède) in Australasian waters. *National Museum of New Zealand Records* **2**(4): 17–34.

Baker, A.N. 1983. *Whales and dolphins of New Zealand and Australia. An identification guide.* Victoria University Press, Wellington. 133 pp.

Baker, A.N. 1985. Pygmy right whale *Caperea marginata* (Gray, 1846). Pp 345–354 in (Ridgway, S.H., and Harrison, R., eds). *Handbook of Marine Mammals 3: The Sirenians and Baleen Whales.* Academic Press, London.

Baker, A.N. 2001. Status, relationships, and distribution of *Mesoplodon bowdoini* Andrews, 1908 (Cetacea: Ziphiidae). *Marine Mammal Science* **17**(3): 473–493.

Baker, A.N., Duignan, P.J., Norman, R.J. de B., and Van Helden, A.L. 2001. A juvenile Hector's beaked whale, *Mesoplodon hectori* (Gray 1871), without functional throat grooves, plus notes on endoparasites (Cetacea: Ziphiidae). *Marine Mammal Science* **17**(1): 171–175.

Balbuena, J.A., and Raga, J.A. 1991. Ecology and host relationships of the whale-louse *Isocyamus delphini* (Amphipoda: Cyamidae) parasitizing long-finned pilot whales (*Globicephala melas*) off the Faroe Islands (Northeast Atlantic). *Canadian Journal of Zoology* **69**: 141–145.

Balcomb (III), K.C. 1989. Baird's beaked whale *Berardius bairdii* Stejneger, 1883; Arnoux's beaked whale *Berardius arnuxii* Duvernoy, 1851. Pp 261–288 in (Ridgway, S.H., and Harrison, R.J., eds). *Handbook of Marine Mammals 4: River dolphins and the larger toothed whales.* Academic Press, London.

Ballance, L.T., Anderson, R.C., Pitman, R.L., Stafford, K., Shaan, A., Waheed, Z., and Brownell Jr, R.L. 2001. Cetacean sightings around the Republic of Maldives, April 1998. *Journal of Cetacean Research and Management* **3**(2): 213–218.

Ballance, L.T., and Pitman, R.L. 1998. Cetaceans of the western tropical Indian Ocean: distribution, relative abundance, and comparisons with cetacean communities of two other tropical ecosystems. *Marine Mammal Science* **14**(3): 429–459.

Ballance, L.T., Pitman, R.L., Reilly, S.B., and Force, M.P. 1996. Report of a cetacean, seabird, marine turtle and flying fish survey of the western tropical Indian Ocean aboard the research vessel *Malcolm Baldrige*, March 21–July 26, 1995. *NOAA Technical Memorandum National Marine Fisheries Service* NOAA-TM-NMFS-SWFSC-224. 132 pp.

Bannister, J.L., and Baker, A. de C. 1967. Observations on food and feeding of baleen whales at Durban. *Norsk Hvalfangsttidende* **56**(4): 78–82.

Bannister, J.L., and Gambell, R. 1965. The succession and abundance of fin, sei and other whales off Durban. *Norsk Hvalfangsttidende* **54**(3): 45–60.

Bannister, J.L., and Grindley, J.R. 1966. Notes on *Balaenophilus unisetus* P.O.C. Aurivillius, 1879, and its occurrence in the Southern Hemisphere (Copepoda, Harpacticoida). *Crustaceana* **10**(3): 296–302.

Bargain, R.-M., Lucas, V., and Thomas, A. 2000. The Seychelles semi-industrial fishery. *Indian Ocean Tuna Commission Proceedings* **3**: 164–168.

Barlow, J. 1985. Variability, trends, and biases in reproductive rates of spotted dolphins, *Stenella attenuata*. *Fishery Bulletin, US* **83**(4): 657–669.

Barnard, K.H. (Undated). *The whale collection.* Unpublished manuscript in South African Museum files. 12 pp.

Barnard, K.H. 1954. *A guide book to South African whales and dolphins.* Guide no. 4, South African Museum, Cape Town. 33 pp.

Barrett-Lennard, L.G., Ford, J.K.B., and Heise, K.A. 1996. The mixed blessing of echolocation: differences in sonar use by fish-eating and mammal-eating killer whales. *Animal Behaviour* **51**: 553–565.

Barros, N.B., and Cockcroft, V.G. 1991. Prey of humpback dolphins (*Sousa plumbea*) stranded in eastern Cape Province, South Africa. *Aquatic Mammals* **17**(3): 134–136.

Bastida, R., Lichtschein, V., and Goodall, R.N.P. 1988. Food habits of *Cephalorhynchus commersonii* off Tierra del Fuego. *Report of the International Whaling Commission* (Special Issue) **9**: 143–160.

Baumgartner, M.F. 1997. The distribution of Risso's dolphin (*Grampus griseus*) with respect to the physiography of the northern Gulf of Mexico. *Marine Mammal Science* **13**(4): 614–638.

Baumgartner, M.F., Mullin, K.D., May, L.N., and Leming, T.D. 2001. Cetacean habitats in the northern Gulf of Mexico. *Fishery Bulletin, US* **99**: 219–239.

Bazúa-Durán, C. 2004. Differences in the whistle characteristics and repertoire of bottlenose and spinner dolphins. *Anais da Academia Brasileira de Ciências* **76**(2): 386–392.

Beal, L.M., and Bryden, H.L. 1997. Observations of an Agulhas Undercurrent. *Deep-Sea Research I* **44**(9–10): 1 715–1 724.

Bearzi, G., Reeves, R.R., Notarbartolo-di-Sciara, G., Politi, E., Cañadas, A., Frantzis, A., and Mussi, B. 2003. Ecology, status and conservation of short-beaked common dolphins *Delphinus delphis* in the Mediterranean Sea. *Mammal Review* **33**(3): 224–252.

Bearzi, M. 2005. Habitat partitioning by three species of dolphins in Santa Monica Bay, California. *Bulletin of the Southern California Academy of Science* **104**(3): 113–124.

Belkin, I.M., and Gordon, A.L. 1996. Southern Ocean fronts from the Greenwich meridian to Tasmania. *Journal of Geophysical Research* **101**(C2): 3 675.

Bell, C.H., Kemper, C.M., and Conran, J.G. 2002. Common dolphins *Delphinus delphis* in southern Australia: a morphometric study. *Australian Mammalogy* **24**: 1–10.

Benirschke, K., Johnston, M.L., and Benirschke, R.J. 1980. Is ovulation in dolphins, *Stenella longirostris* and *Stenella attenuata*, always copulation-induced? *Fishery Bulletin, US* **78**(2): 507–528.

Benjaminsen, T. 1972. On the biology of the bottlenose whale, *Hyperoodon ampullatus* (Forster). *Norwegian Journal of Zoology* **20**(3): 233–241.

Benjaminsen, T., and Christensen, I. 1979. The natural history of the bottlenose whale, *Hyperoodon ampullatus* (Forster). Pp 143–164 in (Winn, H.E., and Olla, B.L., eds) *Behavior of Marine Animals. 3. Cetaceans.* Plenum Publishing Co., New York.

Benoit-Bird, K.J., and Au, W.W.L. 2003. Prey dynamics affect foraging by a pelagic predator (*Stenella longirostris*) over a range of spatial and temporal scales. *Behavioural Ecology and Sociobiology* **53**: 364–373.

Benoit-Bird, K.J., Würsig, B., and McFadden, C.J. 2004. Dusky dolphin (*Lagenorhynchus obscurus*) foraging in two different habitats: active acoustic detection of dolphins and their prey. *Marine Mammal Science* **20**(2): 215–231.

Bernard, H.J., and Hohn, A.A. 1989. Differences in feeding habits between pregnant and lactating spotted dolphins (*Stenella attenuata*). *Journal of Mammalogy* **70**(1): 211–215.

Bernard, H.J., and Reilly, S.B. 1999. Pilot whales *Globicephala* (Lesson, 1828). Pp 245–279 in (Ridgway, S.H., and Harrison, R.J., eds) *Handbook of Marine Mammals 6: The second book of dolphins and the porpoises.* Academic Press, New York.

Bérubé, M. 2002. Hybridism. Pp 596–600 in (Perrin, W.F., Würsig, B., and Thewissen, J.G.M., eds) *Encyclopedia of Marine Mammals.* Academic Press, San Diego.

Bérubé, M., and Aguilar, A. 1998. A new hybrid between a blue whale, *Balaenoptera musculus*, and a fin whale, *B. physalus*: frequency and implications of hybridization. *Marine Mammal Science* **14**(1): 82–98.

Berzin, A.A. 1972. *The sperm whale.* Israel Program for Scientific Translations. 394 pp.

Berzin, A.A., and Vladimirov, V.L. 1983. [A new species of killer whale (Cetacea, Delphinidae) from Antarctic waters.] *Zoologicheskiĭ Zhurnal* **62**(2): 287–295 [In Russian].

Besharse, J.C. 1971. Maturity and sexual dimorphism in the skull, mandible, and teeth of the beaked whale, *Mesoplodon densirostris. Journal of Mammalogy* **52**(2): 297–315.

Best, P.B. 1960. Further information on Bryde's whale (*Balaenoptera edeni* Anderson) from Saldanha Bay, South Africa. *Norsk Hvalfangsttidende* **49**: 201–215.

Best, P.B. 1966. A case of prolonged lactation in the fin whale (*Balaenoptera physalus*). *Norsk Hvalfangsttidende* **55**(6): 119–122.

Best, P.B. 1967. Distribution and feeding habits of baleen whales off the Cape Province. *Investigational Reports of the Division of Sea Fisheries, South Africa* **57**: 1–44.

Best, P.B. 1968. The sperm whale (*Physeter catodon*) off the west coast of South Africa. 2. Reproduction in the female. *Investigational Reports of the Division of Sea Fisheries, South Africa* **66**: 1–32.

Best, P.B. 1969a. A dolphin (*Stenella attenuata*) from Durban, South Africa. *Annals of the South African Museum* **52**: 121–135.

Best, P.B. 1969b. The sperm whale (*Physeter catodon*) off the west coast of South Africa. 3. Reproduction in the male. *Investigational Reports of the Division of Sea Fisheries, South Africa* **72**: 1–20.

Best, P.B. 1969c. The sperm whale (*Physeter catodon*) off the west coast of South Africa. 4. Distribution and movements. *Investigational Reports of the Division of Sea Fisheries, South Africa* **78**: 1–12.

Best, P.B. 1970a. Exploitation and recovery of right whales *Eubalaena australis* off the Cape Province. *Investigational Reports of the Division of Sea Fisheries, South Africa* **80**: 1–20.

Best, P.B. 1970b. Records of the pygmy killer whale, *Feresa attenuata*, from Southern Africa, with notes on behaviour in captivity. *Annals of the South African Museum* **57**: 1–14.

Best, P.B. 1970c. The sperm whale (*Physeter catodon*) off the west coast of South Africa. 5. Age, growth and mortality. *Investigational Reports of the Division of Sea Fisheries, South Africa* **79**: 1–27.

Best, P.B. 1974. The biology of the sperm whale as it relates to stock management. Pp 257–293 in (Schevill, W.E., ed.). *The whale problem. A status report.* Harvard University Press, Cambridge, Mass.

Best, P.B. 1977. Two allopatric forms of Bryde's whale off South Africa. *Report of the International Whaling Commission* (Special Issue) **1**: 10–38.

Best, P.B. 1979. Social organization in sperm whales, *Physeter macrocephalus*. Pp 227–289 in (Winn, H.E., and Olla, B.L., eds) *Behavior of Marine Animals. Current Perspectives in Research. 3. Cetaceans.* Plenum Press, New York.

Best, P.B. 1982. Seasonal abundance, feeding, reproduction, age and growth in minke whales off Durban (with incidental observations from the Antarctic). *Report of the International Whaling Commission* **32**: 759–786.

Best, P.B. 1985. External characters of southern minke whales and the existence of a diminutive form. *Scientific Reports of the Whales Research Institute, Tokyo* **36**: 1–33.

Best, P.B. 1987. Estimates of the landed catch of right (and other whalebone) whales in the American fishery, 1805–1909. *Fishery Bulletin, US* **85**: 403–418.

Best, P.B. 1988a. Right whales (*Eubalaena australis*) at Tristan da Cunha – a clue to the 'non-recovery' of depleted stocks? *Biological Conservation* **46**: 23–51.

Best, P.B. 1988b. The external appearance of Heaviside's dolphin, *Cephalorhynchus heavisidii* (Gray, 1828). *Report of the International Whaling Commission* (Special Issue) **9**: 279–299.

Best, P.B. 1991. The presence of coronuline barnacles on a southern right whale *Eubalaena australis. South African Journal of Marine Science* **11**: 585–587.

Best, P.B. 1994a. A review of the catch statistics for modern whaling in southern Africa, 1908–1930. *Report of the International Whaling Commission* **44**: 467–485.

Best, P.B. 1994b. Seasonality of reproduction and length of gestation in southern right whales *Eubalaena australis. Journal of Zoology, London* **232**: 175–189.

Best, P.B. 1996. Evidence of migration by Bryde's whales from the offshore population in the southeast Atlantic. *Report of the International Whaling Commission* **46**: 315–322.

Best, P.B. 1997. *Whale watching in South Africa. The southern right whale.* Mammal Research Institute, University of Pretoria, Pretoria. 2nd ed. 28 pp.

Best, P.B. 1998. Blue whales off Namibia – a possible wintering ground for the Antarctic population. Paper SC/50/CAWS14 submitted to the Scientific Committee of the International Whaling Commission, Muscat, Oman, April–May 1998.

Best, P.B. 1999. Food and feeding of sperm whales *Physeter macrocephalus* off the west coast of South Africa. *South African Journal of Marine Science* **21**: 393–413.

Best, P.B. 2000. Coastal distribution, movements and site fidelity of right whales *Eubalaena australis* off South Africa, 1969–1998. *South African Journal of Marine Science* **22**: 43–55.

Best, P.B. 2001. Distribution and population separation of Bryde's whale *Balaenoptera edeni* off southern Africa. *Marine Ecology Progress Series* **220**: 277–289.

Best, P.B. 2003. How low did they go? An historical comparison of indices of abundance for some baleen whales on the Durban whaling ground. Paper SC/55/SH18 submitted to the Scientific Committee of the International Whaling Commission, Berlin, May/June 2003.

Best, P.B. 2006. The presence of right whales in summer on the west coast of South Africa: the evidence from historical records. *African Journal of Marine Science* **28**(1): 159–166.

Best, P.B., and Abernethy, R.B. 1994. Heaviside's dolphin *Cephalorhynchus heavisidii*. Pp 289–310, 415–416 in (Ridgway, S.H., and Harrison, R.J., eds) *Handbook of Marine Mammals 5: The first book of dolphins.* Academic Press, New York.

Best, P.B., Brandão, A., and Butterworth, D.S. 2001. Demographic parameters of southern right whales off South Africa. *Journal of Cetacean Research and Management* (Special Issue) **2**: 161–169.

Best, P.B., Butterworth, D.S., and Rickett, L.H. 1984a. An assessment cruise for the South African inshore stock of Bryde's whale (*Balaenoptera edeni*). *Report of the International Whaling Commission* **34**: 403–423.

Best, P.B., Canham, P.A.S., and MacLeod, N. 1984b. Patterns of reproduction in sperm whales, *Physeter macrocephalus. Report of the International Whaling Commission* (Special Issue) **6**: 51–79.

Best, P.B., Findlay, K.P., Sekiguchi, K., Peddemors, V.M., Rakotonirina, B., Rossouw, A., and Gove, D. 1998. Winter distribution and possible migration routes of humpback whales *Megaptera novaeangliae* in the southwest Indian Ocean. *Marine Ecology Progress Series* **162**: 287–299.

Best, P.B., and Gambell, R. 1968a. A comparison of the external characters of sperm whales off South Africa. *Norsk Hvalfangsttidende* **57**(6): 146–164.

Best, P.B., and Gambell, R. 1968b. The abundance of sei whales off South Africa. *Norsk Hvalfangsttidende* **57**: 168–174.

Best, P.B., and Lockyer, C.H. 2002. Reproduction, growth and migrations of sei whales *Balaenoptera borealis* off the west coast of South Africa in the 1960s. *South African Journal of Marine Science* **24**: 111–133.

Best, P.B., Payne, R., Rowntree, V., Palazzo, J.T., and Both, M.C. 1993. Long-range movements of South Atlantic right whales *Eubalaena australis*. *Marine Mammal Science* **9**(3): 227–234.

Best, P.B., Rademeyer, R.A., Burton, C. *et al.* 2003a. The abundance of blue whales on the Madagascar Plateau, December 1996. *Journal of Cetacean Research and Management* **5**(3): 253–260.

Best, P.B., Reeb, D., Morais, M., and Baird, A. 1999. A preliminary investigation of humpback whales off northern Angola. Paper SC51/CAWS33 submitted to the Scientific Committee of the International Whaling Commission, Grenada, May 1999.

Best, P.B., and Ross, G.J.B. 1977. Exploitation of small cetaceans off southern Africa. *Report of the International Whaling Commission* **27**: 494–497.

Best, P.B., and Ross, G.J.B. 1984. Live-capture fishery for small cetaceans in South African waters. *Report of the International Whaling Commission* **34**: 615–618.

Best, P.B., and Ross, G.J.B. 1986. Catches of right whales from shore-based establishments in southern Africa, 1792–1975. *Report of the International Whaling Commission* (Special Issue) **10**: 275–289.

Best, P.B., and Ross, G.J.B. 1989. Whales and whaling. Pp 315–338 in (Payne, A.I.L., and Crawford, R.J.M., eds) *Oceans of Life off southern Africa*. Vlaeberg Publishers, Cape Town.

Best, P.B., and Rüther, H. 1992. Aerial photogrammetry of southern right whales, *Eubalaena australis*. *Journal of Zoology, London* **228**: 595–614.

Best, P.B., Schaeff, C.M., Reeb, D., and Palsbøll, P. 2003b. Composition and possible function of social groupings of southern right whales in South African waters. *Behaviour* **140**: 1 469–1 494.

Best, P.B., and Schell, D.M. 1996. Stable isotopes in southern right whale (*Eubalaena australis*) baleen as indicators of seasonal movements, feeding and growth. *Marine Biology* **124**: 483–494.

Best, P.B., and Scott, H.A. 1993. The distribution, seasonality and trends in abundance of southern right whales *Eubalaena australis* off De Hoop Nature Reserve, South Africa. *South African Journal of Marine Science* **13**: 175–186.

Best, P.B., Sekiguchi, K., and Findlay, K.P. 1995. A suspended migration of humpback whales (*Megaptera novaeangliae*) on the west coast of South Africa. *Marine Ecology Progress Series* **118**: 1–12.

Best, P.B., Sekiguchi, K., Rakotonirina, B., and Rossouw, A. 1996. The distribution and abundance of humpback whales off southern Madagascar, August–September 1994. *Report of the International Whaling Commission* **46**: 323–331.

Best, P.B., and Shaughnessy, P.D. 1981. First record of the melon-headed whale *Peponocephala electra* from South Africa. *Annals of the South African Museum* **83**: 33–47.

Bester, M.N., Keith, M., and Pistorius, P.A. 2002. Killer whales at Marion Island, Southern Ocean. Paper submitted to fourth international orca symposium and workshop, 23–28 September 2002, CEBC-CNRS, Villiers en Bois, France. Pp 35–39.

Bigg, M.A., and Wolman, A.A. 1975. Live-capture killer whale (*Orcinus orca*) fishery, British Columbia and Washington, 1962–73. *Journal of the Fisheries Research Board of Canada* **32**(7): 1 213–1 221.

Birtles, A., and Arnold, P. 2002. *Dwarf minke whales in the Great Barrier Reef. Current state of knowledge*. CRC Reef Research Centre Ltd, Townsville, Australia. 6 pp.

Bloch, D., Desportes, G., Mouritsen, R., Skaaning, S., and Stefansson, E. 1993a. An introduction to studies of the ecology and status of the long-finned pilot whale (*Globicephala melas*) off the Faroe Islands, 1986–88. *Report of the International Whaling Commission* (Special Issue) **14**: 1–32.

Bloch, D., Heide-Jørgensen, M.P., Stefansson, E., Mikkelsen, B., Ofstad, L.H., Dietz, R., and Andersen, L.W. 2003. Short-term movements of long-finned pilot whales *Globicephala melas* around the Faroe Islands. *Wildlife Biology* **9**: 47–58.

Bloch, D., Lockyer, C., and Zachariassen, M. 1993b. Age and growth parameters of the long-finned pilot whale off the Faroe Islands. *Report of the International Whaling Commission* (Special Issue) **14**: 163–207.

Bloch, D., Zachariassen, M., and Zachariassen, P. 1993c. Some external characters of the long-finned pilot whale off the Faroe Islands and a comparison with the short-finned pilot whale. *Report of the International Whaling Commission* (Special Issue) **14**: 117–135.

Boebel, O., Lutjeharms, J.R.E., Schmid, C., Zenk, W., Rossby, T., and Barron, C. 2003. The Cape Cauldron: a regime of turbulent inter-ocean exchange. *Deep-Sea Research II* **50**(1): 57–86.

Boisseau, O. 2005. Quantifying the acoustic repertoire of a population: the vocalizations of free-ranging bottlenose dolphins in Fiordland, New Zealand. *Journal of the Acoustical Society of America* **117**(4): 2 318–2 329.

Bones, M., Neill, B., and Reid, B. 1998. Fraser's dolphin (*Lagenodelphis hosei*) stranded in South Uist: first record in UK waters. *Journal of Zoology, London* **246**: 460–461.

Borobia, M. 1997. Small cetaceans of Eastern, Western and Central African regions: a summary report. Paper SC/49/SM48 submitted to the Scientific Committee of the International Whaling Commission, Bournemouth, UK, September/October 1997, 11 pp.

Borsa, P., and Robineau, D. 2005. Blainville's beaked whale in New Caledonia. *Pacific Science* **59**(3): 467–472.

Boschma, H. 1951. Some smaller whales. *Endeavour* **10**(39): 131–135.

Boucaud, V., and Rota, B. 2006. État des connaissances sur les cétacés à La Réunion (France). In *Compte rendu du séminaire GLOBICE: étude et conservation des cétacés à La Réunion*. University of La Réunion (France), May 2006, pp 2–4.

Bowman, T.E. 1955. A new genus and species of whale-louse (Amphipoda: Cyamidae) from the false killer whale. *Bulletin of Marine Science of the Gulf and Caribbean* **5**(4): 315–320.

Boyd, A.J., Salat, J., and Maso, M. 1987. The seasonal intrusion of relatively saline water on the shelf off northern and central Namibia. *South African Journal of Marine Science* **5**: 107–120.

Branch, T.A., Stafford, K.M., Palacios, D.M. *et al.* 2007. Past and present distribution, densities and movements of blue whales *Balaenoptera musculus* in the Southern Hemisphere and northern Indian Ocean. *Mammal Review* **37**(2): 116–175.

Branch, T.A., Abubaker, E.M.N., Mkango, S., and Butterworth, D.S. In press. Separating southern blue whale subspecies based on length frequencies of sexually mature females. *Marine Mammal Science.*

Branch, T.A., and Butterworth, D.S. 2001a. Estimates of abundance south of 60° S for cetacean species sighted frequently on the 1978/79 to 1997/98 IWC/IDCR-SOWER sighting surveys. *Journal of Cetacean Research and Management* **3**(3): 251–270.

Branch, T.A., and Butterworth, D.S. 2001b. Southern Hemisphere minke whales: standardized abundance estimates from the 1978/79 to 1997/98 IDCR-SOWER surveys. *Journal of Cetacean Research and Management* **3**(2): 143–174.

Branch, T.A., Matsuoka, K., and Miyashita, T. 2004. Evidence for increases in Antarctic blue whales based on Bayesian modelling. *Marine Mammal Science* **20**(4): 726–754.

Breiwick, J.M. 1977. Analysis of the Antarctic fin whale stock in Area I. *Report of the International Whaling Commission* **27**: 124–127.

Briggs, J.C. 1974. *Marine zoogeography.* McGraw-Hill, New York. 475 pp.

Brook, F., Lim, E.H.T., Chua, F.H.C., and Mackay, B. 2004. Assessment of the reproductive cycle of the Indo-Pacific humpback dolphin, *Sousa chinensis*, using ultrasonography. *Aquatic Mammals* **30**(1): 137–148.

Brook, F.M., and Kinoshita, R.E. 2005. Controlled unassisted breeding of captive Indo-Pacific bottlenose dolphins, *Tursiops aduncus*, using ultrasonography. *Aquatic Mammals* **31**(1): 89–95.

Brook, F.M., Kinoshita, R., and Benirschke, K. 2002. Histology of the ovaries of a bottlenose dolphin, *Tursiops aduncus*, of known reproductive history. *Marine Mammal Science* **18**(2): 540–544.

Brown, D.H. 1962. Further observations on the pilot whale in captivity. *Zoologica, NY* **47**(7): 59–64 + 2 pl.

Brown, D.H., Caldwell, D.K., and Caldwell, M.C. 1966. Observations on the behaviour of wild and captive false killer whales, with notes on associated behaviour of other genera of captive delphinids. *Contributions in Science* **95**: 1–32.

Brown, M.W., Corkeron, P.J., Hale, P.T., Schultz, K.W., and Bryden, M.M. 1995. Evidence for a sex-segregated migration in the humpback whale (*Megaptera novaeangliae*). *Proceedings of the Royal Society of London, Series B: Biological Sciences* **259**(1 355): 229–234.

Brown, S.G. 1973. Recent sight records of southern right whale dolphins in the Pacific Ocean. *The Marine Observer* **43**: 78–80.

Brown, S.G. 1977. Whale marking: a short review. Pp 569–581 in (M. Angel, ed.) *A Voyage of Discovery.* Pergamon Press, Oxford.

Brown, S.G. 1982. Southern right whale dolphins off the South West Africa coast. *The Marine Observer* **52**: 33–34.

Brownell Jr, R.L. 1974. Small odontocetes of the Antarctic. Pp 13–19 in (Brown, S.G., Brownell, R.L., Erickson, A.W., Hofman, R.J., Llano, G.A., and Mackintosh, N.A., eds). *Antarctic Mammals. Antarctic Map Folio Series.* 18. American Geophysical Society, New York.

Brownell Jr, R.L., and Cipriano, F. 1999. Dusky dolphin *Lagenorhynchus obscurus* (Gray, 1828). Pp 85–104 in (Ridgway, S.H., and Harrison, R.J., eds) *Handbook of Marine Mammals 6: The second book of dolphins and the porpoises.* Academic Press, New York.

Brownell Jr, R.L., and Clapham, P.J. 1999. Spectacled porpoise *Phocoena dioptrica* Lahille, 1912. Pp 379–391 in (Ridgway, S.H., and Harrison, R.J., eds) *Handbook of Marine Mammals 6: The second book of dolphins and the porpoises.* Academic Press, New York.

Brownell Jr, R.L., Crespo, E.A., and Donahue, M.A. 1999. Peale's dolphin *Lagenorhynchus australis* (Peale, 1848). Pp 105–120 in (Ridgway, S.H., and Harrison, R.J., eds) *Handbook of Marine Mammals 6: The second book of dolphins and the porpoises.* Academic Press, New York.

Brownell Jr, R.L., and Donahue, M.A. 1999. Hourglass dolphin *Lagenorhynchus cruciger* (Quoy and Gaimard, 1824). Pp 121–135 in (Ridgway, S.H., and Harrison, R.J., eds) *Handbook of Marine Mammals 6: The second book of dolphins and the porpoises.* Academic Press, New York.

Brownell Jr, R.L., Heyning, J.E., and Perrin, W.F. 1989. A porpoise, *Australophocaena dioptrica*, previously identified as *Phocoena spinipinnis*, from Heard Island. *Marine Mammal Science* **5**(2): 193–195.

Brownell Jr, R.L., and Ralls, K. 1986. Potential for sperm competition in baleen whales. *Report of the International Whaling Commission* (Special Issue) **8**: 97–112.

Brownlee, S.M., and Norris, K.S. 1994. The acoustic domain. Pp 161–185 in (Norris, K.S., Würsig, B., Wells, R.S., Würsig, M., Brownlee, S.M., Johnson, C.M., and Solow, J., eds) *The Hawaiian spinner dolphin.* University of California Press, Berkeley, Los Angeles.

Brundrit, G.B., De Cuevas, B.A., and Shipley, A.M. 1987. Long-term sea-level variability in the eastern South Atlantic and a comparison with that in the eastern Pacific. *South African Journal of Marine Science* **5**: 73–78.

Bryden, H.L., Beal, L.M., and Duncan, L.M. 2005. Structure and transport of the Agulhas Current and its temporal variability. *Journal of Oceanography* **61**(3): 479–492.

Bryden, M.M. 1976. Observations on a pygmy killer whale, *Feresa attenuata*, stranded on the east coast of Australia. *Australian Wildlife Research* **3**: 21–28.

Bryden, M.M., Dawbin, W.H., Heinsohn, G.E., and Brown, D.H. 1977a. Melon-headed whale, *Peponocephala electra*, on the east coast of Australia. *Journal of Mammalogy* **58**(2): 180–187.

Bryden, M.M., Harrison, R.J., and Lear, R.J. 1977b. Some aspects of the biology of *Peponocephala electra* (Cetacea: Delphinidae) I. General and reproductive biology. *Australian Journal of Marine and Freshwater Research* **28**: 703–715.

Buchholtz, E.A. 2002. Convergent evolution. Pp 297–299 in (Perrin, W.F., Würsig, B., and Thewissen, J.G.M., eds) *Encyclopedia of Marine Mammals.* Academic Press, San Diego.

Buckland, S.T., and Duff, E.I. 1989. Analysis of the Southern Hemisphere minke whale mark-recovery data. *Report of the International Whaling Commission* (Special Issue) **11**: 121–143.

Budillon, G., and Rintoul, S.R. 2003. Fronts and upper ocean thermal variability south of New Zealand. *Antarctic Science* **15**: 141–152.

Budker, P. 1953. Les campagnes baleinières 1949–1952 au Gabon (note préliminaire). *Mammalia* **17**(31): 129–148.

Budker, P., and Collignon, J. 1952. Trois campagnes baleinières au Gabon (1949–1950–1951). *Bulletin de l'Institut d'Études Centrafricaines n.s.* **3**: 75–100.

Budylenko, G.A., Panfilov, B.G., Pakhomova, A.A., and Sazhinov, E.G. 1973. [New data on pygmy right whales *Neobalaena marginata* (Gray, 1848).] *Trudy̌ Atlanticheskiĭ Nauchno-Issledovatel'skiĭ Institut Rýbnogo Khozyaĭstva i Okeanografii* **51**: 122–132.

Burmeister, H. 1867. Preliminary description of a new species of finner whale (*Balaenoptera bonaerensis*). *Proceedings of the Zoological Society of London* **1867**: 707–713.

Busnel, R.-G. 1973. Symbiotic relationship between man and dolphins. *Transactions of the New York Academy of Science* **35**: 112–131.

Butterworth, D.S., Punt, A.E., Geromont, H.F., Kato, H., and Fujise, Y. 1999. Inferences on the dynamics of Southern Hemisphere minke whales from ADAPT analyses of catch-at-age information. *Journal of Cetacean Research and Management* **1**: 11–32.

C

Cadenat, J. 1949. Notes sur les cétacés observés sur les côtes du Sénégal de 1941 à 1948. *Bulletin de l'Institut Français d'Afrique Noire* **11**(1–2): 1–15.

Cadenat, J. 1956. Un *Delphinidae* encore mal connu de la côte occidentale d'Afrique: *Sotalia teüszii* Kükenthal 1892. *Bulletin de l'Institut Français d'Afrique Noire* **18A**(2): 555–566.

Cadenat, J. 1959. Rapport sur les petits Cétacés ouest-africains. Résultats des recherches enterprises sur ces animaux jusqu'au mois de mars 1959. *Bulletin de l'Institut Français d'Afrique Noire* **21A**(4): 1 367–1 409 + 31 pl.

Cadenat, J., and Paraiso, F. 1957. Nouvelle observation de *Sotalia* (Cétacé, Delphinidé) sur les côtes du Sénégal. *Bulletin de l'Institut Français d'Afrique Noire* **19A**(1): 324–332.

Caldwell, D.K., and Brown, D.H. 1964. Tooth wear as a correlate of described feeding behavior by the killer whale, with notes on a captive specimen. *Bulletin of the Californian Academy of Science* **63**(3): 128–140.

Caldwell, D.K., and Caldwell, M.C. 1966. Observations on the distribution, coloration, behavior and audible sound production of the spotted dolphin, *Stenella plagiodon* (Cope). *Los Angeles County Museum Contributions in Science* **104**: 1–28.

Caldwell, D.K., and Caldwell, M.C. 1989. Pygmy sperm whale *Kogia breviceps* (de Blainville, 1838): dwarf sperm whale *Kogia simus* Owen, 1866. Pp 235–260 in (Ridgway, S.H., and Harrison, R., eds) *Handbook of Marine Mammals 4: River dolphins and the larger toothed whales*. Academic Press, London.

Caldwell, D.K., Caldwell, M.C., and Walker, R.V. 1976. First records for Fraser's dolphin (*Lagenodelphis hosei*) in the Atlantic and the melon-headed whale (*Peponocephala electra*) in the western Atlantic. *Cetology* **25**: 1–4.

Caldwell, M.C., Caldwell, D.K., and Tyack, P.L. 1990. Review of the signature-whistle hypothesis for the Atlantic bottlenose dolphin. Pp 199–234 in (Leatherwood, S., and Reeves, R.R., eds) *The bottlenose dolphin*. Academic Press, San Diego.

Calzada, N., Aguilar, A., Lockyer, C., and Grau, E. 1997. Patterns of growth and physical maturity in the western Mediterranean striped dolphin, *Stenella coeruleoalba* (Cetacea: Odontoceti). *Canadian Journal of Zoology* **75**: 632–637.

Cañadas, A., Sagarminaga, R., and García-Tiscar, S. 2002. Cetacean distribution related with depth and slope in the Mediterranean waters off southern Spain. *Deep-Sea Research I* **49**: 2 053–2 073.

Cappozzo, H.L., Negri, M.F., Mahler, B., Lía, V.V., Martínez, P., Gianggiobe, A., and Saubidet, A. 2005. Biological data on two Hector's beaked whales, *Mesoplodon hectori*, stranded in Buenos Aires province, Argentina. *Latin American Journal of Aquatic Mammals* **4**(2): 113–128.

Carrier, D.R., Deban, S.M., and Otterstrom, J. 2002. The face that sunk the *Essex*: potential function of the spermaceti organ in aggression. *Journal of Experimental Biology* **205**: 1 755–1 763.

Carwardine, M. 2002. *Smithsonian Handbooks. Whales dolphins and porpoises*. Dorling Kindersley Ltd, London. 256 pp.

Cassens, I., Van Waerebeek, K., Best, P.B., Tzika, A., Van Helden, A.L., Crespo, E.A., and Milinkovitch, M.C. 2005. Male dispersal along the coasts but no migration in pelagic waters in dusky dolphins (*Lagenorhynchus obscurus*). *Molecular Ecology* **14**: 107–121.

Castro, C. 2004. Encounter with a school of pygmy killer whales (*Feresa attenuata*) in Ecuador, southeast tropical Pacific. *Aquatic Mammals* **30**(3): 441–444.

Chambers, S.L., and James, R.N. 2005. Sonar termination as a cause of mass cetacean strandings in Geographe Bay, south-western Australia. *Proceedings of Acoustics* 9–11 November 2005, Busselton, Western Australia. 8 pp.

Chande, A.I., Mtoka, G.F., and Mhitu, H.A. 1994. Marine mammals and fisheries interactions in Tanzania. Unpublished report of Tanzania Fisheries Research Institute to UNEP. 39 pp.

Chantrapornsyl, S., Kinze, C.C., Leatherwood, S., and Prematunga, W.P. 1991. Notes on the genus *Kogia* in the northern Indian Ocean. *UNEP Marine Mammal Technical Report* **3**: 79–88.

Cheal, A.J., and Gales, N.J. 1992. Growth, sexual maturity and food intake of Australian Indian Ocean bottlenose dolphins, *Tursiops truncatus*, in captivity. *Australian Journal of Zoology* **40**: 215–223.

Chittleborough, R.G. 1955a. Aspects of reproduction in the male humpback whale, *Megaptera nodosa* (Bonnaterre). *Australian Journal of Marine and Freshwater Research* **6**: 1–29.

Chittleborough, R.G. 1955b. Puberty, physical maturity, and relative growth of the female humpback whale, *Megaptera nodosa* (Bonnaterre), on the Western Australian coast. *Australian Journal of Marine and Freshwater Research* **6**: 315–327.

Chittleborough, R.G. 1958. Breeding cycle of the female humpback whale, *Megaptera nodosa* (Bonnaterre). *Australian Journal of Marine and Freshwater Research* **9**: 1–18.

Chittleborough, R.G. 1965. Dynamics of two populations of the humpback whale, *Megaptera novaeangliae* (Borowski). *Australian Journal of Marine and Freshwater Research* **16**: 33–128.

Chivers, S.J., LeDuc, R.G., Robertson, K.M., Barros, N.B., and Dizon, A.E. 2005. Genetic variation of *Kogia* spp with preliminary evidence for two species of *Kogia sima*. *Marine Mammal Science* **21**(4): 619–634.

Chivers, S.J., and Myrick Jr, A.C. 1993. Comparison of age at sexual maturity and other reproductive parameters for two stocks of spotted dolphin, *Stenella attenuata*. *Fishery Bulletin, US* **91**: 611–618.

Cipriano, F. 1997. Antitropical distributions and speciation in dolphins of the genus *Lagenorhynchus*: a preliminary analysis. Pp 305–316 in (Dizon, A.E., Chivers, S.J., and Perrin, W.F., eds) *Molecular genetics of marine mammals*. Special Publication no. 3, The Society of Marine Mammalogy, Lawrence, Kansas, USA.

Cipriano, F., Würsig, B., and Würsig, M. 1989. Diurnal variation in dive-times and movements of dusky dolphins feeding on deep scattering layer-associated prey. *American Zoologist* **29**: 68A.

Clapham, P.J. 1992. Age at attainment of sexual maturity in humpback whales, *Megaptera novaeangliae*. *Canadian Journal of Zoology* **70**: 1 470–1 472.

Clapham, P.J., Berggren, P., Childerhouse, S. *et al.* 2003. Whaling as science. *BioScience* **53**(3): 210–212.

Clapham, P.J., and Palsbøll, P.J. 1997. Molecular analysis of paternity shows promiscuous mating in female humpback whales (*Megaptera novaeangliae*, Borowski). *Proceedings of the Royal Society of London* B **264**: 95–98.

Clapham, P.J., Palsbøll, P.J., Mattila, D.K., and Vasquez, O. 1992. Composition and dynamics of humpback whale competitive groups in the West Indies. *Behaviour* **122** (3–4): 182–194.

Clapham, P.J., Wetmore, S.E., Smith, T.D., and Mead, J.G. 1999. Length at birth and at independence in humpback whales. *Journal of Cetacean Research and Management* **1**(2): 141–146.

Clark, C.W. 1995. Application of US Navy underwater hydrophone arrays for scientific research on whales. *Report of the International Whaling Commission* **45**: 210–212.

Clark, C.W., and Clapham, P.J. 2004. Acoustic monitoring on a humpback whale (*Megaptera novaeangliae*) feeding ground shows continual singing into late spring. *Proceedings of the Royal Society of London* B **271**: 1 051–1 057.

Clark, S.T., and Odell, D.K. 1999a. Allometric relationships and sexual dimorphism in captive killer whales (*Orcinus orca*). *Journal of Mammalogy* **80**(3): 777–785.

Clark, S.T., and Odell, D.K. 1999b. Nursing behavior in captive false killer whales (*Pseudorca crassidens*). *Aquatic Mammals* **25**(3): 183–191.

Clark, S.T., and Odell, D.K. 1999c. Nursing parameters in captive killer whales (*Orcinus orca*). *Zoo Biology* **18**: 373–384.

Clarke, M.R. 1976. Observation on sperm whale diving. *Journal of the Marine Biological Association of the UK* **56**: 809–810.

Clarke, M.R. 1978. Buoyancy control as a function of the spermaceti organ in the sperm whale. *Journal of the Marine Biological Association of the UK* **58**: 27–71.

Clarke, M.R. 1980. Cephalopoda in the diet of sperm whales of the Southern Hemisphere and their bearing on sperm whale biology. *Discovery Reports* **37**: 1–324.

Clarke, M.R. 1986. Cephalopods in the diet of odontocetes. Pp 281–321 in (Bryden, M.M., and Harrison, R., eds) *Research on dolphins*. Clarendon Press, Oxford.

Clarke, M.R. 1996. Cephalopods as prey. III. Cetaceans. *Philosophical Transactions of the Royal Society of London* B **351**: 1 053–1 065.

Clarke, M.R. 2003. Production and control of sound by the small sperm whales, *Kogia breviceps* and *K. sima* and their implications for other Cetacea. *Journal of the Marine Biological Association of the UK* **83**: 241–263.

Clarke, M.R., and Young, R. 1998. Description and analyses of cephalopod beaks from stomachs of six species of odontocete cetaceans stranded on Hawaiian shores. *Journal of the Marine Biological Association of the UK* **78**: 623–641.

Clarke, R. 2004. Pygmy fin whales. *Marine Mammal Science* **20**(2): 329–334.

Clarke, R. 2005. A southern bottlenose whale examined in the Antarctic. *Latin American Journal of Aquatic Mammals* **4**(2): 83–96.

Clarke, R. 2006. The origin of ambergris. *Latin American Journal of Aquatic Mammals* **5**(1): 7–21.

Cockcroft, V.G. 1990. Dolphin catches in the Natal shark nets, 1980–1988. *South African Journal of Wildlife Research* **20**(2): 44–51.

Cockcroft, V.G. 1991. Incidence of shark bites on Indian Ocean hump-backed dolphins (*Sousa plumbea*) off Natal, South Africa. *UNEP Marine Mammal Technical Report* **3**: 277–282.

Cockcroft, V.G. 1999. Organochlorine levels in cetaceans from South Africa: a review. *Journal of Cetaecan Research and Management* (Special Issue) **1**: 169–176.

Cockcroft, V.G. 2002. Natural history of Indo-Pacific humpback dolphins (*Sousa plumbea*) off Southern Africa. Paper SC/54/SM22 submitted to the Scientific Committee of the International Whaling Commission, Shimonoseki, Japan, April–May 2002. 1 p.

Cockcroft, V.G., De Kock, A.C., Lord, D.A., and Ross, G.J.B. 1989. Organochlorines in bottlenose dolphins *Tursiops truncatus* from the east coast of South Africa. *South African Journal of Marine Science* **8**: 207–217.

Cockcroft, V.G., Haschick, S.L., and Klages, N.T.W. 1993. The diet of Risso's dolphin, *Grampus griseus* (Cuvier, 1812), from the east coast of South Africa. *Zeitschrift für Säugetierkunde* **58**: 286–293.

Cockcroft, V.G., and Peddemors, V.M. 1990. Seasonal distribution and identity of common dolphins *Delphinus delphis* off the south-east coast of southern Africa. *South African Journal of Marine Science* **9**: 371–377.

Cockcroft, V.G., Peddemors, V.M., Ryan, P.G., and Lutjeharms, J.R.E. 1990a. Cetaceans associated with Agulhas Current eddies in the Southern Ocean. *South African Journal of Antarctic Research* **20**(2): 64–67.

Cockcroft, V.G., and Ross, G.J.B. 1990a. Age, growth and reproduction of bottlenose dolphins *Tursiops truncatus* from the east coast of southern Africa. *Fishery Bulletin, US* **88**: 289–302.

Cockcroft, V.G., and Ross, G.J.B. 1990b. Food and feeding of the Indian Ocean bottlenose dolphin off southern Natal, Africa. Pp 295–308 in (Leatherwood, S., and Reeves, R.R., eds) *The bottlenose dolphin*. Academic Press, San Diego.

Cockcroft, V.G., and Ross, G.J.B. 1990c. Observations on the early development of a captive bottlenose dolphin calf. Pp 461–478 in (Leatherwood, S., and Reeves, R.R., eds) *The bottlenose dolphin*. Academic Press, San Diego.

Cockcroft, V.G., Ross, G.J.B., and Peddemors, V.M. 1990b. Bottlenose dolphin *Tursiops truncatus* distribution in Natal's coastal waters. *South African Journal of Marine Science* **9**: 1–10.

Cockcroft, V.G., Ross, G.J.B., and Peddemors, V.M. 1991. Distribution and status of bottlenose dolphin *Tursiops truncatus* on the south coast of Natal, South Africa. *South African Journal of Marine Science* **11**: 203–209.

Cockcroft, V.G., Ross, G.J.B., Peddemors, V.M., and Borchers, D.L. 1992. Estimates of abundance and undercounting of bottlenose dolphins off northern Natal, South Africa. *South African Journal of Wildlife Research* **22**(4): 102–109.

Collet, A., and Robineau, D. 1988. Data on the genital tract and reproduction in Commerson's dolphin, *Cephalorhynchus commersonii* (Lacépède, 1804), from the Kerguelen Islands. *Report of the International Whaling Commission* (Special Issue) **9**: 119–141.

Collett, R. 1886. On the external characters of Rudolphi's rorqual (*Balaenoptera borealis*). *Proceedings of the Zoological Society of London* **1886**(18): 243–265 + 2 pl.

Condy, P.R. 1977. Whale observations in the pack ice off the Fimbul Ice Shelf, Antarctica. *South African Journal of Antarctic Research* **7**: 7–9.

Condy, P.R., Van Aarde, R.J., and Bester, M.N. 1978. The seasonal occurrence and behaviour of killer whales *Orcinus orca*, at Marion Island. *Journal of Zoology, London* **184**: 449–464.

Connor, R.C., and Norris, K.S. 1982. Are dolphins reciprocal altruists? *The American Naturalist* **119**(3): 358–374.

Cook, M.L.H., Sayigh, L.S., Blum, J.E., and Wells, R.S. 2004. Signature-whistle production in undisturbed free-ranging bottlenose dolphins (*Tursiops truncatus*). *Proceedings of the Royal Society of London* B **271**: 1 043–1 049.

Cooke, J.G., and De la Mare, W.K. 1984. A note on the estimation of time trends in the age at sexual maturity in baleen whales from transition layer data, with reference to the North Atlantic fin whale (*Balaenoptera physalus*). *Report of the International Whaling Commission* **34**: 701–709.

Corbett, H.D. 1994. The occurrence of cetaceans off Mauritius and in adjacent waters. *Report of the International Whaling Commission* **44**: 393–397.

Corkeron, P.J. 1990. Aspects of the behavioural ecology of inshore dolphins *Tursiops truncatus* and *Sousa chinensis* in Moreton Bay, Australia. Pp 285–293 in (Leatherwood, S., and Reeves, R.R., eds) *The bottlenose dolphin*. Academic Press, San Diego.

Corkeron, P.J., Ensor, P., and Matsuoka, K. 1999. Observations of blue whales feeding in Antarctic waters. *Polar Biology* **22**: 213–215.

Corkeron, P.J., and Van Parijs, S.M. 2001. Vocalizations of eastern Australian Risso's dolphins, *Grampus griseus*. *Canadian Journal of Zoology* **79**: 160–164.

Cornell, L.H., Antrim, J.E., Asper, E.D., and Pincheira, B.J. 1988. Commerson's dolphins (*Cephalorhynchus commersonii*) live-captured from the Strait of Magellan, Chile. *Report of the International Whaling Commission* (Special Issue) **9**: 183–194.

Courtenay-Latimer, M. 1963. Gray's beaked whale *Mesoplodon grayi*. *Annals of the Cape Provincial Museum (Natural History)* **3**: 122.

Cox, T.M., Ragen, T.J., Read, A.J. *et al.* 2006. Understanding the impacts of anthropogenic sound on beaked whales. *Journal of Cetacean Research and Management* **7**(3): 177–187.

Craig, Y. 2001. Amazing whale hunt off Cape St Francis. *Our Times* newspaper, 7 September 2001.

Cranford, T.W. 1999. The sperm whale's nose: sexual selection on a grand scale. *Marine Mammal Science* **15**(4): 1 133–1 157.

Crespo, E.A., Pedraza, S.N., Dans, S.L. *et al.* 1997. Direct and indirect effects of the highseas fisheries on the marine mammal populations in the northern and central Patagonian coast. *Journal of Northwest Atlantic Fishery Science* **22**: 189–207.

Cresswell, G., and Walker, D. 2002. Using visual clues to assess group composition of Cuvier's beaked whales (*Ziphius cavirostris*) in the North Atlantic. *Proceedings of the 16th annual conference of the European Cetacean Society, Liege, Belgium, April 2002.*

Croll, D.A., Acevedo-Gutiérrez, A., Tershy, B.R., and Urbán-Ramírez, J. 2001. The diving behavior of blue and fin whales: is dive duration shorter than expected based on oxygen stores? *Comparative Biochemistry and Physiology* A **129**: 797–809.

Croll, D.A., Clark, C.W., Acevedo, A., Tershy, B., Flores, S., Gedamke, J., and Urban, J. 2002. Only male fin whales sing loud songs. *Nature, London* **417**: 809.

Cruickshank, R.A., and Brown, S.G. 1981. Recent observations and some historical records of southern right-whale dolphins *Lissodelphis peronii*. *Fisheries Bulletin, South Africa* **15**: 109–121.

D

Dahlheim, M.E., and Heyning, J.E. 1999. Killer whale *Orcinus orca* (Linnaeus, 1758). Pp 281–322 in (Ridgway, S.H., and Harrison, R.J., eds) *Handbook of Marine Mammals 6: The second book of dolphins and the porpoises.* Academic Press, New York.

Dalebout, M.L., Baker, C.S., Mead, J.G., Cockcroft, V.G., and Yamada, T.K. 2004a. A comprehensive and validated molecular taxonomy of beaked whales, family Ziphiidae. *Journal of Heredity* **95**(6): 459–473.

Dalebout, M.L., Russell, K.G., Little, M.J., and Ensor, P. 2004b. Observations of live Gray's beaked whales (*Mesoplodon grayi*) in Mahurangi Harbour, North Island, New Zealand, with a summary of at-sea sightings. *Journal of the Royal Society of New Zealand* **34**(4): 347–356.

Dalebout, M.L., Mead, J.G., Baker, C.S., Baker, A.N., and Van Helden, A.L. 2002. A new species of beaked whale *Mesoplodon perrini* sp. n. (Cetacea: Ziphiidae) discovered through phylogenetic analyses of mitochondrial DNA sequences. *Marine Mammal Science* **18**(3): 577–608.

Dalebout, M.L., Robertson, K.M., Frantzis, A., Engelhaupt, D., Mignucci-Giannoni, A.A., Rosario-Delestre, R.J., and Baker, C.S. 2005. Worldwide structure of mtDNA diversity among Cuvier's beaked whales (*Ziphius cavirostris*): implications for threatened populations. *Molecular Ecology* **14**(11): 3 353–3 371.

Dalebout, M.L., Ross, G.J.B., Baker, C.S. *et al.* 2003. Appearance, distribution, and genetic distinctiveness of Longman's beaked whale, *Indopacetus pacificus*. *Marine Mammal Science* **19**(3): 421–461.

Dalebout, M.L., Van Helden, A., Van Waerebeek, K., and Baker, C.S. 1998. Molecular identification of southern hemisphere beaked whales (Cetacea: Ziphiidae). *Molecular Ecology* **7**: 687–694.

Dammerman, K.W. 1924. On Globicephala and some other Delphinidae from the Indo-Australian archipelago. *Treubia* **5**(1–3): 340–352 + 4 pl.

Danil, K. 2004. *Growth and reproduction of female short-beaked common dolphins,* Delphinus delphis, *in the eastern tropical Pacific.* M Sc thesis, University of San Diego, San Diego, USA, i–xiv + 192 pp.

Dans, S.L., Crespo, E.A., Pedraza, S.N., and Alonso, M.K. 1997. Notes on the reproductive biology of female dusky dolphins (*Lagenorhynchus obscurus*) off the Patagonian coast. *Marine Mammal Science* **13**(2): 303–307.

Darling, J., and Bérubé, M. 2001. Interactions of singing humpback whales with other males. *Marine Mammal Science* **17**(3): 570–584.

David, J.H.M., Sekiguchi, K., Meÿer, M.A., and Davis, R. 1998. Investigation of the diving behaviour, home range and movement of Heaviside's dolphins (*Cephalorhynchus heavisidii*) off the west coast of South Africa. *Research Report for Pro Natura Fund* **7**: 153–164.

Davies, J.L. 1960. The southern form of the pilot whale. *Journal of Mammalogy* **41**: 29–34.

Davies, J.L. 1962. The antitropical factor in cetacean evolution. *Evolution* **17**:107–116.

Davis, B., Cliff, G., and Dudley, S.F.J. 1989. The Natal Sharks Board. Pp 209–213 in (Payne, A.I.L., and Crawford, R.J.M., eds) *Oceans of Life off Southern Africa.* Vlaeberg Publishers, Cape Town.

Davis, R.W., Fargion, G.S., May, N. *et al.* 1998. Physical habitat of cetaceans along the continental slope in the North-Central and Western Gulf of Mexico. *Marine Mammal Science* **14**(3): 490–507.

Davis, R.W., Worthy, G.A.J., Würsig, B., Lynn, S.K., and Townsend, F.I. 1996. Diving behavior and at-sea movements of an Atlantic spotted dolphin in the Gulf of Mexico. *Marine Mammal Science* **12**(4): 569–581.

Dawbin, W.H. 1966. The seasonal migratory cycle of humpback whales. Pp 145–170 in (K.S. Norris, ed.) *Whales, dolphins and porpoises.* University of California Press, Berkeley.

Dawbin, W.H., and Cato, D.H. 1992. Sounds of a pygmy right whale (*Caperea marginata*). *Marine Mammal Science* **8**(3): 213–219.

Dawson, S.M. 2002. *Cephalorhynchus* dolphins *C. heavisidii, C. eutropia, C. hectori, C. commersonii*. Pp 200–203 in (Perrin, W.F., Würsig, B., and Thewissen, J.G.M., eds) *Encyclopedia of Marine Mammals.* Academic Press, San Diego.

Deacon, G.E.R. 1937. The hydrology of the Southern Ocean. *Discovery Reports* **15**: 1–124.

Deacon, G.E.R. 1983. Physical and biological zonation in the Southern Ocean. *Deep-Sea Research* **29**(1): 1–15.

De Boer, M.N., Baldwin, R., Burton, C.L.K. *et al.* 2003. Cetaceans in the Indian Ocean Sanctuary: a review. *Whale and Dolphin Conservation Society.* 52 pp.

Debrot, A.O., and Barros, N.B. 1992. Notes on a Gervais' beaked whale, *Mesoplodon europaeus*, and a dwarf sperm whale, *Kogia simus*, stranded in Curaçao, Netherlands Antilles. *Marine Mammal Science* **8**(2): 172–178.

De Bruyn, P.J.N., Hofmeyr, G.J.G., and De Villiers, M.S. 2006. First record of a vagrant Commerson's dolphin, *Cephalorhynchus commersonii*, at the southern African continental shelf. *African Zoology* **41**(1): 131–133.

De Buffrénil, V., Dziedzic, A., and Robineau, D. 1989. Répartition et déplacements des Dauphins de Commerson (*Cephalorhynchus commersonii* (Lacépède, 1804)) dans un golfe des îles Kerguelen; données du marquage individuel. *Canadian Journal of Zoology* **67**: 516–521.

Defran, R.H., and Pryor, K. 1980. The behavior and training of cetaceans in captivity. Pp 319–362 in (Herman, L.M., ed.) *Cetacean behavior: mechanisms and functions*. John Wiley and Sons, New York.

De Gurjão, L.M., Neto, M.A.A.F., Dos Santos, R.A., and Cascon, P. 2004. Análise de conteúdos estomacais de quatro golfinhos (Cetacea: Delphinidae) encalhados em prais no litoral do Estado do Ceará, Brasil. *Revista biociências (Taubaté)* **10**(1–2): 39–45.

De Kock, A.C., Best, P.B., Cockcroft, V., and Bosma, C. 1994. Persistent organochlorine residues in small cetaceans from the east and west coasts of southern Africa. *The Science of the Total Environment* **154**: 153–162.

De Ruijter, W. 1982. Asymptotic analysis of the Agulhas and Brazil Current Systems. *Journal of Physical Oceanography* **12**: 361.

De Ruijter, W.P.M., Biastoch, A., Drijfhout, S.S. *et al.* 1999. Indian-Atlantic interocean exchange: dynamics, estimation and impact. *Journal of Geophysical Research* **104**: 20 885–20 910.

De Ruijter, W.P.M., Ridderinkhof, H., Lutjeharms, J.R.E., Schouten, M.W., and Veth, C. 2002. Observations of the flow in the Mozambique Channel. *Geophysical Research Letters* **29**(10): 1 401–1 403.

De Ruijter, W.P.M., Van Aken, H.M., Beier, E.J., Lutjeharms, J.R.E., Matano, R.P., and Schouten, M.W. 2004. Eddies and dipoles around South Madagascar: formation, pathways and large-scale impact. *Deep-Sea Research I* **51**(3): 383–400.

Desportes, G., and Mouritsen, R. 1993. Preliminary results on the diet of long-finned pilot whales off the Faroe Islands. *Report of the International Whaling Commission* (Special Issue) **14**: 305–324.

Desportes, G., Saboureau, M., and Lacroix, A. 1993. Reproductive maturity and seasonality of male long-finned pilot whales, off the Faroe Islands. *Report of the International Whaling Commission* (Special Issue) **14**: 233–262.

Dias, C.A. 1983a. Note on the evidence of a permanent southward flow of the upper oceanic tropospheric waters off Angola at 12° S. *Collection of Scientific Papers–International Commission for the Southeast Atlantic Fisheries* **10**: 99–102.

Dias, C.A. 1983b. Preliminary report on the physical oceanography off southern Angola, March and July 1971. *Collection of Scientific Papers–International Commission for the Southeast Atlantic Fisheries* **10**: 103–116.

Di Sciara, G.N., Barbaccia, G., and Azzellino, A. 1997. Birth at sea of a false killer whale, *Pseudorca crassidens*. *Marine Mammal Science* **13**(3): 508–511.

Dixon, J.M., Frigo, L., and Moyle, R.L.C. 1994. New information on the southern bottlenose whale, *Hyperoodon planifrons* (Cetacea: Ziphiidae), from a recent stranding in Victoria, Australia. *Australian Mammalogy* **17**: 85–95.

Dohl, T.P., Norris, K.S., and Kang, I. 1974. A porpoise hybrid: *Tursiops* × *Steno*. *Journal of Mammalogy* **55**(1): 217–221.

Dolar, M.L.L., Suarez, P., Ponganis, P.J., and Kooyman, G.L. 1999. Myoglobin in pelagic small cetaceans. *Journal of Experimental Biology* **202**: 227–236.

Dolar, M.L.L., Walker, W.A., Kooyman, G.L., and Perrin, W.F. 2003. Comparative feeding ecology of spinner dolphins (*Stenella longirostris*) and Fraser's dolphins (*Lagenodelphis hosei*) in the Sulu Sea. *Marine Mammal Science* **19**(1): 1–19.

Dos Santos, R.A., and Haimovici, M. 2001. Cephalopods in the diet of marine mammals stranded or incidentally caught along southeastern and southern Brazil (21–34° S). *Fisheries Research* **52**: 99–112.

Dudok van Heel, W.H. 1962. Sound and Cetacea. *Netherlands Journal of Sea Research* **1**: 407–507.

Dudzinski, K.M. 1998. Contact behavior and signal exchange in Atlantic spotted dolphins (*Stenella frontalis*). *Aquatic Mammals* **24**(3): 129–142.

Duffield, D.A., Odell, D.K., McBain, J.F., and Andrews, B. 1995. Killer whale (*Orcinus orca*) reproduction at Sea World. *Zoo Biology* **14**(5): 417–430.

Duguy, R. 1976. Contribution à l'étude des mammiferes marins de la côte nord-ouest Afrique. *Revue des Travaux Institut des Pêches maritimes* **39**(3): 321–332.

Dziedzic, A., and De Buffrénil, V. 1989. Acoustic signals of the Commerson's dolphin, *Cephalorhynchus commersonii*, in the Kerguelen Islands. *Journal of Mammalogy* **70**(2): 449–452.

E

Edds, P.L., Odell, D.K., and Tershy, B.R. 1993. Vocalizations of a captive juvenile and free-ranging adult-calf pairs of Bryde's whales, *Balaenoptera edeni*. *Marine Mammal Science* **9**(3): 269–284.

Eichelberger, L., Fetcher Jr, E.S., Geiling, E.M.K., and Vos Jr, B.J. 1940. The composition of dolphin milk. *Journal of Biological Chemistry* **134**: 171–176.

Elwen, S., and Best, P.B. 2004a. Environmental factors influencing the distribution of southern right whales (*Eubalaena australis*) on the South Coast of South Africa I: Broad scale patterns. *Marine Mammal Science* **20**(3): 567–582.

Elwen, S., and Best, P.B. 2004b. Environmental factors influencing the distribution of southern right whales (*Eubalaena australis*) on the South Coast of South Africa II: Within bay distribution. *Marine Mammal Science* **20**(3): 583–601.

Elwen, S. Ph D thesis draft, University of Pretoria.

Elwen, S., Meÿer, M.A., Best, P.B., Kotze, P.G.H., Thornton, M., and Swanson, S. 2006. Range and movements of female Heaviside's dolphins (*Cephalorhynchus heavisidii*), as determined by satellite-linked telemetry. *Journal of Mammalogy* **87**(5): 866–877.

Eschricht, D.F. 1866. On the species of the genus *Orca* inhabiting the northern seas. Pp 151–188 in (Flower, W.H., ed.) *Recent memoirs on the Cetacea by Professors Eschricht, Reinhardt and Lilljeborg*. The Ray Society, London.

Evans, K., Kemper, C., and Hill, M. 2001. First records of the spectacled porpoise *Phocoena dioptrica* in continental Australian waters. *Marine Mammal Science* **17**(1): 161–170.

Evans, W.E. 1994. Common dolphin, white-bellied porpoise *Delphinus delphis* Linnaeus, 1758. Pp 191–224 in (Ridgway, S.H., and Harrison, R.J., eds) *Handbook of Marine Mammals 5: The first book of dolphins*. Academic Press, New York.

Evans, W.E., Awbrey, F.T., and Hackbarth, H. 1988. High frequency pulses produced by free-ranging Commerson's dolphin (*Cephalorhynchus commersonii*) compared to those of phocoenids. *Report of the International Whaling Commission* (Special Issue) **9**: 173–181.

Eyre, L. 1995. Observations of cetaceans in the Indian Ocean whale sanctuary, May–July 1993. *Report of the International Whaling Commission* **45**: 419–426.

F

Ferguson, M.C., Barlow, J., Reilly, S.B., and Gerrodette, T. 2006. Predicting Cuvier's (*Ziphius cavirostris*) and *Mesoplodon* beaked whale population density from habitat characteristics in the eastern tropical Pacific Ocean. *Journal of Cetacean Research and Management* **7**(3): 287–299.

Fernández, A., Edwards, J.F., Rodríguez, F. *et al.* 2005. 'Gas and fat embolic syndrome' involving a mass stranding of beaked whales (family *Ziphiidae*) exposed to anthropogenic sonar signals. *Veterinary Pathology* **42**: 446–457.

Fernández, M., Berón-Vera, B., García, N.A., Raga, J.A., and Crespo, E.A. 2003. Food and parasites from two hourglass dolphins, *Lagenorhynchus cruciger* (Quoy and Gaimard, 1824), from Patagonian waters. *Marine Mammal Science* **19**(4): 832–836.

Ferreira, I. M Sc thesis, draft, University of Pretoria.

Ferrero, R.C., and Walker, W.A. 1995. Growth and reproduction of the common dolphin, *Delphinus delphis* Linnaeus, in the offshore waters of the North Pacific Ocean. *Fishery Bulletin, US* **93**: 483–494.

Fertl, D., Acevedo-Gutierrez, A., and Darby, F.L. 1996. A report of killer whales (*Orcinus orca*) feeding on a carcharhinid shark in Costa Rica. *Marine Mammal Science* **12**(4): 606–611.

Fertl, D., Barros, N.B., Rowlett, R.A., Estes, S., and Richlen, M. 2004. An update on anomalously white cetaceans, including the first account for the pantropical spotted dolphin (*Stenella attenuata graffmani*). *Latin American Journal of Aquatic Mammals* **3**(2): 163–166.

Fertl, D., Jefferson, T.A., Moreno, I.B., Zerbini, A.N., and Mullin, K.D. 2003. Distribution of the clymene dolphin *Stenella clymene*. *Mammal Review* **33**(3): 253–271.

Fertl, D., and Leatherwood, S. 1997. Cetacean interactions with trawls: a preliminary review. *Journal of Northwest Atlantic Fishery Science* **22**: 219–248.

Fertl, D., Pusser, L.T., and Long, J.J. 1999. First record of an albino bottlenose dolphin (*Tursiops truncatus*) in the Gulf of Mexico, with a review of anomalously white cetaceans. *Marine Mammal Science* **15**(1): 227–234.

Fertl, D., Schiro, A.J., and Peake, D. 1997. Coordinated feeding by Clymene dolphins (*Stenella clymene*) in the Gulf of Mexico. *Aquatic Mammals* **23**(2): 111–112.

Fertl, D., and Würsig, B. 1995. Coordinated feeding by Atlantic spotted dolphins (*Stenella frontalis*) in the Gulf of Mexico. *Aquatic Mammals* **21**(1): 3–5.

Fiedler, P.C., and Reilly, S.B. 1994. Inter-annual variability of dolphin habitats in the eastern tropical Pacific. II: Effects on abundances estimated from tuna vessel sightings, 1975–1990. *Fishery Bulletin, US* **92**: 451–463.

Findlay, K., Meÿer, M., Elwen, S. *et al.* 2004. Distribution and abundance of humpback whales, *Megaptera novaeangliae*, off the coast of Mozambique, 2003. Paper SC56/SH12 submitted to the Scientific Committee of the International Whaling Commission, Sorrento, July 2004.

Findlay, K.P. 1989. *The distribution of cetaceans off the coast of South Africa and South West Africa/Namibia.* M Sc thesis, University of Pretoria. 128 pp + 163 pp of figures.

Findlay, K.P. 1994. *The migrations of east coast humpback whales* (Megaptera novaeangliae). Ph D thesis, University of Pretoria, 221 pp + 14 pl.

Findlay, K.P., and Best, P.B. 1995. Summer incidence of humpback whales on the west coast of South Africa. *South African Journal of Marine Science* **15**: 279–282.

Findlay, K.P., and Best, P.B. 1996a. Assessment of heterogeneity in sighting probabilities of humpback whales within viewing range of Cape Vidal, South Africa. *Marine Mammal Science* **12**(3): 335–353.

Findlay, K.P., and Best, P.B. 1996b. Estimates of the numbers of humpback whales observed migrating past Cape Vidal, South Africa, 1988–1991. *Marine Mammal Science* **12**(3): 354–370.

Findlay, K.P., Best, P.B., Peddemors, V.M., and Gove, D. 1994. The distribution and abundance of humpback whales on their southern and central Mozambique winter grounds. *Report of the International Whaling Commission* **44**: 311–320.

Findlay, K.P., Best, P.B., Ross, G.J.B., and Cockcroft, V.G. 1992. The distribution of small odontocete cetaceans off the coasts of South Africa and Namibia. *South African Journal of Marine Science* **12**: 237–270.

Fish, F.E. 1998. Comparative kinematics and hydrodynamics of odontocete cetaceans: morphological and ecological correlates with swimming performance. *Journal of Experimental Biology* **201**: 2 867–2 877.

Fish, F.E., Nicastro, A.J., and Weihs, D. 2006. Dynamics of the aerial maneuvers of spinner dolphins. *Journal of Experimental Biology* **209**: 590–598.

Fisher, S.R., Ferm, L.M., and Thomas, J.A. 1983. Circumpolar acoustic surveys for marine mammals using a towed hydrophone array. *Antarctic Journal of the United States* **18**(5): 171–173.

Flórez-González, L., Capelía A.J., Haase, B., Bravo, G.A., Félix, F., and Gerrodette, T. 1998. Changes in winter destinations and the northernmost record of southeastern Pacific humpback whales. *Marine Mammal Science* **14**(1): 189–196.

Flower, W.H. 1869. Description of the skeleton of the Chinese white dolphin (*Delphinus sinensis*, Osbeck). *Transactions of the Zoological Society of London* **7**(2): 151–160 + 2 pl.

Flower, W.H. 1872a. On Risso's dolphin, *Grampus griseus* (Cuv.). *Transactions of the Zoological Society of London* **8**(1): 1–21 + 2 pl.

Flower, W.H. 1872b. On the recent ziphoid whales, with a description of the skeleton of *Berardius arnouxi*. *Transactions of the Zoological Society of London* **8**: 203–234.

Ford, J.K.B., and Ellis, G.M. 1999. *Transients: mammal-hunting killer whales of British Columbia, Washington, and southeastern Alaska.* UBC Press, Vancouver; University of Washington Press, Seattle. 96 pp.

Ford, J.K.B., Ellis, G.M., Barrett-Lennard, L.G., Morton, A.B., Palm, R.S., and Balcomb (III), K.C. 1998. Dietary specialization in two sympatric populations of killer whales (*Orcinus orca*) in coastal British Columbia and adjacent waters. *Canadian Journal of Zoology* **76**: 1 456–1 471.

Forney, K.A., and Barlow, J. 1998. Seasonal patterns in the abundance and distribution of California cetaceans, 1991–1992. *Marine Mammal Science* **14**(3): 460–489.

Forrester, D.J., Odell, D.K., Thompson, N.P., and White, J.R. 1980. Morphometrics, parasites, and chlorinated hydrocarbon residues of pygmy killer whales from Florida. *Journal of Mammalogy* **61**(2): 356–360.

Frantzis, A. 1998. Does acoustic testing strand whales? *Nature, London* **392**(6671): 29.

Frantzis, A., and Herzing, D.L. 2002. Mixed-species associations of striped dolphins (*Stenella coeruleoalba*), short-beaked common dolphins (*Delphinus delphis*), and Risso's dolphins (*Grampus griseus*) in the Gulf of Corinth (Greece, Mediterranean Sea). *Aquatic Mammals* **28**(2): 188–197.

Fraser, F.C. 1936. Recent strandings of the false killer whale, *Pseudorca crassidens*, with special reference to those found at Donna Nook, Lincolnshire. *The Scottish Naturalist* **220**: 105–114.

Fraser, F.C. 1940. Three anomalous dolphins from Blacksod Bay, Ireland. *Proceedings of the Royal Irish Academy* **45B**: 413–455 + 7 pl.

Fraser, F.C. 1945. On a specimen of the southern bottlenosed whale, *Hyperoodon planifrons*. *Discovery Reports* **23**: 19–36.

Fraser, F.C. 1950a. Note on a skull of Hector's beaked whale *Mesoplodon hectori* (Gray) from the Falkland Islands. *Proceedings of the Linnean Society of London* **162**: 50–52 + 2 pl.

Fraser, F.C. 1950b. Two skulls of *Globicephala macrorhyncha* (Gray) from Dakar. *Atlantide-Report* **1**: 49–60 + 5 pl.

Fraser, F.C. 1955. The southern right whale dolphin, *Lissodelphis peroni* (Lacépède). *Bulletin of the British Museum (Natural History) Zoology* **2**(11): 341–346 + 1 pl.

Fraser, F.C. 1956. A new Sarawak dolphin. *The Sarawak Museum Journal* **7**(8, n.s., 23, o.s.): 478–503.

Fraser, F.C. 1964. Whales and whaling. Pp 191–205 in (Priestley, R., Adie, R.J., and Robin, G. de Q., eds) *Antarctic Research*. Butterworth's, London.

Fraser, F.C. 1966. Comments on the Delphinoidea. Pp 7–31 in (Norris, K.S., ed.) *Whales, dolphins and porpoises*. University of California Press, Berkeley.

Fraser, F.C. 1968. Notes on a specimen of *Phocoena dioptrica* from South Georgia. *British Antarctic Survey Bulletin* **16**: 51–56 + 1 table.

Fraser, F.C., and Noble, B.A. 1970. Variation of pigmentation pattern in Meyen's dolphin, *Stenella coeruleoalba* (Meyen). *Investigations on Cetacea* **2**: 147–163 + 7 plates.

Fristrup, K.M., and Harbison, G.R. 2002. How do sperm whales catch squids? *Marine Mammal Science* **18**:42–54.

Fullard, K.J., Early, G., Heide-Jørgensen, M.P., Bloch, D., Rosing-Asvid, A., and Amos, W. 2000. Population structure of long-finned pilot whales in the North Atlantic: a correlation with sea surface temperature? *Molecular Ecology* **9**: 949–958.

G

Gabriele, C.M., Straley, J.M., and Neilson, J.L. 2007. Age at first calving of female humpback whales in southeastern Alaska. *Marine Mammal Science* **23**(1): 226–239.

Gaines, C.A., Hare, M.P., Beck, S.E., and Rosenbaum, H. 2005. Nuclear markers confirm taxonomic status and relationships among highly endangered and closely related right whale species. *Proceedings of the Royal Society B* **272**: 533–542.

Gales, N.J., Dalebout, M.L., and Bannister, J.L. 2002. Genetic identification and biological observation of two free-swimming beaked whales: Hector's beaked whale (*Mesoplodon hectori*, Gray, 1871), and Gray's beaked whale (*Mesoplodon grayi*, von Haast, 1876). *Marine Mammal Science* **18**(2): 544–551.

Gambell, R. 1964. A pygmy blue whale from Durban. *Norsk Hvalfangsttidende* **53**(1): 66–68.

Gambell, R. 1966. The dorsal fin of sei and fin whales. *Norsk Hvalfangsttidende* **55**(9): 177–181.

Gambell, R. 1967. Seasonal movements of sperm whales. *Symposium of the Zoological Society of London* **19**: 237–254.

Gambell, R. 1968. Seasonal cycles and reproduction in sei whales of the Southern Hemisphere. *Discovery Reports* **35**: 31–133.

Gambell, R. 1973. Some effects of exploitation on reproduction in whales. *Journal of Reproduction and Fertility, Supplement* **19**: 533–553.

Gambell, R., Best, P.B., and Rice, D.W. 1975. Report on the international Indian Ocean whale marking cruise 24 November 1973–3 February 1974. *Report of the International Whaling Commission* **25**: 240–252.

Gammelsrød, T., Bartholomae, C.H., Boyer, D.C., Filipe, V.L.L., and O'Toole, M.J. 1998. Intrusion of warm surface water along the Angolan–Namibian coast in February–March 1995: The 1995 Benguela Niño. *South African Journal of Marine Science* **19**: 41–56.

Gannier, A. 1998. Seasonal variation of the bathymetric distribution of cetaceans in the Liguro-Provençal basin (Western Mediterranean). *Vie et Milieu* **48**(1): 25–34.

Gannier, A. 1999. Diel variations of the striped dolphin distribution off the French Riviera (northwestern Mediterranean Sea). *Aquatic Mammals* **25**(3): 123–134.

Gannier, A., and West, K.L. 2005. Distribution of the rough-toothed dolphin (*Steno bredanensis*) around the Windward Islands (French Polynesia). *Pacific Science* **59**(1): 17–24.

Gannon, D.P., Read, A.J., Craddock, J.E., Fristrup, K.M., and Nicolas, J.R. 1997. Feeding ecology of long-finned pilot whales *Globicephala melas* in the western North Atlantic. *Marine Ecology Progress Series* **148**: 1–10.

Garzoli, S.L., Gordon, A.L., Kamenkovich, V., Pillsbury, D., and Duncombe-Rae, C. 1996. Variability and sources of the south eastern Atlantic circulation. *Journal of Marine Research* **54**: 1 039–1 071.

Gaskin, D.E. 1968. The New Zealand Cetacea. *Fishery Research Bulletin* **1**(n.s.): 1–92.

Gasparini, J.L., and Sazima, I. 1996. A stranded melon-headed whale, *Peponocephala electra*, in southeastern Brazil, with comments on wounds from the cookie-cutter shark, *Isistius brasiliensis*. *Marine Mammal Science* **12**(2): 308–312.

Gazitúa, F., Gibbons, J., and Cárcamo, J. 1999. Descripcion de un ejemplar de delfin cruzado, *Lagenorhynchus cruciger* (Delphinidae), encontrado en el estrecho de Magallanes. *Anales del Instituto de la Patagonia Serie Ciencias Naturales* **27**: 73–82.

Gedamke, J., Costa, D.P., and Dunstan, A. 2001. Localization and visual verification of a complex minke whale vocalization. *Journal of the Acoustical Society of America* **109**(6): 3 038–3 047.

Geraci, J.R., Harwood, J., and Lounsbury, V.J. 1999. Marine mammal die-offs: causes, investigations, and issues. Pp 367–395 in (Twiss Jr, J.R., and Reeves, R.R., eds) *Conservation and management of marine mammals*. Smithsonian Institution Press, Washington.

Gerrodette, T. 2002. Tuna-dolphin issue. Pp 1 269–1 273 in (Perrin, W.F., Würsig, B., and Thewissen, J.G.M., eds) *Encyclopedia of Marine Mammals*. Academic Press, San Diego.

Gerrodette, T., and Forcada, J. 2005. Non-recovery of two spotted and spinner dolphin populations in the eastern tropical Pacific Ocean. *Marine Ecology Progress Series* **291**: 1–21.

Gianuca, N.M., and Castello, H.P. 1976. First record of the southern bottlenose whale, *Hyperoodon planifrons*, from Brazil. *Scientific Reports of the Whales Research Institute, Tokyo* **28**: 119–126.

Gibson-Hill, C.A. 1950. The whales, porpoises and dolphins known in Sarawak waters. *The Sarawak Museum Journal* **5**(n.s. 2): 288–296.

Gihr, M., Kraus, C., and Pilleri, G. 1982. The manus of *Pseudorca crassidens* (Owen): a study of variability. *Investigations on Cetacea* **13**: 101–124.

Gihr, M., and Pilleri, G. 1969. On the anatomy and biometry of *Stenella styx* Gray and *Delphinus delphis* L. (Cetacea, Delphinidae) of the Western Mediterranean. *Investigations on Cetacea* **1**: 15–65.

Gill, P.C., Ross, G.J.B., Dawbin, W.H., and Wapstra, H. 2000. Confirmed sightings of dusky dolphins (*Lagenorhynchus obscurus*) in southern Australian waters. *Marine Mammal Science* **16**(2): 452–459.

Gill, P.C., and Thiele, D. 1997. A winter sighting of killer whales (*Orcinus orca*) in Antarctic sea ice. *Polar Biology* **17**: 401–404.

Gilmore, R.M. 1962. *Bubbles and other pilot whales*. Barley Brae Printers, Del Mar, California. 16 pp.

Goley, P.D., and Straley, J.M. 1994. Attack on gray whales (*Eschrichtius robustus*) in Monterey Bay, California, by killer whales (*Orcinus orca*) previously identified in Glacier Bay, Alaska. *Canadian Journal of Zoology* **72**: 1 528–1 530.

Golovlev, I.F. 2000. Response to 'mystery of whales'. Pp 11–24 in (Yablokov, A.V., and Zemsky, V.A., eds) *Soviet whaling data (1949–1979)*. Centre for Russian Environmental Policy, Moscow.

Goodall, R.N.P. 1994. Commerson's dolphin, *Cephalorhynchus commersonii* (Lacépède 1804). Pp 241–267 in (Ridgway, S.H., and Harrison, R.J., eds) *Handbook of Marine Mammals 5: The first book of dolphins*. Academic Press, London.

Goodall, R.N.P. 1997. Review of sightings of the hourglass dolphin, *Lagenorhynchus cruciger*, in the South American sector of the Antarctic and Sub-Antarctic. *Report of the International Whaling Commission* **47**: 1 001–1 013.

Goodall, R.N.P. 2002. Spectacled porpoise *Phocaena dioptrica*. Pp 1 158–1 161 in (Perrin, W.F., Würsig, B., and Thewissen, J.G.M., eds) *Encyclopedia of Marine Mammals*. Academic Press, San Diego.

Goodall, R.N.P., Baker, A.N., Best, P.B., Meÿer, M.A., and Miyazaki, N. 1997. On the biology of the hourglass dolphin, *Lagenorhynchus cruciger* (Quoy and Gaimard, 1824). *Report of the International Whaling Commission* **47**: 985–999.

Goodall, R.N.P., and Galeazzi, A.R. 1985. A review of the food habits of the small cetaceans of the Antarctic and sub-antarctic. Pp 566–572 in (Siegfried, W.R., Condy, P.R., and Laws, R.M., eds) *Antarctic Nutrient Cycles and Food Webs*. Springer-Verlag, Berlin.

Goodall, R.N.P., Galeazzi, A.R., Leatherwood, S., Miller, K.W., Cameron, I.S., Kastelein, R.K., and Sobral, A.P. 1988a. Studies of Commerson's dolphins, *Cephalorhynchus commersonii*, off Tierra del Fuego, 1976–1984, with a review of information on the species in the South Atlantic. *Report of the International Whaling Commission* (Special Issue) **9**: 3–70.

Goodall, R.N.P., Galeazzi, A.R., Sobral, A.P., and Cameron, I.S. 1988b. Flipper serration in *Cephalorhynchus commersonii*. *Report of the International Whaling Commission* (Special Issue) **9**: 161–171.

Goodall, R.N.P., and Macnie, S.V. 1998a. A note on the behaviour of pilot whales in a Beagle Channel bay, Tierra del Fuego. *Report of the International Whaling Commission* **48**: 559–563.

Goodall, R.N.P., and Macnie, S.V. 1998b. Sightings of pilot whales off South America south of 30° S: a review of data to 1988. *Report of the International Whaling Commission* **48**: 565–579.

Goodall, R.N.P., and Schiavini, A.C.M. 1995. On the biology of the spectacled porpoise, *Australophocaena dioptrica*. *Report of the International Whaling Commission* (Special Issue) **16**: 411–453.

Goold, J.C., and Jefferson, T.A. 2004. A note on the clicks recorded from free-ranging Indo-Pacific humpback dolphins, *Sousa chinensis*. *Aquatic Mammals* **30**(1): 175–178.

Gordon, A.L., Molinelli, E., and Baker, T. 1978. Large scale relative dynamic topography of the Southern Ocean. *Journal of Geophysical Research* **83**: 3 023–3 032.

Gordon, A.L., Weiss, R.F., Smethie, W.M., and Warner, J. 1992. Thermocline and intermediate water communication between the South Atlantic and Indian Oceans. *Journal of Geophysical Research* **97**: 7 223–7 240.

Gosliner, M.L. 1999. The tuna-dolphin controversy. Pp 120–155 in (Twiss Jr, J.R., and Reeves, R.R., eds) *Conservation and management of marine mammals*. Smithsonian Institution Press, Washington.

Götz, T., Verfuß, U.K., and Schnitzler, H.-U. 2006. 'Eavesdropping' in wild rough-toothed dolphins (*Steno bredanensis*)? *Biological Letters of the Royal Society* **2**(1): 5–7.

Grandi, M.F., Buren, A.D., Crespo, E.A., García, N.A., Svendsen, G.M., and Dans, S.L. 2005. Record of a specimen of Shepherd's beaked whale (*Tasmacetus shepherdi*) from the coast of Santa Cruz, Argentina, with notes on age determination. *Latin American Journal of Aquatic Mammals* **4**(2): 97–100.

Gray, J.E. 1865. Notices of a new genus of delphinoid whales from the Cape of Good Hope, and of other cetaceans from the same seas. *Proceedings of the Zoological Society of London* **1865**: 522–529.

Griffin, M. 1998. The species diversity, distribution and conservation of Namibian mammals. *Biodiversity and Conservation* **7**: 483–494.

Griffin, M., and Coetzee, C.G. 2005. Annotated checklist and provisional national conservation status of Namibian mammals. *Technical Reports of Scientific Services* No. 4. Directorate of Scientific Services, Ministry of Environment and Tourism, Windhoek, Namibia. 207 pp.

Gründlingh, M.L. 1983. On the course of the Agulhas Current. *South African Geographical Journal* **65**(1): 49–57.

Gründlingh, M.L. 1985. An intense cyclonic eddy east of the Mozambique Ridge. *Journal of Geophysical Research* **90**(C4): 7 163–7 167.

Guerrero-Ruiz, M., García-Godos, I., and Urbán R., J. 2005. Photographic match of a killer whale (*Orcinus orca*) between Peruvian and Mexican waters. *Aquatic Mammals* **31**(4): 438–441.

Guiler, E.R., Burton, H.R., and Gales, N.J. 1987. On three odontocete skulls from Heard Island. *Scientific Reports of the Whales Research Institute, Tokyo* **38**: 117–124.

Guinet, C. 1991a. Intentional stranding apprenticeship and social play in killer whales (*Orcinus orca*). *Canadian Journal of Zoology* **69**: 2 712–2 716.

Guinet, C. 1991b. L'orque (*Orcinus orca*) autour de l'archipel Crozet, comparaison avec d'autres localités. *Revue de Ecologie (Terre et la Vie)* **46**: 321–337.

Guinet, C. 1992. Comportement de chasse des orques (*Orcinus orca*) autour des îles Crozet. *Canadian Journal of Zoology* **70**: 1 656–1 667.

Guinet, C., and Bouvier, J. 1995. Development of intentional stranding hunting techniques in killer whale (*Orcinus orca*) calves at Crozet Archipelago. *Canadian Journal of Zoology* **73**: 27–33.

Guissamulo, A., and Cockcroft, V.G. 2004. Ecology and population estimates of Indo-Pacific humpback dolphins (*Sousa chinensis*) in Maputo Bay, Mozambique. *Aquatic Mammals* **30**(1): 94–102.

Gunter, G., Hubbs, C.L., and Beal, M.A. 1955. Records of *Kogia breviceps* from Texas, with remarks on movements and distribution. *Journal of Mammalogy* **36**(2): 263–270.

H

Hale, H.M. 1931. Beaked whales – *Hyperoodon planifrons* and *Mesoplodon layardii* – from South Australia. *Records of the South Australian Museum* **4**(3): 291–311.

Hale, H.M. 1964. Pigmy right whale (*Caperea marginata*) in South Australian waters, Part 2. *Records of the South Australian Museum* **14**(4): 679–694 + 1 pl.

Hale, P.T., Barreto, A.S., and Ross, G.J.B. 2000. Comparative morphology and distribution of the *aduncus* and *truncatus* forms of bottlenose dolphin *Tursiops* in the Indian and Western Pacific Oceans. *Aquatic Mammals* **26**(2): 101–110.

Hammond, D.D., and Leatherwood, S. 1984. Cetaceans live-captured for Ocean Park, Hong Kong April 1974– February 1983. *Report of the International Whaling Commission* **34**: 491–495.

Handley Jr, C.O. 1966. A synopsis of the genus *Kogia* (pygmy sperm whales). Pp 62–69 in (K.S. Norris, ed.) *Whales, dolphins and porpoises*. University of California Press, Berkeley.

Haney, T.A., De Almeida, A.O., and Reiss, M.S.S. 2004. A new species of cyamid (Crustacea: Amphipoda) from a stranded cetacean in southern Bahia, Brazil. *Bulletin of Marine Science* **75**(3): 409–421.

Harlin, A.D., Markowitz, T., Baker, C.S., Würsig, B., and Honeycutt, R.L. 2003. Genetic structure, diversity, and historical demography of New Zealand's dusky dolphin (*Lagenorhynchus obscurus*). *Journal of Mammalogy* **84**(2): 702–717.

Harmer, S.F. 1927. *Report on Cetacea stranded on the British coasts from 1913 to 1926*. London, British Museum (Natural History). No. 10, 91 pp + 7 pl.

Harmer, S.F. 1931. Southern whaling. *Proceedings of the Linnean Society of London* **142**(1929–30): 85–163.

Hart, T.J. 1935. On the diatoms of the skin film of whales, and their possible bearing on problems of whale movements. *Discovery Reports* **10**: 247–282.

Hart, T.J., and Currie, R.I. 1960. The Benguela Current. *Discovery Reports* **31**: 123–298.

Hassel, L.B., Venturotti, A., Magalhães, F.A., De Cuenca, S., Siciliano, S., and Marques, F.F.C. 2003. Summer sightings of dwarf minke whales (*Balaenoptera acutorostrata*) off the eastern coast of Rio de Janeiro state, Brazil. *Latin American Journal of Aquatic Mammals* **2**(1): 47–50.

Heide-Jørgensen, M.P., Bloch, D., Stefansson, E., Mikklesen, B., Ofstad, L.H., and Dietz, R. 2002. Diving behaviour of long-finned pilot whales *Globicephala melas* around the Faroe Islands. *Wildlife Biology* **8**: 307–313.

Hernández-Garcia, V. 2002. Contents of the digestive tract of a false killer whale (*Pseudorca crassidens*) stranded in Gran Canaria (Canary Islands, central east Atlantic). *Bulletin of Marine Science* **71**(1): 367–369.

Herzing, D.L. 1996. Vocalizations and associated underwater behaviour of free-ranging Atlantic spotted dolphins, *Stenella frontalis* and bottlenose dolphins, *Tursiops truncatus*. *Aquatic Mammals* **22**(2): 61–79.

Herzing, D.L. 1997. The life history of free-ranging Atlantic spotted dolphins (*Stenella frontalis*): age classes, color phases, and female reproduction. *Marine Mammal Science* **13**(4): 576–595.

Herzing, D.L., and Brunnick, B.J. 1997. Coefficients of association of reproductively active female Atlantic spotted dolphins, *Stenella frontalis*. *Aquatic Mammals* **23**(3): 155–162.

Herzing, D.L., and Johnson, C.M. 1997. Interspecific interactions between Atlantic spotted dolphins (*Stenella frontalis*) and bottlenose dolphins (*Tursiops truncatus*) in the Bahamas, 1985–1995. *Aquatic Mammals* **23**(2): 85–99.

Herzing, D.L., Moewe, K., and Brunnick, B.J. 2003. Interspecies interactions between Atlantic spotted dolphins, *Stenella frontalis* and bottlenose dolphins, *Tursiops truncatus*, on Great Bahama Bank, Bahamas. *Aquatic Mammals* **29**(3): 335–341.

Heyning, J.E. 1988a. Age determination and reproductive biology of Cuvier's beaked whale, *Ziphius cavirostris*. Paper SC/40/SM22 submitted to the Scientific Committee of the International Whaling Commission, San Diego, USA, May 1988.

Heyning, J.E. 1988b. Presence of solid food in a young calf killer whale (*Orcinus orca*). *Marine Mammal Science* **4**(1): 68–71.

Heyning, J.E. 1989. Cuvier's beaked whale *Ziphius cavirostris* G. Cuvier, 1823. Pp 289–308 in (Ridgway, S.H., and Harrison, R.J., eds). *Handbook of Marine Mammals 4: River dolphins and the larger toothed whales*. Academic Press, London.

Heyning, J.E., and Mead, J.G. 1996. Suction feeding in beaked whales: morphological and observational evidence. *Contributions in Science* **464**: 1–12.

Heyning, J.E., and Perrin, W.F. 1994. Evidence for two species of common dolphins (genus *Delphinus*) from the eastern North Pacific. *Contributions in Science* **442**: 1–35.

Higashi, N., Takemura, A., Kubota, J., and Uchida, S. 1992. Calls of the short-finned pilot whale, *Globicephala macrorhynchus*, in the coastal waters of Okinawa, southern Japan. *Bulletin of the Faculty of Fisheries, Nagasaki University* **72**: 5–10.

Hobson, R.P., and Martin, A.R. 1996. Behaviour and dive times of Arnoux's beaked whales, *Berardius arnuxii*, at narrow leads in fast ice. *Canadian Journal of Zoology* **74**(2): 388–393.

Hoelzel, A.R. 1991. Killer whale predation on marine mammals at Punta Norte, Argentina; food sharing, provisioning and foraging strategy. *Behavioral Ecology and Sociobiology* **29**: 197–204.

Hofmann, E.E. 1985. The large scale horizontal structure of the Antarctic Circumpolar Current from FGGE drifters. *Journal of Geophysical Research* **90**: 7 087–7 097.

Hoffmann, M., and Best, P.B. (subeds). 2005. Suborder Cetacea. Whales and dolphins. Pp 560–615 in (Skinner, J.D., and Chimimba, C.T., eds) *The Mammals of the Southern African Subregion*. 3rd edition. Cambridge University Press, Cape Town.

Hofmeyr-Juritz, L. 2006. *The nature and rate of vocalization by southern right whales* (Eubalaena australis)*, and the evidence for individually distinctive calls*. Ph D thesis, University of Pretoria. 164 pp.

Hohn, A.A., Chivers, S.J., and Barlow, J. 1985. Reproductive maturity and seasonality of male spotted dolphins, *Stenella attenuata*, in the eastern tropical Pacific. *Marine Mammal Science* **1**(4): 273–293.

Hohn, A.A., and Hammond, P.S. 1985. Early postnatal growth of the spotted dolphin, *Stenella attenuata*, in the offshore eastern tropical Pacific. *Fishery Bulletin, US* **83**(4): 553–566.

Holliday, N.P., and Read, J.F. 1998. Surface oceanic fronts between Africa and Antarctica. *Deep-Sea Research I* **45**: 217–238.

Holmes, R.W., Nagasawa, S., and Takano, H. 1993. A re-examination of diatom samples obtained from cetaceans collected off South Africa. *Bulletin of the National Science Museum* Series B (Botany) **19**(4): 127–135.

Hooker, S.K., and Baird, R.W. 1999. Deep-diving behaviour of the northern bottlenose whale, *Hyperoodon ampullatus* (Cetacea: Ziphiidae). *Proceedings of the Royal Society of London* B **266**: 671–676.

Horwood, J. 1987. *The sei whale: population biology, ecology and management*. Croom Helm, London. 375 pp.

Howell, K.M., and Pearson, D.M. 1977. Two records of dolphins from Tanzania. *East African Wildlife Journal* **15**: 167–168.

Hoyt, E. 1992. *The performing orca – why the show must stop*. Whale and Dolphin Conservation Society, Bath. 102 pp.

Hückstädt, L., and Antezana, T. 2001. An observation of parturition in a stranded *Kogia breviceps*. *Marine Mammal Science* **17**(2): 362–365.

Hui, C.A. 1985. Undersea topography and the comparative distributions of two pelagic cetaceans. *Fishery Bulletin, US* **83**(3): 472–475.

Huxley, L. 1947. *Scott's last expedition 1. The journals of Captain R.F. Scott, R.N., C.V.O.* John Murray, London. 497 pp + chart.

I

Ichihara, T. 1963. Identification of the pigmy blue whale in the Antarctic. *Norsk Hvalfangsttidende* **52**(5): 128–130.

Ichihara, T. 1966. The pygmy blue whale, *Balaenoptera musculus brevicauda*, a new subspecies from the Antarctic. Pp 79–113 in (Norris, K.S., ed.) *Whales, dolphins, and porpoises*. University of California Press, Berkeley.

Ichii, T., and Kato, H. 1991. Food and daily food consumption of southern minke whales in the Antarctic. *Polar Biology* **11**: 479–487.

Ingebrigtsen, A. 1929. Whales caught in the North Atlantic and other seas. *Rapports et Procès-Verbaux des Réunions* **56**(2): 1–26.

Iñíguez, M.A., Hevia, M., Gasparrou, C., Tomsin, A.L., and Secchi, E.R. 2003. Preliminary estimate of incidental mortality of Commerson's dolphins (*Cephalorhynchus commersonii*) in an artisanal setnet fishery in La Angelina beach and Ría Gallegos, Santa Cruz, Argentina. *Latin American Journal of Aquatic Mammals* **2**(2): 87–94.

International Whaling Commission. 1980. Report of the workshop. *Report of the International Whaling Commission* (Special Issue) **3**: 1–50.

International Whaling Commission. 1994. Report of the workshop on mortality of cetaceans in passive fishing nets and traps. *Report of the International Whaling Commission* (Special Issue) **15**: 6–71.

International Whaling Commission. 2001a. Report of the workshop on the comprehensive assessment of right whales: a worldwide comparison. *Journal of Cetacean Research and Management* (Special Issue) **2**: 1–60.

International Whaling Commission. 2001b. Report of the Scientific Committee. Annex K. Appendix 4. *Journal of Cetacean Research and Management* **3**(Suppl.): 284–291.

Ito, H., and Miyazaki, N. 1990. Skeletal development of the striped dolphin (*Stenella coeruleoalba*) in Japanese waters. *Journal of the Mammalogical Society of Japan* **14**(2): 79–96.

Ivashin, M.V., Shevchenko, V.I., and Yukov, V.L. 1972. [The pygmy right whale *Caperea marginata* (Cetacea)]. *Zoologicheskiĭ Zhurnal* **51**(11): 1 715–1 723 [In Russian].

J

Jacobs, S.S. 2004. Bottom water production and its links with the thermohaline circulation. *Antarctic Science* **16**(4): 427–437.

Jacobs, S.S., Giulivi, C.F., and Mele, P.A. 2002. Freshening of the Ross Sea during the late 20th century. *Science* **297**(5580): 386–389.

Jansen van Vuuren, B., Best, P.B., Roux, J.-P., and Robinson, T.J. 2002. Phylogeographic population structure in the Heaviside's dolphin (*Cephalorhynchus heavisidii*): conservation implications. *Animal Conservation* **5**: 303–307.

Jefferson, T., and Karczmarski, L. 2001. *Sousa chinensis*. *Mammalian Species* **655**: 1–9.

Jefferson, T.A. 1996. Morphology of the clymene dolphin (*Stenella clymene*) in the northern Gulf of Mexico. *Aquatic Mammals* **22**(1): 35–43.

Jefferson, T.A. 2000. Population biology of the Indo-Pacific humpbacked dolphin in Hong Kong waters. *Wildlife Monographs* **144**: 1–65.

Jefferson, T.A. 2002. Rough-toothed dolphin *Steno bredanensis*. Pp 1 055–1 059 in (Perrin, W.F., Würsig, B., and Thewissen, J.G.M., eds) *Encyclopedia of Marine Mammals*. Academic Press, San Diego.

Jefferson, T.A., and Barros, N.B. 1997. *Peponocephala electra*. *Mammalian Species* **553**: 1–6.

Jefferson, T.A., and Curry, B.E. 2003. *Stenella clymene*. *Mammalian Species* **726**: 1–5.

Jefferson, T.A., Newcomer, M.A., Leatherwood, S., and Van Waerebeek, K. 1994. Right whale dolphins *Lissodelphis borealis* (Peale, 1848) and *Lissodelphis peronii* (Lacépède, 1804). Pp 335–362 in (Ridgway, S.H., and Harrison, R.J., eds) *Handbook of Marine Mammals 5: The first book of dolphins*. Academic Press, New York.

Jefferson, T.A., Odell, D.K., and Prunier, K.T. 1995. Notes on the biology of the clymene dolphin (*Stenella clymene*) in the northern Gulf of Mexico. *Marine Mammal Science* **11**(4): 564–573.

Jefferson, T.A., Pitman, R.L., Leatherwood, S., and Dolar, M.L.L. 1997. Developmental and sexual variation in the external appearance of Fraser's dolphins (*Lagenodelphis hosei*). *Aquatic Mammals* **23**(3): 145–153.

Jefferson, T.A., Stacey, P.J., and Baird, R.W. 1991. A review of killer whale interactions with other marine mammals: predation to co-existence. *Mammal Review* **21**(4): 151–180.

Jefferson, T.A., and Van Waerebeek, K. 2002. The taxonomic status of the nominal dolphin species *Delphinus tropicalis* (van Bree, 1971). *Marine Mammal Science* **18**(4): 787–818.

Jefferson, T.A., and Van Waerebeek, K. 2004. Geographic variation in skull morphology of humpback dolphins (*Sousa* spp.). *Aquatic Mammals* **30**(1): 3–17.

Jepson, P.D., Arbelo, M., Deaville, R. *et al.* 2003. Gas-bubble lesions in stranded cetaceans. *Nature, London* **425**(6958): 575.

Johnson, C., and Norris, K.S. 1994. Social behaviour. Pp 243–286 in (Norris, K.S., Würsig, B., Wells, R.S., Würsig, M., Brownlee, S.M., Johnson, C.M., and Solow, J., eds) *The Hawaiian spinner dolphin*. University of California Press, Berkeley.

Johnson, C.M., and Moewe, K. 1999. Pectoral fin preference during contact in Commerson's dolphins (*Cephalorhynchus commersonii*). *Aquatic Mammals* **25**(2): 73–77.

Johnson, M., Madsen, P.T., Zimmer, W.M.X., De Soto, N.A., and Tyack, P.L. 2004. Beaked whales echolocate on prey. *Proceedings of the Royal Society B: Biological Sciences 271 Biology Letters* Supplement **6**: S383–S386.

Johnston, S.J., and Butterworth, D.S. 2005. A Bayesian assessment of breeding stocks B, C and G of Southern Hemisphere humpback whales using a prior for growth rate from analyses for stocks D and E. Paper SC57/SH16 submitted to the Scientific Committee of the International Whaling Commission, Ulsan, Korea, July 2005.

Jones, E.C. 1971. *Isistius brasiliensis*, a squaloid shark, the probable cause of crater wounds on fishes and cetaceans. *Fishery Bulletin, US* **69**(4): 791–798.

Jonsgård, A. 1951. Studies on the little piked whale or minke whale. *Norsk Hvalfangsttidende* **40**: 209–230.

Joyce, G.G. 1982. Blow patterns as sighting cues for censusing minke whales in Antarctic waters. *Report of the International Whaling Commission* **32**: 787–790.

Joyce, T.M., and Patterson, S.L. 1977. Cyclonic ring formation at the polar front in the Drake Passage. *Nature* **265**: 131–133.

Junge, G.C.A. 1950. On a specimen of the rare fin whale, *Balaenoptera edeni* Anderson, stranded on Pulu Sugi near Singapore. *Zoologische Verhandelingen* **9**: 1–26 + 9 pl.

K

Kamminga, C., and Wiersma, H. 1982. Investigations on cetacean sonar V. The true nature of the sonar sound of *Cephalorhynchus commersonii*. *Aquatic Mammals* **9**(3): 95–104.

Kanda, N., Goto, M., Kato, H., McPhee, M.V., and Pastene, L.A. 2007. Population genetic structure of Bryde's whales (*Balaenoptera brydei*) at the inter-oceanic and trans-equatorial levels.

Conservation Genetics [Online] DOI:19.1007/s10592-006-9232-8, 12pp.

Karczmarski, L. 1999. Group dynamics of humpback dolphins (*Sousa chinensis*) in the Algoa Bay region, South Africa. *Journal of Zoology, London* **249**: 283–293.

Karczmarski, L. 2000. Conservation and management of humpback dolphins: the South African perspective. *Oryx* **34**(3): 207–216.

Karczmarski, L., and Cockcroft, V.G. 1999. Daylight behaviour of humpback dolphins *Sousa chinensis* in Algoa Bay, South Africa. *Zeitschrift für Säugetierkunde* **64**: 19–29.

Karczmarski, L., Cockcroft, V.G., and McLachlan, A. 1999a. Group size and seasonal pattern of occurrence of humpback dolphins *Sousa chinensis* in Algoa Bay, South Africa. *South African Journal of Marine Science* **21**: 89–97.

Karczmarski, L., Cockcroft, V.G., and McLachlan, A. 2000. Habitat use and preferences of Indo-Pacific humpback dolphins *Sousa chinensis* in Algoa Bay, South Africa. *Marine Mammal Science* **16**(1): 65–79.

Karczmarski, L., Thornton, M., and Cockcroft, V.G. 1997. Description of selected behaviours of humpback dolphins *Sousa chinensis*. *Aquatic Mammals* **23**(3): 127–133.

Karczmarski, L., Winter, P.E.D., Cockcroft, V.G., and McLachlan, A. 1999b. Population analyses of Indo-Pacific humpback dolphins *Sousa chinensis* in Algoa Bay, Eastern Cape, South Africa. *Marine Mammal Science* **15**(4): 1 115–1 123.

Karczmarski, L., Würsig, B., Gailey, G., Larson, K.W., and Vanderlip, C. 2005. Spinner dolphins in a remote Hawaiian atoll: social grouping and population structure. *Behavioral Ecology* **16**: 675–685.

Kasamatsu, F., Ensor, P., and Joyce, G.G. 1998. Clustering and aggregations of minke whales in the Antarctic feeding grounds. *Marine Ecology Progress Series* **168**: 1–11.

Kasamatsu, F., Hembree, D., Joyce, G., Tsunoda, L., Rowlett, R., and Nakano, T. 1988. Distribution of cetacean sightings in the Antarctic: results obtained from the IWC/IDCR minke whale assessment cruises, 1978/79 to 1983/84. *Report of the International Whaling Commission* **38**: 449–487.

Kasamatsu, F., and Joyce, G.G. 1995. Current status of odontocetes in the Antarctic. *Antarctic Science* **7**(4): 365–379.

Kasamatsu, F., Nishiwaki, S., and Ishikawa, H. 1995. Breeding areas and southbound migrations of southern minke whales *Balaenoptera acutorostrata*. *Marine Ecology Progress Series* **119**: 1–10.

Kastelein, R.A., Au, W.W.L., and De Haan, D. 2000a. Detection distances of bottom-set gill nets by harbour porpoises (*Phocoena phocoena*) and bottlenose dolphins (*Tursiops truncatus*). *Marine Environmental Research* **49**: 359–375.

Kastelein, R.A., Kershaw, J., Berghout, E., and Wiepkema, P.R. 2003a. Food consumption and suckling in killer whales *Orcinus orca* at Marineland Antibes. *International Zoo Yearbook* **38**: 204–218.

Kastelein, R.A., Macdonald, G.J., and Wiepkema, P.R. 2000b. A note on food consumption and growth of common dolphins (*Delphinus delphis*). *Journal of Cetacean Research and Management* **2**(1): 69–73.

Kastelein, R.A., McBain, J., and Neurohr, B. 1993a. Information on the biology of Commerson's dolphins (*Cephalorhynchus commersonii*). *Aquatic Mammals* **19**(1): 13–19.

Kastelein, R.A., McBain, J., Neurohr, B., Mohri, M., Saijo, S., Wakabayashi, I., and Wiepkema, P.R. 1993b. The food consumption of Commerson's dolphins (*Cephalorhynchus commersonii*). *Aquatic Mammals* **19**(2): 99–121.

Kastelein, R.A., Mosterd, J., Schooneman, N.M., and Wiepkema, P.R. 2000c. Food consumption, growth, body dimensions, and respiration rates of captive false killer whales (*Pseudorca crassidens*). *Aquatic Mammals* **26**(1): 33–44.

Kastelein, R.A., Staal, C., and Wiepkema, P.R. 2003b. Food consumption, food passage time, and body measurements of captive Atlantic bottlenose dolphins (*Tursiops truncatus*). *Aquatic Mammals* **29**(1): 53–66.

Kastelein, R.A., Van der Elst, C.A., Tennant, H.K., and Wiepkema, P.R. 2000d. Food consumption and growth of a female dusky dolphin (*Lagenorhynchus obscurus*). *Zoo Biology* **19**: 131–142.

Kastelein, R.A., and Vaughan, N. 1989. Food consumption, body measurements and weight changes of a female killer whale (*Orcinus orca*). *Aquatic Mammals* **15**(1): 18–21.

Kastelein, R.A., Walton, S., Odell, D., Nieuwstraten, S.H., and Wiepkema, P.R. 2000e. Food consumption of a captive female killer whale (*Orcinus orca*). *Aquatic Mammals* **26**(2): 127–131.

Kasuya, T. 1972. Growth and reproduction of *Stenella caeruleoalba* based on the age determination by means of dentinal growth layers. *Scientific Reports of the Whales Research Institute, Tokyo* **24**: 57–79.

Kasuya, T. 1976. Reconsideration of life history parameters of the spotted and striped dolphins based on cemental layers. *Scientific Reports of the Whales Research Institute, Tokyo* **28**: 73–106.

Kasuya, T. 1985. Effect of exploitation on reproductive parameters of the spotted and striped dolphins off the Pacific coast of Japan. *Scientific Reports of the Whales Research Institute, Tokyo* **36**: 107–138.

Kasuya, T. 1986. False killer whales. Pp 178–187 in (Tamura, T., Ohsumi, S., and Arai, S., eds) *Report of investigation in search of solution for dolphin-fishery conflict in the Iki Island area*. Japan Fisheries Agency [In Japanese, translated by Inagaki, N.].

Kasuya, T. 2002. Giant beaked whales *Berardius bairdii* and *B. arnuxii*. Pp 519–522 in (Perrin, W.F., Würsig, B., and Thewissen, J.G.M., eds) *Encyclopedia of Marine Mammals*. Academic Press, San Diego.

Kasuya, T., Brownell Jr, R.L., and Balcomb (III), K.C. 1997a. Life history of Baird's beaked whales off the Pacific coast of Japan. *Report of the International Whaling Commission* **47**(2): 969–979.

Kasuya, T., Izumisawa, Y., Komyo, Y., Ishino, Y., and Maejima, Y. 1997b. Life history parameters of bottlenose dolphins off Japan. *International Marine Biological Research Institute Reports* **7**: 71–107.

Kasuya, T., and Marsh, H. 1984. Life history and reproductive biology of the short-finned pilot whale, *Globicephala macrorhynchus*, off the Pacific coast of Japan. *Report of the International Whaling Commission* (Special Issue) **6**: 259–310.

Kasuya, T., and Matsui, S. 1984. Age determination and growth of the short-finned pilot whale off the Pacific coast of Japan. *Scientific Reports of the Whales Research Institute, Tokyo* **35**: 57–91.

Kasuya, T., and Miyashita, T. 1997. Distribution of Baird's beaked whales off Japan. *Report of the International Whaling Commission* **47**(2): 963–968.

Kasuya, T., Miyashita, T., and Kasamatsu, F. 1988. Segregation of two forms of short-finned pilot whales off the Pacific coast of Japan. *Scientific Reports of the Whales Research Institute, Tokyo* **39**: 77–90.

Kasuya, T., and Ohsumi, S. 1966. A secondary sexual character of the sperm whale. *Scientific Reports of the Whales Research Institute, Tokyo* **20**: 89–94.

Kasuya, T., and Tai, S. 1993. Life history of short-finned pilot whale stocks off Japan and a description of the fishery. *Report of the International Whaling Commission* (Special Issue) **14**: 439–473.

Kato, H. 1982. Some biological parameters for the Antarctic minke whale. *Report of the International Whaling Commission* **32**: 935–945.

Kato, H. 1984. Observation of tooth scars on the head of male sperm whale, as an indication of intra-sexual fightings. *Scientific Reports of the Whales Research Institute, Tokyo* **35**: 39–46.

Kato, H. 1986. *Studies of biological characteristics for southern minke whales, with special emphasis on the effects of changes over time and with resource condition.* Ph D thesis, Hokkaido University, 145 pp [In Japanese].

Kato, H. 1987. Density dependent changes in growth parameters of the southern minke whale. *Scientific Reports of the Whales Research Institute, Tokyo* **38**: 47–73.

Kato, H., and Fujise, Y. 2000. Dwarf minke whales; morphology, growth and life history with some analyses on morphometric variation among the different forms and regions. Paper SC/52/OS3 submitted to the Scientific Committee of the International Whaling Commission, Adelaide, Australia, June 2000, 30 pp.

Kato, H., Honno, Y., Yoshida, H., Kojima, E., Nomura, A., and Okamura, H. 2002. Further developments on morphological and behavioral key for sub-species discrimination of southern blue whales, analyses from data through 1995/96 to 2001/02 SOWER cruises. Paper SC/54/IA8 submitted to the Scientific Committee of the International Whaling Commission, Shimonoseki, Japan, April–May 2002.

Kato, H., and Miyashita, T. 2005. Japan progress report on cetacean research, May 2004 to April 2005. Document SC/57/ProgRep Japan submitted to the Scientific Committee, International Whaling Commission, Ulsan, Korea, May/June 2005. 14 pp.

Kato, H., Tanaka, E., and Sakuramoto, K. 1993. Movement of southern minke whales in the Antarctic feeding grounds from mark-recapture analyses. *Report of the International Whaling Commission* **43**: 335–342.

Kawai, S., and Fukushima, M. 1981. Relation between the lipid composition and the concentrations of organochlorine compounds in the various organs of striped dolphins *Stenella coeruleoalba*. Pp 85–96 in (Fujiyama, T., ed.) *Studies on the levels of organochlorine compounds and heavy metals in the marine organisms*. Report for the fiscal year of 1980, Grant-in-aid for scientific research, University of the Ryukyus, Okinawa, Japan.

Kawamura, A. 1974. Food and feeding ecology in the southern sei whale. *Scientific Reports of the Whales Research Institute, Tokyo* **26**: 25–144.

Kawamura, A. 1980. A review of food of balaenopterid whales. *Scientific Reports of the Whales Research Institute, Tokyo* **32**: 155–197.

Kawamura, A. 1994. A review of baleen whale feeding in the Southern Ocean. *Report of the International Whaling Commission* **44**: 261–271.

Keith, M., Bester, M.N., Bartlett, P.A., and Baker, D. 2001. Killer whales (*Orcinus orca*) at Marion Island, Southern Ocean. *African Zoology* **36**(2): 163–175.

Keller, R.W., Leatherwood, S., and Holt, S.J. 1982. Indian Ocean cetacean survey, Seychelle Islands, April through June 1980. *Report of the International Whaling Commission* **32**: 503–513.

Kellogg, R. 1928. The history of whales – their adaptation to life in the water. *Quarterly Review of Biology* **3**: 29–208.

Kemper, C., Dutton, J., Foster, B., and McGuire, R. 1997. Sightings and strandings of the pygmy right whale *Caperea marginata* near Port Lincoln, South Australia and a review of other Australasian sightings. *Transactions of the Royal Society of South Australia* **121**(2): 79–82.

Kemper, C.M. 2002a. Distribution of the pygmy right whale, *Caperea marginata*, in the Australasian region. *Marine Mammal Science* **18**(1): 99–111.

Kemper, C.M. 2002b. Pygmy right whale *Caperea marginata*. Pp 1 010–1 012 in (Perrin, W.F., Würsig, B., and Thewissen, J.G.M., eds) *Encyclopedia of Marine Mammals*. Academic Press, San Diego.

Kemper, C.M., and Leppard, P. 1999. Estimating body length of pygmy right whales (*Caperea marginata*) from measurements of the skeleton and baleen. *Marine Mammal Science* **15**(3): 683–700.

Kemper, C.M., and Ling, J.K. 1991. Whale strandings in South Australia (1881–1989). *Transactions of the Royal Society of South Australia* **115**(1): 37–52.

Kerem, D., Goffman, O., and Spanier, E. 2001. Sighting of a single humpback dolphin (*Sousa* sp.) along the Mediterranean coast of Israel. *Marine Mammal Science* **17**(1): 170–171.

Kingston, S.E., and Rosel, P.E. 2004. Genetic differentiation among recently diverged delphinid taxa determined using AFLP markers. *Journal of Heredity* **95**(1): 1–10.

Kiszka, J., Breysse, O., Vely, M., and Boinali, K. 2006a. Marine mammals around the Comoros archipelago (Mozambique Channel): recent records and review of available information. Paper SC/58/O6 submitted to the Scientific Committee of the International Whaling Commission, St Kitts and Nevis, June 2006.

Kiszka, J., Ersts, P.J., and Ridoux, V. 2006b. Cetacean diversity in the lagoon and surrounding waters of Mayotte, Comoros Archipelago. Paper SC/58/O13 submitted to the Scientific Committee of the International Whaling Commission, St Kitts and Nevis, June 2006.

Kita, S., Yoshioka, M., and Kashiwagi, M. 1999. Relationship between sexual maturity and serum and testis testosterone concentrations in short-finned pilot whales *Globicephala macrorhynchus*. *Fisheries Science* **65**(6): 878–883.

Kitchener, D.J., Ross, G.J.B., and Caputi, N. 1990. Variation in skull and external morphology in the false killer whale, *Pseudorca crassidens*, from Australia, Scotland and South Africa. *Mammalia* **54**(1): 119–135.

Kjeld, M., Víkingsson, G.A., Alfredsson, Á., Ólafsson, Ö., and Árnason, A. 2003. Sex hormone concentrations in the blood of sei whales (*Balaenoptera borealis*) off Iceland. *Journal of Cetacean Research and Management* **5**(3): 233–240.

Klinowska, M. 1985a. Cetacean live stranding sites relative to geomagnetic topography. *Aquatic Mammals* **11**(1): 27–32.

Klinowska, M. 1985b. Cetacean live stranding dates relate to geomagnetic disturbances. *Aquatic Mammals* **11**(3): 109–119.

Knox, G.A. 1984. *The biology of the Southern Ocean*. Cambridge University Press, Cambridge. 444 pp.

Kock, K.-H., Purves, M.G., and Duhamel, G. 2006. Interactions between cetacean and fisheries in the Southern Ocean. *Polar Biology* **29**: 379–388.

Kogi, K., Hishii, T., Imamura, A., Iwatani, T., and Dudzinski, K.M. 2004. Demographic parameters of Indo-Pacific bottlenose dolphins (*Tursiops aduncus*) around Mikura Island, Japan. *Marine Mammal Science* **20**(3): 510–526.

Korabelnikov, L.V. 1959. The diet of sperm whales in the Antarctic seas. *Priroda* (3): 103–104.

Kritzler, H. 1952. Observations on the pilot whale in captivity. *Journal of Mammalogy* **33**(3): 321–334.

Kroese, M. 1993. *Age, growth and reproductive biology of striped dolphins,* Stenella coeruleoalba *(Meyen, 1833) off the coast of southern Africa.* M Sc thesis, Faculty of Science, University of Port Elizabeth. 73 pp.

Kruse, S., Caldwell, D.K., and Caldwell, M.C. 1999. Risso's dolphin *Grampus griseus* (G. Cuvier, 1812). Pp 183–212 in (Ridgway, S.H., and Harrison, R.J., eds) *Handbook of Marine Mammals 6: The second book of dolphins and the porpoises.* Academic Press, New York.

Krützen, M., Barré, L.M., Connor, R.C., Mann, J., and Sherwin, W.B. 2004. 'O father: where art thou?' – paternity assessment in an open fission-fusion society of wild bottlenose dolphins (*Tursiops* sp.) in Shark Bay, Western Australia. *Molecular Ecology* **13**(7): 1 975–1 990.

Kükenthal, W. 1892. *Sotalia teuszii* n. sp. Eine planzenfressender (?) Delphin aus Kamerun. *Zoologische Jahrbucher, Abteilungen Systematik, Ökologie und Geographie der Tiere* VI: 442–446 + 1 pl.

L

Lagerquist, B.A., Stafford, K.M., and Mate, B.R. 2000. Dive characteristics of satellite-monitored blue whales (*Balaenoptera musculus*) off the central Californian coast. *Marine Mammal Science* **16**(2): 375–391.

Lahille, F. 1912. Nota preliminary sobre una nueva especie de marsopa del rio de la Plata. *Anales del Museo Nacional de Historia Natural de Buenos Aires* **23**: 269–278 + 4 pl.

Laist, D.W., Coe, J.M., and O'Hara, K.J. 1999. Marine debris pollution. Pp 342–366 in (Twiss Jr, J.R., and Reeves, R.R., eds) *Conservation and management of marine mammals.* Smithsonian Institution Press, Washington.

Lambertsen, R.H. 1983. The internal mechanism of rorqual feeding. *Journal of Mammalogy* **64**(1): 76–88.

Lambertsen, R.H., Ulrich, N., and Straley, J. 1995. Frontomandibular stay of Balaenopteridae: a mechanism for momentum recapture during feeding. *Journal of Mammalogy* **76**(3): 877–899.

Lang, T.G., and Pryor, K. 1966. Hydrodynamic performance of porpoises (*Stenella attenuata*). *Science* **152** (3721): 531–533.

Laporta, P., Praderi, R., Little, V., and Le Bas, A. 2005. An Andrew's beaked whale *Mesoplodon bowdoini* (Cetacea, Ziphiidae) stranded on the Atlantic coast of Uruguay. *Latin American Journal of Aquatic Mammals* **4**(2): 101–111.

Lass, H.U., Schmidt, M., Mohrholz, V., and Nausch, G. 2000. Hydrographic and current measurements in the area of the Angola–Benguela front. *Journal of Physical Oceanography* **30**: 2 589–2 609.

Laws, R.M. 1961. Reproduction, growth and age of southern fin whales. *Discovery Reports.* **31**: 327–486 + 4 pl.

Layard, E.L. 1862. *The catalogue of the South African Museum. Part I.* Published at the Cape Argus office, Cape Town. 83 pp.

Layne, J.N. 1965. Observations on marine mammals in Florida waters. *Bulletin of the Florida State Museum* **9**(4): 131–181.

Le Roux, J.-P., Best, P.B., and Stander, P.E. 2001. Sightings of southern right whales (*Eubalaena australis*) in Namibian waters, 1971–1999. *Journal of Cetacean Research and Management* (Special Issue) **2**: 181–185.

Leaper, R., Gillespie, D., and Papastavrou, V. 2000. Results of passive acoustic surveys for odontocetes in the Southern Ocean. *Journal of Cetacean Research and Management* **2**(3): 187–196.

Leaper, R., and Scheidat, M. 1998. An acoustic survey for cetaceans in the Southern Ocean Sanctuary conducted from the German government research vessel *Polarstern*. *Report of the International Whaling Commission* **48**: 431–437.

Leatherwood, S. 1986. Whales, dolphins and porpoises of the Indian Ocean Cetacean Sanctuary. A catalogue of available information. *Hubbs Marine Research Center Technical Report* 87–197. 207 pp.

Leatherwood, S., Grove, J.S., and Zuckerman, A.E. 1991a. Dolphins of the genus *Lagenorhynchus* in the tropical South Pacific. *Marine Mammal Science* **7**(2): 194–197.

Leatherwood, S., Kastelein, R.A., and Hammond, P.S. 1988. Estimate of numbers of Commerson's dolphins in a portion of the northeastern Strait of Magellan, January–February 1984. *Report of the International Whaling Commission* (Special Issue) **9**: 93–102.

Leatherwood, S., and Ljungblad, D.K. 1979. Nighttime swimming and diving behavior of a radio-tagged spotted dolphin, *Stenella attenuata*. *Cetology* **34**: 1–6.

Leatherwood, S., McDonald, D., Prematunga, W.P., Girton, P., Bangakoon, A., and McBrearty, D. 1991b. Records of the 'blackfish' (killer, false killer, pilot, pygmy killer and melon-headed whales) in the Indian Ocean, 1772–1986. UNEP *Marine Mammal Technical Report* **3**: 33–65.

Leatherwood, S., Perrin, W.F., Kirby, V.L., Hubbs, C.L., and Dahlheim, M. 1980. Distribution and movements of Risso's dolphin, *Grampus griseus*, in the Eastern North Pacific. *Fishery Bulletin, US* **77**(4): 951–963.

Leatherwood, S., Reeves, R.R., Perrin, W.F., and Evans, W.E. 1982. Whales, dolphins and porpoises of the Eastern North Pacific and adjacent Arctic waters. *NOAA Technical Report NMFS Circular* **444**: 1–245.

Leatherwood, S., Thomas, J.A., and Awbrey, F.T. 1981. Minke whales off northwestern Ross Island. *Antarctic Journal* **16**: 154–156.

Leatherwood, S., and Walker, W.A. 1979. The northern right whale dolphin *Lissodelphis borealis* Peale in the eastern North Pacific. Pp 85–141 in (Winn, H.E., and Olla, B.L., eds) *Behavior of Marine Animals. Current Perspectives in Research. 3. Cetaceans.* Plenum Press, New York.

LeDuc, R.G., Perrin, W.F., and Dizon, A.E. 1999. Phylogenetic relationships among the delphinid cetaceans based on full cytochrome *b* sequences. *Marine Mammal Science* **15**(3): 619–648.

Leighton, T.G., Richards, S.D., and White, P.R. 2004. Trapped within a 'wall of sound'. A possible mechanism for the bubble nets of humpback whales. *Acoustics Bulletin* **29**(1): 24–29.

Lescrauwaet, A.C., Gibbons, J., Guzman, L., and Schiavini, A. 2000. Abundance estimation of Commerson's dolphin in the eastern area of the Strait of Magellan – Chile. *Revista Chilena de Historia Natural* **73**(3): 473–478.

Leslie, M.S., Perrin, W.F., and Rosenbaum, H.C. 2005. Evaluating cetacean hybrids: the utility of multiple sources of data to investigate the origin of an intermediate species, Stenella clymene. 16th Biennial Conference on the Biology of Marine Mammals, San Diego, California, December 12–16, 2005. Abstracts: **166**.

Leung, Y.-M. 1967. An illustrated key to the species of whale-lice (Amphipoda, Cyamidae), ectoparasites of cetacea, with a guide to the literature. *Crustaceana* **12**(3): 279–291.

Lichter, A.A. 1986. Records of beaked whales (Ziphiidae) from the western South Atlantic. *Scientific Reports of the Whales Research Institute, Tokyo* **37**: 109–127.

Lichter, A.A., Fraga, F., and Castello, H.P. 1990. First record of the pygmy killer whale, *Feresa attenuata*, in the southwest Atlantic. *Marine Mammal Science* **6**(1): 85–86.

Lillie, D.G. 1915. Cetacea. *British Antarctic Terra Nova Expedition, 1910. Natural History Reports. (Zoology)* **1**(3): 85–124 + 8 pl.

Lincoln, R.J., and Hurley, D.E. 1981. A new species of the whale-louse *Syncyamus* (Crustacea: Amphipoda: Cyamidae) ectoparasitic on dolphins from South Africa. *Annals of the Cape Provincial Museums (Natural History)* **13**(13):187–194.

Ljungblad, D.K., Clark, C.W., and Shimada, H. 1998. A comparison of sounds attributed to pygmy blue whales (*Balaenoptera musculus brevicauda*) recorded south of the Madagascar plateau and those attributed to 'true' blue whales (*Balaenoptera musculus*) recorded off Antarctica *Report of the International Whaling Commission* **48**: 439–442.

Llido, J., Garçon, V., Lutjeharms, J.R.E., and Sudre, J. 2005. Event-scale blooms drive enhanced primary productivity at the Subtropical Convergence. *Geophysical Research Letters* **32**(15): L5611, doi. 10.1029/2005GL022880.

Lockerby, W. 1925. *The journal of William Lockerby sandalwood trader in the Fijian islands during the years 1808–1809* (Im Thurn, E., and Wharton, L.C., eds) London. The Hakluyt Society 2nd ser. 52. 250 pp.

Lockyer, C. 1976. Body weights of some species of large whales. *Journal du Conseil pour l'Exploration de la Mer* **36**(3): 259–273.

Lockyer, C. 1977. Some estimates of growth in the sei whale, *Balaenoptera borealis*. *Report of the International Whaling Commission* (Special Issue) **1**: 58–62.

Lockyer, C. 1978. A theoretical approach to the balance between growth and food consumption in fin and sei whales, with special reference to the female reproductive cycle. *Report of the International Whaling Commission* **28**: 243–249.

Lockyer, C. 1981a. Growth and energy budgets of large baleen whales from the Southern Hemisphere. Pp 379–487 in Mammals in the Seas 3. General papers and large cetaceans. *FAO Fisheries Series* (5) 3, 504 pp.

Lockyer, C. 1981b. Estimates of growth and energy budget for the sperm whale, *Physeter catodon*. Pp 489–504 in Mammals in the Seas 3. General papers and large cetaceans. *FAO Fisheries Series* (5) 3, 504 pp.

Lockyer, C. 1981c. Estimation of the energy costs of growth, maintenance and reproduction in the female minke whale, (*Balaenoptera acutorostrata*), from the Southern Hemisphere. *Report of the International Whaling Commission* **31**: 337–343.

Lockyer, C. 1990. Review of incidents involving wild, sociable dolphins, worldwide. Pp 337–353 in (Leatherwood, S., and Reeves, R.R., eds) *The bottlenose dolphin*. Academic Press, San Diego.

Lockyer, C. 1993. Seasonal changes in body fat condition of Northeast Atlantic pilot whales, and their biological significance. *Report of the International Whaling Commission* (Special Issue) **14**: 325–350.

Lockyer, C., Goodall, R.N.P., and Galeazzi, A.R. 1988. Age and body length characteristics of *Cephalorhynchus commersonii* from incidentally-caught specimens off Tierra del Fuego. *Report of the International Whaling Commission* (Special Issue) **9**: 103–118.

Lodi, L., and Hetzel, B. 1999. Rough-toothed dolphin, *Steno bredanensis*, feeding behaviors in Ilha Grande Bay, Brazil. *Biociências* **7**(1): 29–42.

Lodi, L., Siciliano, S., and Capistrano, L. 1990. Mass stranding of *Peponocephala electra* (Cetacea, Globicephalinae) on Piracanga Beach, Bahia, northeastern Brazil. *Scientific Reports of Cetacean Research* **1**: 79–84.

Longhurst, A.R. 2006. *Ecological geography of the sea*. Elsevier, Amsterdam. Second edition. 560 pp.

Lönnberg, E. 1906. Contributions to the fauna of South Georgia. 1. Taxonomic and biological notes on vertebrates. *Kungliga Svenska Vetenskapsakademiens Handlingar* **40**(5): 1–104 + 12 pl.

Lutjeharms, J.R.E. 1985. Location of frontal systems between Africa and Antarctica: some preliminary results. *Deep-Sea Research* **32**: 1 499–1 509.

Lutjeharms, J.R.E. 2001. The Agulhas Current. Pp. 104–113 in (Steele, J., Thorpe, S., and Turekian, K., eds) *Encyclopedia of Ocean Science*. Volume 1, Academic Press, London.

Lutjeharms, J.R.E. 2006a. The coastal oceans of south-eastern Africa. Pp. 783–834 in (Robinson, A.R., and Brink, K.H., eds) *The Sea*. Volume 14B, Harvard University Press, Cambridge, MA, USA.

Lutjeharms J.R.E. 2006b. *The Agulhas Current*. Springer, Heidelberg. 329 pp.

Lutjeharms, J.R.E., and Ansorge, I.J. 2001. The Agulhas Return Current. *Journal of Marine Systems* **30**(1/2): 115–138.

Lutjeharms, J.R.E., Catzel, R., and Valentine, H.R. 1989. Eddies and other border phenomena of the Agulhas Current. *Continental Shelf Research* **9**(7): 597–616.

Lutjeharms, J.R.E., and Cooper, J. 1996. Interbasin leakage through Agulhas Current filaments. *Deep-Sea Research I* **43**(2): 213–238.

Lutjeharms, J.R.E., Cooper, J., and Roberts, M. 2000b. Upwelling at the inshore edge of the Agulhas Current. *Continental Shelf Research* **20**(7): 737–761.

Lutjeharms, J.R.E., and Jorge da Silva, A. 1988. The Delagoa Bight eddy. *Deep-Sea Research* **35**(4): 619–634.

Lutjeharms, J.R.E., and Machu, E. 2000. An upwelling cell inshore of the East Madagascar Current. *Deep-Sea Research I* **47**(12): 2 405–2 411.

Lutjeharms, J.R.E., and Meeuwis, J.M. 1987. The extent and variability of South-East Atlantic upwelling. *South African Journal of Marine Science* **5**: 51–62.

Lutjeharms, J.R.E., and Roberts, H.R. 1988. The Natal pulse; an extreme transient on the Agulhas Current. *Journal of Geophysical Research* **93**(C1): 631–645.

Lutjeharms, J.R.E., and Stockton, P.L. 1987. Kinematics of the upwelling front off southern Africa. *South African Journal of Marine Science* **5**: 35–50.

Lutjeharms, J.R.E., and Valentine, H.R. 1984. Southern Ocean thermal fronts south of Africa. *Deep-Sea Research* **31**: 1 461–1 476.

Lutjeharms, J.R.E., Valentine, H.R., and Van Ballegooyen, R.C. 2000a. The hydrography and water masses of the Natal Bight, South Africa. *Continental Shelf Research* **20**(14): 1 907–1 939.

Lutjeharms, J.R.E., and Van Ballegooyen, R.C. 1988. The retroflection of the Agulhas Current. *Journal of Physical Oceanography* **18**(12): 1 570–1 583.

Lynas, E.M., and Sylvestre, J.P. 1988. Feeding techniques and foraging strategies of minke whales (*Balaenoptera acutorostrata*) in the St Lawrence River estuary. *Aquatic Mammals* **14**(1): 21–32.

M

Mackintosh, N.A. 1942. The southern stocks of whalebone whales. *Discovery Reports* **22**: 197–300.

Mackintosh, N.A. 1966. The distribution of southern blue and fin whales. Pp 125–144 in (Norris, K.S., ed.) *Whales, Dolphins and Porpoises*. University of California Press, Berkeley.

Mackintosh, N.A., and Wheeler, J.F.G. 1929. Southern blue and fin whales. *Discovery Reports* **1**: 257–540.

MacLeod, C.D. 1998. Intraspecific scarring in odontocete cetaceans: an indicator of male 'quality' in aggressive social interactions? *Journal of Zoology, London* **244**: 71–77.

MacLeod, C.D. 2000. Review of the distribution of *Mesoplodon* species (order Cetacea, family Ziphiidae) in the North Atlantic. *Mammal Review* **30** (1): 1–8.

MacLeod, C.D. 2002. Possible functions of the ultradense bone in the rostrum of Blainville's beaked whale (*Mesoplodon densirostris*). *Canadian Journal of Zoology* **80**: 178–184.

MacLeod, C.D. 2006. How big is a beaked whale? A review of body length and sexual size dimorphism in the family Ziphiidae. *Journal of Cetacean Research and Management* **7**(3): 301–308.

MacLeod, C.D., and Bennett, E. 2007. Pan-tropical spotted dolphins (*Stenella attenuata*) and other cetaceans around St Helena in the tropical south-eastern Atlantic. *Journal of the Marine Biological Association of the UK* **87**(1): 339–344.

MacLeod, C.D., and D'Amico, A. 2006. A review of beaked whale behaviour and ecology in relation to assessing and mitigating impacts of anthropogenic noise. *Journal of Cetacean Research and Management* **7**(3): 211–221.

MacLeod, C.D., Hauser, N., and Peckham, H. 2004. Diversity, relative density and structure of the cetacean community in summer months east of Great Abaco, Bahamas. *Journal of the Marine Biological Association of the UK* **84**: 469–474.

MacLeod, C.D., Perrin, W.F., Pitman, R. *et al.* 2006. Known and inferred distributions of beaked whale species (Cetacea: Ziphiidae). *Journal of Cetacean Research and Management* **7**(3): 271–286.

MacLeod, C.D., Santos, M.B., and Pierce, G.J. 2003. Review of data on diets of beaked whales: evidence of niche separation and geographic segregation. *Journal of the Marine Biological Association of the UK* **83**: 651–665.

MacLeod, C.D., and Zuur, A.F. 2005. Habitat utilization by Blainville's beaked whales off Great Abaco, northern Bahamas, in relation to seabed topography. *Marine Biology* **147**: 1–11.

Madsen, P.T., Johnson, M., De Soto, N.A., Zimmer, W.M.X., and Tyack, P. 2005. Biosonar performance of foraging beaked whales (*Mesoplodon densirostris*). *Journal of Experimental Biology* **208**: 181–194.

Madsen, P.T., Kerr, I., and Payne, R. 2004a. Echolocation clicks of two free-ranging, oceanic delphinids with different food preferences: false killer whales *Pseudorca crassidens* and Risso's dolphins *Grampus griseus*. *Journal of Experimental Biology* **207**: 1 811–1 823.

Madsen, P.T., Kerr, I., and Payne, R. 2004b. Source parameter estimates of echolocation clicks from wild pygmy killer whales (*Feresa attenuata*) (L). *Journal of the Acoustical Society of America* **116**(4) Pt 1: 1 909–1 912.

Mahnken, T., and Gilmore, R.M. 1960. Suckerfish on porpoise. *Journal of Mammalogy* **41**(1): 134.

Maia-Nogueira, R., and Nunes, J.A.C.C. 2005. Record of the Layard's beaked whale, *Mesoplodon layardii* (Gray, 1856), in northeastern Brazil. *Latin American Journal of Aquatic Mammals* **4**(2): 137–139.

Maigret, J. 1980. Données nouvelles sur l'écologie du *Sousa teuszii* (Cetacea: Delphinidae) de la côte ouest africaine. *Bulletin de l'Institut Français d'Afrique Noire* **42A**(3): 619–633.

Mangel, M. 1993. Effects of high-seas driftnet fisheries on the northern right whale dolphin *Lissodelphis borealis*. *Ecological Applications* **3**(2): 221–229.

Mann, J., Connor, R.C., Barre, L.M., and Heithaus, M.R. 2000. Female reproductive success in bottlenose dolphins (*Tursiops* sp.): life history, habitat, provisioning, and group-size effects. *Behavioural Ecology* **11**(2): 210–219.

Mann, J., and Smuts, B. 1999. Behavioral development in wild bottlenose dolphin newborns (*Tursiops* sp). *Behaviour* **136**: 529–566.

Manzanilla, S.R. 1989. The 1982–1983 El Niño event recorded in dentinal growth layers in teeth of Peruvian dusky dolphins (*Lagenorhynchus obscurus*). *Canadian Journal of Zoology* **67**: 2 120–2 125.

Marine Conservation Society Seychelles. 2004. Cetacean sightings around the inner Islands. *Marine Conservation News* **2**(1): 1–2.

Markowitz, T.M. 2004. *Social organization of the New Zealand dusky dolphin*. Ph D thesis, Texas A and M University.

Markowitz, T.M., Harlin, A.D., Würsig, B., and McFadden, C.J. 2004. Dusky dolphin foraging habitat: an overlap with aquaculture in New Zealand. *Aquatic Conservation: Marine and Freshwater Ecosystems* **14**: 133–149.

Marloth, R. 1900. Notes on the mode of growth of *Tubicinella trachealis*, the barnacle of the southern right whale. *Transactions of the South African Philosophical Society*. **11**: 1–6.

Marsh, H., and Kasuya, T. 1984. Changes in the ovaries of the short-finned pilot whale, *Globicephala macrorhynchus*, with age and reproductive activity. *Report of the International Whaling Commission* (Special Issue) **6**: 311–335.

Marsh, H., and Kasuya, T. 1986. Evidence for reproductive senescence in female cetaceans. *Report of the International Whaling Commission* (Special Issue) **8**: 57–74.

Marten, K. 2000. Ultrasonic analysis of pygmy sperm whale (*Kogia breviceps*) and beaked whale (*Mesoplodon carlhubbsi*) clicks. *Aquatic Mammals* **26**: 45–48.

Martin, A.R., and Rothery, P. 1993. Reproductive parameters of female long-finned pilot whales (*Globicephala melas*) around the Faroe Islands. *Report of the International Whaling Commission* (Special Issue) **14**: 263–304.

Martin, J.W., and Heyning, J.E. 1999. First record of *Isocyamus kogiae* Sedlak-Weinstein, 1992 (Crustacea, Amphipoda, Cyamidae) from the eastern Pacific, with comments on morphological characters, a key to the genera of the Cyamidae, and a checklist of cyamids and their hosts. *Bulletin of the Southern California Academy of Science* **98**(1): 26–38.

Martinez, D.R., and Klinghammer, E. 1970. The behavior of the whale *Orcinus orca*: a review of the literature. *Zeitschrift für Tierpsychologie* **27**: 829–839.

Martins, A.M.A., Alves Jr, T.T., Furtado Neto, M.A.A., and Lien, J. 2004. The most northern record of Gervais' beaked whale, *Mesoplodon europaeus* (Gervais, 1855), for the Southern Hemisphere. *Latin American Journal of Aquatic Mammals* **3**(2): 151–155.

Martuscelli, P., Milanelo, M., and Olmos, F. 1995. First record of Arnoux's beaked whale (*Berardius arnuxii*) and southern right-whale dolphin (*Lissodelphis peronii*) from Brazil. *Mammalia* **59**(2): 274–275.

Mate, B.R., Lagerquist, B.A., and Calambokidis, J. 1999. Movements of North Pacific blue whales during the feeding season off southern California and their southern fall migration. *Marine Mammal Science* **15**(4): 1 246–1 257.

Mate, B.R., Lagerquist, B.A., Winsor, M., Geraci, J., and Prescott, J.H. 2005. Movements and dive habits of a satellite-monitored longfinned pilot whale (*Globicephala melas*) in the northwest Atlantic. *Marine Mammal Science* **21**(1): 136–144.

Mate, B.R., Rossbach, K.A., Nieukirk, S.L., Wells, R.S., Irvine, A.B., Scott, M.D., and Read, A.J. 1995. Satellite-monitored movements and dive behaviour of a bottlenose dolphin (*Tursiops truncatus*) in Tampa Bay, Florida. *Marine Mammal Science* **11**(4): 452–463.

Matsuoka, K., Fujise, Y., and Pastene, L.A. 1996. A sighting of a large school of the pygmy right whale, *Caperea marginata*, in the southeast Indian Ocean. *Marine Mammal Science* **12**(4): 594–597.

Matsuoka, K., Pitman, R.L., and Marquez, F.F.C. 2005. A note on a pygmy right whale (*Caperea marginata*) sighting in the southwestern Pacific Ocean. *Journal of Cetacean Research and Management* **7**(1): 71–73.

Matsushita, T. 1955. Daily rhythmic activity of the sperm whales in the Antarctic Ocean. *Bulletin of the Japanese Society of Scientific Fisheries* **20**(9): 770–773.

Matthews, L.H. 1938a. The humpback whale, *Megaptera nodosa*. *Discovery Reports* **17**: 7–92.

Matthews, L.H. 1938b. The sperm whale, *Physeter catodon*. *Discovery Reports* **17**: 93–168.

Matthews, L.H. 1938c. Notes on the southern right whale, *Eubalaena australis*. *Discovery Reports* **17**: 169–182.

Matthews, L.H. 1938d. The sei whale, *Balaenoptera borealis*. *Discovery Reports* **17**: 183–290 + 2 pl.

Mattila, D., Bérubé, M., Bowman, R. *et al.* 2001. Humpback whale habitat use on the West Indies breeding grounds. Paper SC53/NAH3 submitted to the Scientific Committee of the International Whaling Commission, London, July 2001.

Maxwell, C. 2002. The Natal sardine run: 'sardine fever'. [Online]. Available: http://www.underwatervideo.co.za/dive_locations/south_africa/sardine_run/natal_sardine_run.asp [August 2006].

Mazeika, P.A. 1967. Thermal domes in the eastern Atlantic Ocean. *Limnology and Oceanography* **12**: 537–539.

McAlpine, D.F., and Murison, L.D. 1997. New records for the pygmy sperm whale, *Kogia breviceps* (Physeteridae) from Atlantic Canada with notes on diet and parasites. *Marine Mammal Science* **13**(4): 701–704.

McAuliffe, K., and Whitehead, H. 2005. Eusociality, menopause and information in matrilineal whales. *Trends in Ecology and Evolution* **20**(12): 650.

McCann, C. 1963. Occurrence of Blainville's beaked-whale [*Mesoplodon densirostris* (Blainville)] in the Indian Ocean. *Journal of the Bombay Natural History Society* **60**(3): 727–731 + 1 pl.

McCann, C. 1975. A study of the genus *Berardius* Duvernoy. *Scientific Reports of the Whales Research Institute, Tokyo* **27**: 111–137.

McCann, C., and Talbot, F.H. 1963. The occurrence of True's beaked whale (*Mesoplodon mirus* True) in South African waters, with a key to South African species of the genus. *Proceedings of the Linnean Society of London* **175**(2): 137–144 + 6 pl.

McCowan, B., and Reiss, D. 2001. The fallacy of 'signature whistles' in bottlenose dolphins: a comparative perspective of 'signature information' in animal vocalizations. *Animal Behaviour* **62**: 1 151–1 162.

McDonald, M.A. 2006. An acoustic survey of baleen whales off Great Barrier Island, New Zealand. *New Zealand Journal of Marine and Freshwater Research* **40**: 519–529.

McDonald, M.A., Hildebrand, J.A., Wiggins, S.M., Thiele, D., Glasgow, D., and Moore, S.E. 2005. Sei whale sounds recorded in the Antarctic. *Journal of the Acoustical Society of America* **118**(6): 3 941–3 945.

McDonald, M.A., Mesnick, S.L., and Hildebrand, J.A. 2006. Biogeographic characterisation of blue whale song worldwide: using song to identify populations. *Journal of Cetacean Research and Management* **8**(1): 55–65.

Mead, J.G. 1981. First records of *Mesoplodon hectori* (Ziphiidae) from the Northern Hemisphere and a description of the adult male. *Journal of Mammalogy* **62**(2): 430–432.

Mead, J.G. 1984. Survey of reproductive data for the beaked whales (Ziphiidae). *Report of the International Whaling Commission* (Special Issue) **6**: 91–96.

Mead, J.G. 1989a. Shepherd's beaked whale *Tasmacetus shepherdi* Oliver, 1937. Pp 309–320 in (Ridgway, S.H., and Harrison, R.J., eds). *Handbook of Marine Mammals 4: River dolphins and the larger toothed whales*. Academic Press, London.

Mead, J.G. 1989b. Bottlenose whales *Hyperoodon ampullatus* (Forster, 1770) and *Hyperoodon planifrons* Flower, 1882. Pp 321–348 in (Ridgway, S.H., and Harrison, R.J., eds) *Handbook of Marine Mammals 4. River dolphins and the larger toothed whales*. Academic Press, London.

Mead, J.G. 1989c. Beaked whales of the genus *Mesoplodon*. Pp 349–430 in (Ridgway, S.H., and Harrison, R.J., eds) *Handbook of Marine Mammals 4: River dolphins and the larger toothed whales*. Academic Press, London.

Mead, J.G. 2002. Shepherd's beaked whale *Tasmacetus shepherdi*. Pp 1 078–1 081 in (Perrin, W.F., Würsig, B., and Thewissen, J.G.M., eds) *Encyclopedia of Marine Mammals*. Academic Press, San Diego.

Mead, J.G., and Baker, A.N. 1987. Notes on the rare beaked whale, *Mesoplodon hectori* (Gray). *Journal of the Royal Society of New Zealand* **17**(3): 303–312.

Mead, J.G., Odell, D.K., Wells, R.S., and Scott, M.D. 1980. Observations on a mass stranding of spinner dolphin, *Stenella longirostris*, from the west coast of Florida. *Fishery Bulletin, US* **78**(2): 353–360.

Mead, J.G., and Payne, R.S. 1975. A specimen of the Tasman beaked whale, *Tasmacetus shepherdi*, from Argentina. *Journal of Mammalogy* **56**(1): 213–218.

Mead, J.G., and Potter, C.W. 1995. Recognizing two populations of the bottlenose dolphin (*Tursiops truncatus*) off the Atlantic coast of North America – morphologic and ecologic considerations. *International Marine Biological Research Institute Reports* **5**: 31–44.

Meeuwis, J.M., and Lutjeharms, J.R.E. 1990. Surface thermal characteristics of the Angola–Benguela front. *South African Journal of Marine Science* **9**: 261–280.

Melville, H. 1851. *Moby-Dick; or, the whale*. Harper and Bros, New York. 634 pp.

Mendolia, C. 1989. *Reproductive biology of common dolphins (*Delphinus delphis*, Linnaeus) off the south east coast of southern Africa*. MSc thesis, University of Port Elizabeth, 111 pp.

Mercado (III), E., Frazer, L.N., and Herman, L.M. 1996. Humpback whale sonar. *Journal of the Acoustical Society of America* **100**(4): 2 644.

Michel, C., and Van Bree, P.J.H. 1976. On two strandings of the beaked whale *Mesoplodon densirostris* (de Blainville, 1817) on Mauritius. *Zeitschrift für Säugetierkunde* **41**: 194–196.

Mignucci-Giannoni, A.A., Montoya-Ospina, R., Pérez-Zayas, J.J., Rodgríguez-López, M.A., and Williams Jr, E.H. 1999a. New records of Fraser's dolphin (*Lagenodelphis hosei*) for the Caribbean. *Aquatic Mammals* **25**(1): 15–19.

Mignucci-Giannoni, A.A., Toyos-González, G.M., Pérez-Padilla, J., Rodríguez-López, M.A., and Overing, J. 1999b. Mass stranding of pygmy killer whales (*Feresa attenuata*) in the British Virgin Islands. *Journal of the Marine Biological Association of the UK* **80**: 759–760.

Mikhalev, Y.A. 2000. Whaling in the Arabian Sea by the whaling fleets *Slava* and *Sovetskaya Ukraina*. Pp 141–181 in (Yablokov, A.V., and Zemsky, V.A., eds) *Soviet whaling data (1949–1979)*. Centre for Russian Environmental Policy, Moscow.

Mikhalev, Y.A., Ivashin, M.V., Savusin, V.P., and Zelenaya, F.E. 1981. The distribution and biology of killer whales in the

Southern Hemisphere. *Report of the International Whaling Commission* **31**: 551–566.

Miklosovic, D.S., Murray, M.M., Howle, L.E., and Fish, F.E. 2004. Leading-edge tubercles delay stall on humpback whale (*Megaptera novaeangliae*) flippers. *Physics of Fluids* **16**(5): L39–L42.

Miles, J.A., and Herzing, D.L. 2003. Underwater analysis of the behavioural development of free-ranging Atlantic spotted dolphin (*Stenella frontalis*) calves (birth to 4 years of age). *Aquatic Mammals* **29**(3): 363–377.

Miller, P.J.O., Johnson, M.P., and Tyack, P.L. 2004a. Sperm whale behaviour indicates the use of echolocation click buzzes 'creaks' in prey capture. *Proceedings of the Royal Society of London.* B **271**: 2 239–2 247.

Miller, P.J.O., Johnson, M.P., Tyack, P.L., and Terray, E.A. 2004b. Swimming gaits, passive drag and buoyancy of diving sperm whales *Physeter macrocephalus*. *Journal of Experimental Biology* **207**: 1 953–1 967.

Minamikawa, S., Iwasaki, T., Noji, S. *et al.* 2003. Diurnal pattern in diving behaviour of a striped dolphin. *Abstracts of the 22nd Annual Meeting of the Ethological Society of Japan*, October 2003: 86.

Mitchell, E.D. 1975. Porpoise, dolphin and small whale fisheries of the world. *IUCN Monograph* **3**: 1–129.

Miyazaki, N. 1977. Growth and reproduction of *Stenella coeruleoalba* off the Pacific coast of Japan. *Scientific Reports of the Whales Research Institute, Toyko* **29**: 21–48.

Miyazaki, N. 1980. Preliminary note on age determination and growth of the rough-toothed dolphin, *Steno bredanensis*, off the Pacific coast of Japan. *Report of the International Whaling Commission* (Special Issue) **3**: 171–179.

Miyazaki, N. 1984. Further analyses of reproduction in the striped dolphin, *Stenella coeruleoalba*, off the Pacific coast of Japan. *Report of the International Whaling Commission* (Special Issue) **6**: 343–353.

Miyazaki, N. 1989. Notes on the school composition of killer whales in the Southern Hemisphere. *Bulletin of the National Science Museum, Tokyo* Series A **15**(1): 53–59.

Miyazaki, N., Fujise, Y., and Iwata, K. 1998. Biological analysis of a mass stranding of melon-headed whales (*Peponocephala electra*) at Aoshima, Japan. *Bulletin of the National Science Museum, Tokyo* Series A **24**(1): 31–60.

Miyazaki, N., Kasuya, T., and Nishiwaki, M. 1974. Distribution and migration of two species of *Stenella* in the Pacific coast of Japan. *Scientific Reports of the Whales Research Institute, Tokyo* **26**: 227–243.

Miyazaki, N., and Kato, H. 1988. Sighting records of small cetaceans in the Southern Hemisphere. *Bulletin of the National Science Museum, Tokyo* Series A **14**(1): 47–65.

Miyazaki, N., and Nishiwaki, M. 1978. School structure of the striped dolphin off the Pacific coast of Japan. *Scientific Reports of the Whales Research Institute, Tokyo* **30**: 65–115.

Miyazaki, N., and Perrin, W.F. 1994. Rough-toothed dolphin *Steno bredanensis* (Lesson, 1828). Pp 1–21 in (Ridgway, S.H., and Harrison, R.J., eds) *Handbook of Marine Mammals 5: The first book of dolphins*. Academic Press, New York.

Miyazaki, N., and Wada, S. 1978a. Observation of Cetacea during whale marking cruise in the western tropical Pacific, 1976. *Scientific Reports of the Whales Research Institute, Tokyo* **30**: 179–195.

Miyazaki, N., and Wada, S. 1978b. Fraser's dolphin, *Lagenodelphis hosei* in the western North Pacific. *Scientific Reports of the Whales Research Institute, Tokyo* **30**: 231–244.

Mizroch, S.A., and York, A.E. 1984. Have pregnancy rates of southern hemisphere fin whales, *Balaenoptera physalus*, increased? *Report of the International Whaling Commission* (Special Issue) **6**: 401–410.

Mizue, K., and Yoshida, K. 1962. Studies on the little toothed whales in the west sea area of Kyusyu. VIII. About *Grampus griseus* caught in Goto Is., Nagasaki Pref. *Bulletin of the Faculty of Fisheries Nagasaki University* **12**: 45–52 + 2 pl.

Monteiro Neto, C., Alves Jr, T.T., Avila, F.J.C., Campos, A.A., Costa, A.F., Silva, C.P.N., and Furtado Neto, M.A.A. 2000. Impact of fisheries on the tucuxi (*Sotalia fluviatilis*) and rough-toothed dolphin (*Steno bredanensis*) populations off Ceara State, northeastern Brazil. *Aquatic Mammals* **26**(1): 49–56.

Moore, J.C. 1960. New records of the Gulf-stream beaked whale, *Mesoplodon gervaisi*, and some taxonomic considerations. *American Museum Novitates* **1993**: 1–35.

Moore, J.C. 1966. Diagnoses and distributions of beaked whales of the genus *Mesoplodon* known from North American waters. Pp 32–61 in (Norris, K.S., ed.) *Whales, dolphins and porpoises*. University of California Press, Berkeley.

Moore, J.C., and Gilmore, R.M. 1965. A beaked whale new to the Western Hemisphere. *Nature, London* **205**(4 977): 1 239–1 240.

Moore, J.C., and Wood Jr, F.G. 1957. Differences between the beaked whales *Mesoplodon mirus* and *Mesoplodon gervaisi*. *American Museum Novitates* **1831**: 1–25.

Moore, M.J., and Early, G.A. 2004. Cumulative sperm whale bone damage and the bends. *Science* **306**(5705): 2 215.

Moore, M.J., Knowlton, A.R., Kraus, S.D., McLellan, W.A., and Bonde, R.K. 2004. Morphometry, gross morphology and available histopathology in North Atlantic right whale (*Eubalaena glacialis*) mortalities (1970–2002). *Journal of Cetacean Research and Management* **6**(3): 199–214.

Moore, S.E. 1990. Spotted dolphin bioacoustics. *NOAA Southwest Fisheries Science Center Administrative Report* LJ-90-24C. 61 pp.

Morell, V. 2007. Killing whales for science? *Science* **316**(5824): 532–534.

Moreno, F.P. 1895. Nota sobre los restos de Hyperoodontes conservados en el Museo de la Plata. *Anales de Museo de la Plata* (Zool) **3**: 3–8 + 1 pl.

Moreno, I.B., Zerbini, A.N., Danilewicz, D., Santos, M.C. de O., Simões-Lopes, P.C., Lailson-Brito Jr, J., and Azevedo, A.F. 2005. Distribution and habitat characteristics of dolphins of the genus *Stenella* (Cetacea: Delphinidae) in the southwest Atlantic Ocean. *Marine Ecology Progress Series* **300**: 229–240.

Morimitsu, T., Nagai, T., Ide, M., Kawano, H., Naichuu, A., Koono, M., and Ishii, A. 1987. Mass stranding of odontoceti caused by parasitogenic eighth cranial neuropathy. *Journal of Wildlife Diseases* **23**(4): 586–590.

Morisaka, T., Shinohara, M., Nakahara, F., and Akamatsu, T. 2005. Geographic variations in the whistles among three Indo-Pacific bottlenose dolphin *Tursiops aduncus* populations in Japan. *Fisheries Science* **71**: 568–576.

Moroshkin, K.V., Bunov, V.A., and Bulatov, R.P. 1970. Water circulation in the eastern South Atlantic Ocean. *Oceanology* **10**: 27–34.

Morris, R.A., and Mowbray, L.S. 1966. An unusual barnacle attachment on the teeth of the Hawaiian spinning dolphin. *Norsk Hvalfangsttidende* **55**(1): 15–16.

Mörzer Bruyns, W.F.J. 1971. *Field guide of whales and dolphins*. C.A. Mees, Amsterdam, 258 pp.

MRI Whale Unit data. Data (published and unpublished) collected or compiled by staff at the Whale Unit, Mammal Research Institute,

University of Pretoria, C/o Iziko South African Museum, P.O. Box 61, Cape Town, 8000, South Africa.

Mullin, K.D., Higgins, L.V., Jefferson, T.A., and Hansen, L.J. 1994a. Sightings of the clymene dolphin (*Stenella clymene*) in the Gulf of Mexico. *Marine Mammal Science* **10**(4): 464–470.

Mullin, K.D., Hoggard, W., Roden, C.L., Lohoefener, R.R., Rogers, C.M., and Taggart, B. 1994b. Cetaceans on the upper continental slope in the north-central Gulf of Mexico. *Fishery Bulletin, US* **92**(4): 773–786.

Mullin, K.D., Hoggard, W., and Hansen, L.J. 2004. Abundance and seasonal occurrence of cetaceans in outer continental shelf and slope waters of the North-Central and Northwestern Gulf of Mexico. *Gulf of Mexico Science* **22**(1): 62–73.

Mullin, K.D., Jefferson, T.A., Hansen, L.J., and Hoggard, W. 1994c. First sightings of melon-headed whales (*Peponocephala electra*) in the Gulf of Mexico. *Marine Mammal Science* **10**(3): 342–348.

Murphy, R.C. 1947. *Logbook for Grace. Whaling brig Daisy, 1912–1913.* The Macmillan Co., New York. 290 pp.

Murphy, S., Collet, A., and Rogan, E. 2005. Mating strategy in the male common dolphin (*Delphinus delphis*): what gonadal analysis tells us. *Journal of Mammalogy* **86**(6): 1 247–1 258.

Murphy, S., Herman, J.S., Pierce, G.J., Rogan, E., and Kitchener, A.C. 2006. Taxonomic status and geographical cranial variation of common dolphins (*Delphinus*) in the eastern North Atlantic. *Marine Mammal Science* **22**(3): 573–599.

Murray, S.O., Mercado, E., and Roitblat, H.L. 1998. Characterizing the graded structure of false killer whale (*Pseudorca crassidens*) vocalizations. *Journal of the Acoustical Society of America* **104**(3): 1 679–1 688.

Mussi, B., Miragliuolo, A., and Bearzi, G. 2002. Short-beaked common dolphins around the island of Ischia, Italy (southern Tyrrhenian Sea). *European Research on Cetaceans 16*, 3 pp.

Myrick, A.C., Hohn, A.A., Barlow, J., and Sloan, P.A. 1986. Reproductive biology of female spotted dolphins, *Stenella attenuata*, from the eastern tropical Pacific. *Fishery Bulletin, US* **84**(2): 247–259.

N

Nachtigall, P.E., Pawloski, J.L., Schroeder, J.P., and Sinclair, S. 1990. Successful maintenance and research with a formerly stranded Risso's dolphin (*Grampus griseus*). *Aquatic Mammals* **16**(1): 8–13.

Nachtigall, P.E., Yuen, M.M.L., Mooney, T.A., and Taylor, K.A. 2005. Hearing measurements from a stranded infant Risso's dolphin, *Grampus griseus*. *Journal of Experimental Biology* **208**: 4 181–4 188.

Nagorsen, D. 1985. *Kogia simus*. *Mammalian Species* **239**: 1–6.

Nakajima, M., and Nishiwaki, M. 1965. The first occurrence of a porpoise (*Electra electra*) in Japan. *Scientific Reports of the Whales Research Institute, Tokyo* **19**: 91–104 + 7 pl.

Natoli, A., Cañadas, A., Peddemors, V.M., Aguilar, A., Vaquero, C., Fernández-Piqueras, P., and Hoelzel, A.R. 2005. Phylogeography and alpha taxonomy of the common dolphin (*Delphinus* sp.). *Journal of Evolutionary Biology* **19** (3): 943–954.

Natoli, A., Peddemors, V.M., and Hoelzel, A.R. 2004. Population structure and speciation in the genus *Tursiops* based on microsatellite and mitochondrial DNA analyses. *Journal of Evolutionary Biology* **17**: 363–375.

Nawojchik, R., St Aubin, D.J., and Johnson, A. 2003. Movements and dive behavior of two stranded, rehabilitated long-finned pilot whales (*Globicephala melas*) in the northwest Atlantic. *Marine Mammal Science* **19**(1): 232–239.

Nemoto, T. 1959. Food of baleen whales with reference to whale movements. *Scientific Reports of the Whales Research Institute, Tokyo* **14**: 149–290 + 1 pl.

Nemoto, T. 1962a. A secondary sexual character of fin whales. *Scientific Reports of the Whales Research Institute, Tokyo* **16**: 29–34 + 2 pl.

Nemoto, T. 1962b. Food of baleen whales collected in recent Japanese Antarctic whaling expeditions. *Scientific Reports of the Whales Research Institute, Tokyo* **16**: 89–103.

Nemoto, T. 1973. Feeding patterns of baleen whales in the ocean. Pp 241–252 in (J.H. Steele, ed.) *Marine food chains*. Oliver and Boyd, Edinburgh.

Nemoto, T., Best, P.B., Ishimaru, K., and Takano, H. 1980. Diatom films on whales in South African waters. *Scientific Reports of the Whales Research Institute, Tokyo* **32**: 97–103.

Nester, A.E., Thomas, J.A., and Acevedo-Guiterrez, A. 2001. The underwater whistle repertoire of wild false killer whales (*Pseudorca crassidens*). *Journal of the Acoustical Society of America* **109**(5): 2 390.

Neumann, D.R. 2001a. Seasonal movements of short-beaked common dolphins (*Delphinus delphis*) in the north-western Bay of Plenty, New Zealand: influence of sea surface temperature and El Niño/La Niña. *New Zealand Journal of Marine and Freshwater Research* **35**: 371–374.

Neumann, D.R. 2001b. The activity budget of free-ranging common dolphins (*Delphinus delphis*) in the northwestern Bay of Plenty, New Zealand. *Aquatic Mammals* **27**(2): 121–136.

Neumann, D.R., Leitenberger, A., and Orams, M.B. 2002a. Photo-identification of short-beaked common dolphins (*Delphinus delphis*) in north-east New Zealand: a photo-catalogue of recognizable individuals. *New Zealand Journal of Marine and Freshwater Research* **36**: 593–604.

Neumann, D.R., and Orams, M.B. 2003. Feeding behaviours of short-beaked common dolphins, *Delphinus delphis*, in New Zealand. *Aquatic Mammals* **29**(1): 137–149.

Neumann, D.R., Russell, K., Orams, M.B., Baker, C.S., and Duignan, P. 2002b. Identifying sexually mature, male short-beaked common dolphins (*Delphinus delphis*) at sea, based on the presence of a postanal hump. *Aquatic Mammals* **28**(2): 181–187.

Newcomer, M.W., Jefferson, T.A., and Brownell Jr, R.L. 1996. *Lissodelphis peronii*. *Mammalian Species* **531**: 1–5.

Nichols, J.T. 1908. Notes on two porpoises captured on a voyage into the Pacific Ocean. *Bulletin of the American Museum of Natural History* **24**: 217–219 + 1 pl.

Nielsen, R., Mattila, D.K., Clapham, P.J., and Palsbøll, P.J. 2001. Statistical approaches to paternity analysis in natural populations and applications to the North Atlantic humpback whale. *Genetics* **157**: 1 673–1 682.

Nilsson-Cantell, C.A. 1930. Thoracic cirripedes collected in 1925–27. *Discovery Reports* **2**: 223–260 + 1 pl.

Nishiwaki, M. 1950. On the body weight of whales. *Scientific Reports of the Whales Research Institute, Tokyo* **4**: 184–209.

Nishiwaki, M. 1963. Taxonomical consideration on genera of Delphinidae. *Scientific Reports of the Whales Research Institute, Tokyo* **17**: 93–103.

Nishiwaki, M., and Hayashi, K. 1950. Biological survey of fin and blue whales taken in the Antarctic season 1947/48 by the Japanese fleet. *Scientific Reports of the Whales Research Institute, Tokyo* **3**: 132–190.

Nishiwaki, M., and Kamiya, T. 1958. A beaked whale *Mesoplodon* stranded at Oiso Beach, Japan. *Scientific Reports of the Whales Research Institute, Tokyo* **13**: 53–83 + 17 pl.

Nishiwaki, M., Kasuya, T., Kamiya, T., Tobayama, T., and Nakajima, M. 1965. *Feresa attenuata* captured at the Pacific coast of Japan in 1963. *Scientific Reports of the Whales Research Institute, Tokyo* **19**: 65–90 + 5 pl.

Nishiwaki, M., Kasuya, T., Kureha, K., and Oguro, N. 1972. Further comments on *Mesoplodon ginkgodens*. *Scientific Reports of the Whales Research Institute, Tokyo* **24**: 43–56 + 10 pl.

Nishiwaki, M., and Oguro, N. 1972. Catch of the Cuvier's beaked whales off Japan in recent years. *Scientific Reports of the Whales Research Institute, Tokyo* **24**: 35–41.

Nolan, C.P., Liddle, G.M., and Elliot, J. 2000. Interactions between killer whales (*Orcinus orca*) and sperm whales (*Physeter macrocephalus*) with a longline fishing vessel. *Marine Mammal Science* **16**: 658–664.

Norman, S.A., and Mead, J.G. 2001. *Mesoplodon europaeus*. *Mammalian Species* **688**: 1–5.

Norris, K.S. 1976. *The porpoise watcher. A naturalist's experiences with porpoises and whales*. John Murray, London. 250 pp.

Norris, K.S., Baldwin, A., and Samson, D.J. 1965. Open ocean diving test with a trained porpoise (*Steno bredanensis*). *Deep-Sea Research* **12**: 505–509.

Norris, K.S., and Evans, W.E. 1967. Directionality of echolocation clicks in the rough-tooth porpoise, *Steno bredanensis* (Lesson). *Marine Bio-acoustics* **2**: 305–316.

Norris, K.S., and Mohl, B. 1983. Can odontocetes debilitate prey with sound? *American Naturalist* **122**: 85–104.

Norris, K.S., Würsig, B., and Wells, R.S. 1994. Aerial behaviour. Pp 103–121 in (Norris, K.S., Würsig, B., Wells, R.S., Würsig, M., Brownlee, S.M., Johnson, C.M., and Solow, J., eds) *The Hawaiian spinner dolphin*. University of California Press, Berkeley.

Nøttestad, L., Fernö, A., and Axelsen, B.E. 2002. Digging in the deep: killer whales' advanced hunting tactic. *Polar Biology* **25**(12): 939–941.

Nowlin, W.D., and Clifford, M. 1982. The kinematic and thermohaline zonation of the Antarctic Circumpolar Current at the Drake Passage. *Journal of Marine Research* **40** Supplement: 481–507.

Nowlin, W.D., and Klinck, J.M. 1986. The physics of the Antarctic Circumpolar Current. *Review of Geophysics* **24**: 469–491.

O

Odell, D.K., Asper, E.D., Baucom, J., and Cornell, L.H. 1980. A recurrent mass stranding of the false killer whale, *Pseudorca crassidens*, in Florida. *Fishery Bulletin, US* **78**(1): 171–177.

Odell, D.K., and McClune, K.M. 1999. False killer whale *Pseudorca crassidens* (Owen, 1846). Pp 213–243 in (Ridgway, S.H., and Harrison, R.J., eds) *Handbook of Marine Mammals 6: The second book of dolphins and the porpoises*. Academic Press, New York.

Oftedal, O.T. 1997. Lactation in whales and dolphins: evidence of divergence between baleen- and toothed-species. *Journal of Mammary Gland Biology and Neoplasia* **2**(3): 205–230.

Ohizumi, H., Isoda, T., Kishiro, T., and Kato, H. 2003. Feeding habits of Baird's beaked whale *Berardius bairdii*, in the western North Pacific and Sea of Okhotsk off Japan. *Fisheries Science* **69**(1): 11–20.

Ohizumi, H., Yoshioka, M., Mori, K., and Miyazaki, N. 1998. Stomach contents of common dolphins (*Delphinus delphis*) in the pelagic western North Pacific. *Marine Mammal Science* **14**(4): 835–844.

Ohsumi, S. 1978. Provisional report on the Bryde's whales caught under special permit in the Southern Hemisphere. *Report of the International Whaling Commission* **28**: 281–287.

Ohsumi, S. 1979. Interspecies relationships among some biological parameters in cetaceans and estimation of the natural mortality coefficient of the Southern Hemisphere minke whale. *Report of the International Whaling Commission* **29**: 397–406.

Ohsumi, S., and Masaki, Y. 1975. Biological parameters of the Antarctic minke whale at the virginal population level. *Journal of the Fisheries Research Board of Canada* **32**: 995–1 004.

Ohsumi, S., and Masaki, Y. 1978. Age-length keys and growth curves of the Southern Hemisphere sei whale. *Report of the International Whaling Commission* **28**: 431–436.

Ohsumi, S., Masaki, Y., and Kawamura, A. 1970. Stock of the Antarctic minke whale. *Scientific Reports of the Whales Research Institute, Tokyo* **22**: 75–125.

Ohsumi, S., Nishiwaki, M., and Hibiya, T. 1958. Growth of fin whale in the northern Pacific. *Scientific Reports of the Whales Research Institute, Tokyo* **13**: 97–133.

Olesiuk, P.F., Bigg, M.A., and Ellis, G.M. 1990. Life history and population dynamics of resident killer whales (*Orcinus orca*) in the coastal waters of British Columbia and Washington State. *Report of the International Whaling Commission* (Special Issue) **12**: 209–243.

Oleson, E.M., Barlow, J., Gordon, J., Rankin, S., and Hildebrand, J.A. 2003. Low frequency calls of Bryde's whales. *Marine Mammal Science* **19**(2): 407–419.

Oliver, W.R.B. 1922. A review of the Cetacea of the New Zealand seas – 1. *Proceedings of the Zoological Society of London* **1922**: 557–585.

Oliver, W.R.B. 1937. *Tasmacetus shepherdi*: a new genus and species of beaked whale from New Zealand. *Proceedings of the Zoological Society of London, Series* **B**: 371–381 + 5 pl.

Olsen, Ø. 1913. On the external characters and biology of Bryde's whale (*Balaenoptera brydei*), a new rorqual from the coast of South Africa. *Proceedings of the Zoological Society of London* **1913**: 1 073–1 090 + 5 pl.

Olsen, Ø. 1914. Hvaler og hvalfangst i Sydafrika. *Bergens Museum Aarbok* **1914–15**(5): 1–56.

Omura, H. 1953. Biological study on the humpback whales in the Antarctic whaling areas IV and V. *Scientific Reports of the Whales Research Institute, Tokyo* **8**: 81–102.

Omura, H. 1959. Bryde's whale from the coast of Japan. *Scientific Reports of the Whales Research Institute, Tokyo* **14**: 1–33 + 6 pl.

Omura, H. 1972. An osteological study of the Cuvier's beaked whale, *Ziphius cavirostris*, in the northwest Pacific. *Scientific Reports of the Whales Research Institute, Tokyo* **24**: 1–34.

Omura, H. 1975. Osteological study of the minke whale from the Antarctic. *Scientific Reports of the Whales Research Institute, Tokyo* **27**: 1–36.

Omura, H., Fujino, K., and Kimura, S. 1955. Beaked whale *Berardius bairdii* off Japan, with notes on *Ziphius cavirostris*. *Scientific Reports of the Whales Research Institute, Tokyo* **10**: 89–132.

Omura, H., Ichihara, T., and Kasuya, T. 1970. Osteology of pygmy blue whale with additional information on external and other characteristics. *Scientific Reports of the Whales Research Institute, Tokyo* **22**: 1–27.

Omura, H., and Kasuya, T. 1976. Additional information on skeleton of the minke whale from the Antarctic. *Scientific Reports of the Whales Research Institute, Tokyo* **28**: 57–68.

Omura, H., Kasuya, T., Kato, H., and Wada, S. 1981. Osteological study of the Bryde's whale from the central South Pacific and eastern Indian Ocean. *Scientific Reports of the Whales Research Institute, Tokyo* **33**: 1–26 + 8 pl.

Omura, H., Ohsumi, S., Nemoto, T., Nasu, K., and Kasuya, T. 1969. Black right whales in the North Pacific. *Scientific Reports of the Whales Research Institute, Tokyo* **21**: 1–78 + 18 pl.

Orams, M.B., and Schuetze, C. 1998. Seasonal and age/size-related occurrence of a barnacle (*Xenobalanus globicipitis*) on bottlenose dolphins (*Tursiops truncatus*). *Marine Mammal Science* **14**(1): 186–189.

Orsi, A.H., Whitworth (III), T., and Nowlin Jr, W.D. 1995. On the meridional extent and fronts of the Antarctic Circumpolar Current. *Deep-Sea Research I* **42**: 641–673.

Oswald, J.N., Barlow, J., and Norris, T.F. 2003. Acoustic identification of nine delphinid species in the eastern tropical Pacific Ocean. *Marine Mammal Science* **19**(1): 20–37.

O'Toole, M.J. 1980. Seasonal distribution of temperature and salinity in the surface waters off South West Africa, 1972–1974. *Investigational Report South Africa Sea Fisheries Institute* **121**: 1–25.

Ottensmeyer, C.A., and Whitehead, H. 2003. Behavioural evidence for social units in long-finned pilot whales. *Canadian Journal of Zoology* **81**: 1 327–1 338.

Ozharovskaya, L.V. 1990. The female reproductive cycle of Black Sea bottlenose dolphins as revealed by analysis of plasma progesterone levels. *Report of the International Whaling Commission* **40**: 481–485.

P

Pack, A.A., Salden, D.R., Ferrari, M.J., Glockner-Ferrari, D.A., Herman, L.M., Stubbs, H.A., and Straley, J.M. 1998. Male humpback whale dies in competitive group. *Marine Mammal Science* **14**(4): 861–873.

Palacios, D.M., and Mate, B.R. 1996. Attack by false killer whales (*Pseudorca crassidens*) on sperm whales (*Physeter macrocephalus*) in the Galápagos Islands. *Marine Mammal Science* **12**(4): 582–587.

Palmer, R.S. 1948. A school of whales, Kogia? *Journal of Mammalogy* **29**: 421.

Panigada, S., Zanardelli, M., Canese, S., and Jahoda, M. 1999. How deep can baleen whales dive? *Marine Ecology Progress Series* **187**: 309–311.

Park, Y.-H., Gamberoni, L., and Charriaud, E. 1991. Frontal structure and transport of the Antarctic Circumpolar Current in the South Indian Ocean sector, 40°–80° E. *Marine Chemistry* **35**: 45–62.

Park, Y.-H., Gamberoni, L., and Charriaud, E. 1993. Frontal structure, water masses and circulation in the Crozet Basin. *Journal of Geophysical Research* **98**(C7): 12 361–12 385.

Parks, S.E. 2003. Response of North Atlantic right whales (*Eubalaena glacialis*) to playback of calls recorded from surface active groups in both the North and South Atlantic. *Marine Mammal Science* **19**(3): 563–580.

Parrish, R.H., Bakun, A., Husby, D.M., and Nelson, C.S. 1983. Comparative climatology of selected environmental processes in relation to eastern boundary current pelagic fish reproduction. Pp 731–777 in (Sharp, G.D., and Csirke, J., eds) *Proceedings of the Expert Consultation to Examine Changes in Abundance and Species Composition of Neritic Fish Resources*, San Jose, Costa Rica, April, 1982. *F.A.O. Fisheries Reports* **291**.

Pastene, L.A., Fujise, Y., and Numachi, K. 1994. Differentiation of mitochondrial DNA between ordinary and dwarf forms of southern minke whale. *Report of the International Whaling Commission* **44**: 277–281.

Pastene, L.A., Goto, M., Kanda, N. *et al.* 2007. Radiation and speciation of pelagic organisms during periods of global warming: the case of the common minke whale, *Balaenoptera acutorostrata*. *Molecular Ecology* **16**(7): 1 481–1 495.

Patenaude, N.J., Portway, V.A., Schaeff, C.M. *et al.* 2007. MtDNA diversity and population structure among southern right whales (*Eubalaena australis*). *Journal of Heredity* **98**(2): 147–157.

Paterson, R.A., and Parker, A.E. 1994. Aerial observations of large ziphiid whales, possibly *Berardius arnuxii*, off the southern coast of New South Wales. *Memoirs of the Queensland Museum* **37**(1): 301–306.

Patterson, I.A.P., Reid, R.J., Wilson, B., Grellier, K., Ross, H.M., and Thompson, P.M. 1998. Evidence for infanticide in bottlenose dolphins: an explanation for violent interactions with harbour porpoises? *Proceedings of the Royal Society of London* B **265**: 1 167–1 170.

Pavey, C.R. 1992. The occurrence of the pygmy right whale, *Caperea marginata* (Cetacea: Neobalaenidae), along the Australian coast. *Australian Mammalogy* **15**: 1–6.

Payne, K.B., Tyack, P., and Payne, R.S. 1983. Progressive changes in the songs of humpback whales. Pp 9–57 in (Payne, R. ed.) *Communication and behavior of whales*. AAAS Selected Symposia Series 76, Westview Press, Boulder, Colorado.

Payne, P.M., and Heinemann, D.W. 1993. The distribution of pilot whales (*Globicephala* spp.) in shelf/shelf-edge and slope waters of the northeastern United States, 1978–1988. *Report of the International Whaling Commission* (Special Issue) **14**: 51–68.

Payne, R., Brazier, O., Dorsey, E.M., Perkins, J.S., Rowntree, V.J., and Titus, A. 1983. External features in southern right whales (*Eubalaena australis*) and their use in identifying individuals. Pp 371–445 in (Payne, R., ed.) *Communication and behavior of whales*. AAAS Selected Symposia Series 76, Westview Press, Boulder, Colorado.

Payne, R., and Webb, D. 1971. Orientation by means of long range acoustic signaling in baleen whales. *Annals of the New York Academy of Science* **188**: 110–142.

Payne, R.S., and McVay, S. 1971. Songs of humpback whales. *Science* **173**: 585–597.

Peddemors, V.M. 1989. Minimum age at sexual maturation of a female south-east Atlantic bottlenose dolphin *Tursiops truncatus*. *South African Journal of Marine Science* **8**: 345–347.

Peddemors, V.M. 1992. Feeding and growth in a captive-born bottlenose dolphin *Tursiops truncatus*. *South African Journal of Zoology* **27**(2): 74–80.

Peddemors, V.M. 2004. *Tursiops aduncus* (migratory subpopulation). Pp 668–669 in (Friedmann, Y., and Daly, B., eds) *Red Data Book of the mammals of South Africa: a conservation assessment*. CBSG Southern Africa, Conservation breeding specialist group (SSC/IUCN), Endangered Wildlife Trust, Parkview, South Africa.

Peddemors, V.M., Best, P.B., Findlay, K.P., Gove, D., Rakotonirina, B., Rossouw, A., and Sekiguchi, K. In press. Small cetaceans of the south-west Indian Ocean. *Journal of Cetacean Research and Management*.

Peddemors, V.M., Cockcroft, V.G., and Best, P.B. 1998. Exploitation of small cetaceans off South Africa: 1978–1996. Paper SC/49/SM34 submitted to the Scientific Committee of the International Whaling Commission, Bournemouth, UK, September/October 1997.

Peddemors, V.M., De Muelenaere, H.J.H., and Devchand, K. 1989. Comparative milk composition of the bottlenosed dolphin (*Tursiops truncatus*), humpback dolphin (*Sousa plumbea*) and common dolphin (*Delphinus delphis*) from southern African waters. *Comparative Biochemistry and Physiology* **94**A(4): 639–641.

Peddemors, V.M., and Ross, G.J.B. 1988. First record of the melon-headed whale *Peponocephala electra* (Gray, 1846) for the East African coast. *African Journal of Ecology* **26**: 345–346.

Peddemors, V.M., and Thompson, G. 1994. Beaching behaviour during shallow water feeding by humpback dolphins *Sousa plumbea*. *Aquatic Mammals* **20**(2): 65–67.

Perkins, J.D., and Miller, G.W. 1983. Mass stranding of *Steno bredanensis* in Belize. *Biotropica* **15**(3): 235–236.

Perrin, W.F. 1969a. Color pattern of the eastern Pacific spotted porpoise *Stenella graffmani* Lönnberg (Cetacea, Delphinidae). *Zoologica, New York* **54**(4): 135–142 + 7 pl.

Perrin, W.F. 1969b. The barnacle, *Conchoderma auritum*, on a porpoise (*Stenella graffmani*). *Journal of Mammalogy* **50**: 149–151.

Perrin, W.F. 1975. Variation of spotted and spinner porpoise (genus *Stenella*) in the eastern Pacific and Hawaii. *Bulletin of the Scripps Institution of Oceanography* **21**: 1–206.

Perrin, W.F. 1976. First record of the melon-headed whale, *Peponocephala electra*, in the eastern Pacific, with a summary of world distribution. *Fishery Bulletin, US* **74**(2): 457–458.

Perrin, W.F. 1985. The former dolphin fishery at St Helena. *Report of the International Whaling Commission* **35**: 423–428.

Perrin, W.F. 1990. Subspecies of *Stenella longirostris* (Mammalia: Cetacea: Delphinidae). *Proceedings of the Biological Society of Washington* **103**(2): 453–463.

Perrin, W.F. 1998. *Stenella longirostris*. *Mammalian Species* **599**: 1–7.

Perrin, W.F. 2001a. *Stenella attenuata*. *Mammalian Species* **683**: 1–8.

Perrin, W.F. 2001b. *Stenella frontalis*. *Mammalian Species* **702**: 1–6.

Perrin, W.F. 2002. Common dolphins *Delphinus delphis, D. capensis,* and *D. tropicalis*. Pp 245–248 in (Perrin, W.F., Würsig, B., and Thewissen, J.G.M., eds) *Encyclopedia of Marine Mammals*. Academic Press, San Diego.

Perrin, W.F., Aquino, M.T., Dolar, M.L.L., and Alava, M.N.R. 2007. External appearance of the dwarf spinner dolphin *Stenella longirostris roseiventris*. *Marine Mammal Science* **23**(2): 464–467.

Perrin, W.F., Armstrong, W.A., Baker, A.N. *et al.* 1995. An anomalously pigmented form of the short-beaked common dolphin (*Delphinus delphis*) from the southwestern Pacific, eastern Pacific, and eastern Atlantic. *Marine Mammal Science* **11**(2): 240–247.

Perrin, W.F., Best, P.B., Dawbin, W.H., Balcomb, K.C., Gambell, R., and Ross, G.J.B. 1973. Rediscovery of Fraser's dolphin *Lagenodelphis hosei*. *Nature, London* **241**(5 388): 345–350.

Perrin, W.F., Caldwell, D.K., and Caldwell, M.C. 1994a. Atlantic spotted dolphin *Stenella frontalis* (G. Cuvier, 1829). Pp 173–190 in (Ridgway, S.H., and Harrison, R.J., eds) *Handbook of Marine Mammals 5: The first book of dolphins*. Academic Press, New York.

Perrin, W.F., Coe, J.M., and Zweifel, J.R. 1976. Growth and reproduction of the spotted porpoise, *Stenella attenuata*, in the offshore eastern tropical Pacific. *Fishery Bulletin, US* **74**(2): 229–269.

Perrin, W.F., Dolar, M.L.L., and Robineau, D. 1999. Spinner dolphins (*Stenella longirostris*) of the western Pacific and southeast Asia: pelagic and shallow-water forms. *Marine Mammal Science* **15**(4): 1 029–1 053.

Perrin, W.F., Dolar, M.L.L., Chan, C.M., and Chivers, S.J. 2005. Length-weight relationships in the spinner dolphin (*Stenella longirostris*). *Marine Mammal Science* **21**(4): 765–778.

Perrin, W.F., and Gilpatrick Jr, J.W. 1994. Spinner dolphin *Stenella longirostris* (Gray, 1828). Pp 99–128 in (Ridgway, S.H., and Harrison, R.J., eds) *Handbook of Marine Mammals 5: The first book of dolphins*. Academic Press, New York.

Perrin, W.F., Goodall, R.N.P., and Cozzuol, M.A. 2000. Osteological variation in the spectacled porpoise (*Phocoena dioptrica*). *Journal of Cetacean Research and Management* **2**(3): 211–215.

Perrin, W.F., and Hubbs, C.L. 1969. Observations on a young pygmy killer whale (*Feresa attenuata* Gray) from the eastern tropical Pacific Ocean. *Transactions of the San Diego Society of Natural History* **15**(18): 297–308.

Perrin, W.F., Leatherwood, S., and Collet, A. 1994b. Fraser's dolphin *Lagenodelphis hosei* Fraser, 1956. Pp 225–240 in (Ridgway, S.H., and Harrison, R.J., eds) *Handbook of Marine Mammals 5: The first book of dolphins*. Academic Press, New York.

Perrin, W.F., and Mead, J.G. 1994. Clymene dolphin *Stenella clymene* (Gray, 1846). Pp 161–171 in (Ridgway, S.H., and Harrison, R.J., eds) *Handbook of Marine Mammals 5: The first book of dolphins*. Academic Press, New York.

Perrin, W.F., Mitchell, E.D., Mead, J.G., Caldwell, D.K., and Van Bree, P.J.H. 1981. *Stenella clymene*, a rediscovered tropical dolphin of the Atlantic. *Journal of Mammalogy* **62**(3): 583–598.

Perrin, W.F., Mitchell, E.D., Mead, J.G., Caldwell, D.K., Caldwell, M.C., Van Bree, P.J.H., and Dawbin, W.H. 1987. Revision of the spotted dolphins, *Stenella* spp. *Marine Mammal Science* **3**(2): 99–170.

Perrin, W.F., Mitchell, E.D, and Van Bree, P.J.H. 1978. Historical zoogeography of tropical pelagic dolphins. *II. Congressus Theriologicus Internationalis, Brno, June 20–27, 1978. Abstracts*: 73.

Perrin, W.F., and Walker, W.A. 1975. The rough-toothed porpoise, *Steno bredanensis*, in the eastern tropical Pacific. *Journal of Mammalogy* **56**(4): 905–907.

Perrin, W.F., Wilson, C.E., and Archer (II), F.I. 1994c. Striped dolphin *Stenella coeruleoalba* (Meyen, 1833). Pp 129–159 in (Ridgway, S.H., and Harrison, R.J., eds) *Handbook of Marine Mammals 5: The first book of dolphins*. Academic Press, New York.

Perryman, W.L., Au, D.W.K., Leatherwood, S., and Jefferson, T.A. 1994. Melon-headed whale *Peponocephala electra* (Gray, 1846). Pp 363–386 in (Ridgway, S.H., and Harrison, R.J., eds) *Handbook of Marine Mammals 5: The first book of dolphins*. Academic Press, New York.

Perryman, W.L., and Foster, T.C. 1980. Preliminary report on predation by small whales, mainly the false killer whale, *Pseudorca crassidens*, on dolphins (*Stenella* spp., and *Delphinus delphis*) in the eastern tropical Pacific. *NOAA Administrative Report* **LJ-80-05**. 9 pp.

Perryman, W.L., and Lynn, M.S. 1993. Identification of geographic forms of common dolphin (*Delphinus delphis*) from aerial photogrammetry. *Marine Mammal Science* **9**(2): 119–137.

Peschak, T.P. 2005. The sardine run: the greatest shoal on earth. In: *Currents of Contrast: Life in southern Africa's two oceans*. Struik Publishers, Cape Town. 200 pp.

Peters, W. 1876. S.M.S. Gazelle gesammelten Säugethiere aus den Abtheilungen der Nager, Hufthiere, Sirenen, Cetaceen und Beutelthiere. *Monatsberichte der k. Academie der Wissenschaften zu Berlin* **1876**: 355–366 + 3 pl.

Peterson, R.G., and Stramma, L. 1991. Upper-level circulation in the South Atlantic Ocean. *Progress in Oceanography* **26**: 1–73.

Philippi, R.A. 1893. Die Delphine an der Südspitze Südamerikas. *Anales del Museo Nacional de Chile*: 1–16 + 5 pl.

Philips, J.D., Nachtigall, P.E., Au, W.W.L., Pawloski, J.L., and Roitblat, H.L. 2003. Echolocation in the Risso's dolphin, *Grampus griseus*. *Journal of the Acoustical Society of America* **113**(1): 605–616.

Pichler, F.B., Robineau, D., Goodall, R.N.P., Meÿer, M.A., Olivarria, C., and Baker, C.S. 2001. Origin and radiation of

Southern Hemisphere coastal dolphins (genus *Cephalorhynchus*). *Molecular Ecology* **10**: 2 215–2 223.

Pierpoint, C., Gordon, J., and Frizell, J. 1995. Incidental sightings of cetaceans in the Southern Ocean sanctuary. Paper SC/47/O12 submitted to the Scientific Committee of the International Whaling Commission, Dublin, Ireland, 8–20 May 1995. 10 pp.

Pike, G.C. 1953. Two records of *Berardius bairdi* from the coast of British Columbia. *Journal of Mammalogy* **34**(1): 98–104.

Pillar, S.C., and Barange, M. 1993. Feeding selectivity of juvenile Cape hake, *Merluccius capensis* in the southern Benguela. *South African Journal of Marine Science* **13**: 255–268.

Pilleri, G. 1964. Morphologie des Gehirnes des Southern Right Whale, *Eubalaena australis* Desmoulins 1822 (Cetacea, Mysticeti, Balaenidae). *Acta Zoologica, Stockholm* **46**: 245–272.

Pilleri, G., and Pilleri, O. 1982. Cetacean records in the Mediterranean Sea. *Investigations on Cetacea* **14**: 49–63.

Pilson, M.E.Q., and Waller, D.W. 1970. Composition of milk from spotted and spinner porpoises. *Journal of Mammalogy* **51**(1): 74–79.

Pinedo, M.C., Barreto, A.S., Lammardo, M.P., Andrade, A.L.V., and Geracitan, L. 2002. Northernmost records of the spectacled porpoise, Layard's beaked whale, Commerson's dolphin and Peale's dolphin in the southwestern Atlantic Ocean. *Aquatic Mammals* **28**(1): 32–37.

Pistorius, P.A., Taylor, F.E., Louw, C. *et al.* 2002. Distribution, movement, and estimated population size of killer whales at Marion Island, December 2000. *South African Journal of Wildlife Research* **32**(1): 86–92.

Pitman, R.L., and Dutton, P.H. 2004. Killer whale predation on a leatherback turtle in the northeast Pacific. *Pacific Science* **58**(3): 497–498.

Pitman, R.L., and Ensor, P. 2003. Three forms of killer whales (*Orcinus orca*) in Antarctic waters. *Journal of Cetacean Research and Management* **5**(2): 131–139.

Pitman, R.L., Palacios, D.M., Brennan, P.L.R., Brennan, B.J., Balcomb (III), K.C., and Miyashita, T. 1999. Sightings and possible identity of a bottlenose whale in the tropical Indo-Pacific: *Indopacetus pacificus*? *Marine Mammal Science* **15**(2): 531–549.

Pitman, R.L., Perryman, W.L., LeRoi, D., and Eilers, E. 2007. A dwarf form of killer whale in Antarctica. *Journal of Mammalogy* **88**(1): 43–48.

Pitman, R.L., and Stinchcomb, C. 2002. Rough-toothed dolphins (*Steno bredanensis*) as predators of mahimahi (*Coryphaena hippurus*). *Pacific Science* **56**(4): 447–450.

Pitman, R.L., Van Helden, A., Best, P.B., and Pym, A. 2006. Shepherd's beaked whale (*Tasmacetus shepherdi*): information on appearance and biology based on strandings and at-sea observations. *Marine Mammal Science* **22**(3): 744–755.

Pivorunas, A. 1979. The feeding mechanisms of baleen whales. *American Scientist* **67**: 432–440.

Plön, S. 2004. *The status and natural history of pygmy (*Kogia breviceps*) and dwarf (*K. sima*) sperm whales off southern Africa*. Ph D thesis, Rhodes University, South Africa. 551 pp.

Pomilla, C., and Rosenbaum, H.C. 2005. Against the current: an inter-oceanic whale migration event. *Biology Letters (Royal Society)* **1**: 476–479.

Pomilla, C., Best, P.B., Findlay, K.P. *et al.* 2005. Population structure and sex-biased gene flow in humpback whales from wintering regions A, B, C and X based on nuclear microsatellite variation. Paper SC57/SH13 submitted to the Scientific Committee of the International Whaling Commission, Ulsan, Korea, July 2005.

Poncelet, E., Barbraud, C., and Guinet, C. In press. Population dynamics of killer whales in Crozet Archipelago, southern Indian Ocean: exploiting opportunistic and protocol-based photographs in a mark-recapture study. *Journal of Cetacean Research and Management.*

Ponganis, P.J., Kooyman, G.L., and Castellini, M.A. 1995. Multiple sightings of Arnoux's beaked whales along the Victoria Land coast. *Marine Mammal Science* **11**(2): 247–250.

Popper, A.N. 1980. Sound emission and detection by delphinids. Pp 1-52 in (Herman, L.M., ed.) *Cetacean behaviour: mechanisms and functions*. John Wiley, New York.

Porter, J.W. 1977. *Pseudorca* stranding. *Oceans* **10**(4): 8–15.

Praetsch, F. 1994. Second year progress report. The ecology and population biology of bottlenose dolphins (*Tursiops truncatus*, Montagu 1821) off the central Namibian coast. Report on data collection as part of Ph D dissertation, University of Cambridge, UK.

Praetsch, F. 1995. The ecology and population biology of bottlenose dolphins (*Tursiops truncatus*, Montagu 1821) off the central Namibian coast. Report on data collection as part of Ph D dissertation, University of Cambridge, UK.

Progress Report(s), South Africa, International Whaling Commission. 1998–2004.

Pryor, K. 1975. *Lads before the wind. Adventures in porpoise training*. Harper and Row, New York. 278 pp.

Pryor, K., Haag, R., and O'Reilly, J. 1969. The creative porpoise: training for novel behavior. *Journal of the Experimental Analysis of Behavior* **12**: 653–661.

Pryor, K., and Shallenberger, I.K. 1991. Social structure in spotted dolphins (*Stenella attenuata*) in the tuna purse seine fishery in the eastern tropical Pacific. Pp 161–196 in (Pryor, K., and Norris, K.S., eds) *Dolphin societies. Discoveries and puzzles*. University of California Press, Berkeley.

Pryor, T., Pryor, K., and Norris, K.S. 1965. Observations on a pygmy killer whale (*Feresa attenuata* Gray) from Hawaii. *Journal of Mammalogy* **46**(3): 450–461.

Psarakos, S., Herzing, D.L., and Marten, K. 2003. Mixed-species associations between pantropical spotted dolphins (*Stenella attenuata*) and Hawaiian spinner dolphins (*Stenella longirostris*) off Oahu, Hawaii. *Aquatic Mammals* **29**(3): 390–395.

Purves, P.E., and Pilleri, G. 1978. The functional anatomy and general biology of *Pseudorca crassidens* (Owen) with a review of the hydrodynamics and acoustics in Cetacea. *Investigations on Cetacea* **9**: 67–227 + 13 pl.

R

Raga, J.A., and Balbuena, J.A. 1993. Parasites of the long-finned pilot whale, *Globicephala melas* (Traill, 1809), in European waters. *Report of the International Whaling Commission* (Special Issue) **14**: 391–406.

Ralls, K., Brownell Jr, R.L., and Ballou, J. 1980. Differential mortality by sex and age in mammals, with special reference to the sperm whale. *Report of the International Whaling Commission* (Special Issue) **2**: 233–243.

Ramp, C., Bérubé, M., Hagen, W., and Sears, R. 2006. Survival of adult blue whales *Balaenoptera musculus* in the Gulf of St. Lawrence, Canada. *Marine Ecology Progress Series* **319**: 287–295.

Raven, H.C. 1937. Notes on the taxonomy and osteology of two species of *Mesoplodon* (*M. europaeus* Gervais, *M. mirus* True). *American Museum Novitates* **905**: 1–30.

Raven, H.C. 1942. On the structure of *Mesoplodon densirostris*, a rare beaked whale. *Bulletin of the American Museum of Natural History* **80**: 23–50.

Rayner, G.W. 1940. Whale marking: progress and results to December 1939. *Discovery Reports* **19**: 245–284.

Razafindrakoto, Y., Andrianarivelo, N., and Rosenbaum, H.C. 2004. Sightings, catches, and other records of Indo-Pacific humpback dolphins in the coastal waters of Madagascar. *Aquatic Mammals* **30**(1): 103–110.

Read, A.J. 2002. Porpoises, overview. Pp 982–984 in (Perrin, W.F., Würsig, B., and Thewissen, J.G.M., eds) *Encyclopedia of Marine Mammals*. Academic Press, San Diego.

Read, J.F., and Pollard, R.T. 1993. Structure and transport of the Antarctic Circumpolar Current and the Agulhas Return Current at 40° E. *Journal of Geophysical Research* **98**: 12 281–12 295.

Reeb, D., and Best, P.B. 1999. Anatomy of the laryngeal apparatus of the pygmy right whale, *Caperea marginata* (Gray 1846). *Journal of Morphology* **242**: 67–81.

Reeb, D., Duffield, M., and Best, P.B. 2005. Evidence of postnatal ecdysis in southern right whales, *Eubalaena australis*. *Journal of Mammalogy* **86**(1): 131–138.

Reeves, R.R., Perrin, W.F., Taylor, B.L., Baker, C.S., and Mesnick, S.L. 2004. Report of the Workshop on Shortcomings of Cetacean Taxonomy in Relation to Needs of Conservation and Management, April 30–May 2, 2004, La Jolla, California. *NOAA Technical Memorandum NMFS NOAA-TM-SWFSC-***363**. 94 pp.

Reeves, R.R., Stewart, B.S., Clapham, P.J., and Powell, J.A. 2002. *National Audubon Society guide to marine mammals of the world*. Alfred A. Knopf, Inc., New York. 527 pp.

Reilly, S.B. 1990. Seasonal changes in distribution and habitat differences among dolphins in the eastern tropical Pacific. *Marine Ecology Progress Series* **66**: 1–11.

Reinhardt, J. 1866. *Pseudorca crassidens*, a cetacean hitherto unknown in the Danish fauna. Pp 189–218 in (Flower, W.H., ed.) *Recent memoirs on the Cetacea by Professors Eschricht, Reinhardt and Lilljeborg*. The Ray Society, London.

Rendell, L.E., Matthews, J.N., Gill, A., Gordon, J.C.D., and Macdonald, D.W. 1999. Quantitative analysis of tonal calls from five odontocete species, examining interspecific and intraspecific variation. *Journal of Zoology, London* **249**: 403–410.

Reyes, J.C. 1990. Gray's beaked whale *Mesoplodon grayi* in the south east Pacific. *Zeitschrift für Säugetierkunde* **55**(2): 139–141.

Reyes, J.C. 1996. A possible case of hybridism in wild dolphins. *Marine Mammal Science* **12**(2): 301–307.

Rice, D.W. 1989. Sperm whale *Physeter macrocephalus* Linnaeus, 1758. Pp 177–233 in (Ridgway, S.H., and Harrison, R.J., eds) *Handbook of Marine Mammals 4: River dolphins and the larger toothed whales*. Academic Press, London.

Rice, D.W. 1998. Marine mammals of the world. Systematics and distribution. *The Society for Marine Mammalogy Special Publication* **4**: 1–231.

Rice, D.W. 2002. Ambergris. Pp 20–21 in (Perrin, W.F., Würsig, B., and Thewissen, J.G.M., eds) *Encyclopedia of Marine Mammals*. Academic Press, San Diego.

Rice, D.W., and Caldwell, D.K. 1961. Observations on the habits of the whalesucker (*Remilegia australis*). *Norsk Hvalfangsttidende* **50**(5): 181–189.

Rice, F.H., and Saayman, G.S. 1984. Movements and behaviour of Heaviside's dolphins (*Cephalorhynchus heavisidii*) off the western coasts of southern Africa. *Investigations on Cetacea* **16**: 49–63.

Rice, F.H., and Saayman, G.S. 1987. Distribution and behaviour of killer whales (*Orcinus orca*) off the coasts of southern Africa. *Investigations on Cetacea* **20**: 231–250.

Richards, R. 1994. *Into the South Seas: the southern whale fishery comes of age on the Brazil Banks 1765–1812*. The Paremata Press, Wellington, New Zealand.

Ridgway, S.H., and Carder, D.A. 2001. Assessing hearing and sound production in cetaceans not available for behavioral audiograms: experiences with sperm, pygmy sperm, and gray whales. *Aquatic Mammals* **27**(3): 267–276.

Ringelstein, J., Pusineri, C., Hassani, S., Meynier, L., Nicolas, R., and Ridoux, V. 2006. Food and feeding ecology of the striped dolphin, *Stenella coeruleoalba*, in the oceanic waters of the north-east Atlantic. *Journal of the Marine Biological Association of the UK* **86**: 909–918.

Risting, S. 1928. Whales and whale fetuses. Statistics of catch and measurements collected from the Norwegian Whalers' Association 1922–25. *Rapports et Procès-Verbaux des Réunions* **50**: 1–122.

Ritter, F. 2002. Behavioural observations of rough-toothed dolphins (*Steno bredanensis*) off La Gomera, Canary Islands (1995–2000) with special reference to their interactions with humans. *Aquatic Mammals* **28**(1): 46–59.

Ritter, F., and Brederlau, B. 1999. Behavioural observations of dense-beaked whales (*Mesoplodon densirostris*) off La Gomera, Canary Islands (1995–1997). *Aquatic Mammals* **25**(2): 55–61.

Robeck, T.R., Schneyer, A.L., McBain, J.F., Dalton, L.M., Walsh, M.T., Czekala, N.M., and Kraemer, D.C. 1993. Analysis of urinary immunoreactive steroid metabolites and gonadotropins for characterization of the estrous cycle, breeding period, and seasonal estrous activity of captive killer whales (*Orcinus orca*). *Zoo Biology* **12**: 173–187.

Robertson, K.M., and Chivers, S.J. 1997. Prey occurrence in pantropical spotted dolphins, *Stenella attenuata* from the eastern tropical Pacific. *Fishery Bulletin, US* **95**: 334–348.

Robineau, D. 1975. Échouage d'un *Ziphius cavirostris* Cuvier, 1823 (Cetacea, Hyperoodontidae) dans l'Archipel des Comores (Océan Indien). *Mammalia* **39**(2): 513–514.

Robineau, D. 1984. Morphologie externe et pigmentation du dauphin de Commerson, *Cephalorhynchus commersonii* (Lacépède, 1804), en particulier celui des îles Kerguelen. *Canadian Journal of Zoology* **62**: 2 465–2 475.

Robineau, D. 1985. Données preliminaries sur la repartition du dauphin de Commerson *Cephalorhynchus commersonii* aux îles Kerguelen, en particulier dans le Golfe du Morbihan. *Biological Conservation* **31**: 85–93.

Robineau, D. 1986. Valeur adaptive des caractères morphologiques distinctifs (taille et pigmentation) d'une population isolée d'un dauphin subantarctique, *Cephalorhynchus commersonii* (Lacépède, 1804), en particulier celui des îles Kerguelen. *Mammalia* **50**(3): 357–368.

Robineau, D. 1989. Les cétacés des îles Kerguelen. *Mammalia* **53**(2): 265–278.

Robineau, D., and De Buffrénil, V. 1985. Données ostéologiques et ostéométriques sur le dauphin de Commerson, *Cephalorhynchus commersonii* (Lacépède, 1804), en particulier celui des îles Kerguelen. *Mammalia* **49**(1): 109–123.

Robineau, D., and Duhamel, G. 1984. Régime alimentaire du dauphin de Commerson *Cephalorhynchus commersonii* (Lacépède, 1804) aux îles Kerguelen, pendant l'été austral. *Bulletin du Muséum d' Histoire Naturelle, Paris, 4e sér.* **6A**(2): 551–559.

Robineau, D., and Vely, M. 1993. Stranding of a specimen of Gervais' beaked whale (*Mesoplodon europaeus*) on the coast of west Africa (Mauritania). *Marine Mammal Science* **9**(4): 438–440.

Robison, B.H., and Craddock, J.E. 1983. Mesopelagic fishes eaten by Fraser's dolphin, *Lagenodelphis hosei*. *Fishery Bulletin, US* **81**(2): 283–289.

Robson, F.D. 1975. On vestigial and normal teeth in the scamperdown beaked whale, *Mesoplodon grayi*. *Tuatara* **21**(3): 105–107.

Rodríguez-López, M.A., and Mignucci-Giannoni, A.A. 1999. A stranded pygmy killer whale (*Feresa attenuata*) in Puerto Rico. *Aquatic Mammals* **25**(2): 119–121.

Rogers, T.L., and Brown, S.M. 1999. Acoustic observations of Arnoux's beaked whale (*Berardius arnuxii*) off Kemp Land, Antarctica. *Marine Mammal Science* **15**(1): 192–198.

Rohr, J.J., Fish, F.E., and Gilpatrick Jr, J.W. 2002. Maximum swim speeds of captive and free-ranging delphinids: critical analysis of extraordinary performance. *Marine Mammal Science* **18**(1): 1–19.

Rommel, S. 1990. Osteology of the bottlenose dolphin. Pp 29–49 in (Leatherwood, S., and Reeves, R.R., eds) *The bottlenose dolphin*. Academic Press, San Diego.

Rosario-Delestre, R.J., Rodríguez-López, M.A., Mignucci-Giannoni, A.A., and Mead, J.G. 1999. New records of beaked whales (*Mesoplodon* spp.) for the Caribbean. *Caribbean Journal of Science* **35**(1–2): 144–148.

Rose, B., and Payne, A.I.L. 1991. Occurrence and behavior of the southern right whale dolphin *Lissodelphis peronii* off Namibia. *Marine Mammal Science* **7**(1): 25–34.

Rosel, P.E., Dizon, A.E., and Heyning, J.E. 1994. Genetic analysis of sympatric morphotypes of common dolphins (genus *Delphinus*). *Marine Biology* **119**: 159–167.

Rosenbaum, H., and Collins, T. 2006. The ecology, population characteristics and conservation efforts for humpback whales (*Megaptera novaeangliae*) on their wintering grounds in the coastal waters of Gabon. In (Alonso, F., Lee, M.E., Campbell, P., Pauwels, O.S.G., and Dallmeier, F., eds) Gamba, Gabon: biodiversity of an equatorial African rainforest. *Bulletin of the Biological Society of Washington* **12**: 425–433.

Rosenbaum, H.C. 2003. Marine Mammals. Pp 213–216 in (Goodman, S., and Bengston, J., eds) *Natural History of Madagascar*. University of Chicago Press, Chicago.

Rosenbaum, H.C., Brownell Jr, R.L., Brown, M.W. *et al.* 2000. World-wide genetic differentiation of *Eubalaena*: questioning the number of right whale species. *Molecular Ecology* **9**: 1 793–1 802.

Rosenbaum, H.C., Ersts, P.J., Razafindrakoto, Y. *et al.* 2002a. Population characteristics, distribution, and relative abundance of humpback whales off the coasts of Madagascar and Gabon: an update on recent and planned research. Paper SC54/H20 submitted to the Scientific Committee of the International Whaling Commission, Shimonoseki, Japan, April–May 2002.

Rosenbaum, H.C., Glaberman, S., Jefferson, T., Collins, T., Minton, G., Peddemors, V., and Baldwin, R. 2002b. Phylogenetic relationships and population structure among humpback dolphins based on mtDNA variation. Paper SC/54/SM34 submitted to the Scientific Committee of the International Whaling Commission, Shimonoseki, Japan, April–May 2002.

Rosenbaum, H.C., Pomilla, C., Leslie, M. *et al.* 2004a. Mitochondrial DNA diversity and population structure of humpback whales from their wintering areas in the Indian and South Atlantic Oceans (wintering regions A, B, C and X). Paper SC56/SH3 submitted to the Scientific Committee of the International Whaling Commission, Sorrento, Italy, July 2004.

Rosenbaum, H.C., Razafindrakoto, Y., Vahoavy, J., and Pomilla, C. 2001. A note on recent sightings of southern right whales (*Eubalaena australis*) along the east coast of Madagascar. *Journal of Cetacean Research and Management* (Special Issue) **2**: 177–180.

Rosenbaum, H.C., Strindberg, S., and Ersts, P.J. 2004b. Initial estimates abundance and distribution of humpback whales on their wintering grounds in the coastal waters of Gabon (southeastern Atlantic Ocean, Area B) based on aerial surveys. Paper SC56/SH2 submitted to the Scientific Committee of the International Whaling Commission, Sorrento, Italy, July 2004.

Ross, G.J.B. 1969. Evidence for a southern breeding population of True's beaked whale. *Nature, London* **222**(5193): 585.

Ross, G.J.B. 1970. The occurrence of Hector's beaked whale *Mesoplodon hectori* (Gray) in South African waters. *Annals of the Cape Provincial Museums (Natural History)* **8**(13): 195–204.

Ross, G.J.B. 1977. The taxonomy of bottlenosed dolphins *Tursiops* species in South African waters, with notes on their biology. *Annals of the Cape Provincial Museums (Natural History)* **11**(9): 135–194.

Ross, G.J.B. 1979a. Records of pygmy and dwarf sperm whales, genus *Kogia*, from southern Africa, with biological notes and some comparisons. *Annals of the Cape Provincial Museums (Natural History)* **11**(14): 259–327.

Ross, G.J.B. 1979b. *The smaller cetaceans of the south east coast of southern Africa*. Ph D thesis, University of Port Elizabeth, South Africa. 415 pp.

Ross, G.J.B. 1984. The smaller cetaceans of the south east coast of southern Africa. *Annals of the Cape Provincial Museums (Natural History)* **15**(2): 173–410.

Ross, G.J.B. 2002. Humpback dolphins *Sousa chinensis, S. plumbea* and *S. teuszii*. Pp 585–589 in (Perrin, W.F., Würsig, B., and Thewissen, J.G.M., eds) *Encyclopedia of Marine Mammals*. Academic Press, San Diego.

Ross, G.J.B., Best, P.B., and Donnelly, B.G. 1975. New records of the pygmy right whale (*Caperea marginata*) from South Africa, with comments on distribution, migration, appearance and behaviour. *Journal of the Fisheries Research Board of Canada* **32**: 1 005–1 017.

Ross, G.J.B., Best, P.B., Baker, A.N., and Mead, J.G. 1988. A review of colour patterns and their ontogenetic variation in beaked whales (Ziphiidae, Cetacea). Paper SC/40/SM6 submitted to the Scientific Committee of the International Whaling Commission, San Diego, USA, May 1988.

Ross, G.J.B., and Cockcroft, V.G. 1990. Comments on Australian bottlenose dolphins and the taxonomic status of *Tursiops aduncus* (Ehrenberg, 1832). Pp 101–128 in (Leatherwood, S., and Reeves, R.R., eds) *The bottlenose dolphin*. Academic Press, San Diego.

Ross, G.J.B., Cockcroft, V.G., and Butterworth, D.S. 1987. Offshore distribution of bottlenosed dolphins in Natal coastal waters and Algoa Bay, eastern Cape. *South African Journal of Zoology* **22**(1): 50–56.

Ross, G.J.B., Cockcroft, V.G., and Cliff, G. 1985. Additions to the marine mammal faunas of South Africa and Natal. *Lammergeyer* **35**: 36–40.

Ross, G.J.B., Cockcroft, V.G., Melton, D.A., and Butterworth, D.S. 1989. Population estimates for bottlenose dolphins *Tursiops truncatus* in Natal and Transkei waters. *South African Journal of Marine Science* **8**: 119–129.

Ross, G.J.B., Heinsohn, G.E., and Cockcroft, V.G. 1994. Humpback dolphins *Sousa chinensis* (Osbeck, 1765), *Sousa plumbea* (G. Cuvier, 1829) and *Sousa teuszii* (Kükenthal, 1892). Pp 23–42 in (Ridgway, S.H., and Harrison, R.J., eds) *Handbook of Marine Mammals 5: The first book of dolphins.* Academic Press, New York.

Ross, G.J.B., and Leatherwood, S. 1994. Pygmy killer whale *Feresa attenuata* Gray, 1874. Pp 387–404 in (Ridgway, S.H., and Harrison, R.J., eds) *Handbook of Marine Mammals 5: The first book of dolphins.* Academic Press, New York.

Roux, J.P. 1986. Le cycle annuel d'abondance des orques, *Orcinus orca*, aux îles Saint-Paul et Amsterdam. *Mammalia* **50**(1): 5–8.

Rowntree, V.J. 1996. Feeding, distribution, and reproductive behavior of cyamids (Crustacea: Amphipoda) living on humpback and right whales. *Canadian Journal of Zoology* **74**: 103–109.

S

Saayman, G.S., Bower, D., and Tayler, C.K. 1972. Observations on inshore and pelagic dolphins on the south-eastern Cape coast of South Africa. *Koedoe* **15**: 1–24.

Saayman, G.S., and Tayler, C.K. 1973. Social organisation of inshore dolphins (*Tursiops aduncus* and *Sousa*) in the Indian Ocean. *Journal of Mammalogy* **54** (4): 993–996.

Saayman, G.S., and Tayler, C.K. 1979. The socioecology of humpback dolphins (*Sousa* sp.). Pp 165–226 in (Winn, H.E., and Olla, B.L., eds) *Behavior of Marine Animals. Current Perspectives in Research. 3. Cetaceans.* Plenum Press, New York.

Saayman, G.S., Tayler, C.K., and Bower, D. 1973. Diurnal activity cycles in captive and free-ranging Indian Ocean bottlenose dolphins (*Tursiops aduncus* Ehrenburg). *Behaviour* **44**: 212–233.

Saint Thomas Productions. 2000. *Killer whales. Up close and personal.* St Thomas Productions, Aix en Provence, France [video recording].

Samaai, T., Best, P.B., and Gibbons, M.J. 2005. The taxonomic status of common dolphins *Delphinus* spp. in South African waters. *African Journal of Marine Science* **27**(2): 449–458.

Santos, M.C. de O. 1997. Lone sociable bottlenose dolphin in Brazil: human fatality and management. *Marine Mammal Science* **13**(2): 355–356.

Santos, M.C. de O., Zampirolli, É., Castro, A.F.V. de, and Alvarenga, F.S. 2003. A Gervais' beaked whale (*Mesoplodon europaeus*) washed ashore in southeastern Brazil: extra limital record? *Aquatic Mammals* **29**(3): 404–410.

Sargeant, B.L., Mann, J., Berggren, P., and Krützen, M. 2005. Specialization and development of beach hunting, a rare foraging behavior, by wild bottlenose dolphins (*Tursiops* sp.). *Canadian Journal of Zoology* **83**: 1 400–1 410.

Sasaki, S., Komura, H., Hashikata, M. *et al.* 2004. Composition and characteristics of ginkgo-toothed beaked whale (*Mesoplodon ginkgodens*) milk. *Milk Science* **53**(1): 15–18.

Sasaki, T., Nikaido, M., Wada, S., Yamada, T.K., Cao, Y., Hasegawa, M., and Okada, N. 2006. *Balaenoptera omurai* is a newly discovered baleen whale that represents an ancient evolutionary lineage. *Molecular Phylogenetics and Evolution* **41**: 40–52.

Sayigh, L.S., Tyack, P.L., Wells, R.S., Solow, A.R., Scott, M.D., and Irvine, A.B. 1998. Individual recognition in wild bottlenose dolphins: a field test using playback experiments. *Animal Behaviour* **57**: 41–50.

Sazima, I., Sazima, C., and Silva-Jr, J.M. 2003. The cetacean offal connection: feces and vomits of spinner dolphins as a food source for reef fishes. *Bulletin of Marine Science* **72**(1): 151–160.

Scarff, J.E. 1986. Occurrence of the barnacles *Coronula diadema, C. reginae* and *Cetopirus complanatus* (Cirripedia) on right whales. *Scientific Reports of the Whales Research Institute, Tokyo* **37**: 129–153.

Schaeff, C.M., Best, P.B., Rowntree, V.J., Payne, R., Jarvis, C., and Portway, V.A. 1999. Dorsal skin color patterns among southern right whales (*Eubalaena australis*): genetic basis and evolutionary significance. *Journal of Heredity* **90**(4): 464–471.

Schevill, W.E., and Watkins, W.A. 1972. Intense low frequency sounds from an Antarctic minke whale, *Balaenoptera acutorostrata. Breviora* **388**: 1–8.

Schnell, G.D., Douglas, M.E., and Hough, D.J. 1982. Geographic variation in morphology of spotted and spinner dolphins (*Stenella attenuata* and *S. longirostris*) from the eastern tropical Pacific. *NOAA National Marine Fisheries Service Southwest Fisheries Center Administrative Report* LJ-82-15C. 213 pp.

Scott, E.M., Mann, J., Watson-Capps, J.J., Sargeant, B.L., and Connor, R.C. 2005. Aggression in bottlenose dolphins: evidence for sexual coercion, male-male competition, and female tolerance through analysis of tooth-rake marks and behaviour. *Behaviour* **142**: 21–44.

Scott, M.D., and Chivers, S.J. 1990. Distribution and herd structure of bottlenose dolphins in the eastern tropical Pacific Ocean. Pp 387–402 in (Leatherwood, S., and Reeves, R.R., eds) *The bottlenose dolphin.* Academic Press, San Diego.

Scott, M.D., and Cordaro, J.G. 1987. Behavioral observations of the dwarf sperm whale, *Kogia simus. Marine Mammal Science* **3**(4): 353–354.

Scott, M.D., Hohn, A.A., Westgate, A.J., Nicolas, J.R., Whitaker, B.R., and Campbell, W.B. 2001. A note on the release and tracking of a rehabilitated pygmy sperm whale (*Kogia breviceps*). *Journal of Cetacean Research and Management* **3**(1): 87–94.

Secchi, E.R., Barcellos, L., Zerbini, A.N., and Dalla Rosa, L. 2003. Biological observations on a dwarf minke whale, *Balaenoptera acutorostrata*, caught in southern Brazilian waters, with a new record of prey for the species. *Latin American Journal of Aquatic Mammals* **2**(2): 109–115.

Secchi, E.R., and Siciliano, S. 1995. Comments on the southern range of the spinner dolphin (*Stenella longirostris*) in the western south Atlantic. *Aquatic Mammals* **21**(2): 105–108.

Secretariat of the International Whaling Commission. 1988. *International Whaling Statistics* XCV and XCVI. Cambridge. 68 pp.

Sedlak-Weinstein, E. 1991. Three new records of cyamids (Amphipoda) from Australian cetaceans. *Crustaceana* **60**(1): 90–104.

Sedlak-Weinstein, E. 1992. The occurrence of a new species of *Isocyamus* (Crustacea, Amphipoda) from Australian and Japanese pilot whales, with a key to species of *Isocyamus. Journal of Natural History* **26**: 937–946.

Sekiguchi, K. 1994. *Studies on feeding habits and dietary analytical methods for smaller odontocete species along the southern African coast.* Ph D thesis, University of Pretoria. 259 pp.

Sekiguchi, K., Best, P.B., and Kaczmaruk, B.Z. 1992a. New information on the feeding habits and baleen morphology of the pygmy right whale *Caperea marginata. Marine Mammal Science* **8**(3): 288–293.

Sekiguchi, K., Klages, N.T.W., and Best, P.B. 1992b. Comparative analysis of the diets of smaller odontocete cetaceans along the coast of southern Africa. *South African Journal of Marine Science* **12**: 843–861.

Sekiguchi, K., Klages, N., Findlay, K., and Best, P.B. 1993. Feeding habits and possible movements of southern bottlenose whales (*Hyperoodon planifrons*). *Proceedings of the National Institute of Polar Research Symposium on Polar Biology* **6**: 84–97.

Sekiguchi, K., Klages, N.T.W., and Best, P.B. 1996. The diet of strap-toothed whales (*Mesoplodon layardii*). *Journal of Zoology, London* **239**: 453–463.

Sekiguchi, K., Olavarría, C., Morse, L. *et al.* In press. The spectacled porpoise *Phocoena dioptrica* in Antarctic waters. *Journal of Cetacean Research and Management.*

Sergeant, D.E. 1962. The biology of the pilot or pothead whale *Globicephala melaena* (Traill) in Newfoundland waters. *Bulletin of the Fisheries Research Board of Canada* **132**: 1–84.

Sergeant, D.E. 1982. Mass strandings of toothed whales (Odontoceti) as a population phenomenon. *Scientific Reports of the Whales Research Institute, Tokyo* **34**: 1–47.

Shane, S. 1995a. Behavior patterns of pilot whales and Risso's dolphins off Santa Catalina Island, California. *Aquatic Mammals* **21**(3): 195–197.

Shane, S.H. 1994. Pilot whales carrying dead sea lions. *Mammalia* **58**(3): 494–498.

Shane, S.H. 1995b. Relationship between pilot whales and Risso's dolphins at Santa Catalina Island, California, USA. *Marine Ecology Progress Series* **123**(1–3): 5–11.

Shane, S.H., Tepley, L., and Costello, L. 1993. Life-threatening contact between a woman and a pilot whale captured on film. *Marine Mammal Science* **9**(3): 331–336.

Shane, S.H., Wells, R.S., and Würsig, B. 1986. Ecology, behaviour and social organization of the bottlenose dolphin: a review. *Marine Mammal Science* **2**(1): 34–63.

Shannon, L.V. 1985. The Benguela Ecosystem. Part I. Evolution of the Benguela physical features and processes. *Oceanography and Marine Biology, Annual Review* **23**: 105.

Shannon, L.V., and Agenbag, J.J. 1990. A large-scale perspective on interannual variability in the environment in the South-East Atlantic. *South African Journal of Marine Science* **9**: 161–168.

Shannon, L.V., Agenbag, J.J., and Buys, M.E.L. 1987. Large- and mesoscale features of the Angola–Benguela front. *South African Journal of Marine Science* **5**: 11–34.

Shannon, L.V., Boyd, A.J., Brundrit, G.B., and Taunton-Clark, J. 1986. On the existence of an El Niño type phenomenon in the Benguela system. *Journal of Marine Research* **44**: 495–520.

Shannon, L.V., Crawford, R.J.M., Brundrit, G.B., and Underhill, L.G. 1988. Responses of fish populations in the Benguela ecosystem to environmental change. *Journal du Conseil pour l'Exploration de la Mer* **45**(1): 5–12.

Shannon, L.V., Crawford, R.J.M., Pollock, D.E. *et al.* 1992. The 1980s – a decade of change in the Benguela ecosystem. *South African Journal of Marine Science* **12**: 271–296.

Shannon, L.V., Lutjeharms, J.R.E., and Agenbag, J.J. 1989. Episodic input of Subantarctic water into the Benguela region. *South African Journal of Science* **85**(5): 317–322.

Shannon, L.V., and Nelson, G. 1996. The Benguela: large scale features and processes and system variability. Pp 163–210 In (Wefer, G., Berger, W.H., Siedler, G., and Webb, D.J., eds) *The South Atlantic: present and past circulation*. Springer-Verlag, Berlin.

Shevchenko, V.I. 1970. [Puzzling white scars on whale body]. *Priroda* **6**: 72–73.

Shevchenko, V.I. 1975. [The nature of the interrelationships between killer whales and other cetaceans]. *Morskie Mlekopitayushchie Chast'* **2**: 173–174 [In Russian – English translation by Canadian Fisheries and Marine Service Translation Series no. 3839, 1976, 4 pp].

Shevchenko, V.I. 1977. Application of white scars to the study of the location and migrations of sei whale populations in Area III of the Antarctic. *Report of the International Whaling Commission* (Special Issue) **1**: 130–134.

Signorini, S.R., Murtuguddo, R.G., McClain, C.R., Christian, J.R., Picaut, J., and Busalacchi, A.J. 1999. Biological and physical signatures in the tropical and subtropical Atlantic. *Journal of Geophysical Research* **104**: 18 367–18 382.

Silva-Jr, J.M., Pandolfo, L.J., and Sazima, I. 2004. Vomiting behavior of the spinner dolphin (*Stenella longirostris*) and squid meals. *Aquatic Mammals* **30**(2): 271–274.

Silva-Jr, J.M., and Sazima, I. 2003. Whalesuckers and a spinner dolphin bonded for weeks: does host fidelity pay off? *Biota Neotropica* **3**(2): 1–5.

Silva-Jr, J.M., and Sazima, I. 2006. Whalesuckers on spinner dolphins: an underwater view. *JMBA2 – Biodiversity Records*: 1–6.

Silva-Jr, J.M., Silva, F.J.L., and Sazima, I. 2005a. Rest, nurture, sex, release, and play: diurnal underwater behaviour of the spinner dolphin at Fernando de Noronha Archipelago, SW Atlantic. *Aqua, Journal of Ichthyology and Aquatic Biology* **9**(4): 161–176.

Silva-Jr, J.M., Silva, F.J.L., and Sazima, I. 2005b. Two presumed interspecific hybrids in the genus *Stenella* (Delphinidae) in the tropical west Atlantic. *Aquatic Mammals* **31**(4): 468–472.

Simmonds, M.P., and Lopez-Jurado, L.F. 1991. Whales and the military. *Nature, London* **351**(6326): 448.

Sinclair, E.H. 1992. Stomach contents of four short-finned pilot whales (*Globicephala macrorhynchus*) from the southern California Bight. *Marine Mammal Science* **8**(1): 76–81.

Širović, A., Hildebrand, J., Wiggins, S., McDonald, M., Moore, S., and Thiele, D. 2004. Seasonality of blue and fin whale calls, and the influence of sea ice in the Western Antarctic Peninsula. *Deep-Sea Research II* **51**: 2 327–2 344.

Slip, D.J., Moore, G.J., and Green, K. 1995. Stomach contents of a southern bottlenose whale, *Hyperoodon planifrons*, stranded at Heard Island. *Marine Mammal Science* **11**(4): 575–584.

Small, J.A., and Small, G.J. 1991. Cetacean observations from the Somali Democratic Republic, September 1985 through May 1987. UNEP *Marine Mammal Technical Report* **3**: 179–210.

Smith, T.G., Siniff, D.B., Reichle, R., and Stone, S. 1981. Coordinated behavior of killer whales, *Orcinus orca*, hunting a crabeater seal, *Lobodon carcinophagus*. *Canadian Journal of Zoology* **59**: 1 185–1 189.

Smithers, R.H.N. 1938. Notes on the stranding of a school of *Pseudorca crassidens* at the Berg River Mouth, December 27, 1936. *Transactions of the Royal Society of South Africa* **25**(4): 403–411.

Smolker, R.A., Richards, A.F., Connor, R.C., and Pepper, J.W. 1992. Sex differences in patterns of association among Indian Ocean bottle-nosed dolphins. *Behaviour* **123**: 38–69.

Smythe-Wright, D., Chapman, P., Duncombe Rae, C., Shannon, L.V., and Boswell, S.M. 1998. Characteristics of the South Atlantic Subtropical Frontal Zone between 15° W and 5° E. *Deep-Sea Research I* **45**:167–192.

Southall, B.L., Braun, R., Gulland, F.M.D., Heard, A.D., Baird, R.W., Wilkin, S.M., and Rowles, T.K. 2006. Hawaiian melon-headed whale (*Peponocephala electra*) mass stranding event of July 3–4, 2004. *NOAA Technical Memorandum* NMFS-OPR-31. 73 pp.

Souza, S.P., Siciliano, S., Cuenca, S., and Sanctis, B. 2005. A True's beaked whale (*Mesoplodon mirus*) on the coast of Brazil: adding a new beaked whale species to the western tropical Atlantic and South America. *Latin American Journal of Aquatic Mammals* **4**(2): 129–136.

Speich, S., Blanke, B., and Madec, G. 2001. Warm and cold water paths of a GCM thermohaline conveyor belt. *Geophysical Research Letters* **28**: 311–314.

Stacey, P.J., Leatherwood, S., and Baird, R.W. 1994. *Pseudorca crassidens. Mammalian Species* **456**: 1–6.

Stafford, K.M., Bohnenstiehl, D.R., Tolstoy, M., Chapp, E., Mellinger, D.K., and Moore, S.E. 2004. Antarctic-type blue whale calls recorded at low latitudes in the Indian and eastern Pacific Oceans. *Deep-Sea Research I* **51**: 1 337–1 346.

Stahl, J.-Cl. 1982. Observations de cétacés dans le sud-ouest de l'Océan Indien. *Annales de la Société des Sciences naturelles de la Charente-Maritime* Suppl. Décembre: 59–63.

Stander, G.H. 1964. The pilchard of South West Africa (*Sardinops ocellata*). The Benguela Current off South West Africa. *Investigational reports – Marine Research Laboratory of South West Africa* **11**: 1–42.

Steiner, L. 1995. Rough-toothed dolphin, *Steno bredanensis*: a new species record for the Azores, with some notes on behaviour. *Arquipélago. Boletim da Universidade dos Açores. Ciencias Biologicas e Marinhas.* (13A): 125–127.

Steiner, W.W., Hain, J.W., Winn, H.E., and Perkins, P.J. 1979. Vocalizations and feeding behaviour of the killer whale (*Orcinus orca*). *Journal of Mammalogy* **60**(4): 823–827.

Stensland, E., Carlén, I., Särnblad, A., Bignert, A., and Berggren, P. 2006. Population size, distribution, and behaviour of Indo-Pacific bottlenose (*Tursiops aduncus*) and humpback (*Sousa chinensis*) dolphins off the south coast of Zanzibar. *Marine Mammal Science* **22**(3): 667–682.

Stramma, L., and England, M.H. 1999. On the water masses and mean circulation of the South Atlantic Ocean. *Journal of Geophysical Research* **104**: 20 863–29 883.

Summers, R.F.H. 1975. *A history of the South African Museum 1825–1975.* A.A. Balkema, Cape Town. 245 pp.

Svane, I. 2005. Occurrence of dolphins and seabirds and their consumption of by-catch during prawn trawling in Spencer Gulf, South Australia. *Fisheries Research* **76**: 317–327.

Swart, S., Speich, S., Ansorge, I.J., Lutjeharms, J.R.E., Gladyshev, S., and Goni, G. (Submitted). Monitoring the variability of the Antarctic Circumpolar Current south of Africa. *Journal of Geophysical Research.*

Sylvestre, J.P. 1988. On a specimen of pygmy sperm whale *Kogia breviceps* (Blainville 1838) from New-Caledonia. *Aquatic Mammals* **14**(2): 76–77.

Sylvestre, J.P., and Tasaka, S. 1985. On the intergeneric hybrids in cetaceans. *Aquatic Mammals* **11**(3): 101–108.

T

Taruski, A.G. 1979. The whistle repertoire of the North Atlantic pilot whale (*Globicephala melaena*) and its relationship to behavior and environment. Pp 345–368 in (Winn, H.E., and Olla, B.L., eds) *Behavior of marine animals. Current perspectives in research. 3. Cetaceans.* Plenum Press, New York.

Taunton-Clark, J., and Shannon, L.V. 1988. Annual and interannual variability in the South-East Atlantic during the 20th century. *South African Journal of Marine Science* **6**: 97–106.

Tayler, C.K., and Saayman, G.S. 1972. The social organization and behaviour of dolphins (*Tursiops aduncus*) and baboons (*Papio ursinus*): some comparisons and assessments. *Annals of the Cape Provincial Museums (Natural History)* **9**(2): 11–49.

Taylor, H., Gordon, A.L., and Molinelli, E. 1978. Climatic characteristics of the Antarctic Polar Front Zone. *Journal of Geophysical Research* **83**: 4 572–4 578.

Taylor, R.J.F. 1957. An unusual record of three species of whale being restricted to pools in Antarctic sea-ice. *Proceedings of the Zoological Society of London* **129**: 325–331 + 7 pl.

Taylor, W.R. 1986. Family no. 59: Ariidae. Pp 211–213 in (Smith, M.M., and Heemstra, P.C., eds) *Smith's Sea Fishes.* Macmillan South Africa (Publishers) (Pty) Ltd., Johannesburg, South Africa.

Tershy, B.R., and Wiley, D.N. 1992. Asymmetrical pigmentation in the fin whale: a test of two feeding related hypotheses. *Marine Mammal Science* **8**(3): 315–318.

Thiele, D., and Gill, P.C. 1999. Cetacean sightings during a winter voyage into Antarctic sea ice south of Australia. *Antarctic Science* **11**(1): 48–53.

Thomas, J.A., Moore, P.W.B., Nachtigall, P.E., and Gilmartin, W.G. 1990. A new sound from a stranded pygmy sperm whale. *Aquatic Mammals* **16**(1): 28–30.

Thomas, J., and Turl, C.W. 1990. Echolocation characteristics and range detection threshold of a false killer whale (*Pseudorca crassidens*). Pp 321–333 in (Thomas, J., and Kastelein, R., eds) *Sensory abilities of cetaceans.* De Spil Publishers, New York.

Tietz, R.M. 1966. The southern bottlenose whale *Hyperoodon planifrons* from Humewood, Port Elizabeth. *Annals of the Cape Provincial Museums (Natural Sciences)* **5**: 101–107.

Tietz, R.M., and Tayler, C.K. 1964. Dolphin talk. *Scientific South Africa* **1**(10): 385–390.

Tobayama, T., Nishiwaki, M., and Yang, H.C. 1973. Records of the Fraser's Sarawak dolphin (*Lagenodelphis hosei*) in the western North Pacific. *Scientific Reports of the Whales Research Institute, Tokyo* **25**: 251–263.

Tobayama, T., Uchida, S., and Nishiwaki, M. 1969. A white bellied right whale dolphin caught in the waters off Ibaragi, Japan. *Journal of the Mammalogical Society of Japan* **4**(4–6): 112–120.

Tomilin, A.G. 1957. Cetacea. In (Heptner, V.G., ed.) *Mammals of the USSR and adjacent countries.* Vol. 9. Akad. Nauk SSSR, Moscow. [Translated by Israel Program for Scientific Translations, Jerusalem, 1967].

Tomilin, A.G., and Latyshev, V.M. 1967. New data on the flathead bottlenose whale (*Hyperoodon planifrons* Flower, 1882). *Byulletin' Moskovskogo Obshchestva Ispŷtateleĭ Prirodŷ. Biologiya* **72**(3): 119–122.

Tønnessen, J.N., and Johnsen, A.O. 1982. *The history of modern whaling.* C. Hurst and Co., London. 798 pp.

Tormosov, D.D., Mikhaliev, Y.A., Best, P.B., Zemsky, V.A., Sekiguchi, K., and Brownell Jr, R.L. 1998. Soviet catches of southern right whales *Eubalaena australis*, 1951–1971. Biological data and conservation implications. *Biological Conservation* **86**: 185–197.

Torres N.D., and Aguayo L., A. 1979. Habitos alimentarios de *Lissodelphis peronii* (Lacépède, 1804) en Chile central (Cetacea: Delphinidae). *Revista de Biologia Marina* **16**(3): 221–224.

Tove, M. 1995. Live sightings of *Mesoplodon* cf *M. mirus*, True's beaked whale. *Marine Mammal Science* **11**(1): 80–85.

Townsend, C.H. 1935. The distribution of certain whales as shown by logbook records of American whaleships. *Zoologica, NY* **19**(1): 1–50.

Treshnikov, A.F., and Baranov, G.I. 1976. Circulation of the world ocean: global studies. *Problemy Arktiki i Antarktiki* **47**: 7–22.

True, F.W. 1889. Contributions to the natural history of the cetaceans: a review of the family *Delphinidae*. *Bulletin of the US National Museum* **36**: 1–191.

True, F.W. 1904. The whalebone whales of the western North Atlantic compared with those occurring in European waters with some observations on the species in the North Pacific. *Smithsonian Contributions to Knowledge* **33**: 1–332 + 50 pl.

Turner, W. 1880. Report on the bones of Cetacea collected during the voyage of H.M.S. *Challenger* in the years 1873–1876. *Report of the scientific results of the voyage of H.M.S.* Challenger *during the years 1873–76*. Zoology **1**: 1–45 + 3 pl.

Tyack, P. 1981. Interactions between singing Hawaiian humpback whales and conspecifics nearby. *Behavioral Ecology and Sociobiology* **13**: 49–55.

Tyack, P. 2000. Functional aspects of cetacean communication. Pp 270–307 in (Mann, J., Connor, R.C., Tyack, P.L., and Whitehead, H., eds) *Cetacean societies. Field studies of dolphins and whales*. University of Chicago Press, Chicago.

Tyack, P.L., Johnson, M., Soto, N.A., Sturlese, A., and Madsen, P.T. 2006. Extreme diving of beaked whales. *Journal of Experimental Biology* **209**: 4 238–4 253.

U

Ulmer Jr, F.A. 1941. *Mesoplodon mirus* in New Jersey, with additional notes on the New Jersey *M. densirostris*, and a list and key to the ziphoid whales of the Atlantic coast of North America. *Proceedings of the Academy of Natural Sciences of Philadelphia* **93**: 107–122 + 2 pl.

Umezaki, K., Mihara, S., and Koike, H. 2003. Isotope analysis for the melon-headed whale (*Peponocephala electra*). *Honyurui-Kagaku [Mammalian Science]* **3**(Suppl): 79–82.

US Dept Commerce and US Navy. 2001. Joint interim report Bahamas marine mammal stranding event of 15–16 March 2000. 66 pp.

V

Van Beneden, [P.-J.], and Gervais, P. 1880. *Ostéographie des Cétacés vivants et fossils*. Arthus Bertrand, Paris. 634 pp + 64 pl.

Van Bree, P.J.H. 1971. The rabbit-eared barnacle, *Conchoderma auritum*, on the teeth of the dolphin *Stenella frontalis*. *Zeitschrift für Säugetierkunde* **36**(5): 316–317.

Van Bree, P.J.H., Best, P.B., and Ross, G.J.B. 1978. Occurrence of the two species of pilot whales (genus *Globicephala*) on the coast of South Africa. *Mammalia* **4**(3): 323–328.

Van Bree, P.J.H., Collet, A., Desportes, G., Hussenot, E., and Raga, J.A. 1986. Le dauphin de Fraser, *Lagenodelphis hosei* (Cetacea, Odontoceti), espèce nouvelle pour la faune d'Europe. *Mammalia* **50**(1): 57–86.

Van Bree, P.J.H., and Duguy, R. 1965. Sur un crane de *Sotalia teuszii* Kükenthal, 1892 (Cetacea, Delphinidae). *Zeitschrift für Säugetierkunde* **30**(5): 311–314.

Van Bree, P.J.H., and Gallagher, M.D. 1978. On the taxonomic status of *Delphinus tropicalis* van Bree, 1971 (Notes on Cetacea, Delphinoidea IX). *Beaufortia* **28**(342): 1–8.

Van Helden, A.L., Baker, A.N., Dalebout, M.L., Reyes, J.C., Van Waerebeek, K., and Baker, C.S. 2002. Resurrection of *Mesoplodon traversii* (Gray, 1874), senior synonym of *M. bahamondi* Reyes, Van Waerebeek, Cárdenas and Yáñez, 1995 (Cetacea: Ziphiidae). *Marine Mammal Science* **18**(3): 609–621.

Van Parijs, S.M., and Corkeron, P.J. 2001a. Boat traffic affects the acoustic behaviour of Pacific humpback dolphins, *Sousa chinensis*. *Journal of the Marine Biological Association of the UK* **81**: 533–538.

Van Parijs, S.M., and Corkeron, P.J. 2001b. Vocalizations and behaviour of Pacific humpback dolphins *Sousa chinensis*. *Ethology* **107**: 701–716.

Van Waerebeek, K. 1993a. External features of the dusky dolphin *Lagenorhynchus obscurus* (Gray, 1828) from Peruvian waters. *Estudios Oceanológicos* **12**: 37–53.

Van Waerebeek, K. 1993b. Presumed *Lagenorhynchus* skull at Tasmanian Museum reidentified as *Lissodelphis peronii*. *Australian Mammalogy* **16**(1): 41–43.

Van Waerebeek, K. 1997. Long-beaked and short-beaked common dolphins sympatric off central-west Africa. Paper SC/49/SM46 submitted to the Scientific Committee of the International Whaling Commission, Bournemouth, UK, September–October 1997.

Van Waerebeek, K., Barnett, L., Camara, A. ***et al.*** 2004. Distribution, status, and biology of the Atlantic humpback dolphin, *Sousa teuszii* (Kükenthal, 1892). *Aquatic Mammals* **30**(1): 56–83.

Van Waerebeek, K., Canto, J., Gonzalez, J., Oporto, J., and Brito, J.L. 1991. Southern right whale dolphins, *Lissodelphis peronii* off the Pacific coast of South America. *Zeitschrift für Säugetierkunde* **56**: 284–295.

Van Waerebeek, K., and De Smet, W.M.A. 1996. A second confirmed record of the false killer whale *Pseudorca crassidens* (Owen, 1846) (Cetacea, Delphinidae) from West Africa. *Mammalia* **60**(2): 319–322.

Van Waerebeek, K., Gallagher, M., Baldwin, R., Papastavrou, V., and Al-Lawati, S.M. 1999. Morphology and distribution of the spinner dolphin, *Stenella longirostris*, rough-toothed dolphin, *Steno bredanensis* and melon-headed whale, *Peponocephala electra*, from waters off the Sultanate of Oman. *Journal of Cetacean Research and Management* **1**(2): 167–177.

Van Waerebeek, K., Goodall, R.N.P., and Best, PB. 1997. A note on evidence for pelagic warm-water dolphins resembling *Lagenorhynchus*. *Report of the International Whaling Commission* **47**: 1 015–1 017.

Van Waerebeek, K., and Read, A.J. 1994. Reproduction of dusky dolphins, *Lagenorhynchus obscurus*, from coastal Peru. *Journal of Mammalogy* **75**(4): 1 054–1 062.

Van Waerebeek, K., Reyes, J.C., and Alfaro, J. 1993. Helminth parasites and phoronts of dusky dolphins *Lagenorhynchus obscurus* (Gray, 1828) from Peru. *Aquatic Mammals* **19**(3): 159–169.

Van Waerebeek, K., Reyes, J.C., Read, A.J., and McKinnon, J.S. 1990. Preliminary observations of bottlenose dolphins from the Pacific coast of South America. Pp 143–154 in (Leatherwood, S., and Reeves, R.R., eds) *The bottlenose dolphin*. Academic Press, San Diego.

Van Waerebeek, K., Van Bree, P.J.H., and Best, P.B. 1995. On the identity of *Prodelphinus petersii* Lutken, 1889 and records of dusky dolphins *Lagenorhynchus obscurus* (Gray, 1828) from the southern Indian and Atlantic Oceans. *South African Journal of Marine Science* **16**: 25–35.

Venegas, C.C. 1996. Estimation of population density by aerial line transects of Commerson's dolphin *Cephalorhynchus commersonii* in the Strait of Magellan, Chile. *Anales del Institut de la Patagonia Ciencias Naturales* **24**: 41–48.

Vidal, O., Findley, L.T., Turk, P.J., and Boyer, R.E. 1987. Recent records of pygmy sperm whales in the Gulf of Mexico. *Marine Mammal Science* **3**(4): 354–356.

Visser, I.N. 1998. Prolific body scars and collapsing dorsal fins on killer whales (*Orcinus orca*) in New Zealand waters. *Aquatic Mammals* **24**(2): 71–81.

Visser, I.N., Fertl, D., and Pusser, L.T. 2004. Melanistic southern right-whale dolphins (*Lissodelphis peronii*) off Kaikoura, New Zealand, with records of other anomalously all-black cetaceans. *New Zealand Journal of Marine and Freshwater Research* **38**: 833–836.

Visser, I.N., and Mäkeläinen, P. 2000. Variation in eye-patch shape of killer whales (*Orcinus orca*) in New Zealand waters. *Marine Mammal Science* **16**(2): 459–469.

Voisin, J.-F. 1972. Notes on the behaviour of the killer whale, *Orcinus orca* (L.). *Norwegian Journal of Zoology* **20**(1): 93–96.

Von Haast, J. 1876. On a new ziphioid whale. *Proceedings of the Zoological Society of London* **1876**: 7–13.

W

Wada, S., and Numachi, K. 1991. Allozyme analyses of genetic differentiation among the populations and species of the *Balaenoptera*. *Report of the International Whaling Commission* (Special Issue) **13**: 125–154.

Wada, S., Oishi, M., and Yamada, T.K. 2003. A newly discovered species of living baleen whale. *Nature, London* **426**(6964): 278–281.

Wade, P.R., and Gerrodette, T. 1993. Estimates of cetacean abundance and distribution in the eastern tropical Pacific. *Report of the International Whaling Commission* **43**: 477–493.

Walker, M.M., Kirschvink, J.L., Ahmed, G., and Dizon, A.E. 1992. Evidence that fin whales respond to the geomagnetic field during migration. *Journal of Experimental Biology* **171**: 67–78.

Walker, W.A. 1981. Geographical variation in morphology and biology of bottlenose dolphins (*Tursiops*) in the Eastern North Pacific. *NOAA Administrative Report. LJ-81-03C*. 54 pp.

Waller, G.N.H. 1989. Two new species of whale lice (Cyamidae) from the ziphoid whale *Berardius bairdii*. *Investigations on Cetacea* **22**: 292–297.

Wang, D., Würsig, B., and Evans, W.E. 1995a. Comparisons of whistles among seven odontocete species. Pp 299–323 in (Kastelein, R.A., Thomas, J.A., and Nachtigall, P.E., eds) *Sensory systems of aquatic mammals*. De Spil Publishers, Woerden, The Netherlands.

Wang, D., Würsig, B., and Evans, W.E. 1995b. Whistles of bottlenose dolphins: comparisons among populations. *Aquatic Mammals* **19**: 20–37.

Wang, M-.C., Walker, W.A., Shao, K-.T., and Chou, L.S. 2003. Feeding habits of the pantropical spotted dolphin, *Stenella attenuata*, off the eastern coast of Taiwan. *Zoological Studies* **42**(2): 368–378.

Wardle, W.J., Haney, T.A., and Worthy, G.A.J. 2000. New host record for the whale louse *Isocyamus delphinii* (Amphipoda, Cyamidae). *Crustaceana* **73**(5): 639–641.

Waring, G.T., Palka, D.L., Clapham, P.J. *et al.* 1999. US Atlantic and Gulf of Mexico marine mammal stock assessments – 1999. *NOAA Technical Memorandum* NMFS-NE-153. 187 pp.

Watkins, W.A. 1983. Current status of whale tagging. Pp 81–97 in (Albert, T.F., Kelley, J.J., and Dronenburg, R., eds) *Proceedings of the first conference on the biology of the bowhead whale,* Balaena mysticetus, *25–29 January 1982, Anchorage, Alaska, USA*.

Watkins, W.A., Daher, M.A., DiMarzio, N.A. *et al.* 2002. Sperm whale dives tracked by radio telemetry. *Marine Mammal Science*. **18**(1): 55–68.

Watkins, W.A., Daher, M.A., Fristrup, K., and Di Sciara, G.N. 1994. Fishing and acoustic behavior of Fraser's dolphin (*Lagenodelphis hosei*) near Dominica, Southeast Caribbean. *Caribbean Journal of Science* **30**(1–2): 76–82.

Watkins, W.A., Daher, M.A., Fristrup, K.M., Howald, T.J., and Di Sciara, G.N. 1993. Sperm whales tagged with transponders and tracked underwater by sonar. *Marine Mammal Science* **9**(1): 55–67.

Watkins, W.A., Daher, M.A., Samuels, A., and Gannon, D.P. 1997. Observations of *Peponocephala electra*, the melon-headed whale, in the southeastern Caribbean. *Caribbean Journal of Science* **33**(1–2): 34–40.

Watkins, W.A., Moore, K.E., and Tyack, P. 1985. Sperm whale acoustic behaviors in the southeast Caribbean. *Cetology* **49**: 1–15.

Watkins, W.A., Di Sciara, G.N., and Moore, K.E. 1979. Observations and radio tagging of *Balaenoptera edeni* near Puerto la Cruz, Venezuela. *Woods Hole Oceanographic Institution Technical Report WHO1-79-89*. 8 pp.

Watkins, W.A., and Schevill, W.E. 1979. Aerial observation of feeding behavior in four baleen whales: *Eubalaena glacialis*, *Balaenoptera borealis*, *Megaptera novaeangliae*, and *Balaenoptera physalus*. *Journal of Mammalogy* **60**(1): 155–163.

Watkins, W.A., Schevill, W.E., and Best, P.B. 1977. Underwater sounds of *Cephalorhynchus heavisidii* (Mammalia: Cetacea). *Journal of Mammalogy* **58**: 316–320.

Watkins, W.A., Tyack, P., and Moore, K.E. 1987. *Steno bredanensis* in the Mediterranean Sea. *Marine Mammal Science* **3**(1): 78–82.

Watson, L. 1981. *Sea Guide to Whales of the World*. E.P. Dutton, New York. 302 pp.

Watwood, S.L., Miller, P.J.O., Johnson, M.O., Madsen, P.T., and Tyack, P.L. 2006. Deep-diving foraging behaviour of sperm whales (*Physeter macrocephalus*). *Journal of Animal Ecology* **75**: 814–825.

Wedekin, L.L., Freitas, A., Engel, M.H., and Sazima, I. 2004. Rough-toothed dolphins (*Steno bredanensis*) catch diskfishes while interacting with humpback whales (*Megaptera novaeangliae*) off Abrolhos Bank breeding ground, Southwest Atlantic. *Aquatic Mammals* **30**(2): 327–329.

Wedepohl, P.M., Lutjeharms, J.R.E., and Meeuwis, J.M. 2000. Surface drift in the South-East Atlantic Ocean. *South African Journal of Marine Science* **22**: 71–79.

Weeks, S.J., and Shillington, F.A. 1996. Phytoplankton pigment distribution and frontal structure in the subtropical convergence region south of Africa. *Deep-Sea Research I* **43**: 739–741.

Weilgart, L.S., and Whitehead, H. 1990. Vocalizations of the North Atlantic pilot whale (*Globicephala melas*) as related to behavioural contexts. *Behavioral Ecology and Sociobiology* **26**: 399–402.

Weir, C.R. 2006a. First confirmed records of clymene dolphin, *Stenella clymene* (Gray, 1850), from Angola and Congo, south-east Atlantic Ocean. *African Zoology* **41**(2): 297–300.

Weir, C.R. 2006b. Sightings of beaked whales (Cetacea: Ziphiidae) including first confirmed Cuvier's beaked whales *Ziphius cavirostris* from Angola. *African Journal of Marine Science* **28**(1): 173–175.

Weir, C.R. In press. The distribution and seasonal occurrence of cetaceans off northern Angola, Africa. *Journal of Cetacean Research and Management*.

Weir, C.R., Stokes, J., Martin, C., and Cermeño, P. 2004. Three sightings of *Mesoplodon* species in the Bay of Biscay: first confirmed True's beaked whales (*M. mirus*) for the north-east Atlantic? *Journal of the Marine Biological Association of the UK* **84**: 1 095–1 099.

Weller, D.W., Würsig, B., Whitehead, H. *et al.* 1996. Observations of an interaction between sperm whales and short-finned pilot whales in the Gulf of Mexico. *Marine Mammal Science* **12**(4): 588–594.

Wells, R.S. 1984. Reproductive behavior and hormonal correlates in Hawaiian spinner dolphins, *Stenella longirostris*. *Report of the International Whaling Commission* (Special Issue) **6**: 465–472.

Wells, R.S., and Gannon, J.G. 2005. Release and follow-up monitoring of rehabilitated rough-toothed dolphins. Pp 4–18 in (Manire, C.A., and Wells, R.S., eds) *Rough-toothed dolphin* *rehabilitation and post-release monitoring. Final report for J.H. Prescott Marine Mammal Rescue Assistance Grant Program.* Award no. (FL) 2005-0162-001.

Wells, R.S., Hansen, L.J., Baldridge, A., Dohl, T.P., Kelly, D., and Defran, R.H. 1990. Northward extension of the range of bottlenose dolphins along the California coast. Pp 421–431 in (Leatherwood, S., and Reeves, R.R., eds) *The bottlenose dolphin.* Academic Press, San Diego.

Wells, R.S., Rhinehart, H.L., Cunningham, P., Whaley, J., Baran, M., Koberna, C., and Costa, D.P. 1999. Long distance offshore movements of bottlenose dolphins. *Marine Mammal Science* **15**(4): 1 098–1 114.

Wells, R.S., and Scott, M.D. 2002. Bottlenose dolphins *Tursiops truncatus* and *T. aduncus.* Pp 122–128 in (Perrin, W.F., Würsig, B., and Thewissen, J.G.M., eds) *Encyclopedia of Marine Mammals.* Academic Press, San Diego.

Werth, A. 2000. A kinematic study of suction feeding and associated behavior in the long-finned pilot whale, *Globicephala melas* (Traill). *Marine Mammal Science* **16**(2): 299–314.

Werth, J. 2001. How do mysticetes remove prey trapped in baleen? *Bulletin of the Museum of Comparative Zoology* **156**(1): 189–203.

Weru, S. 2001. Marine conservation programmes in Kenya: a statement on critical species. Paper presented to International Workshop on Cooperative Research on humpback whales in the southwest Indian Ocean, Cape Town, April 2001.

Wheeler, J.F.G. 1946. Observations on whales in the South Atlantic Ocean in 1943. *Proceedings of the Zoological Society, London* **16**: 221–224.

White, J.R. 1976. A pygmy killer whale found on the east coast of Florida. *Florida Science* **39**(1): 31–36.

White, W.B., and Peterson, R.G. 1996. An Antarctic Circumpolar Wave in surface pressure, wind, temperature and sea-ice extent. *Nature* **380**(6576): 699–702.

Whitehead, H. 2002. Estimates of the current global population size and historical trajectory for sperm whales. *Marine Ecology Progress Series* **242**: 295–304.

Whitehead, H. 2003. *Sperm whales. Social Evolution in the Ocean.* University of Chicago Press, Chicago. 431 pp.

Whittow, G.C., Hampton, F.G., and Ohata, C.A. 1978. Body temperature of the rough-toothed dolphin. *Journal of Wildlife Management* **42**(1): 184–185.

Whitworth (III), T. 1980. Zonation and geostrophic flow of the Antarctic Circumpolar Current at the Drake Passage. *Deep-Sea Research* **27**: 497–507.

Whitworth (III), T., and Nowlin, W.D. 1987. Water masses and currents of the Southern Ocean at the Greenwich Meridian. *Journal of Geophysical Research* **92**: 6 462–6 476.

Wiggins, S.M., Oleson, E.M., McDonald, M.A., and Hildebrand, J.A. 2005. Blue whale (*Balaenoptera musculus*) diel call patterns offshore of southern California. *Aquatic Mammals* **31**(2): 161–168.

Williams, A.D., Williams, R., and Brereton, T. 2002. The sighting of pygmy killer whales (*Feresa attenuata*) in the southern Bay of Biscay and their association with cetacean calves. *Journal of the Marine Biological Association of the UK* **82**: 509–511.

Williams, A.J., Dyer, B.M., Randall, R.M., and Komen, J. 1990. Killer whales *Orcinus orca* and seabirds: 'play', predation and association. *Marine Ornithology* **18**: 37–41.

Willis, P.M., and Baird, R.W. 1998. Status of the dwarf sperm whale, *Kogia simus*, with special reference to Canada. *Canadian Field-Naturalist* **112**(1): 114–125.

Wilson, C.E., Perrin, W.F., Gilpatrick Jr, J.W., and Leatherwood, S. 1987. Summary of worldwide locality records of the striped dolphin *Stenella coeruleoalba. US Dept of Commerce, NOAA Technical Memorandum* NMFS-TM-NMFS-SWFC-90. 65 pp.

Winn, H.E., Perkins, P.J., and Winn, L. 1970. Sounds and behavior of the northern bottle-nosed whale. *Seventh Annual Conference on Biological Sonar and Diving Mammals*: 53–59.

Würsig, B., and Bastida, R. 1986. Long-range movement and individual associations of two dusky dolphins (*Lagenorhynchus obscurus*) off Argentina. *Journal of Mammalogy* **67**(4): 773–774.

Würsig, B., Jefferson, T., and Schmidly, D.J. 2000. *The marine mammals of the Gulf of Mexico.* Texas A and M University Press, College Station, Texas. 232 pp.

Würsig, B., Wells, R.S., and Norris, K.S. 1994a. Food and feeding. Pp 216–231 in (Norris, K.S., Würsig, B., Wells, R.S., Würsig, M., Brownlee, S.M., Johnson, C.M., and Solow, J., eds) *The Hawaiian spinner dolphin.* University of California Press, Berkeley.

Würsig, B., Wells, R.S., Norris, K.S., and Würsig, M. 1994b. A spinner dolphin's day. Pp 65–102 in (Norris, K.S., Würsig, B., Wells, R.S., Würsig, M., Brownlee, S.M., Johnson, C.M., and Solow, J., eds) *The Hawaiian spinner dolphin.* University of California Press, Berkeley.

Würsig, B., and Würsig, M. 1980. Behavior and ecology of the dusky dolphin, *Lagenorhynchus obscurus*, in the South Atlantic. *Fishery Bulletin, US* **77**(4): 871–890.

Y

Yamada, M. 1954a. Some remarks on the pygmy sperm whale, *Kogia. Scientific Reports of the Whales Research Institute, Tokyo* **9**: 37–58 + 1 pl.

Yamada, M. 1954b. An account of a rare porpoise, *Feresa* Gray from Japan. *Scientific Reports of the Whales Research Institute, Tokyo* **9**: 59–88.

Yamada, T.K. 2002. *On an unidentified beaked whale found stranded in Kagoshima.* [Online] Available: http://svrsh1.kahaku.go.jp/sendai/indexE.html [June 2007].

Yamada, Y. 1956. An analysis in mass osteology of the false killer whale *Pseudorca crassidens* (Owen) Part I. *Okajimas Folia Anatomica Japonica* **28**: 453–463.

Yazdi, P. 2002. A possible hybrid between the dusky dolphin (*Lagenorhynchus obscurus*) and the southern right whale dolphin (*Lissodelphis peronii*). *Aquatic Mammals* **28**(2): 211–217.

Yin, S.E. 1999. *Movement patterns, behaviors, and whistle sounds of dolphin groups off Kaikoura, New Zealand.* M Sc thesis, Texas A and M University. 117 pp.

Yonekura, M., Matsui, S., and Kasuya, T. 1980. On the external characters of *Globicephala macrorhynchus* off Taiji, Pacific coast of Japan. *Scientific Reports of the Whales Research Institute, Tokyo* **32**: 67–95.

Yoshida, H., and Kato, H. 1999. Phylogenetic relationships of Bryde's whales in the western North Pacific and adjacent waters inferred from mitochondrial DNA sequences. *Marine Mammal Science* **15**(4): 1 269–1 286.

Young, D.D., and Cockcroft, V.G. 1994. Diet of common dolphins (*Delphinus delphis*) off the south-east coast of southern Africa: opportunism or specialization? *Journal of Zoology, London* **234**: 41–53.

Young, D.D., and Cockcroft, V.G. 1995. Stomach contents of stranded common dolphins *Delphinus delphis* from the south-east coast of southern Africa. *Zeitschrift für Säugetierkunde* **60**: 343–351.

Z

Zbinden, K., Pilleri, G., Kraus, C., and Bernath, O. 1977. Observations on the behaviour and the underwater sounds of the plumbeous dolphin (*Sousa plumbea* G. Cuvier, 1829) in the Indus Delta region. *Investigations on Cetacea* **8**: 259–286 + 4 pl.

Zemsky, V.A., Berzin, A.A., Mikhalev, Y.A., and Tormosov, D.D. 1995. *Soviet Antarctic whaling data (1947–1972).* Centre for Russian Environmental Policy, Moscow. 320 pp.

Zemsky, V.A., and Budylenko, G.A. 1970. Flatheaded bottlenose whales in the Antarctic. *Trudÿ Atlanticheskiĭ Nauchno-Issledovatel'skiĭ Institut Rÿbnogo Khozyaĭstva i Okeanografii* **29**: 193–201.

Zemsky, V.A., and Sazhinov, E.G. 1982. Distribution and abundance of the pygmy blue whale. *Morskiye Mlekopitayushchiye* VNIRO: 53-70. [Translated by NOAA Southwest Fisheries Science Center Administrative Report LJ-94-02, 1994].

Zerbini, A.N., and Secchi, E.R. 2001. Occurrence of Hector's beaked whale, *Mesoplodon hectori*, in southern Brazil. *Aquatic Mammals* **27**(2): 149–153.

Zerbini, A.N., Secchi, E.R., Siciliano, S., and Simões-Lopes, P.C 1996. The dwarf form of the minke whale, *Balaenoptera acutorostrata* Lacépède, 1804, in Brazil. *Report of the International Whaling Commission* **46**: 333–340.

Zerbini, A.N., Secchi, E.R., Siciliano, S., and Simões-Lopes, P.C. 1997. A review of the occurrence and distribution of whales of the genus *Balaenoptera* along the Brazilian coast. *Report of the International Whaling Commission* **47**: 407–417.

Zerbini, A.N., and Simões-Lopes, P.C. 2000. Morphology of the skull and taxonomy of Southern Hemisphere minke whales. Paper SC/52/OS10 submitted to the Scientific Committee of the International Whaling Commission, Adelaide, Australia, June 2000, 28 pp.

Zhang, T., Yang, G., Zhou, K., and Wei, F. 2005. An initial study on the mitochondrial DNA control region sequences of false killer whales (*Pseudorca crassidens*). *Acta Theriologica Sinica* **25**(3): 219–223.

Zimmer, W.M.X., Johnson, M.P., Madsen, P.T., and Tyack, P.L. 2005. Echolocation clicks of free-ranging Cuvier's beaked whales (*Ziphius cavirostris*). *Journal of the Acoustical Society of America* **117**(6): 3 919–3 927.

Zioupos, P., Currey, J.D., Casinos, A., and De Buffrénil, V. 1997. Mechanical properties of the rostrum of the whale *Mesoplodon densirostris*, a remarkably dense bony tissue. *Journal of Zoology, London* **241**: 725–737.

ILLUSTRATION ACKNOWLEDGEMENTS

Photographs

Page 27	Tim Collins, Wildlife Conservation Society (tcollins@wcs.org; www.wcs.org/international/marine/ccrp/ccrpgabon) and Indian Ocean Research and Conservation Association (http://iorca.org)
Fig. 7 A-E	Desray Reeb, Mammal Research Institute, University of Pretoria, South Africa
Fig. 9 A-G	Peter B. Best, Mammal Research Institute, University of Pretoria, South Africa
Fig. 10	Robert L. Pitman
Fig. 12 A-H	Peter B. Best
Fig. 13	Leonard J.V. Compagno, Shark Research Centre, Iziko South African Museum, Cape Town
Fig. 14	Leonard J.V. Compagno
Fig. 16	Keiko Sekiguchi, University of Hawai'i, Hilo, Hawai'i, USA (ksekiguc@hawaii.edu)
Fig. 17 A-H	Peter B. Best
Fig. 18	Patricia Evans, Kommetjie, Cape Town (patevans@global.co.za)
Fig. 19	Keiko Sekiguchi
Fig. 20 A-H	Peter B. Best
Fig. 23 A-D	Peter B. Best
Fig. 23 E-F	Stephen Van Dyck, Queensland Museum, Australia
Fig. 24	Peter B. Best
Fig. 25	Gwenith S. Penry, Centre for Dolphin Studies, South Africa, and St Andrews University, Scotland (info@dolphinstudies.co.za; www.dolphinstudies.co.za)
Fig. 27 A-H	Peter B. Best
Fig. 28 A-B	Paul Ensor, IWC/SOWER cruise programme (paulensor@xtra.co.nz)
Fig. 29 A-H	Peter B. Best
Fig. 31	Peter B. Best
Fig. 32 A-B	Keiko Sekiguchi
Fig. 34 A-H	Peter B. Best
Fig. 35	Stephanie Plön, Bayworld, Box 13147, Port Elizabeth, South Africa (stephanie@bayworld.co.za)
Fig. 36	Keiko Sekiguchi
Fig. 37	Paula Olson, Southwest Fisheries Science Center, NOAA National Marine Fisheries Service, USA (Paula.Olson@noaa.gov)
Fig. 40 A-H	Peter B. Best
Fig. 41	From Anderson, L. 1935. *Lis sails the Atlantic.* George Routledge & Sons.
Fig. 42	Caroline R. Weir, Ketos Ecology, UK (www.ketosecology.co.uk)
Fig. 44 A-E	Peter B. Best
Fig. 45	Peter B. Best
Fig. 47 A-F	Peter B. Best
Fig. 48	Michael Newcomer (deceased)
Fig. 49	Barbara Taylor, Southwest Fisheries Science Center, USA http://swfsc.noaa.gov/textblock.aspx?Division=PRD&ParentMenuId=229&id=1415
Fig. 51 A-F	Peter B. Best
Fig. 52 A-B	L. Todd Pusser (TPusser@nc.rr.com)
Fig. 53 A-F	Donald Hurlbert, Smithsonian Institution, Washington, DC, USA
Fig. 53 G-I	Peter B. Best
Fig. 55 A-B	Michael A. Meÿer, Marine and Coastal Management, Cape Town, South Africa (Mmeyer@deat.gov.za)
Fig. 57 A-F	Peter B. Best
Fig. 57 G-J	Museum of New Zealand Te Papa Tongarewa, Wellington, New Zealand
Fig. 58	Peter B. Best
Fig. 60 A-F	Donald Hurlbert
Fig. 60 G-I	Peter B. Best
Fig. 62 A-B	Christopher Johnson, earthOCEAN (www.earthocean.tv; cj@earthocean.tv)
Fig. 64 A-G	Peter B. Best
Fig. 65 A	Robert L. Pitman

Fig. 65 B	Paula Olson
Fig. 67 A–I	Peter B. Best
Fig. 68 A–H	Museum of New Zealand Te Papa Tongarewa
Fig. 70	John L. Bannister
Fig. 72 A–F	Donald Hurlbert
Fig. 72 G–I	Peter B. Best
Fig. 73	Sergio Hanquet (info@sergiohanquet.com; www.sergiohanquet.com)
Fig. 75 A–F	Donald Hurlbert
Fig. 77 A–B	Paula Olson
Fig. 78 A–I	Peter B. Best
Fig. 79 A–H	Museum of New Zealand Te Papa Tongarewa
Fig. 81 A	Isabel Beasley, IWC/SOWER cruise 2006/7 (Isabel.Beasley@jcu.edu.au)
Fig. 81 B	Paula Olson
Fig. 83 A–F	Donald Hurlbert
Fig. 83 G–H	Peter B. Best
Fig. 84	Iziko South African Museum, Box 61, Cape Town, South Africa
Fig. 85	L. Todd Pusser
Fig. 87 A–F	Donald Hurlbert
Fig. 87 G–I	Peter B. Best
Fig. 88	Peter B. Best
Fig. 89 A–F	Donald Hurlbert
Fig. 89 G–H	Charles Anderson, Maldives (anderson@dhivehinet.net.mv)
Fig. 92	Simon Elwen, Mammal Research Institute, University of Pretoria, South Africa (www.simonelwen.net)
Fig. 93 A–F	Peter B. Best
Fig. 95	Sascha Hooker, Sea Mammal Research Unit, University of St Andrews, Scotland (s.hooker@st-andrews.ac.uk).
Fig. 96 A–F	Peter B. Best
Fig. 98	Robert L. Pitman
Fig. 99 A–G	Peter B. Best
Fig. 100	Tim Collins
Fig. 102 A–F	Jamie Owen, Natural History Museum, London, UK
Fig. 104 A	Gwenith S. Penry
Fig. 104 B	Leszek Karczmarski, Mammal Research Institute, University of Pretoria, South Africa, and Institute of Marine Life Sciences, Texas A&M at Galveston, USA (leszek@iziko.org.za; karczmal@tamug.edu)
Fig. 105 A–F	Peter B. Best
Fig. 106	Bayworld, Port Elizabeth, South Africa
Fig. 107	Graham Saayman, Department of Psychology, University of Victoria, BC, Canada (gsaayman@uvic.ca)
Fig. 108 A	Rod Haestier, P.O. Box 50, Port St Johns, South Africa (rod.h@mweb.co.za)
Fig. 108 B	Desray Reeb
Fig. 110 A–F	Peter B. Best
Fig. 111	Peter B. Best
Fig. 113 A	Gwenith S. Penry
Fig. 113 B	Peter B. Best
Fig. 114 A–F	Peter B. Best
Fig. 117	Rod Haestier
Fig. 118 A–F	Peter B. Best
Fig. 119 A–B	Caroline R. Weir
Fig. 121 A–F	Peter B. Best
Fig. 122 A–B	Leszek Karczmarski
Fig. 124 A–F	Peter B. Best
Fig. 125	Leszek Karczmarski
Fig. 127	Robert L. Pitman
Fig. 128 A–F	Peter B. Best
Fig. 130	Caroline R. Weir
Fig. 131 A–F	Peter B. Best
Fig. 133	Paul Ensor
Fig. 134 A–F	Peter B. Best

Fig. 135	Gwenith S. Penry
Fig. 137 A–F	Peter B. Best
Fig. 138	Rod Haestier
Fig. 140	The Royal Society
Fig. 141	Robert L. Pitman
Fig. 142 A–F	Peter B. Best
Fig. 144	Paula Olson
Fig. 145 A–F	Peter B. Best
Fig. 148	Paul Ensor
Fig. 149 A–F	Peter B. Best
Fig. 151 A–F	Peter B. Best
Fig. 152 A–B	Paula Olson
Fig. 154 A–F	Peter B. Best
Fig. 155 A–B	Caroline R. Weir
Fig. 157	Robert L. Pitman
Fig. 158 A–F	Peter B. Best
Fig. 160 A–F	Peter B. Best
Fig. 161	Louella Dolar, Tropical Marine Research for Conservation, San Diego, CA, USA (ldolar@san.rr.com)
Fig. 163	Margaret A. Best
Fig. 164 A–F	Peter B. Best
Fig. 165	Robert L. Pitman
Fig. 166	Paula Olson
Fig. 168 A	Paula Olson
Fig. 168 B	Robert L. Pitman
Fig. 169 A–F	Peter B. Best
Fig. 170	Paula Olson
Fig. 172 A–F	Peter B. Best
Fig. 174 A–B	Caroline R. Weir
Fig. 175 A–F	Peter B. Best
Fig. 177 A–F	Peter B. Best
Fig. 178 A	Paula Olson
Fig. 178 B	Keiko Sekiguchi
Fig. 179 A–B	Peter B. Best
Page 289	Ian Sorby

Illustrations

Figs 5, 6	Val Myburgh
Fig. 115	Peter B. Best

Frontispiece, pp 18–23, 29, 36, 41, 48, 54, 58, 64, 70, 76, 83, 90, 94, 98, 103, 108, 111, 114, 119, 122, 125, 128, 132, 135, 140, 145, 146, 148, 153, 158, 163, 167, 174, 181, 188, 193, 198, 204, 208, 213, 218, 223, 228, 234, 238, 242, 247, 252, 257, 263, 272, 278, 284 Pieter A. Folkens

Maps

Figs 1, 2, 3, 4, 8, 10, 15, 21, 22, 26, 31, 34, 39, 40, 44, 47, 51, 55, 57, 60, 62, 66, 69, 71, 74, 76, 80, 82, 86, 90, 91, 94, 97, 101, 103, 109, 112, 116, 120, 123, 126, 129, 132, 136, 139, 143, 147, 150, 153, 156, 159, 162, 167, 171, 173, 176 John Hall

Figs 2, 3, 4 adapted by John Hall from originals by Anne Westoby

GLOSSARY

THE TERMS BELOW that refer to specific bones in the cetacean skull are not defined here. Instead, the reader is referred to the section of this book entitled 'Anatomy of cetacean skulls' on pages 24 and 25.

A

advection
Horizontal transport (of mass, heat, etc.) effected by horizontal exchange of air or water.

albedo
The measure of the reflectivity of a surface such as ice or snow.

alisphenoid
See Anatomy of cetacean skulls.

alloparental care
Parental care shown by individuals other than the biological parents ('baby-sitting').

alveolus (pl. alveoli)
Cavity or socket for a tooth.

amphipod
Small crustacean of the Order Amphipoda. A good terrestrial example would be the so-called sand flea.

annelid
A segmented worm, such as an earthworm or leech.

anoestrus
The period of the mammalian reproductive cycle when females are not receptive to males.

anthropogenic
Arising from human activities or behaviour.

antorbital notch
See Anatomy of cetacean skulls.

antorbital process
See Anatomy of cetacean skulls.

apical teeth
Teeth at the very front of the lower jaw.

ascending process
See Anatomy of cetacean skulls.

ascidian
Sometimes known as a sea squirt, this is a sac-like invertebrate belonging to the Subphylum Tunicata.

asdic
A sonar device used in fish-finders or depth-sounders (originally an acronym for the Anti-Submarine Detection Investigation Committee).

asymptote (adj. asymptotic)
Refers to that part of a growth curve where growth ceases (size stays more or less constant irrespective of age).

axilla (pl. axillae)
The 'armpit' or region where the trailing edge of the flipper meets the body.

B

balanoid
An acorn barnacle, or one with an external shell, as opposed to a stalked barnacle such as the goose barnacle.

baleen
The filtering apparatus used by baleen whales, consisting of a rack-like series of plates mounted in the upper jaw. The plates are keratinous, like hair or horn, and were commercially known as whalebone.

basi-occipital
See Anatomy of cetacean skulls.

bathymetry (adj. bathymetric)
The spatial distribution of the depth of the ocean floor from the sea surface. Also known as bottom topography.

bathypelagic
Living in the depths of the ocean, usually defined as between 1 000 and 4 000 m from the surface.

benthic
Living on or close to the sea floor.

bifurcated
Forked into two (like a snake's tongue).

biota
Living things (plant or animal), usually of a region.

brotulid
A fish of the Family Brotulidae, related to the kingklip.

C

cancellous bone
Spongy bone of low density, such as that which fills long bones.

caudal peduncle
That part of the cetacean body between the dorsal fin and the tail.

centrum (pl. centra)
The main body of each vertebra, to which an epiphysis or disc becomes fused at either end.

cephalopod
Member of the Order Cephalopoda, or squid, cuttlefish and octopus.

cetacean
Member of the Order Cetacea, or whales, dolphins and porpoises.

cetology
The study of whales, dolphins and porpoises.

commensals
Different species that live in association without much mutual influence on each other.

condyle
The rounded part of a jointed area that fits into a socket.

condylobasal length (CBL)
The length of the skull in a straight line from the hindmost margins of the occipital condyles to the tip of the rostrum.

condylo-premaxillary length (CPL)
For whales and dolphins, this is essentially the same measurement as the CBL.

congener
A member of the same genus.

copepod
A tiny crustacean that is one of the most abundant multicellular organisms in the aquatic kingdom. The Greek word from which the name is derived means 'oar-footed', referring to a pair of swimming legs on the same body segment that are moved together like someone sculling.

Coriolis force
A conceptual force that is used to explain the deflection of a body to the right in the northern hemisphere and to the left in the southern hemisphere, when that body moves for a long period. It is exerted on a parcel of air or water as a result of the rotation of the earth.

corpus luteum
A temporary glandular body formed in the mammalian ovary following the rupture of a follicle at ovulation. It is one of the principal sources of oestrogen and progesterone and is essential for the establishment and maintenance of pregnancy.

cranchiid
A squid belonging to the Family Cranchiidae, whose members typically accumulate ammonia solution in the body cavity as a means to aid flotation.

crustacean
A member of the Subphylum Crustacea, including lobsters, crabs, shrimps, crayfish and barnacles.

cyamid
Small amphipod crustacean (also known as a whale louse) that spends its entire life on the surface of a cetacean host.

cytochrome b
An electron-transporting protein found in the mitochondria of cells and frequently used in genetic analysis.

D

delphinid
Member of the cetacean Family Delphinidae, or oceanic dolphins.

demersal
Living near to the coast or sea floor (in contrast to pelagic).

denticle
A small tooth-like protuberance.

diel pattern
That which takes place over 24 hours. It is different to diurnal (the opposite of nocturnal), or that which takes place only during daylight.

dimorphism
Existing in two forms.

dorsal peduncle
The upper part of the caudal peduncle.

E

ecdysis
A drastic type of moult in which the entire outer layer of skin is shed either as a unit or in large sheets.

ecotype
A group of animals within a species that adapt themselves genetically to a particular environment, but are able to breed freely with other ecotypes of the same species.

eddy
A circular movement of water in the ocean, the dimensions of which may vary from metres to hundreds of kilometres. Ocean eddies may extend all the way to the sea floor.

elasmobranch
A cartilaginous (as opposed to a bony) fish, such as a shark, skate or ray.

endemic
Belonging to a particular region.

enoploteuthid
A squid belonging to the clade Enoploteuthidae. They are generally small in size, carry many light-producing organs and live in the midwater zone.

epimeletic behaviour
Assistance provided by a healthy individual to one incapacitated through sickness or injury.

epipelagic
Living in the upper layers of the ocean (usually defined as above 200 m).

euphausiid
Shrimp-like crustacean, the best-known being krill.

exoccipital process
See Anatomy of cetacean skulls.

F

falcate
Curved like a sickle.

foramen (pl. foramina)
A natural perforation of bone through which nerves or blood vessels pass.

front, frontal band
A region where there is a relatively marked horizontal change in properties, such as sea surface temperature.

frontal bone
See Anatomy of cetacean skulls.

G

genome
The entire hereditary information of an individual as contained in its DNA.

gonatid
A squid belonging to the Family Gonatidae. They are generally medium-sized and have few light-producing organs, but many carry hooks as well as suckers on their arms.

gyre
A large rotating ocean current system. The term is usually used when describing an ocean-wide movement in the subtropical ocean basins.

H

hamular process
See Anatomy of cetacean skulls.

harpacticoid copepod
A copepod belonging to the Order Harpacticoida, many of the members of which live benthically or in association with other animals, rather than as members of the plankton community.

histioteuthid
A squid belonging to the Family Histioteuthidae, members of which generally have very long, thick arms and short bodies, and in which the two eyes are of very different size and shape.

homoplasy
Similarity between a feature of two different animals that does not arise from their having a common ancestor, but because the feature has developed independently in response to a common environmental influence.

hyoid bones
A horseshoe-shaped set of bones in the throat region that acts as a support for the tongue.

hyperiid
A crustacean belonging to the amphipod Suborder Hyperiidae, many of which live in association with gelatinous zooplankton such as jellyfish and salps.

I

inferior process
See Anatomy of cetacean skulls.

insolation
Warming caused by radiation from the sun received at the earth's surface.

introgression
The movement of a gene from one species into the gene pool of another by back-crossing an interspecific hybrid with one of its parents.

isotherm
Line of equal temperature.

J

jugal
See Anatomy of cetacean skulls.

K

kogiid
A member of the genus *Kogia*, or a short-headed sperm whale.

L

labial
Used here to indicate the side adjacent to the lips.

lachrymal
See Anatomy of cetacean skulls.

lingual
Used here to indicate the side adjacent to the tongue.

lobate
Resembling a lobe in shape.

lunate
Half-moon shaped.

M

mandible (adj. mandibular)
Lower jaw.

maxilla (pl. maxillae)
See Anatomy of cetacean skulls.

melon
A fatty structure in the forehead of all cetaceans, but particularly well developed in toothed whales and

dolphins, where it is believed to be used principally as an acoustic lens in sound production.

meridional
Along a meridian, or in a north-south rather than east-west direction.

mesethmoid cartilage
A cartilage forming part of the nasal septum, that extends along the centre of the rostrum within the groove formed by the vomer. In some species (e.g. beaked whales) it can become partially or completely ossified.

mesopelagic
Living in the midwater of the ocean, usually defined as between 200 m and 1 000 m below the surface.

mesoscale
A meteorological and oceanographic term to describe phenomena that have horizontal scales ranging from a few to several hundred kilometres.

microsatellite DNA
DNA that is usually derived from the nucleus of a cell that is highly variable in nature. Widely used in kinship, paternity and population studies.

mitochondrial DNA
DNA derived from cell mitochondria, outside the cell nucleus, and so normally passed directly from mother to offspring.

molecular phylogeny
The use of genetic data to gain information on an animal's evolutionary relationships.

morphology
The study of an animal's form or shape.

morphometrics
The measurement of an animal's form or shape.

morphotype (or morph)
A subpopulation that differs in appearance or behaviour from the rest of the population of the same species.

myctophid
Fish belonging to the Family Myctophidae. They are small, highly abundant deep-sea fish that get their common name of lantern fish from the numerous light-producing organs on their body.

myoglobin
The respiratory pigment found in muscle, or the equivalent of haemoglobin in the blood.

mysid
A shrimp-like crustacean of the Order Mysidacea, sometimes called an opossum shrimp because the females carry their eggs in a pouch beneath their thorax.

mysticete
A whale belonging to the Suborder Mysticeti, with baleen instead of teeth in the mouth. Otherwise known as baleen or whalebone whales.

N

naris (pl. nares)
See Anatomy of cetacean skulls.

nasal
See Anatomy of cetacean skulls.

nematode
Non-segmented worm, one of the most numerous multicellular organisms, many of which are parasitic in marine and other mammals.

neonate
A newly born animal.

neritic habitat
A zone, close to shore, often defined as from the low tide mark to the edge of the continental shelf.

neurocranium
The part of the skull containing the brain.

O

odontocete
A whale belonging to the Suborder Odontoceti, with teeth instead of baleen in the mouth. Otherwise known as toothed whales.

ommastrephid
Squid belonging to the Family Ommastrephidae. They are small to large muscular animals frequently dominating oceanic squid communities, with some species forming the basis of commercial fisheries.

otoliths
Small chalky stones found in the inner ear of fish and used for balance and orientation rather than hearing. Lines of calcium carbonate accretion in the otolith can be used to estimate the age of the fish.

P

palatine
See Anatomy of cetacean skulls.

pantropical
Occurring in the tropical areas of both the northern and the southern hemispheres.

parietal, interparietal
See Anatomy of cetacean skulls.

peduncle
See caudal peduncle.

pelagic
Living in the open sea, not near the coast or sea floor (in contrast to demersal).

phocoenid
Member of the Family Phocoenidae, or porpoises.

phylogeny (adj. phylogenetic)
The origin and evolution of a set of species – sometimes known as an evolutionary tree.

physiography (adj. physiographic)
The study of the physical features of the earth's surface.

phytoplankton
Members of the plankton community that are plants, not animals (the latter being known as zooplankton).

pinnipeds
Members of the Suborder Pinnipedia ('fin-foots'), or seals, sea lions and walruses.

piscivorous
Fish-eating.

polychaetes
Segmented worms (mostly marine) in which each segment has a pair of fleshy protrusions that carry many bristles – hence the alternative name bristle worms. The bloodworm is a well-known South African example.

polyoestrous
Undergoing several ovulatory cycles in one breeding season.

postorbital process
See Anatomy of cetacean skulls.

premaxilla (pl. premaxillae)
See Anatomy of cetacean skulls.

prenarial triangle
See Anatomy of cetacean skulls.

pterygoid
See Anatomy of cetacean skulls.

R

remora
Commonly known as sucker fish.

retroflection
A term used to describe when an ocean current turns back on itself.

rorqual
A term that generally refers to all baleen whales with throat grooves, i.e. excluding bowhead, right and pygmy right whales.

rostrum (adj. rostral)
The front portion of the head or skull, corresponding to the 'beak' if one is present.

rugosity
An area that appears roughened, in contrast to the normally smooth cetacean skin.

S

salp
A barrel-shaped, gelatinous tunicate that moves by pumping water through its body.

sepiids
Members of the cephalopod Family Sepiidae, commonly known as cuttlefish.

serranid
A member of the fish Family Serranidae, or rock cods and sea basses.

sessile
Sedentary, unable to move about.

sinusoidal
S-shaped.

speciation
The process by which a new species is formed.

squamosal
See Anatomy of cetacean skulls.

stenellid
A member of the dolphin genus *Stenella*.

stomatopod
A crustacean belonging to the Order Stomatopoda, or mantis shrimps.

succourant behaviour
Care and attention directed to individuals in distress, not necessarily kin.

supra-occipital
See Anatomy of cetacean skulls.

supra-orbital process
See Anatomy of cetacean skulls.

sympatric species
Species that share areas of distribution but do not interbreed.

symphysis
A fixed articulation between two bones usually involving a fibrous or cartilaginous pad.

T

taxonomy
The process of classifying animals into species, genera, families, etc.

teleost
A fish with a bony skeleton, as opposed to an elasmobranch, which has a cartilaginous skeleton.

temporal fossa (pl. fossae)
See Anatomy of cetacean skulls.

thermocline
The layer in the water column where the vertical temperature gradient is exceptionally strong. The permanent thermocline between warm surface and cold deep water is at a depth of about 1 000 m, while the seasonal thermocline may be at a depth of a few hundred metres.

thermohaline
Thermohaline circulation describes the density-driven circulation of the oceans. Derived from thermo- for heat and -haline for salt.

topography (adj. topographic)
A depiction of the earth's surface elevation, usually by means of contour lines. In the ocean this is also called bathymetry.

tunicate
A member of the Subphylum Tunicata, sometimes known as sea squirts.

U

umbilicus
Navel.

V

vertex
Summit, or highest point.

vicariant species
Species, usually very similar, that share a common ancestor but occupy adjacent areas separated by a barrier.

vomer
See Anatomy of cetacean skulls.

W

water mass
A body of oceanic water with a distinctive narrow range of temperature and salinity indicating a specific origin.

wind stress curl
The tendency for the wind field to rotate. In the oceans the wind stress curl gives rise to large-scale ocean currents.

Y

y chromosome
The sex chromosome that in mammals is found only in males.

Z

ziphiid
A whale belonging to the Family Ziphiidae, or beaked and bottlenose whales.

zooplankton
Members of the plankton community that are animals, not plants (the latter being known as phytoplankton).

zygomatic process
See Anatomy of cetacean skulls.

INDEX

Please note: Page numbers in *italics* refer to maps.

LIST OF SUBSCRIBERS

Sponsors' Edition

Ismail Hussain
Michael Clacey
Philip E Read
Cyril Ramaphosa

Collectors' Edition

Gordon and Janet Hall
Catherine and Alec Grant
T C A Wadeson
Rick and Caroline Menell
The Brenthurst Library
Hannah Parsraman
Valdon Smith
R G Jeffery
Jessica Hayward
Alex Barrell
J S Gericke Library, Stellenbosch University
G H Tudor
Margaret Best
Sibuya Game Reserve
Paul Larsen

Standard Edition

Bobby Godsell
Mevagh Glyn
O J Mackenzie
Lorna Ramsden
Jean de Lange
Richard and Stephanie Harris
Bobby and Carol Savosnick
Brian W Fargher
Christo Fabricus
Michael Fabricus
Ryan-Moloney Family
Natal Sharks Board
Nigel Willis
George Branch
Elizabeth Baines
Graham Kerley
Henk Bouwman Prof - NWU
South African Institute for Aquatic Biodiversity
University of Johannesburg Library and Information Centre
Trevor and Ruth Branch
Emma and Reinhard le Roux
J S Gericke Library, Stellenbosch University
Walter Stuart
Oceanographic Research Institute
G H Tudor
Jerome O'Regan and Alison Dyer
John B Dyer
Per Berggren

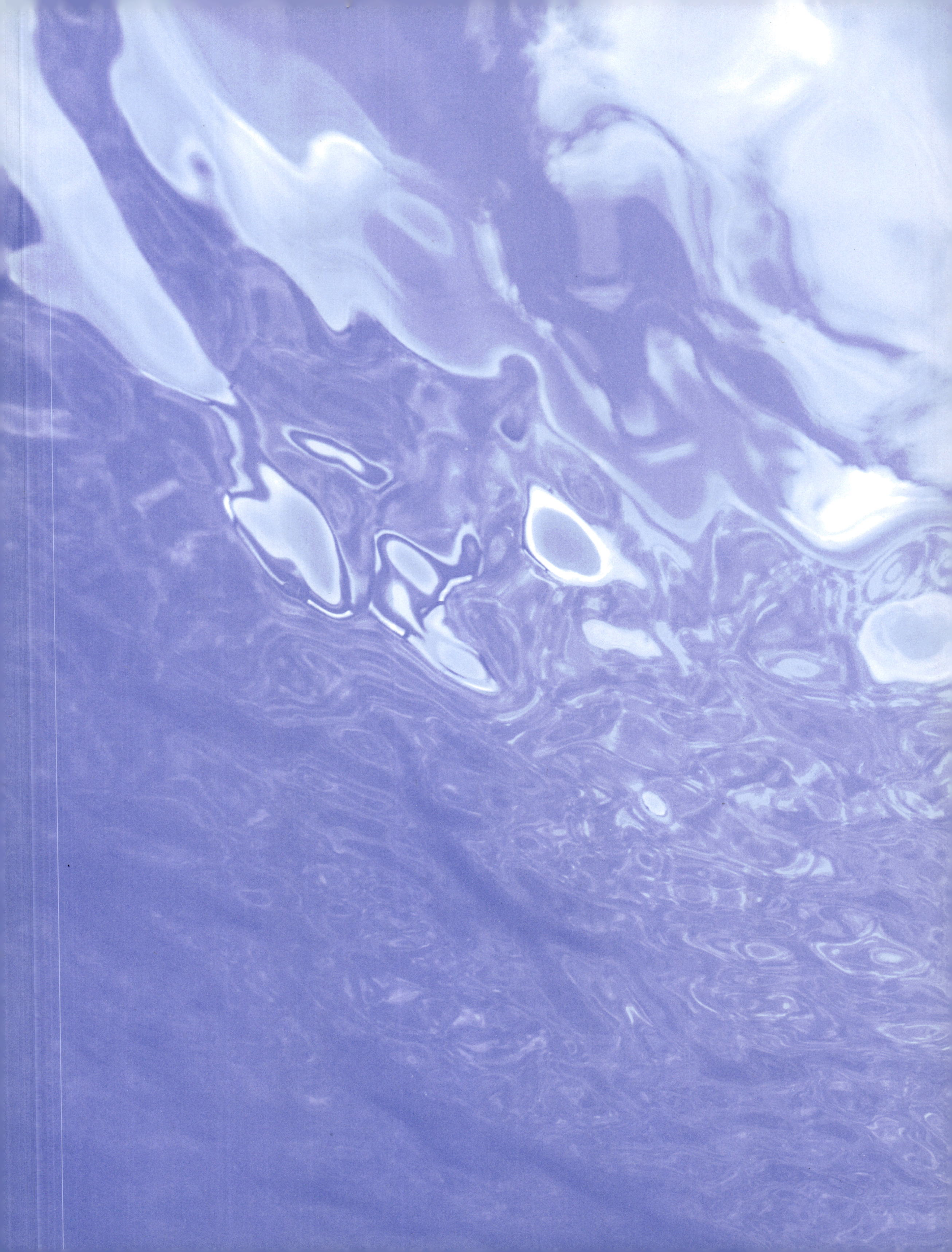